LONDON MATHEMATICAL SOCIETY STUDENT TEXTS

Managing Editor: Ian J. Leary,
Mathematical Sciences, University of Southampton, UK

61 An introduction to noncommutative Noetherian rings (2nd Edition), K. R. GOODEARL & R. B. WARFIELD, JR
62 Topics from one-dimensional dynamics, KAREN M. BRUCKS & HENK BRUIN
63 Singular points of plane curves, C. T. C. WALL
64 A short course on Banach space theory, N. L. CAROTHERS
65 Elements of the representation theory of associative algebras I, IBRAHIM ASSEM, DANIEL SIMSON & ANDRZEJ SKOWROŃSKI
66 An introduction to sieve methods and their applications, ALINA CARMEN COJOCARU & M. RAM MURTY
67 Elliptic functions, J. V. ARMITAGE & W. F. EBERLEIN
68 Hyperbolic geometry from a local viewpoint, LINDA KEEN & NIKOLA LAKIC
69 Lectures on Kähler geometry, ANDREI MOROIANU
70 Dependence logic, JOUKU VÄÄNÄNEN
71 Elements of the representation theory of associative algebras II, DANIEL SIMSON & ANDRZEJ SKOWROŃSKI
72 Elements of the representation theory of associative algebras III, DANIEL SIMSON & ANDRZEJ SKOWROŃSKI
73 Groups, graphs and trees, JOHN MEIER
74 Representation theorems in Hardy spaces, JAVAD MASHREGHI
75 An introduction to the theory of graph spectra, DRAGOŠ CVETKOVIĆ, PETER ROWLINSON & SLOBODAN SIMIĆ
76 Number theory in the spirit of Liouville, KENNETH S. WILLIAMS
77 Lectures on profinite topics in group theory, BENJAMIN KLOPSCH, NIKOLAY NIKOLOV & CHRISTOPHER VOLL
78 Clifford algebras: An introduction, D. J. H. GARLING
79 Introduction to compact Riemann surfaces and dessins d'enfants, ERNESTO GIRONDO & GABINO GONZÁLEZ–DIEZ
80 The Riemann hypothesis for function fields, MACHIEL VAN FRANKENHUIJSEN
81 Number theory, Fourier analysis and geometric discrepancy, GIANCARLO TRAVAGLINI
82 Finite geometry and combinatorial applications, SIMEON BALL
83 The geometry of celestial mechanics, HANSJÖRG GEIGES
84 Random graphs, geometry and asymptotic structure, MICHAEL KRIVELEVICH *et al*
85 Fourier analysis: Part I - Theory, ADRIAN CONSTANTIN
86 Dispersive partial differential equations, M. BURAK ERDOĞAN & NIKOLAOS TZIRAKIS
87 Riemann surfaces and algebraic curves, R. CAVALIERI & E. MILES
88 Groups, languages and automata, DEREK F. HOLT, SARAH REES & CLAAS E. RÖVER
89 Analysis on Polish spaces and an introduction to optimal transportation, D. J. H. GARLING
90 The homotopy theory of $(\infty, 1)$-categories, JULIA E. BERGNER
91 The block theory of finite group algebras I, MARKUS LINCKELMANN
92 The block theory of finite group algebras II, MARKUS LINCKELMANN
93 Semigroups of linear operators, DAVID APPLEBAUM
94 Introduction to approximate groups, MATTHEW C. H. TOINTON
95 Representations of finite groups of Lie type (2nd Edition), FRANÇOIS DIGNE & JEAN MICHEL
96 Tensor products of C^*-algebras and operator spaces, GILLES PISIER
97 Topics in cyclic theory, DANIEL G. QUILLEN & GORDON BLOWER
98 Fast track to forcing, MIRNA DŽAMONJA
99 A gentle introduction to homological mirror symmetry, RAF BOCKLANDT
100 The calculus of braids, PATRICK DEHORNOY

London Mathematical Society Student Texts 101

Classical and Discrete Functional Analysis with Measure Theory

MARTIN BUNTINAS
Loyola University Chicago

CAMBRIDGE
UNIVERSITY PRESS

University Printing House, Cambridge CB2 8BS, United Kingdom

One Liberty Plaza, 20th Floor, New York, NY 10006, USA

477 Williamstown Road, Port Melbourne, VIC 3207, Australia

314–321, 3rd Floor, Plot 3, Splendor Forum, Jasola District Centre,
New Delhi – 110025, India

103 Penang Road, #05–06/07, Visioncrest Commercial, Singapore 238467

Cambridge University Press is part of the University of Cambridge.

It furthers the University's mission by disseminating knowledge in the pursuit of education, learning, and research at the highest international levels of excellence.

www.cambridge.org
Information on this title: www.cambridge.org/9781107034143
DOI: 10.1017/9781139524445

First published 2022

A catalogue record for this publication is available from the British Library.

Library of Congress Cataloging-in-Publication Data
Names: Buntinas, Martin, author.
Title: Classical and discrete functional analysis with measure theory / Martin Buntinas.
Description: Cambridge ; New York, NY : Cambridge University Press, 2022.|
Series: London Mathematical Society student texts ; 101 |
Includes bibliographical references and index.
Identifiers: LCCN 2021029831 (print) | LCCN 2021029832 (ebook) |
ISBN 9781107034143 (hardback) | ISBN 9781107634886 (paperback) |
ISBN 9781139524445 (epub)
Subjects: LCSH: Functional analysis. | Measure theory. |
BISAC: MATHEMATICS / Algebra / Abstract
Classification: LCC QA320 .B83 2022 (print) | LCC QA320 (ebook) |
DDC 515/.7–dc23
LC record available at https://lccn.loc.gov/2021029831
LC ebook record available at https://lccn.loc.gov/2021029832

ISBN 978-1-107-03414-3 Hardback
ISBN 978-1-107-63488-6 Paperback

Contents

*Sections marked with * may be skipped on first reading.*

Preface

This book grew out of two analysis courses taught to both undergraduate and first-year graduate students at Loyola University Chicago. One of them covered measure theory and integration, the other functional analysis. Both assumed background knowledge of undergraduate-level real analysis and linear algebra, but neither was a prerequisite for the other.

Although there are other excellent books on functional analysis and measure theory, this book can be used for the first analysis course after real analysis and linear algebra. It is suitable for a course spanning from one quarter to an entire year, or for two independent courses, or for a capstone undergraduate course. Topology and Lebesgue integration are not prerequisites. This book is also suitable for self-study. Any section marked with an asterisk ($*$) may be skipped on the first reading.

The book starts with a chapter titled "Preliminaries." The following chapters are grouped into three parts:

- Much of the Preliminaries chapter is for reference and may be read only as needed. It is therefore labeled Chapter 0.
- Part I deals with Lebesgue measure and integration.
- Part II deals with elements of metric spaces, normed spaces, and linear operators.
- Part III deals with discrete functional analysis. The term *discrete functional analysis* refers to the parts of functional analysis that are in a discrete setting, such as Fourier series, sequence spaces, matrix maps, and summability. These topics do not require a background in general topology, and include areas that are especially rich in applications to the sciences and engineering.

The modern theory of measure and integration, as covered in Part I, was developed a little over a century ago by the French mathematician Henri Lebesgue (1875–1941). It remains the standard form of integration used in mathematical analysis and its applications today.

Functional analysis (Part II) is an important topic for every mathematical scientist. There are many reasons for studying it.

- Functional analysis is the gateway to all higher analysis. It is the subject to be studied after real analysis.
- Functional analysis has applications in probability theory, statistics, economics, physics, chemistry, engineering, as well as in pure mathematics.
- Functional analysis is one of the triumphs of twentieth-century mathematics. It deals with infinite-dimensional vector spaces and is one of the most beautiful areas of mathematics.
- Studying this subject is a training of the mind. After studying functional analysis, one sees the world with different eyes. To obtain this benefit, it is important for the reader to proceed through the rigorous development of the subject as presented in this book.

Discrete functional analysis (Part III) includes a study of Fourier series, sequence spaces, matrix maps, multipliers, summability theory, and a variety of applications. The topics and applications chosen for study can depend on the interests of the reader or instructor.

Courses: For self-study, reading may start with Chapter 0 (Preliminaries), and then continue through all of the chapters in order. Any starred ($*$) section may be skipped on first reading.

A one-term course could start with either Part I (Measure and Integration) or Part II (Elements of Classical Functional Analysis).

If starting with Part I, the prior reading of the sections Euclidean Measure (Section 0.9) and Sets of Measure Zero (Section 0.10) from Chapter 0 (Preliminaries) is recommended.

If starting with Part II, the prior reading of the section Overview of Integration (Section 0.11) is recommended in addition to Sections 0.9 and 0.10. These three sections cover sufficient measure and integration theory for the understanding of Part II and beyond.

A one-year sequence would include most of Part I (Lebesgue Measure and Integration), Part II (Elements of Functional Analysis), and selections from Part III (Discrete Functional Analysis).

Exercises: This book has 760 exercises intended to help the reader understand the content and to develop skills in measure theory and functional analysis.

Symbols: A list of symbols in this book is given at the beginning of the index.

Acknowledgments: I thank my former students for finding typographical and mathematical errors in the manuscript. I thank my dear daughters-in-law, Karen, Tina, and Tracey, for thoroughly proofreading the manuscript. I also thank Nancy and Vojin Drenovac for their superb graphics. And I am indebted to John Wenger for his constructive comments and criticisms.

0
Preliminaries

It is assumed that you, the reader, have a background in undergraduate real analysis and linear algebra. In this preliminary chapter we collect some of the basic ideas about sets, real and complex numbers, linear algebra, Euclidean space $\mathbb{R}^n$, sequences and series, and inequalities. If you are self-studying this book, you may wish to start here and read all of the chapters in order. In an undergraduate or graduate course, it is reasonable to start with Chapter 1 (or Chapter 5, depending on the syllabus) and refer to the various sections in this chapter only as needed.

0.1 Sets

Here we give definitions, notation, and results on sets used in the book.

Sets are usually denoted by upper case letters, and their elements by lower case letters. If A is a set, then we write $a \in A$ to denote that a is an element of A, and $a \notin A$ to denote that a is not an element of A.

These are some standard sets that we take for granted:

$\emptyset$, the empty set,
$\mathbb{N} = \{1, 2, 3, \ldots\}$, the set of positive integers,
$\mathbb{Z} = \{0, +1, -1, +2, -2, \ldots\}$, the set of all integers,
$\mathbb{Q}$, the set of all rational numbers,
$\mathbb{R}$, the set of all real numbers,
$\mathbb{C}$, the set of all complex numbers,
$\mathbb{R}^n$, the set of all n-tuples $(a_1, \ldots, a_n)$ with $a_k \in \mathbb{R}$ for $k = 1, 2, \ldots, n$,
$\mathbb{C}^n$, the same except that $a_k \in \mathbb{C}$ for $k = 1, 2, \ldots, n$.

If A and B are sets, and if every element of A is an element of B, we say that A is a **subset** of B and write $A \subset B$, or equivalently $B \supset A$. We have $A = B$ if and only if $A \subset B$ and $B \subset A$.

If A and B are sets, the **union** of A and B is the set

$$A \cup B = \{x \mid x \in A \text{ or } x \in B\},$$

the **intersection** of A and B is the set

$$A \cap B = \{x \mid x \in A \text{ and } x \in B\},$$

and the **complement** of A with respect to B (or simply the **difference**) is

$$B - A = \{x \mid x \in B \text{ and } x \notin A\}.$$

Note that A need not be a subset of B in the definition of complement. If $B = U$ is understood to be a fixed **universe of discourse,** we write $A^c = U - A$. For example, in number theory, the universe of discourse is understood to be $U = \mathbb{N}$; thus, for $A = \{1,2,3,4\}$, the complement is $A^c = \{5,6,7,\ldots\}$, and its complement is $(A^c)^c = A = \{1,2,3,4\}$. For real analysis, $U = \mathbb{R}$, and for complex analysis, $U = \mathbb{C}$.

We can generalize the notions of union and intersection to any arbitrary collection of sets. Let $\mathcal{A}$ be a collection of sets, which may be finite, countably infinite, or uncountable. Such a collection $\mathcal{A}$ is sometimes referred to as a **family** of sets. Then the **union** of these sets is the set

$$\bigcup_{A \in \mathcal{A}} A = \left\{x \,\middle|\, x \in A \text{ for at least one } A \in \mathcal{A}\right\},$$

and similarly the **intersection** is the set

$$\bigcap_{A \in \mathcal{A}} A = \left\{x \,\middle|\, x \in A \text{ for all } A \in \mathcal{A}\right\}.$$

If the sets in $\mathcal{A}$ can be indexed by some set $\mathcal{I}$ (that is, the sets of $\mathcal{A}$ are all of the form A_κ for $\kappa \in \mathcal{I}$; and for each $\kappa \in \mathcal{I}$, there corresponds a set $A_\kappa \in \mathcal{A}$), then the union and intersection may be written

$$\bigcup_{\kappa \in \mathcal{I}} A_\kappa \quad \text{and} \quad \bigcap_{\kappa \in \mathcal{I}} A_\kappa, \text{ respectively.}$$

The set $\mathcal{I}$ is called the **index set** of the family $\mathcal{A}$. If $\mathcal{I} = \mathbb{N} = \{1,2,3,\ldots\}$, we write $\bigcup_{k=1}^{\infty} A_k$ and $\bigcap_{k=1}^{\infty} A_k$, respectively.

A collection of sets $\mathcal{A} = \{A_\kappa\}_{\kappa \in \mathcal{I}}$ is **pairwise disjoint** if

$$\kappa \neq \kappa' \quad \Rightarrow \quad A_\kappa \cap A_{\kappa'} = \emptyset.$$

The collection of *all* subsets of a set A is called the **power set** of A and is denoted by either 2^A or $P(A)$.

Example 0.1 Let $\mathcal{I} = (1, \infty) = \{x \in \mathbb{R} \mid x > 1\}$, be the open interval of reals with left end point 1. For each $x \in \mathcal{I}$, let A_x be the interval $A_x = [0, x) = \{y \in \mathbb{R} \mid 0 \leq y < x\}$. Then

$$\bigcap_{x \in \mathcal{I}} A_x = [0, 1] = \{x \in \mathbb{R} \mid 0 \leq x \leq 1\}, \text{ and} \tag{0.1}$$

$$\bigcup_{x \in \mathcal{I}} A_x = [0, \infty). \tag{0.2}$$

Often we need De Morgan's laws, which relate unions, intersections, and complements. We do not prove the theorem.

Theorem 0.1 De Morgan's laws: *For all families of sets, relative to some universal set U, we have*

$$\left(\bigcup_{\kappa \in \mathcal{I}} A_\kappa\right)^c = \bigcap_{\kappa \in \mathcal{I}} A_\kappa^c,$$

$$\left(\bigcap_{\kappa \in \mathcal{I}} A_\kappa\right)^c = \bigcup_{\kappa \in \mathcal{I}} A_\kappa^c.$$

We say that sets A and B are **conumerous** (or we say that A and B have the same **cardinality**) if there exists a one-to-one correspondence between the two sets. That is, there exists a function $f : A \longrightarrow B$ which is one-to-one and onto. Such a function is also called a **bijection**. A set A is called **countable** if it is either conumerous with $\mathbb{N}$ or with a subset of $\mathbb{N}$. Otherwise, A is called **uncountable**. Examples of countable sets are $\emptyset$, $\{5, 6, 7, 8\}$, $\{2, 4, 6, 8, \ldots\}$, $\mathbb{N}$, $\mathbb{Z}$, and $\mathbb{Q}$. The following theorem shows that $\mathbb{R}$ is uncountable.

Theorem 0.2 Cantor's theorem: *The set $\mathbb{R}$ is uncountable.*

Proof Assume $\mathbb{R}$ is countable. Then the interval $[0, 1)$ is countable as well. The set can thus be written in a sequence $x_1, x_2, x_3, \ldots$ of reals which lists *all* of the elements in the interval $[0, 1)$. Write out the decimal expansion of each x_k for $k = 1, 2, 3, \ldots$. Note that decimal expansions are *not* unique. For example, $0.26999\ldots$ is the same real number as $0.27000\ldots$. To make the expansion unique for each x_k, do not use any decimal expansion ending with all 9s but instead use the equivalent one ending with all 0s. The list looks like this.

$$
\begin{array}{lll}
x_1 & = 0.a_{11}a_{12}a_{13}a_{14}\cdots & = \dfrac{a_{11}}{10} + \dfrac{a_{12}}{10^2} + \dfrac{a_{13}}{10^3} + \dfrac{a_{14}}{10^4} + \cdots \\
x_2 & = 0.a_{21}a_{22}a_{23}a_{24}\cdots & = \dfrac{a_{21}}{10} + \dfrac{a_{22}}{10^2} + \dfrac{a_{23}}{10^3} + \dfrac{a_{24}}{10^4} + \cdots \\
x_3 & = 0.a_{31}a_{32}a_{33}a_{34}\cdots & = \dfrac{a_{31}}{10} + \dfrac{a_{32}}{10^2} + \dfrac{a_{33}}{10^3} + \dfrac{a_{34}}{10^4} + \cdots \\
. & \cdots\cdots\cdots\cdots & \cdots\cdots\cdots\cdots\cdots \\
. & \cdots\cdots\cdots\cdots & \cdots\cdots\cdots\cdots\cdots \\
. & \cdots\cdots\cdots\cdots & \cdots\cdots\cdots\cdots\cdots
\end{array}
$$

Now define a new decimal number $y = 0.b_1b_2b_3\ldots$ according to the rule

$$
b_k = \begin{cases} 1 & \text{if } a_{kk} = 0 \\ 0 & \text{if } a_{kk} \neq 0. \end{cases}
$$

Clearly $y \in [0,1)$ but y is different from all of the x_k in the list because $b_k \neq a_{kk}$ for all $k = 1, 2, \ldots$. This contradicts the statement that *all* of the elements of $[0,1)$ are in the list, which means that the assumption of the countability of $\mathbb{R}$ is false. ∎

Exercises for Section 0.1

0.1.1 Show that $A \subset B \iff A \cup B = B \iff A \cap B = A$.

0.1.2 Prove that $(A \cap B) \cup C = A \cap (B \cup C)$ if and only if $C \subset A$.

0.1.3 Prove Identity (0.1) of Example 0.1 by showing both inclusions:

$$
\bigcap_{x \in \mathcal{I}} A_x \subset [0,1] \quad \text{and} \quad [0,1] \subset \bigcap_{x \in \mathcal{I}} A_x.
$$

0.1.4 For each $x \in \mathcal{I} = (1, \infty)$, consider the open interval $A_x = (x, \infty) = \{y \in \mathbb{R} \mid x < y < \infty\}$. Find $\bigcap_{x \in \mathcal{I}} A_x$ and $\bigcup_{x \in \mathcal{I}} A_x$.

0.1.5 For each $x \in \mathcal{I} = (1, \infty)$, consider the open interval $A_x = (0, x) = \{y \in \mathbb{R} \mid 0 < y < x\}$. Find $\bigcap_{x \in \mathcal{I}} A_x$ and $\bigcup_{x \in \mathcal{I}} A_x$.

0.1.6 Give an example of a decreasing sequence of *nonempty* sets $A_1 \supset A_2 \supset \cdots$ such that $\bigcap_{k=1}^{\infty} A_k = \emptyset$. Give a second example where $\bigcap_{k=1}^{\infty} A_k$ has exactly one element.

0.1.7 Prove the identities of De Morgan laws in Theorem 0.1.

0.1.8 For any index set $\mathcal{I}$ prove this distributive property:

$$A \cap \left(\bigcup_{\kappa \in \mathcal{I}} A_\kappa \right) = \bigcup_{\kappa \in \mathcal{I}} (A \cap A_\kappa).$$

0.1.9 For any index set $\mathcal{I}$ prove this distributive property:

$$A \cup \left(\bigcap_{\kappa \in \mathcal{I}} A_\kappa \right) = \bigcap_{\kappa \in \mathcal{I}} (A \cup A_\kappa).$$

0.1.10 Show that the union of two countable sets is countable.

0.1.11 Show that if $A_1, A_2, \ldots$ is a countable collection of countable sets, then the union $A = \bigcup_{k=1}^{\infty} A_k$ is also countable. Hint: If $A_k = \{a_{k1}, a_{k2}, \ldots\}$, create the list $\{a_{11}, a_{12}, a_{21}, a_{13}, a_{22}, a_{31}, a_{14}, a_{23}, a_{32}, a_{41}, a_{15}, a_{24}, \ldots\}$.

0.1.12 Show that $\mathbb{Q}$ is countable. Hint: Use the preceding Exercise 0.1.11.

0.1.13 Determine whether the set of irrational numbers in $\mathbb{R}$ is countable.

0.1.14 Let $A_1, A_2, \ldots$ be a sequence of sets. For each $k = 1, 2, \ldots$, define the set $B_k = \bigcup_{i=k}^{\infty} A_i$. Note that $B_1 \supset B_2 \supset B_3 \supset \cdots$. Then define the set $\limsup A_i = \bigcap_{k=1}^{\infty} B_k$. Show that $\limsup A_i = \{x \mid x \in A_i$ for infinitely many $i\}$.

0.1.15 Let $A_1, A_2, \ldots$ be a sequence of sets. For each $k = 1, 2, \ldots$, define the set $C_k = \bigcap_{i=k}^{\infty} A_i$. Note that $C_1 \subset C_2 \subset C_3 \subset \cdots$. Then define $\liminf A_i = \bigcup_{k=1}^{\infty} C_k$. Show that $\liminf A_i = \{x \mid x \in A_i$ for all but finitely many $i\}$.

0.1.16 Referring to Exercises 0.1.14 and 0.1.15, show that $\liminf A_i \subset \limsup A_i$.

0.1.17 Let $A_1 \supset A_2 \supset \cdots$ be a decreasing sequence of sets. Referring to Exercises 0.1.14 and 0.1.15, show that

$$\liminf A_i = \limsup A_i = \bigcap_{i=1}^{\infty} A_i.$$

0.1.18 Referring to Exercise 0.1.17, state and prove a corresponding result for an increasing sequence of sets $A_1 \subset A_2 \subset \cdots$.

0.1.19 Show that any open interval $(a, b,) \subset \mathbb{R}$ is conumerous with the interval $(0, 1)$.

0.1.20 Show that the open interval $(0, 1)$ is conumerous with the interval $[0, 1)$.

0.1.21 Show that the open interval $(0, 1)$ is conumerous with $\mathbb{R}$.

0.1.22 Can an uncountable union of distinct sets be countable? If yes, give an example. If no, prove it.

0.2 Properties of $\mathbb{R}$ and $\mathbb{C}$

We say that a set of reals $A \subset \mathbb{R}$ has an **upper bound** $b \in \mathbb{R}$ if $x \le b$ for all $x \in A$ (or we say A is **bounded above**). For a set $A \subset \mathbb{R}$ which is bounded above, a **supremum**, written $m = \sup A$, is an upper bound of A satisfying the condition $m \le b$ for all other upper bounds b of A. An important property of the real numbers is the **completeness property**, which we assume.

Axiom Completeness property of $\mathbb{R}$: Every nonempty set $A \subset \mathbb{R}$ which is bounded above has a unique supremum $\sup A$.

An equivalent formulation of the completeness property is the following nested interval theorem.

Theorem 0.3 Nested interval: *Let $I_k = [a_k, b_k]$, $k = 1, 2, \ldots$ be a sequence of closed, bounded, and nonempty intervals in $\mathbb{R}$ such that $I_{k+1} \subset I_k$ for each k. Then $\bigcap_{k=1}^{\infty} I_k \neq \emptyset$. Furthermore, if* $\operatorname{diam} I_k = b_k - a_k$ *tends to zero, then this intersection consists of a single point.*

Functional analysis is usually undertaken in the setting of complex numbers. For this reason, we now give a short introduction to complex numbers $\mathbb{C}$. However, most results in this book *could* be restricted to real numbers $\mathbb{R}$.

A **complex number** z is a number expressed in two dimensions $z = (x, y)$ with $x, y \in \mathbb{R}$. We often use symbols $a, b, c, d, r, s, t, u, v, x, y$ for real numbers, and z, w for complex numbers. If we write $\mathbf{1} = (1, 0)$ and $i = (0, 1)$, then we have the representation

$$z = (x, y) = x(1, 0) + y(0, 1) = x\mathbf{1} + yi.$$

Generally, the symbol $\mathbf{1} = (1, 0)$ is suppressed and often we write iy instead of yi. So

$$z = (x, y) = x + iy.$$

For $z = (x, y)$ and $w = (u, v)$, addition is coordinatewise:

$$z + w = (x, y) + (u, v) = (x + u, y + v),$$

and multiplication is defined as follows:

$$z \cdot w = (xu - yv, xv + yu) = (xu - yv) + i(xv + yu).$$

As a consequence,

$$\mathbf{1} \cdot z = z, \quad x\mathbf{1} + u\mathbf{1} = (x + u)\mathbf{1}, \quad x\mathbf{1} \cdot u\mathbf{1} = xu\mathbf{1}, \quad \text{and} \quad i^2 = i \cdot i = -\mathbf{1}.$$

We write $\operatorname{Re}(x + iy) = x$ and $\operatorname{Im}(x + iy) = y$. We say z is **real** if $\operatorname{Im}(z) = y = 0$, and we say z is **imaginary** if $\operatorname{Re}(z) = x = 0$.

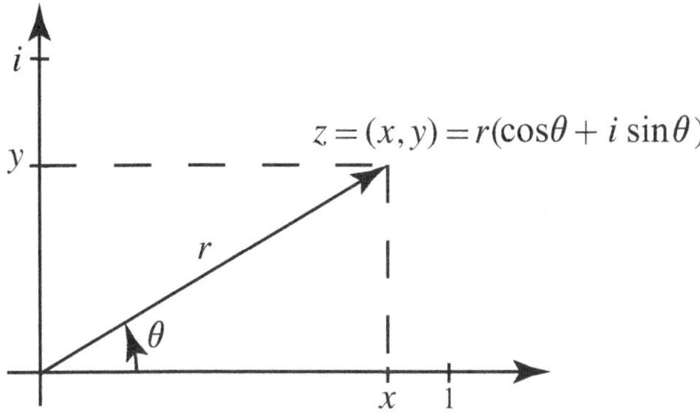

Figure 0.1 The complex number $z = x + iy = r(\cos\theta + i\sin\theta)$.

For $z = x + iy \in \mathbb{C}$, the **complex conjugate** of z is

$$\overline{z} = x - iy.$$

We easily obtain

$$\mathrm{Re}(z) = \frac{z + \overline{z}}{2}, \qquad \mathrm{Im}(z) = \frac{z - \overline{z}}{2i}, \qquad \text{and} \qquad z \cdot \overline{z} = x^2 + y^2.$$

The **absolute value** or **length** or **magnitude** of $z = x + iy$ is

$$|z| = \sqrt{x^2 + y^2} = \sqrt{z \cdot \overline{z}}.$$

When $z = x$ is real, $|z|$ is the usual absolute value of the real number x. Note also

$$|\mathrm{Re}(z)| \leq |z| \quad \text{and} \quad |\mathrm{Im}(z)| \leq |z|. \tag{0.3}$$

Definition 0.1 The **argument** (or **amplitude**) of the complex number $z = x + iy$ is the angle θ formed by the two-dimensional vector z. If we write $r = |z|$ and $\theta = \arg(z)$, then $x = r\cos\theta$ and $y = r\sin\theta$ (see Figure 0.1). Thus $z = r(\cos\theta + i\sin\theta)$. Note that the argument of z is not unique. The arguments θ and $\theta + 2\pi$ define the same complex number z. For example, $z = 2\big(\cos\frac{\pi}{6} + i\sin\frac{\pi}{6}\big) = 2\big(\cos\frac{13\pi}{6} + i\sin\frac{13\pi}{6}\big) = \sqrt{3} + i$.

If $z = x + iy = r(\cos\theta + i\sin\theta)$ and $w = u + iv = s(\cos\phi + i\sin\phi)$, then

$$\begin{aligned} z \cdot w &= r \cdot s(\cos\theta + i\sin\theta)(\cos\phi + i\sin\phi) \\ &= r \cdot s\{(\cos\theta\cos\phi - \sin\theta\sin\phi) + i(\cos\theta\sin\phi + \sin\theta\cos\phi)\}. \end{aligned}$$

By the trigonometric addition formulas

$$\cos(\alpha+\beta) = \cos\alpha\cos\beta - \sin\alpha\sin\beta, \tag{0.4}$$

$$\sin(\alpha+\beta) = \cos\alpha\sin\beta + \sin\alpha\cos\beta, \tag{0.5}$$

we have

$$z \cdot w = r \cdot s\{\cos(\theta+\phi) + i\sin(\theta+\phi)\}.$$

Thus the product of complex numbers multiplies the absolute value and adds the arguments. For this reason, Euler's notation,

$$e^{i\theta} = \cos\theta + i\sin\theta, \tag{0.6}$$

is used to obtain

$$z \cdot w = re^{i\theta} \cdot se^{i\phi} = rse^{i(\theta+\phi)}.$$

Euler's formula may be extended to all complex z as follows. The complex exponential function e^z may be defined as

$$e^z = e^{x+iy} = e^x \cdot e^{iy} = e^x(\cos y + i\sin y).$$

Then e^z has magnitude e^x and amplitude y.

Exercises for Section 0.2

0.2.1 For any real x and y with $y > 0$, show $-y \le x \le y \iff |x| \le y$.

0.2.2 For any real x and y, prove $\max\{x, y\} = \frac{1}{2}(x + y + |x - y|)$.

0.2.3 Use the completeness property of $\mathbb{R}$ to prove the nested interval theorem (Theorem 0.3).

0.2.4 Use the nested interval theorem (Theorem 0.3) to prove the completeness property of $\mathbb{R}$.

0.2.5 For any complex numbers z and w, show that $\overline{z+w} = \overline{z} + \overline{w}$ and $\overline{z \cdot w} = \overline{z} \cdot \overline{w}$.

0.2.6 For any nonzero complex number $z = x + iy$, show that $z^{-1} = \frac{\overline{z}}{|z|^2} = \frac{x-iy}{x^2+y^2}$.

0.2.7 For $z = 2 - i5$ and $w = 1 + i2$, find $z + w$, $z - w$, $z \cdot w$, $\frac{z}{w}$, $\frac{1-z}{w+1}$. Write each in Cartesian form $a + ib$.

0.2.8 For $z = 1 + i\sqrt{3}$ and $w = 1 + i(\sqrt{3}/3)$, find the magnitude and amplitude of $z, w, z \cdot w$, and z/w. Write each in polar form $re^{i\theta} = r(\cos\theta + i\sin\theta)$.

0.2.9 For any complex z and w, show $|z \cdot \overline{w}| = |\overline{z} \cdot w| = |z \cdot w|$.

0.2.10 For any complex z and w, show $|z + w| \leq |z| + |w|$.

0.2.11 Show $\big||z| - |w|\big| \leq |z - w|$. Hint: First use Exercise 0.2.10 to show that $|z| - |w| \leq |z - w|$. Then interchange z and w to obtain $-|z - w| \leq |z| - |w|$.

0.2.12 Prove that for any $z = x + iy$, we have $\max\{|x|, |y|\} \leq |z| \leq \sqrt{2}\max\{|x|, |y|\}$.

0.2.13 Let $p \in \mathbb{R}$ with $1 \leq p < \infty$. Use the inequality $|z + w| \leq 2\max\{|z|, |w|\}$ to show that for complex numbers z, w, we have $|z + w|^p \leq 2^p(|z|^p + |w|^p)$.

0.3 Linear Spaces

We use the terms **vector space** and **linear space** interchangeably. In such a space we have addition of **vectors** as well as multiplication of vectors by **scalars.** For the study of algebra, the scalars are in some algebraic field. For analysis, the field will be either $\mathbb{R}$ or $\mathbb{C}$.[1] That is, scalars are always assumed to range over either the real or complex numbers. We will use the same symbol for the zero vector and the 0 scalar. Generally, there is little risk of confusion. Here is a formal definition.

Definition 0.2 A **linear space** consists of a set V with two operations, vector addition and scalar multiplication. The operations have to satisfy the following conditions for all $x, y, z \in V$ and all scalars $\alpha, \beta \in \mathbb{R}$ (or $\mathbb{C}$).

(VA0) $x + y \in V$
(VA1) $x + (y + z) = (x + y) + z$
(VA2) $x + y = y + x$
(VA3) $x + 0 = x$
(VA4) For each $x \in V$, there exists $-x \in V$ such that $x + (-x) = 0$
(VM0) $\alpha x \in V$
(VM1) $(\alpha\beta)x = \alpha(\beta x)$
(VM2) $(\alpha + \beta)x = \alpha x + \beta x$
(VM3) $\alpha(x + y) = \alpha x + \alpha y$
(VM4) $1x = x$.

[1] Just as the completeness property is basic to real analysis, the completeness property for finite-dimensional linear spaces (Corollary 5.47) is basic to functional analysis. For example, the Cauchy criterion (Theorem 0.9) for convergence holds in $\mathbb{R}^n$ but not in $\mathbb{Q}^n$.

The algebraic closure conditions **(VA0)** and **(VM0)** are implicit in the term "operation."

Theorem 0.4 *Suppose that V is known to be a linear space. Then a nonempty subset $W \subset V$ will be a vector space if the following two conditions are satisfied:*

(SS1) *If $x, y \in W$, then $x + y \in W$. We say W is* ***algebraically closed*** *under vector addition.*

(SS2) *If $x \in W$ and α is a scalar, then $\alpha x \in W$. W is* ***algebraically closed*** *under scalar multiplication.*

In this case, we say that W is a ***subspace*** *of V.*

You are familiar with finite-dimensional vector spaces such as $\mathbb{R}^n$ and $\mathbb{C}^n$. In functional analysis, vector spaces are usually infinite-dimensional. Examples of infinite-dimensional vector spaces include **sequence spaces** such as ω, ℓ_∞, c, c_0, cs, bs, and ℓ_p $(1 \le p < \infty)$, defined later by Equations (0.7) through (0.15). Infinite-dimensional **function spaces** include the following (here $a < b$):

$P(\mathbb{R})$, the space of all real valued polynomials,
$C[a,b]$, the space of all continuous functions on the interval $[a,b]$,
$C^k[a,b]$, the space of all functions with k continuous derivatives on $[a,b]$,
$C^\infty[a,b]$, the space of functions f for which with the kth derivative $f^{(k)}$ exists on $[a,b]$ for *all* $k = 1, 2, 3, \ldots$,
$\mathcal{R}[a,b]$, the space of all Riemann integrable functions on $[a,b]$.

For any nonempty subset A of a vector space V, a **linear combination** of A is a vector of the form $\alpha_1 x_1 + \cdots + \alpha_n x_n$, where $x_1, \ldots, x_n \in A$ and $\alpha_1, \ldots, \alpha_n$ are scalars. The collection of all linear combinations of A is called the **span** or **linear hull** of A and denoted $\operatorname{span}(A)$. The span of A is a vector subspace of V and is the smallest subspace that contains A. If $\operatorname{span}(A) = W$, we say A **spans** W.

A set of vectors A, whether finite or infinite, is **linearly independent** if for any *finite* set of distinct elements $x_1, \ldots, x_n$ in A, we have the implication

$$\alpha_1 x_1 + \cdots + \alpha_n x_n = 0 \implies \alpha_1 = \cdots = \alpha_n = 0.$$

A subset $B \subset V$ is an **algebraic basis** of V (also known as a **Hamel basis**) if B is linearly independent and B spans V. In this case, every nonzero vector in V is a unique linear combination of B; that is, for each nonzero $x \in V$, there exist unique (except for rearrangement) vectors $x_1, \ldots, x_n$ in B with unique nonzero scalars $\alpha_1, \ldots, \alpha_n \in \mathbb{R}$ (or $\mathbb{C}$) such that

$$x = \alpha_1 x_1 + \cdots + \alpha_n x_n.$$

A vector space V is **finite-dimensional** if it has a finite algebraic basis, otherwise V is said to be **infinite-dimensional.**

The space $\mathbb{R}^n$ has an algebraic basis $B = \{e^1, \ldots, e^k, \ldots, e^n\}$, where

$$e^k = (0, \ldots, 0, 1, 0, \ldots, 0) \text{ with 1 in the } k\text{th position.}$$

An important result in linear algebra is that for a given finite-dimensional vector space V, every algebraic basis has the same number of elements. This number is called the **dimension** of the vector space. Indeed, even for a given infinite-dimensional vector space, any two algebraic bases are conumerous. Thus $\mathbb{R}^n$ has dimension n. Similarly, $\mathbb{C}^n$ has dimension n, when considered over the field of scalars $\mathbb{C}$ (and has dimension $2n$ when considered over the field $\mathbb{R}$).

The space $P(\mathbb{R})$ of all real valued polynomials is infinite-dimensional with a *countable* algebraic basis of polynomials $B = \{1, x, x^2, x^3, \ldots\}$.

Similarly, Φ, the sequence space of all sequences with only a finite number of nonzero terms, has a countable algebraic basis $B = \{e^k \mid k = 1, 2, 3 \ldots\}$, where

$$e^k = (0, \ldots, 0, 1, 0, \ldots) \text{ with 1 in the } k\text{th position.}$$

However, the spaces $C[0,1]$, $C^k[0,1]$, and $\mathcal{R}[0,1]$ are infinite-dimensional with no countable algebraic basis. The same holds for the following spaces, defined in the next section ω, c, c_0, cs, bs, ℓ_p $(1 \leq p \leq \infty)$.

Definition 0.3 Let V and W be linear spaces. A function $T\colon V \longrightarrow W$ is a **linear operator** or **linear map** if for all $x, y \in V$ and all scalars α, β, we have

$$T(\alpha x + \beta y) = \alpha T(x) + \beta T(y).$$

If the codomain W is the same as the domain V, we say that T is a **linear operator on** V. If the codomain W is the field of scalars $\mathbb{C}$ or $\mathbb{R}$, we say that T is a **linear functional on** V. By convention we sometimes write Tx instead of $T(x)$.

Exercises for Section 0.3

0.3.1 Let V be a vector space. Use the axioms for a linear space to prove that $0x = 0$ and $(-1)x = -x$ for all $x \in V$.

0.3.2 Let V be a vector space. Show that the intersection of any family of subspaces of V is a subspace of V.

0.3.3 Let W_1 and W_2 be subspaces of V, and define $W_1 + W_2 = \{x_1 + x_2 \mid x_1 \in W_1, x_2 \in W_2\}$. Show that $W_1 + W_2$ is a subspace of V.

0.3.4 Show that $C^k[a,b]$, as defined in Section 0.3, is a linear space. Also show that it is infinite-dimensional.

0.3.5 Show that the set of power functions $\{1, x, x^2, \ldots\}$ is linearly independent in the space $P(\mathbb{R})$ of polynomials.

0.3.6 Prove Theorem 0.4.

0.3.7 Suppose B is a (finite or infinite) algebraic basis of V. Show that for each nonzero $x \in V$, there exist *unique* (up to rearrangement) vectors $x_1, \ldots, x_n \in B$ and *unique* nonzero scalars $\alpha_1, \ldots, \alpha_n$ such that

$$x = \alpha_1 x_1 + \cdots + \alpha_n x_n.$$

0.3.8 For any linear operator $T\colon V \longrightarrow W$, prove that $T0 = 0$, where the first 0 is the zero vector of V and the second 0 is the zero vector of W.

0.3.9 Consider the space $V = C^\infty[0,1]$ as defined in Section 0.3. Show that the derivative operator $Df = f'$ is a linear operator on V.

0.3.10 Consider the linear space $\mathbb{R}^n$. For any fixed scalars $\alpha_1, \ldots, \alpha_n \in \mathbb{R}$, show that the function $Tx = \alpha_1 x_1 + \cdots + \alpha_n x_n$ defines a linear functional on $\mathbb{R}^n$. Additionally, show that every linear functional on $\mathbb{R}^n$ can be expressed in this form.

0.4 Sequences and Series

A **real sequence** x is a function $x\colon \mathbb{N} \longrightarrow \mathbb{R}$. That is, each $k \in \mathbb{N}$ is associated with a real number $x(k)$, written conveniently as x_k. Also, we write

$$x = (x_k) = (x_1, x_2, \ldots).$$

Occasionally, a sequence starts with x_0

$$x = (x_0, x_1, x_2, \ldots).$$

A **complex sequence** is one where, for all k, $x_k \in \mathbb{C}$ instead of $x_k \in \mathbb{R}$.

Denote the set of *all* real (or complex) sequences by

$$\omega = \{x = (x_k) \mid \forall k,\ x_k \in \mathbb{R} \ \ (\text{or } x_k \in \mathbb{C})\}. \tag{0.7}$$

If we define the operation of vector addition in ω by the identity

$$x + y = (x_k) + (y_k) = (x_k + y_k) \quad \text{for} \quad x, y \in \omega,$$

and the operation of scalar multiplication by

$$\alpha x = \alpha(x_k) = (\alpha x_k) \quad \text{for} \quad x \in \omega \quad \text{and} \quad \alpha \in \mathbb{R} \text{ (or } \mathbb{C}),$$

then ω forms a vector space.

We say that a sequence $x \in \omega$ is **bounded** if there exists $M \geq 0$ such that $|x_k| \leq M$ for all $k \in \mathbb{N}$. The set of all bounded sequences forms a subspace of ω which is denoted by

$$\ell_\infty = \{x \in \omega \mid \exists\, M \geq 0 \text{ such that } |x_k| \leq M\ \forall k\}. \tag{0.8}$$

We say that a sequence $x \in \omega$ is **convergent** if for some $L \in \mathbb{R}$ (or $\mathbb{C}$) and for every given $\epsilon > 0$, there exists $N > 0$ such that $|x_k - L| < \epsilon$ whenever $k > N$. We write $L = \lim_{k\to\infty} x_k$ or $x_k \to L$ as $k \to \infty$. The set of all convergent sequences is another subspace of ω denoted by

$$c = \{x \in \omega \mid \exists\, L \text{ such that } x_k \to L \text{ as } k \to \infty\}. \tag{0.9}$$

If $L = 0$, we say that x is a **null sequence**. The set of all null sequences is a subspace of c denoted by

$$c_0 = \{x \in c \mid x_k \to 0 \text{ as } k \to \infty\}. \tag{0.10}$$

A sequence $x \in \omega$ is a **Cauchy sequence** if for every given $\epsilon > 0$, there exists $N > 0$ such that $|x_k - x_j| < \epsilon$ whenever $k, j > N$.

The inequality

$$|x_k - x_j| = |x_k - L + L - x_j| \leq |x_k - L| + |L - x_j| = |x_k - L| + |x_j - L|,$$

which is less than ϵ whenever $|x_k - L|$ and $|x_j - L|$ are each less than $\epsilon/2$, shows that every convergent sequence is a Cauchy sequence. The converse is the **Cauchy criterion for convergence** in $\mathbb{R}$ (or $\mathbb{C}$) stated below, which is equivalent to the completeness axiom as stated in Section 0.2. We do not prove this Cauchy criterion for convergence but refer you to undergraduate-level real analysis.

Theorem 0.5 Cauchy criterion for convergence: *In $\mathbb{R}$ or $\mathbb{C}$ every Cauchy sequence is convergent.*

An **infinite series** of a sequence $x \in \omega$, written $\sum x_k$, is the sequence $s = (s_n)$ of partial sums $s_n = x_1 + x_2 + \cdots + x_n$,

$$\sum x_k = (s_1, s_2, \ldots).$$

If the series converges to L, that is, $\lim_{n\to\infty} s_n = L$, then we write $\sum_{k=1}^{\infty} x_k = L$. The set of all sequences $x \in \omega$ which have a convergent series forms a subspace of ω denoted by

$$cs = \left\{ x \in \omega \,\middle|\, \sum x_k = (s_1, s_2, \ldots) \text{ converges} \right\}. \tag{0.11}$$

The sequence space cs is a proper subset of c_0. For example, the sequence $x = \left(1, \frac{1}{2}, \frac{1}{3}, \ldots, \frac{1}{k}, \ldots\right)$ is in c_0 because $\lim_{k\to\infty} \frac{1}{k} = 0$. Yet this series of x is the harmonic series, which does not converge (that is, $x \notin cs$).

The set of all sequences $x \in \omega$ which have a **bounded series** forms a subspace of ω denoted by

$$bs = \left\{ x \in \omega \,\middle|\, \sum x_k = (s_1, s_2, \ldots) \text{ is a bounded sequence} \right\}. \tag{0.12}$$

For example, the sequence $x = (+1, -1, +1, -1, \ldots)$ has partial sums

$$s_n = x_1 + \cdots + x_n = \begin{cases} 1 & \text{if } n = 1, 3, 5, \ldots \\ 0 & \text{if } n = 2, 4, 6, \ldots. \end{cases}$$

Since this (s_n) is bounded but not convergent, the sequence x is in bs but not in cs.

The space of all sequences $x \in \omega$ whose series are absolutely convergent is denoted by ℓ or ℓ_1,

$$\ell = \ell_1 = \left\{ x \in \omega \,\middle|\, \sum_{k=1}^{\infty} |x_k| < \infty \right\}. \tag{0.13}$$

We have

$$\ell_1 \subset cs \subset c_0 \subset c \subset \ell_\infty \quad \text{and} \quad cs \subset bs \subset \ell_\infty. \tag{0.14}$$

Finally, for any $1 \le p < \infty$, we define

$$\ell_p = \left\{ x \in \omega \,\middle|\, \sum_{k=1}^{\infty} |x_k|^p < \infty \right\}. \tag{0.15}$$

We can use the inequality $|z + w|^p \le 2^p(|z|^p + |w|^p)$ of Exercise 0.2.13 to show that ℓ_p is a vector space for any $1 \le p < \infty$. We can also use Jensen's inequality (Theorem 0.58) to show that the ℓ_p spaces become larger with p. That is,

$$\ell_1 \subset \ell_q \subset \ell_p \subset \ell_\infty \quad \text{for} \quad 1 < q < p < \infty. \tag{0.16}$$

Exercises for Section 0.4

0.4.1 Give an example of a proper infinite-dimensional subspace of ℓ_1.

0.4.2 Let $y \in \ell_\infty$. Show that for $x \in \ell_1$, $T_y x = (y_1x_1, y_2x_2, y_3x_3, \ldots)$ defines a linear operator on ℓ_1.

0.4.3 Let $y \in \ell_\infty$. Show that for $x \in c$, $T_y x = (y_1x_1, y_2x_2, y_3x_3, \ldots)$ defines a linear operator $T : c \longrightarrow \ell_\infty$.

0.4.4 Show that a sequence can have at most one limit.

0.4.5 Let $x_k = \frac{k}{k+1}$ for all $k = 1, 2, \ldots$. Use the definition of limit to formally prove $\lim_{k\to\infty} x_k = 1$.

0.4.6 Let $x_k = 1/k$ for all $k = 1, 2, \ldots$. Formally prove that $x \in c_0$ but $x \notin cs$.

0.4.7 Let $x_k = i^k$ for all $k = 1, 2, \ldots$. Prove that the sequence has no limit.

0.4.8 Let $x_k = (i/2)^k$ for all $k = 1, 2, \ldots$. Prove that $x \in c_0$.

0.4.9 Let $x_k = (i/2)^k$ for all $k = 1, 2, \ldots$. Prove $x \in cs$ and find $\sum_{k=1}^{\infty} x_k = L$.

0.4.10 Let $x_k = \frac{i}{k(k+1)}$ for all $k = 1, 2, \ldots$. Show that $\sum_{k=1}^{\infty} x_k = i$.

0.4.11 Suppose $x, y \in c$ and $x_k < y_k$ for all k. Show by example that $\lim_{k\to\infty} x_k$ need not be strictly less than $\lim_{k\to\infty} y_k$.

0.4.12 Suppose x is a real sequence with $\lim_{k\to\infty} x_k = L > 0$. Prove that the sequence is eventually positive.

0.4.13 Suppose x is a complex sequence and $\lim_{k\to\infty} x_k = L \in \mathbb{C}$. Prove that $\lim_{k\to\infty} |x_k| = |L| \in \mathbb{R}$.

0.4.14 **Squeezing theorem:** Suppose x, y, z are real sequences with $x_k \leq y_k \leq z_k$ for all k. Show that if $\lim_{k\to\infty} x_k = \lim_{k\to\infty} z_k = L \in \mathbb{R}$, then $\lim_{k\to\infty} y_k = L$.

0.4.15 In the definition of $\lim_{k\to\infty} x_k = L$, show that the condition $|x_k - L| < \epsilon$ may be replaced by $|x_k - L| < c\epsilon$ for any fixed $c > 0$, or even $|x_k - L| \leq \epsilon$.

0.4.16 Verify that ℓ_∞ (Section 0.4) is a linear space. Also show that it is infinite-dimensional.

0.4.17 Show by example that all of the inclusions below are strict.

$$\ell_1 \subset cs \subset c_0 \subset c \subset \ell_\infty \subset \omega.$$

0.4.18 Show that $cs \subset bs \subset \ell_\infty$ and that these inclusions are strict.

0.4.19 Show that $L = \lim_{k\to\infty} x_k$ if and only if $L = x_1 + \sum_{k=1}^{\infty}(x_{k+1} - x_k)$.

0.4.20 Show ℓ_p is a linear space for any $1 \leq p < \infty$. Hint: Use Exercise 0.2.13.

0.4.21 Find an example of a sequence in ℓ_∞ which is not in any ℓ_p for any $p > 1$.

0.4.22 Find an example of a sequence which is in every ℓ_p for $p > 1$, but which is not in ℓ_1.

0.5 Euclidean Space

Let $x \in \mathbb{R}^n$. The **Euclidean norm** (sometimes called the **quadratic norm**) of x is defined to be

$$\|x\|_2 = \sqrt{x_1^2 + \cdots + x_n^2}.$$

Often we suppress the subscript $\|\cdot\|_2$ and simply write

$$\|x\| = \|x\|_2.$$

The space $\mathbb{R}^n$ along with the Euclidean norm is called **Euclidean space** or **Euclidean n-space**.

Theorem 0.6 *The Euclidean norm has the following properties for all x, $y \in \mathbb{R}^n$ and all $\alpha \in \mathbb{R}$:*

(E0) $\|x\| \geq 0$;
(E1) $\|x\| = 0$ *if and only if* $x = 0$;
(E2) $\|\alpha x\| = |\alpha| \|x\|$ (absolute homogeneity);
(E3) $\left|\sum_{k=1}^n x_k y_k\right| \leq \|x\| \|y\|$ (Cauchy's inequality);
(E4) $\|x + y\| \leq \|x\| + \|y\|$ (triangle inequality).

Proof The properties **(E0)**, **(E1)**, and **(E2)** are easy to prove.

Cauchy's inequality **(E3)** can be proven as follows. If either $x = 0$ or $y = 0$, then both sides of **(E3)** are zero; so the inequality holds. Next consider the case $\|x\| = \|y\| = 1$. Using the fact that for $x_k, y_k \in \mathbb{R}$ we have $0 \leq (|x_k| - |y_k|)^2$, it follows that $|x_k y_k| \leq \frac{1}{2}(x_k^2 + y_k^2)$. Summing from $k = 1$ to n, we obtain

$$\left|\sum_{k=1}^n x_k y_k\right| \leq \sum_{k=1}^n |x_k y_k| \leq \frac{1}{2}\left(\|x\|^2 + \|y\|^2\right) = \frac{1}{2}(1^2 + 1^2) = 1. \quad (0.17)$$

Then, for general nonzero x and y, the norms of $\dfrac{x}{\|x\|}$ and $\dfrac{y}{\|y\|}$ are equal to 1, so by (0.17) we have

$$\left|\sum_{k=1}^n \left(\frac{x_k}{\|x\|}\right)\left(\frac{y_k}{\|y\|}\right)\right| \leq 1.$$

Multiplying by $\|x\|$ and $\|y\|$, results in the Cauchy inequality **(E3)**.

Finally, the triangle inequality **(E4)** can be proven from Cauchy's inequality **(E3)** as follows.

$$\begin{aligned}\|x+y\|^2 &= \sum_{k=1}^{n}(x_k+y_k)^2 \le \sum_{k=1}^{n}(x_k^2+2|x_k y_k|+y_k^2)\\ &\le \|x\|^2+2\|x\|\|y\|+\|y\|^2 = (\|x\|+\|y\|)^2.\end{aligned}$$ ∎

The **Euclidean metric** or **Euclidean distance** is the function

$$d(x,y) = \|x-y\| \quad \text{for} \quad x,y \in \mathbb{R}^n.$$

It has the following properties for all $x,y,z \in \mathbb{R}^n$:

(M0) $d(x,y) \ge 0$;
(M1) $d(x,y) = 0$ if and only if $x = y$;
(M2) $d(x,y) = d(y,x)$ (symmetry);
(M3) $d(x,y) \le d(x,z) + d(z,y)$ (triangle inequality).

Sequences of elements of $\mathbb{R}^n$ are denoted with superscripts $x^1, x^2, \ldots$ to distinguish them from coordinates. The kth element of the sequence is thus $x^k = (x_1^k, x_2^k, \ldots, x_n^k)$. Convergent and Cauchy sequences in $\mathbb{R}^n$ are defined as in $\mathbb{R}$.

Definition 0.4 A sequence $x^1, x^2, \ldots, x^k, \ldots$ in $\mathbb{R}^n$, also written (x^k), is said to be **convergent** if there exists an $x \in \mathbb{R}^n$ such that $d(x^k, x) \to 0$ as $k \to \infty$. A **Cauchy** sequence is one for which $d(x^k, x^j) \to 0$ as $k, j \to \infty$.

Theorem 0.7 *A sequence (x^k) in $\mathbb{R}^n$ is Cauchy if and only if for each fixed coordinate $m = 1, 2, \ldots, n$, the sequence x_m^k is Cauchy in $\mathbb{R}$.*

Proof For each fixed coordinate $m = 1, 2, \ldots, n$, we have

$$\begin{aligned}|x_m^k - x_m^j| &\le d(x^k, x^j)\\ &= \sqrt{(x_1^k - x_1^j)^2 + \cdots + (x_m^k - x_m^j)^2 + \cdots + (x_n^k - x_n^j)^2}.\end{aligned}$$

Thus $d(x^j, x^k) \to 0$ if and only if we have $|x_m^k - x_m^j| \to 0$ for each m. ∎

Corollary 0.8 *A sequence x^k in $\mathbb{R}^n$ is convergent if and only if it is* **coordinatewise convergent**.

The Cauchy criterion for convergence (Theorem 0.5) in $\mathbb{R}$ can thus be extended to $\mathbb{R}^n$ as follows.

Theorem 0.9 *Cauchy criterion for $\mathbb{R}^n$: A sequence in $\mathbb{R}^n$ is Cauchy if and only if it is convergent.*

Definition 0.5 Suppose $x^1, x^2, \ldots$ is a sequence and $k_1 < k_2 < k_3 \cdots$ is a strictly increasing sequence of natural numbers. Then the sequence $x^{k_1}, x^{k_2}, x^{k_3}, \ldots$ is called a **subsequence** of $x^1, x^2, \ldots$.

Exercises for Section 0.5

0.5.1 Prove the properties **(M0)** through **(M4)** of the Euclidean metric.

0.5.2 Prove the inequality $\big|\, \|x\| - \|y\| \,\big| \leq \|x - y\|$ for all $x, y \in \mathbb{R}^n$.

0.5.3 Prove the inequality $|d(x,z) - d(z,y)| \leq d(x,y)$ for all $x, y, z \in \mathbb{R}^n$.

0.5.4 Suppose $x^1, x^2, \ldots$ is a Cauchy sequence in $\mathbb{R}^n$. Show that the set $A = \{x^1, x^2, \ldots\} \subset \mathbb{R}^n$ is bounded. That is, there exists a real number $M > 0$ such that $\|x^k\| \leq M$ for all k.

0.5.5 Suppose $x^1, x^2, \ldots$ and $y^1, y^2, \ldots$ are Cauchy sequences in $\mathbb{R}^n$. Show that $d(x^k, y^k)$ converges in $\mathbb{R}$.

0.5.6 Show that any subsequence of a convergent sequence in $\mathbb{R}^n$ izs convergent to the same limit.

0.5.7 Show that any subsequence of a Cauchy sequence in $\mathbb{R}^n$ is a Cauchy sequence.

0.5.8 Suppose $x^1, x^2, \ldots$ is a Cauchy sequence in $\mathbb{R}^n$ with a subsequence that is convergent. Show that $x^1, x^2, \ldots$ is convergent.

0.6 Topology of Euclidean Space

In this chapter, the term **topology**[2] refers to the study of closeness and continuity in Euclidean space $\mathbb{R}^n$. This includes topics such as open and closed sets, boundaries and limits, continuity, convergence, and compactness.

Definition 0.6 For every $x \in \mathbb{R}^n$ and $0 < r < \infty$ we define the **open sphere** with **center** x and **radius** r to be the set

$$S(x,r) = \{y \in \mathbb{R}^n \mid d(x,y) < r\}.$$

This is often called an n-dimensional **open ball**. It includes its interior but not its surface.

[2] The term **topology** was introduced by the German mathematician Johann Benedict Listing (1808–1882) in a paper referring to the study of properties that are invariant under continuous deformations such as bending, twisting, and stretching but without tearing or gluing (*Vorstudien zur Topologie*, Vandenhoeck und Ruprecht, Göttingen 1848, p. 67).

For $n = 1$, the open sphere $S(x, r)$ is the open interval $(x - r, x + r)$.
For $n = 2$, it is the open disk with center $x = (x_1, x_2)$ and radius r.
For $n = 3$, it is the open ball with center at $x = (x_1, x_2, x_3)$ and radius r.

Definition 0.7 For any given set $A \subset \mathbb{R}^n$, a point $x \in \mathbb{R}^n$ is an **interior point** of A if there exists an open sphere $S(x, r)$ with $r > 0$ such that

$$S(x, r) \subset A.$$

A point x is **exterior** to a set A if there exists an open sphere $S(x, r)$ with $r > 0$ such that

$$S(x, r) \subset A^c = \mathbb{R}^n - A.$$

A point x is a **boundary point** of A if it is neither an interior point of A nor exterior to A. That is, for a boundary point x, *every* open sphere $S(x, r)$ intersects both A and A^c.

Each set $A \subset \mathbb{R}^n$ is associated with three sets, A°, $\text{Ext}(A)$, and ∂A, defined as follows:

(1) A° is the set of all interior points of A, called the **interior** of A;
(2) $\text{Ext}(A)$ is the set of points exterior to A, called the **exterior** of A;
(3) ∂A is the set of boundary points of A, called the **boundary** of A.

Clearly $A^\circ \subset A$ but it may be a proper subset. Similarly, $\text{Ext}(A) \subset A^c$ but it may be a proper subset. The set A may include none, part, or all of the boundary ∂A.

Theorem 0.10 *For any set A, we have the following:*

(1) *the sets A°, $\text{Ext}(A)$, and ∂A are pairwise disjoint;*
(2) $A^\circ \cup \partial A \cup \text{Ext}(A) = \mathbb{R}^n$;
(3) $\partial A = \partial(A^c)$;
(4) $\text{Ext}(A) = (A^c)^\circ$;
(5) $A^\circ = A - \partial A$.

The proofs of these properties follow easily from the definitions.

Definition 0.8 A set $G \subset \mathbb{R}^n$ is said to be an **open set** in $\mathbb{R}^n$ if all its points are interior points of G; that is, $G = G^\circ$.

Let $A \subset \mathbb{R}^n$. A set B with $B \subset A$ is said to be **open relative to** A if there exists an open set $G \subset \mathbb{R}^n$ such that $B = A \cap G$.

Let $x \in A \subset \mathbb{R}^n$. If there exists an open set $G \subset \mathbb{R}^n$ with $x \in G \subset A$, then we say that A is a **neighborhood** of the point x.

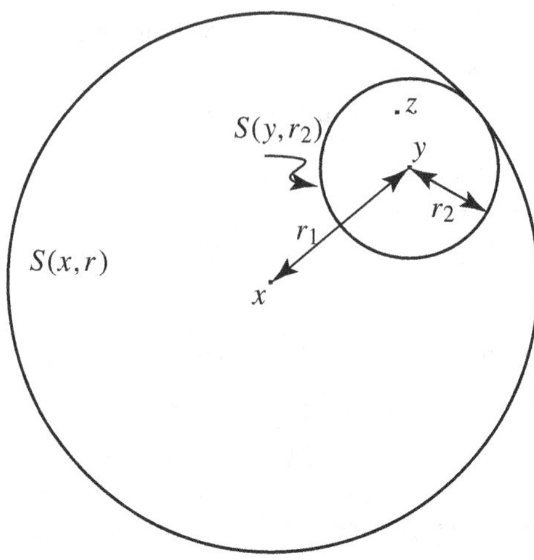

Figure 0.2 An open sphere is an open set.

We follow tradition by denoting an arbitrary open set with the letter G.

Theorem 0.11 *A set G is open if and only if it contains none of its boundary points; that is,*

$$G \cap \partial G = \emptyset.$$

Proof This follows from Property **(1)** of Theorem 0.10. ∎

Theorem 0.12 *Every open sphere $S(x,r)$ is an open set; that is, $\left(S(x,r)\right)^{\circ} = S(x,r)$.*

Proof We show that every point y in $S(x,r)$ is an interior point; that is, we find an open sphere about y which is contained in $S(x,r)$. See Figure 0.2. Let $d(x,y) = r_1$. Since $y \in S(x,r)$, we have $r_1 < r$. Let $r_2 = r - r_1 > 0$. We claim $S(y,r_2) \subset S(x,r)$. For if $z \in S(y,r_2)$, then $d(x,z) \leq d(x,y) + d(y,z) < r_1 + r_2 = r$. Hence, $z \in S(x,r)$. This shows $S(y,r_2) \subset S(x,r)$, which proves the result. ∎

Theorem 0.13 *The open sets of $\mathbb{R}^n$ have the following properties.*

(T1) *The sets $\emptyset$ and $\mathbb{R}^n$ are open.*
(T2) *The union of* any *collection (countable or not) of open sets is open.*
(T3) *The intersection of a* finite *collection of open sets is open.*

Proof **(T1)**: Since $\emptyset^\circ = \emptyset$, the empty set is open. For each $x \in \mathbb{R}^n$, $S(x,1) \subset \mathbb{R}^n$, so $(\mathbb{R}^n)^\circ = \mathbb{R}^n$, and hence $\mathbb{R}^n$ is open.

(T2): Consider a family of open sets $\boldsymbol{\mathcal{A}}$. Suppose x is in the union of this family of open sets; that is, suppose

$$x \in \bigcup_{G \in \boldsymbol{\mathcal{A}}} G.$$

Then $x \in G'$ for some open set G' in the family $\boldsymbol{\mathcal{A}}$. Since G' is open, we have $x \in S(x,r) \subset G'$, for some $r > 0$. This means that there is an open sphere about x in the union

$$x \in S(x,r) \subset G' \subset \bigcup_{G \in \boldsymbol{\mathcal{A}}} G.$$

(T3): Consider two open sets G_1 and G_2, and a point x in their intersection $G_1 \cap G_2$. Since x is in both open sets G_1 and G_2, there exist open spheres $x \in S(x,r_1) \subset G_1$ and $x \in S(x,r_2) \subset G_2$. Let $r = \min\{r_1, r_2\}$. Then we have both $x \in S(x,r) \subset S(x,r_1) \subset G_1$ and $x \in S(x,r) \subset S(x,r_2) \subset G_2$. Hence, $x \in S(x,r) \subset G_1 \cap G_2$, which means that x is in the interior of the intersection $G_1 \cap G_2$. The same argument works for a finite family of open sets $G_1, \ldots, G_N$ and its intersection $\bigcap_{k=1}^{N} G_k$. ∎

Definition 0.9 A set F is said to be **closed** if its complement $G = F^c$ is open.

We follow tradition by denoting an arbitrary closed set with the letter F.

Theorem 0.14 *A set F is closed if and only if it contains all of its boundary; that is, $\partial F \subset F$.*

Proof This follows from Theorem 0.11 and Property **(3)** of Theorem 0.10. ∎

Theorem 0.15 *The closed sets of $\mathbb{R}^n$ have the following properties.*

(TC1) *The sets $\emptyset$ and $\mathbb{R}^n$ are closed.*
(TC2) *The intersection of any collection (countable or not) of closed sets is closed.*
(TC3) *The union of a* finite *collection of closed sets is closed.*

The proof, which involves Theorem 0.13 and De Morgan's laws (Theorem 0.1), is left as Exercise 0.6.7.

Theorem 0.16 *Let A be any subset of $\mathbb{R}^n$. Then $A^\circ = \bigcup\{G \mid G$ is open and $G \subset A\}$. This shows that the interior A° is an open set, and that it is the largest open set contained in A.*

Proof The set A° is open and $A^\circ \subset A$. So $A^\circ \subset \cup\{G \mid G \text{ is open and } G \subset A\}$. Conversely, if G is open and $G \subset A$, then every point of G is an interior point of G, and hence also an interior point of A. Then $G \subset A^\circ$. Thus $\cup\{G \mid G \text{ is open and } G \subset A\} \subset A^\circ$. ∎

Definition 0.10 The **closure** of a set A is $\overline{A} = A \cup \partial A$.

Theorem 0.17 *Let A be any subset of $\mathbb{R}^n$. Then the closure $\overline{A}$ is equal to the set*

$$\bigcap\{F \mid F \text{ is closed and } A \subset F\}.$$

This shows that the closure $\overline{A}$ is a closed set and it is the smallest closed set containing A.

The proof is left as Exercise 0.6.15.

Definition 0.11 Let A be any subset of $\mathbb{R}^n$. A point $x \in \mathbb{R}^n$, not necessarily a point of A itself, is said to be a **limit point** of A if, for every $r > 0$, there exists a point $y \neq x$ such that $y \in A \cap S(x,r)$.

Theorem 0.18 *A point x is a limit point of a set A if and only if for every $r > 0$, the open sphere $S(x,r)$ contains infinitely many points of A.*

The proof is left as Exercise 0.6.17.

Theorem 0.19 *A set F is closed if and only if it contains all of its limit points.*

Proof ($\Rightarrow$): Suppose F is closed. If a given limit point is not a point of F, then it is clearly a boundary point. But a closed set contains all of its boundary points. So it must contain all of its limit points.

($\Leftarrow$): Consider a boundary point x of F. Either x is an element of F or, if not, every $S(x,r)$ must contain a point $y \in A$ (necessarily, $y \neq x$). That is, either $x \in F$ or, if not, x is a limit point of F. In either case, each boundary point is an element of F, which means that F is closed. ∎

Definition 0.12 A set $A \subset \mathbb{R}^n$ is said to be **bounded** if A is contained in some n-dimensional (bounded) interval I of the form

$$I = \{(x_1,\dots,x_n) \in \mathbb{R}^n \mid -\infty < a_k \leq x_k \leq b_k < \infty, \text{ for } k = 1,2,\dots,n\}.$$

Clearly a set is bounded if and only if it is a subset of an open sphere.

Definition 0.13 The **diameter** of a nonempty set $A \subset \mathbb{R}^n$ is $\operatorname{diam}(A) = \sup_{x,y \in A} d(x,y)$.

Clearly a nonempty set is bounded if and only if its diameter is finite.

Exercises for Section 0.6

0.6.1 Give an example of a set in $\mathbb{R}^2$ which is neither open nor closed.

0.6.2 Prove Property **(T3)** of Theorem 0.13 in detail for the intersection of a finite number of open sets $G_1 \dots, G_N$.

0.6.3 Give an example to show that the word 'finite' is needed in Property **(T3)** of Theorem 0.13.

0.6.4 Give an example to show that the word 'finite' is needed in Property **(TC3)** of Theorem 0.15.

0.6.5 Show that for any subset A of $\mathbb{R}^n$, we have $(A^\circ)^\circ = A^\circ$.

0.6.6 A **closed sphere** is a set of the form $S[x,r] = \{y \in \mathbb{R}^n \mid d(x,y) \leq r\}$. Show that a closed sphere is a closed set.

0.6.7 Prove Theorem 0.15.

0.6.8 Show that if $A \subset B$, then $A^\circ \subset B^\circ$.

0.6.9 Show that $(A \cap B)^\circ = A^\circ \cap B^\circ$.

0.6.10 Show that $A^\circ \cup B^\circ \subset (A \cup B)^\circ$, but that equality is not true in general.

0.6.11 Show that if $A \subset B$, then $\overline{A} \subset \overline{B}$.

0.6.12 Give an example of a set in $\mathbb{R}^2$ which has a boundary point that is not a limit point of that set.

0.6.13 Give an example of a set in $\mathbb{R}^2$ which has a limit point that is not a boundary point.

0.6.14 Prove that for any set A, we have $\partial A = \overline{A} - A^\circ$.

0.6.15 Prove Theorem 0.17.

0.6.16 Show that in $\mathbb{R}$, the only sets that are both open and closed are $\emptyset$ and $\mathbb{R}$.

0.6.17 Prove Theorem 0.18.

0.6.18 Prove the assertion that a set A is bounded if and only if it is a subset of an open sphere.

0.6.19 Prove the assertion that a nonempty set A is bounded if and only if its diameter is finite.

0.6.20 Show that the union of two bounded sets is a bounded set.

0.6.21 Let $\emptyset \neq A \subset B \subset \mathbb{R}^n$. Show that $\text{diam}(A) \leq \text{diam}(B)$.

0.6.22 What can you say about a nonempty set A for which $\text{diam}(A) = 0$?

0.7 Compact Sets in Euclidean Space

An important result of real analysis is the fact that a continuous real-valued function defined on a closed and bounded interval $[a,b] \subset \mathbb{R}$ attains its maximum (and minimum) value on the interval. This **maximum value property** is not true for open sets, nor for unbounded sets. To extend the maximum value property to $\mathbb{R}^n$ we define the concept of a **compact set**. Compact sets, we will show in Theorem 0.27, have the maximum value property.

Definition 0.14 Let K be a subset of $\mathbb{R}^n$. A collection of sets $\mathcal{A}$ is called a **cover** of K if K is contained in the union of the sets in $\mathcal{A}$; that is, $K \subset \bigcup_{A \in \mathcal{A}} A$. If the sets in $\mathcal{A}$ are indexed by some index set $\mathcal{I}$ (that is, the sets of $\mathcal{A}$ are all of the form A_κ for $\kappa \in \mathcal{I}$ and, further, for each $\kappa \in \mathcal{I}$ there corresponds a set $A_\kappa \in \mathcal{A}$), then we can also write $K \subset \bigcup_{\kappa \in \mathcal{I}} A_\kappa$.

If every set A in $\mathcal{A}$ is open, then we say that $\mathcal{A}$ is an **open cover** of K.

If a subcollection of $\mathcal{A}$ still covers K, we say the subcollection is a **subcover** of K.

Definition 0.15 We say that a set $K \subset \mathbb{R}^n$ is **compact** if every open cover $\mathcal{A}$ of K contains a finite subcover. In the case where the open cover $\mathcal{A}$ is indexed by some index set $\mathcal{I}$, say $\mathcal{A} = \{G_\kappa\}_{\kappa \in \mathcal{I}}$, compactness means there exists a finite subset $\{G_{\kappa_1}, G_{\kappa_2}, \ldots, G_{\kappa_N}\}$ such that

$$K \subset G_{\kappa_1} \bigcup G_{\kappa_2} \bigcup \cdots \bigcup G_{\kappa_N}.$$

Clearly every finite set is compact. Also a finite union of compact sets is compact.

Theorem 0.20 *If K is a compact subset of $\mathbb{R}^n$, then K is closed and bounded.*

Proof This proof is known as Cantor's diagonalization. **A compact set K must be bounded:** Consider the concentric open spheres about the origin $S(0,1) \subset S(0,2) \subset \cdots \subset S(0,k) \subset \cdots$. Since $\bigcup_{k=1}^{\infty} S(0,k) = \mathbb{R}^n$, the open spheres cover K. Since there exists a finite subcover, K is contained in the largest of them, say $K \subset S(0,N)$. This means that K is bounded.

A compact set K must be closed: We do this by showing that the complement of K is open. Consider a fixed $x \in K^c$. For each $y \in K$, let $r_y = \frac{1}{2}d(x,y)$, so that the open spheres $S(x,r_y)$ and $S(y,r_y)$ are disjoint. The spheres $\{S(y,r_y) \mid y \in K\}$ cover K. Since K is compact, there exists a finite subcover, say

$$S(y_1,r_{y_1}) \bigcup S(y_2,r_{y_2}) \bigcup \cdots \bigcup S(y_N,r_{y_N}) \supset K.$$

Now we turn our attention to the fixed $x \in K^c$. The intersection of the corresponding open spheres about x, namely $\bigcap_{k=1}^{N} S(x,r_{y_k})$, is an open sphere with center x. This open sphere is disjoint from each $S(y_k,r_{y_k})$, and hence disjoint from the union $\bigcup_{k=1}^{N} S(y_k,r_{y_k})$ which contains K. Thus each $x \in K^c$ is an interior point of K^c, which makes K^c open. ∎

Theorem 0.21 *Let K be a compact set in $\mathbb{R}^n$. Every closed subset F of K is compact.*

Proof Consider an open cover of the closed subset F. This might not be a cover of the larger compact set K. However, if we adjoin the open set $G = F^c = \mathbb{R}^n - F$, then it will become an open cover of $\mathbb{R}^n$, hence, an open cover of K. Since K is compact, there exists a finite subcover of K. Since G is disjoint from F, by removing G from this finite subcover, we are still left with a finite cover of F. ∎

Theorem 0.22 *Every closed and bounded n-dimensional interval of the form*

$$I = \{(x_1, \ldots, x_n) \in \mathbb{R}^n \mid -\infty < a_k \le x_k \le b_k < \infty, \text{ for } k = 1, 2, \ldots, n\}$$

is compact.

Proof Let $d = \text{diam}(I)$. Assume that I is not compact. That is, assume that there exists an open cover $\mathcal{A}$ of I with no finite subcover. Use the n midpoints $c_k = \frac{1}{2}(a_k + b_k)$ for $k = 1, 2, \ldots, n$ to bisect the sides of I. For dimension $n = 2$, we obtain 4 closed subintervals. For $n = 3$, we obtain 8 closed subintervals. For general n, we obtain 2^n closed subintervals, all covered by $\mathcal{A}$. If each of the 2^n closed subintervals had a finite subcover, then there would be a finite subcover of the entire interval I. Since it is assumed that this is not the case, then for at least one of the 2^n closed subintervals, $\mathcal{A}$ has no finite subcover. Denote one of these (necessarily noncompact) subinterval by I_1. Note that $\text{diam}(I_1) = d/2$. Again bisect each side of I_1 to obtain a smaller closed subinterval I_2 of I_1 for which $\mathcal{A}$ has no finite subcover. Note $\text{diam}(I_2) = d/4$, and so on. We obtain a nest of closed intervals $I \supset I_1 \supset I_2 \supset \cdots$, each with no finite subcover, and $\text{diam}(I_k) = d/2^k$. By applying the completeness property of $\mathbb{R}$ (Section 0.2), the intersection of this nest consists of a single point, say $x \in \bigcap_{k=1}^{\infty} I_k$. Since $x \in I$, one of the open sets G of the open cover $\mathcal{A}$ contains x. Since G is open, x is an interior point of G, say $x \in S(x,r) \subset G$. Choose k sufficiently large so that $\text{diam}(I_k) = d/2^k < r$. Then $x \in I_k \subset S(x,r) \subset G$, which means that some element I_k of the nest is contained in G. This shows that I_k is covered by a *single* open set G of $\mathcal{A}$. This single open set $G \in \mathcal{A}$ is a finite subcover of I_k, which is a contradiction. ∎

Theorem 0.23 Heine–Borel: *A subset K of $\mathbb{R}^n$ is compact if and only if it is closed and bounded.*[3]

[3] We shall see later in Section 5.10 that this characterization of compactness fails for all infinite-dimensional spaces.

Proof (⇒): is Theorem 0.20. (⇐): If K is bounded, then K is a subset of a closed and bounded n-dimensional interval I. Since I is compact by Theorem 0.22, and K is closed, the result follows from Theorem 0.21. ∎

Exercises for Section 0.7

0.7.1 Show that a finite union of compact sets is compact.

0.7.2 Show that an infinite intersection of compact sets is compact.

0.7.3 Find an open cover of $\mathbb{R}^n$ which has no finite subcover.

0.7.4 Consider the interval $I = (0,1] \subset \mathbb{R}$. Find an open cover of I which has no finite subcover.

0.7.5 Give an example of a closed set $F \subset \mathbb{R}^n$ and an open cover of F which has no finite subcover.

0.7.6 Give an example of a bounded set $A \subset \mathbb{R}^n$ and an open cover of A which has no finite subcover.

0.7.7 Give an example of a closed and bounded set $A \subset \mathbb{R}^n$ and a collection of sets that cover A but which cannot be covered by any finite subcollection.

0.7.8 Consider the open interval $I = (0,1) \subset \mathbb{R}$. Show that I is a countable union of closed intervals.

0.7.9 A set A in $\mathbb{R}^n$ is **totally bounded** if for each $\epsilon > 0$, there exists a finite collection of open ϵ-spheres $S(x^k, e)$, $k = 1, 2, \ldots, N$, which cover A. Show that in $\mathbb{R}^n$, a set is bounded if and only if it is totally bounded. Note: we will show later in Section 5.11 that this property does carry over to infinite-dimensional linear spaces.

0.7.10 Show that any open cover $\mathcal{A}$ of any set $A \subset \mathbb{R}^n$ has a *countable* subcover. The set A need not be closed nor bounded. Hint: Each $x \in A$ is an element of some open set G of the open cover $\mathcal{A}$. Find an open sphere $S(y,r) \subset G$, where the center $y = (r_1, r_2, \ldots, r_n) \in \mathbb{R}^n$ has rational coordinates, the radius r is rational, and $x \in S(y,r)$.

0.8 Continuity in Euclidean Space

Definition 0.16 Let $A \subset \mathbb{R}^n$ and consider a function $f \colon A \longrightarrow \mathbb{R}^m$. We say that f is **continuous at** $x_0 \in A$ if for every $\epsilon > 0$, there exists $\delta = \delta(\epsilon, x_0) > 0$ such that $d(f(x_0), f(x)) < \epsilon$ whenever $x \in A$ and $d(x_0, x) < \delta$.

If f is continuous at x_0 for all $x_0 \in A$, then we say f is **continuous on** A.

If $\delta = \delta(\epsilon)$ (that is, although δ depends on ϵ, the same δ works for *all* $x_0 \in A$), we say that f is **uniformly continuous on** A.

We will now characterize continuity in terms of open sets. This characterization completely avoids ϵ and δ used in the definition above.

Definition 0.17 Let $f: A \longrightarrow \mathbb{R}^m$, where $A \subset \mathbb{R}^n$. For $B \subset A$, the **image** of B is the following subset of $\mathbb{R}^m$:

$$f(B) = \{f(x) \mid x \in B\} = \{y \in \mathbb{R}^m \mid \exists x \in A \ni y = f(x)\}.$$

For $C \subset \mathbb{R}^m$, the **inverse image** of C is the following subset of A

$$f^{-1}(C) = \{x \in A \mid f(x) \in C\}.$$

Theorem 0.24 *Let $f: \mathbb{R}^n \longrightarrow \mathbb{R}^m$. The function f is continuous on $\mathbb{R}^n$ if and only if, for every open set G in $\mathbb{R}^m$, the inverse image $f^{-1}(G)$ is open in $\mathbb{R}^n$.*

Proof **($\Rightarrow$):** Suppose $f: \mathbb{R}^n \longrightarrow \mathbb{R}^m$ is continuous and let G be an open subset of $\mathbb{R}^m$. We show that any $x_0 \in f^{-1}(G)$ is an interior point. Since G is open and $f(x_0) \in G$, there exists an open sphere $S(f(x_0), r) \subset G$. Let $\epsilon = r > 0$. By continuity of f, there exists $\delta > 0$ such that, whenever $x \in S(x_0, \delta)$, we have $f(x) \in S(f(x_0), \epsilon) \subset G$. Thus the image of $S(x_0, \delta)$ is a subset of G, or $S(x_0, \delta) \subset f^{-1}(G)$. So every $x_0 \in f^{-1}(G)$ is an interior point.

($\Leftarrow$): Suppose that the inverse image of every open set in $\mathbb{R}^m$ is open in $\mathbb{R}^n$. Let $x_0 \in \mathbb{R}^n$ and $\epsilon > 0$ be given. Since $S(f(x_0), \epsilon)$ is an open subset of $\mathbb{R}^m$, the inverse image $f^{-1}(S(f(x_0), \epsilon))$ must be open in $\mathbb{R}^n$. Then $x_0 \in f^{-1}(S(f(x_0), \epsilon))$ must be an interior point. That is, there exists an open sphere $S(x_0, r) \subset f^{-1}(S(f(x_0), \epsilon))$, or $f(x) \in S(f(x_0), \epsilon)$, whenever $x \in S(x_0, r)$. Letting $\delta = r > 0$, we see that $d(f(x_0), f(x)) < \epsilon$ whenever $d(x_0, x) < \delta$. This completes the proof. ∎

The result can easily be extended to continuity on subsets A of $\mathbb{R}^n$.

Corollary 0.25 *Let $A \subset \mathbb{R}^n$ and $f: A \longrightarrow \mathbb{R}^m$. Then f is continuous on A if and only if, for every open set $G \subset \mathbb{R}^m$, the inverse image $f^{-1}(G)$ is open relative to A (that is, there exists an open set $G_1 \subset \mathbb{R}^n$ such that $f^{-1}(G) = A \cap G_1$).*

Theorem 0.26 *Let $f: \mathbb{R}^n \longrightarrow \mathbb{R}^m$ be continuous. If K is a compact subset of $\mathbb{R}^n$, then the image $f(K)$ is compact in $\mathbb{R}^m$.*

Proof Suppose $f: \mathbb{R}^n \longrightarrow \mathbb{R}^m$ is continuous and K is compact in $\mathbb{R}^n$. Consider an open cover $\bigcup_k G_k$ of $f(K)$. By Theorem 0.24, the inverse image $f^{-1}(G_k)$ of every open G_k is open in $\mathbb{R}^n$. The inverse images $f^{-1}(G_k)$

cover K. Since K is compact, there exists an open subcover of K, say $f^{-1}(G_{k_1}), f^{-1}(G_{k_2}), \ldots, f^{-1}(G_{k_N})$. Taking the direct images of these N open sets

$$f(f^{-1}(G_{k_1})) \subset G_{k_1},\ f(f^{-1}(G_{k_2})) \subset G_{k_2},\ \ldots,\ f(f^{-1}(G_{k_N})) \subset G_{k_N},$$

we see that sets G_{k_j} $(j = 1, \ldots, N)$ form a finite open subcover of $f(K)$. ∎

Theorem 0.27 *Suppose f is a continuous real valued function on a compact subset K of $\mathbb{R}^n$. Then f is bounded on K and attains its bounds.*

Proof Suppose $f : \mathbb{R}^n \longrightarrow \mathbb{R}$ is continuous and let $K \subset \mathbb{R}^n$ be compact. By Theorem 0.26, the image $f(K)$ is a compact subset of $\mathbb{R}$, which is closed and bounded by Theorem 0.20. So f is bounded on K, say, $a = \inf f(K)$ and $b = \sup f(K)$. We first show that there exists $x_0 \in K$ such that $f(x_0) = b$, or $b \in f(K)$.

Claim: The point b is a boundary point of $f(K)$. That is, for each $\epsilon > 0$, the open sphere $S(b, \epsilon) = (b - \epsilon, b + \epsilon)$ intersects both $f(K)$ and its complement. By definition of the least upper bound, $b - \epsilon$ cannot be an upper bound of $f(K)$. Thus there exists $f(x) \in f(K)$ such that $b - \epsilon < f(x) \leq b = \sup f(K)$. Additionally, any y with $\sup f(K) = b < y < b + \epsilon$ cannot be in $f(K)$, so y must be in its complement. This proves the claim.

Since $f(K)$ is closed, it contains all of its boundary points; thus $b \in f(K)$. That is, there exists $x_0 \in K$ such that $f(x_0) = b$.

The same argument applies to show $a = \inf f(K) \in f(K)$. ∎

Theorem 0.28 *Suppose $f : K \longrightarrow \mathbb{R}^m$ is a continuous function from a compact set K in $\mathbb{R}^n$ to $\mathbb{R}^m$. Then f is uniformly continuous.*

Proof Let $\epsilon > 0$ be given. For each $x \in K$, there exists $\delta_x > 0$ such that, $d\big(f(y), f(x)\big) < \epsilon/2$ whenever $y \in S(x, \delta_x)$. The collection of open spheres $S(x, \frac{1}{2}\delta_x)$ is an open cover of K. Since K is compact, we can extract a finite subcover, say $S\big(x^1, \frac{1}{2}\delta_{x^1}\big), \ldots, S\big(x^N, \frac{1}{2}\delta_{x^N}\big)$. Let δ be the minimum of the radii $\big\{\frac{1}{2}\delta_{x^1}, \ldots, \frac{1}{2}\delta_{x^N}\big\}$. Now let $x, y \in K$ with $d(x, y) < \delta$. Then for some k, we have $x \in S\big(x^k, \frac{1}{2}\delta_{x^k}\big)$. Furthermore $y \in S\big(x^k, \delta_{x^k}\big)$, since

$$d(y, x^k) \leq d(y, x) + d(x, x^k) < \delta + \frac{\delta_{x^k}}{2} \leq \delta_{x^k} \text{ because } \delta \leq \frac{1}{2}\delta_{x^k}.$$

Thus $d\big(f(x), f(y)\big) \leq d\big(f(x), f(x^k)\big) + d\big(f(x^k), f(y)\big) < \frac{\epsilon}{2} + \frac{\epsilon}{2} = \epsilon$. ∎

Definition 0.18 The **distance between a point** x **and a set** A is defined to be

$$d(x, A) = \inf_{y \in A} d(x, y).$$

The **distance between two sets** A **and** B is

$$d(A, B) = \inf_{x \in A,\ y \in B} d(x, y).$$

The following theorem will be used to prove an important result in measure theory (Theorem 1.29).

Theorem 0.29 *If F is a closed set and K is a compact set which is disjoint from F, then $d(K, F) > 0$.*

Proof The set $G = F^c$ is open. Each point $x \in K \subset G$ is an interior point of G, so there exists an open sphere about x entirely contained in G, say $x \in S(x, 2r_x) \subset G = F^c$. Take halves of these radii and, for each $x \in K$, denote them $G_x = S(x, r_x)$. This collection of open spheres is an open cover of K, and since K is compact, it has a finite subcover of open spheres, say $G_{x^1}, \ldots, G_{x^N}$. Let r be the smallest of their radii, then $r = \min\{r_{x^1}, r_{x^2}, \ldots r_{x^N}\}$. For each $x \in K$, there exists x^k such that $x \in G_{x^k}$.

Claim: $S(x,r) \subset S(x, r_{x^k}) \subset S(x^k, 2r_{x^k}) \subset G$.

- The first inclusion is clear since $r \leq r_{x^k}$ for all k.
- For the second inclusion, let $y \in S(x, r_{x^k})$. Since $x \in G_{x^k}$, we have $d(y, x^k) \leq d(y, x) + d(x, x^k) < r_{x^k} + r_{x^k} = 2r_{x^k}$. Thus $y \in S(x^k, 2r_{x^k})$. This shows the second inclusion $S(x, r_{x^k}) \subset S(x^k, 2r_{x^k})$.
- The third inclusion is clear since $x \in S(x, 2r_x) \subset G$ for all $x \in K \subset G$.

So for all $x \in K$ we have $d(x, F) > r$. This shows that $d(K, F) \geq r > 0$. ∎

Theorem 0.30 *For any nonempty subset A of $\mathbb{R}^n$, the real-valued function $f(x) = d(x, A)$ is uniformly continuous on $\mathbb{R}^n$.*

Proof Let $x, y \in \mathbb{R}^n$. For any $z \in A$, $d(x, A) \leq d(x, z) \leq d(x, y) + d(y, z)$, or

$$d(x, A) - d(x, y) \leq d(y, z).$$

Taking the infimum over $z \in A$ results in $d(x, A) - d(x, y) \leq d(y, A)$, or

$$f(x) - f(y) = d(x, A) - d(y, A) \leq d(x, y).$$

By symmetry we also have, $d(y, A) - d(x, A) \leq d(y, x) = d(x, y)$. Thus

$$|f(x) - f(y)| = |d(x, A) - d(y, A)| \leq d(x, y).$$

Given $\epsilon > 0$, let $\delta = \epsilon$. We have $|f(x) - f(y)| < \epsilon$ whenever $d(x, y) < \delta$. ∎

Theorem 0.31 Urysohn lemma: *For every closed set F and open set G with $F \subset G \subset \mathbb{R}^n$, there exists a continuous function $f : \mathbb{R}^n \longrightarrow \mathbb{R}$ such that*

(a) $0 \le f(x) \le 1$ *for all* $x \in \mathbb{R}^n$,
(b) $f(x) = 1$ *for all* $x \in F$, *and*
(c) $f(x) = 0$ *for all* $x \in G^c$.

Proof If either F or G^c are empty, the result is trivial. Otherwise, it is easy to verify that the function

$$f(x) = \frac{d(x, G^c)}{d(x, G^c) + d(x, F)}$$

satisfies all of the conditions. ∎

Exercises for Section 0.8

0.8.1 Show that the Euclidean norm on $\mathbb{R}^n$ is continuous. That is, show that the function $x \longrightarrow \|x\|_2$ is a continuous function from $\mathbb{R}^n$ to $\mathbb{R}$.

0.8.2 Consider a fixed $y \in \mathbb{R}^n$ and the function $f(x) = d(y, x)$ from $\mathbb{R}^n$ to $\mathbb{R}$. Show that f is continuous.

0.8.3 Give an example of a continuous function f and an open set G for which $f(G)$ is not open.

0.8.4 Let $A = [0, 1) \subset \mathbb{R}$ and $f(x) = \frac{x}{1-x}$. Show that $f : A \longrightarrow \mathbb{R}$ is continuous.

0.8.5 If a bounded function $f : [0, 1] \longrightarrow \mathbb{R}$ is uniformly continuous on the interval $[\epsilon, 1]$ for all $\epsilon > 0$, is it necessarily true that f is uniformly continuous on $[0, 1]$? Either prove or find a counterexample.

0.8.6 Find examples of disjoint closed sets A and B in $\mathbb{R}^2$ for which $d(A, B) = 0$.

0.8.7 Let A be a nonempty subset of $\mathbb{R}^n$ and let $x, y \in \mathbb{R}^n$. Prove that

$$|d(x, A) - d(y, A)| \le d(x, y).$$

0.8.8 Let $A \subset \mathbb{R}^n$ be nonempty. Show that $x \in \overline{A}$ if and only if $d(x, A) = 0$.

0.8.9 Let $f : \mathbb{R}^n \longrightarrow \mathbb{R}^n$. We say that f is a **contraction** function if there exists a constant $0 \le c < 1$ such that

$$d\big(f(x), f(y)\big) \le c d(x, y) \quad \text{for all} \quad x, y \in \mathbb{R}^n.$$

Show that a contraction function is uniformly continuous.

0.8.10 Suppose that F_0, F_1, F_2 are closed subsets of $\mathbb{R}^n$ which are pairwise disjoint. Show that there exists a continuous function $f : \mathbb{R}^n \longrightarrow \mathbb{R}$ such that $f(x) = k$ for all $x \in F_k$, for each $k = 0, 1, 2$.

0.9 Euclidean Measure

To measure the size of subsets of $\mathbb{R}^n$ we begin with intervals, the most elementary sets for measure theory.

In $\mathbb{R}$, an **open interval** is of the form $I = (a,b) = \{x \in \mathbb{R} \mid a < x < b\}$. The **length** of I is the number $b - a$. We write $|I| = b - a$. The same value of length $b - a$ applies to intervals of the half-open types $(a,b]$, $[a,b)$, and the closed type $[a,b]$.

In $\mathbb{R}^2$, a (two-dimensional) open interval is of the form

$$I = \left\{(x,y) \in \mathbb{R}^2 \mid a_1 < x < b_1, a_2 < y < b_2\right\}$$

for real numbers $a_1 < b_1, a_2 < b_2$, where any or all of the signs $<$ may be replaced by $\leq$. The **area** of I is $|I| = (b_1 - a_1)(b_2 - a_2)$.

We now extend this concept to n-dimensional intervals.

Definition 0.19 An n-dimensional **open interval** (or **open n-interval**) is a set in $\mathbb{R}^n$ of the form

$$I = \left\{(x_1, \ldots, x_n) \in \mathbb{R}^n \mid a_k < x_k < b_k \text{ for } k = 1,2,\ldots,n\right\} \tag{0.18}$$

for real numbers $a_1 < b_1, \ldots, a_k < b_k, \ldots, a_n < b_n$. An n-dimensional interval is an n-dimensional **rectangle** with sides parallel to the axes. If all of the sides $(b_k - a_k)$ are equal, the interval is an n-dimensional **cube**.

The **(Euclidean) measure** of the interval I is the product

$$|I| = \prod_{k=1}^{n}(b_k - a_k) = (b_1 - a_1)(b_2 - a_2)\cdots(b_n - a_n). \tag{0.19}$$

If $a_k = -\infty$ or $b_k = \infty$ for any k, then we say that I is an **unbounded open interval** and we define $|I| = \infty$. Note that $\mathbb{R}^n$ is an unbounded open interval. For $n = 1$, the Euclidean measure of an interval is called its **length**, for $n = 2$, its **area**, and for $n = 3$, its **volume**. Any or all of the $<$ signs in (0.18) can be replaced by $\leq$ to obtain other types of n-dimensional intervals.

A **degenerate interval** is one in which we have equality of some bounds $a_k = b_k$. Although the interior of a nondegenerate interval is an open interval with the same Euclidean measure, the interior of degenerate interval, whether bounded or unbounded, is empty and always has Euclidean measure 0. In this case, at least one of the factors in Equation (0.19) is 0. The (n-dimensional) Euclidean measure of a degenerate interval is always understood to be 0, even if another factor is ∞. This leads to the convention $0 \cdot \infty = 0$ in measure theory.

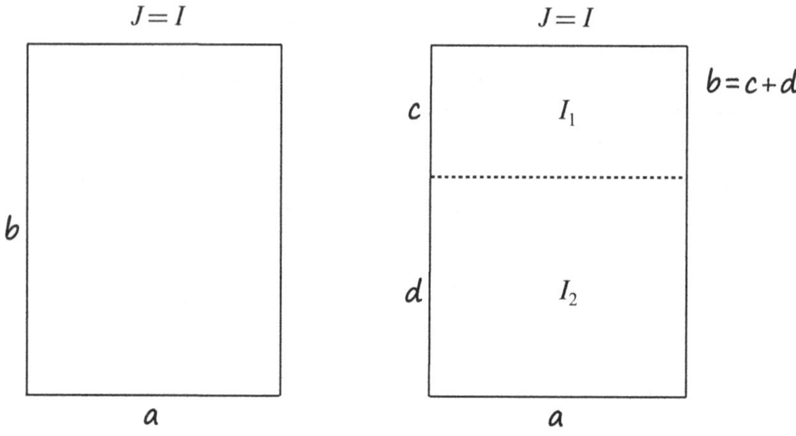

Figure 0.3 Partition of an interval I horizontally into I_1 and I_2.

We say that intervals are **nonoverlapping** if their interiors are disjoint. If J is a finite union of *nonoverlapping* intervals, say $J = \bigcup_{k=1}^{N} I_k$, then we define the **Euclidean measure** of J to be $|J| = \sum_{k=1}^{N} |I_k|$. We now prove that this is well defined.

Theorem 0.32 Finite additivity for intervals: *If J is a finite union of nonoverlapping intervals $J = \bigcup_{k=1}^{N} I_k$, then the Euclidean measure $|J| = \sum_{k=1}^{N} |I_k|$ is well defined. That is, any two representations of J as finite unions of nonoverlapping intervals $J = \bigcup_{i=1}^{N} I_i = \bigcup_{k=1}^{M} I'_k$, will yield the same measure$|J| = \sum_{i=1}^{N} |I_i| = \sum_{k=1}^{M} |I'_k|$.*

Proof Although the result is intuitively clear, it can be tedious to prove formally in detail. Here we illustrate the proof for dimension $n = 2$.

We start with the case where J is a *bounded interval* I. Partition J horizontally into two intervals I_1 and I_2 as shown in Figure 0.3.

We can use the distributive property of $\mathbb{R}$ to show that the two representations $J = I$ and $J = I_1 \cup I_2$ give the same Euclidean measure.

$$|I| = a \cdot b = a(c + d) = a \cdot c + a \cdot d = |I_1| + |I_2|.$$

Clearly this holds for a vertical partition as well. The procedure can be repeated for multiple (but finite) horizontal and vertical partitions, each yielding the same Euclidean measure.

The case of a *degenerate* interval $J = I$ works the same, and any representation yields the measure 0. Similarly, the case of an *unbounded nondegenerate* interval will yield infinite measure for any representation.

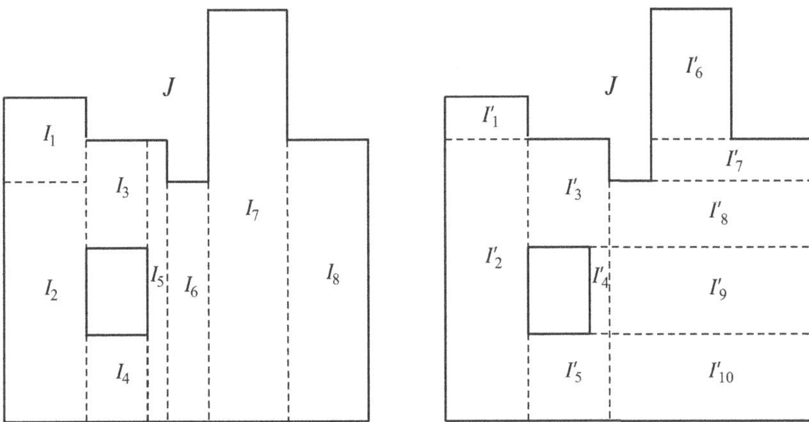

Figure 0.4 Decompositions of J nonoverlapping intervals where $N = 8$ and $M = 10$.

Now consider (still for dimension $n = 2$) a *finite union of nonoverlapping nondegenerate bounded intervals* with representations $J = \bigcup_{i=1}^{N} I_i = \bigcup_{k=1}^{M} I'_k$. For the example in Figure 0.4, we have $N = 8$ and $M = 10$. Note that J has a rectangular hole.

By the preceding case, whenever we add a horizontal or vertical partition within J, the number of nonoverlapping intervals increases but the sum determining the Euclidean measure remains unchanged. To each representation we add horizontal and vertical partitions for *all* of the distinct boundaries of the finite nonoverlapping intervals. Both representations of our example lead to the same refinement shown in Figure 0.5, with 25 nonoverlapping intervals. The sum determining the Euclidean measure remains constant.

This procedure can easily be extended to include degenerate and unbounded intervals. Additionally, this procedure can be generalized to any dimension n to obtain a rigorous proof of the theorem. ∎

Corollary 0.33 Finite additivity: *Suppose each of $J_1, J_2, \ldots, J_N$ is a finite union of nonoverlapping intervals. If they are nonoverlapping with each other, then the Euclidean measures add so that*

$$\left| \bigcup_{k=1}^{N} J_k \right| = \sum_{k=1}^{N} |J_k|.$$

Theorem 0.34 *Any finite union of intervals J, whether overlapping or not, can be written as a finite union of nonoverlapping intervals. Any such J thus has a Euclidean measure $|J|$.*

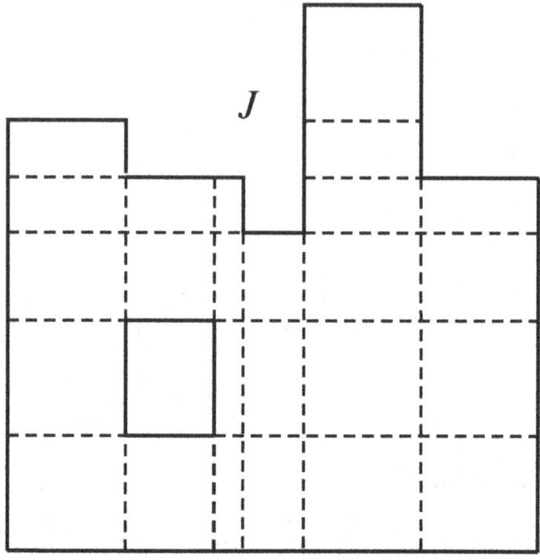

Figure 0.5 Common refinement of J into 25 nonoverlapping intervals.

Proof Let J be the finite union of intervals $I_1, I_2, \ldots, I_N$. Write

$$J_1 = I_1,\ J_2 = I_2 - J_1,\ J_3 = I_3 - (J_1 \cup J_2),\ \ldots, J_N$$
$$= I_N - (J_1 \cup J_2 \cup \cdots \cup J_{N-1}).$$

Clearly $J = \bigcup_{k=1}^N I_k = \bigcup_{k=1}^N J_k$ and the sets J_k are nonoverlapping with each other. Also each J_k is a union of nonoverlapping *intervals*. Thus every finite union of intervals J can be written as a finite union of nonoverlapping intervals. We can then use Theorem 0.33 to obtain $|J| = \sum_{k=1}^N |J_k|$. ∎

Here is another intuitive result. The proof is left as Exercise 0.9.3.

Theorem 0.35 *Suppose J and J' are both finite unions of intervals. If $J \subset J'$, then $|J| \le |J'|$.*

Euclidean measure of triangles: A two-dimensional bounded interval can be bisected into two congruent right triangles with adjacent sides parallel to the axes. The Euclidean measure of each such triangle is given as half of the two-dimensional interval.

Next we can define the Euclidean measure of any triangle T in $\mathbb{R}^2$. We can adjoin right triangles and intervals to T to form a circumscribed interval I, as shown in Figure 0.6. The Euclidean measure $|T|$ will then be the measure of the

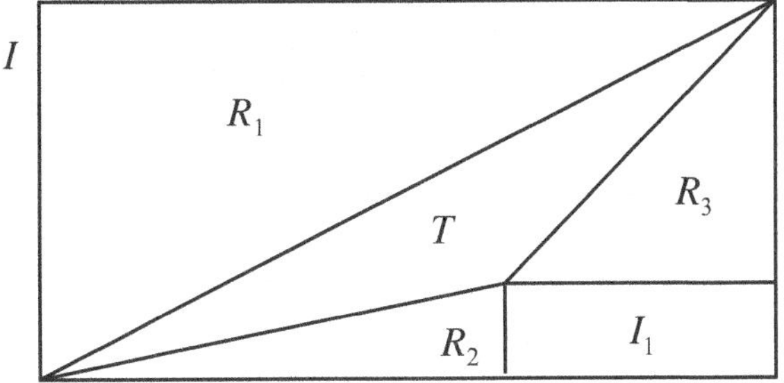

Figure 0.6 Measure of a triangle $|T| = |I| - |R_1| - |R_2| - |R_3| - |I_1|$.

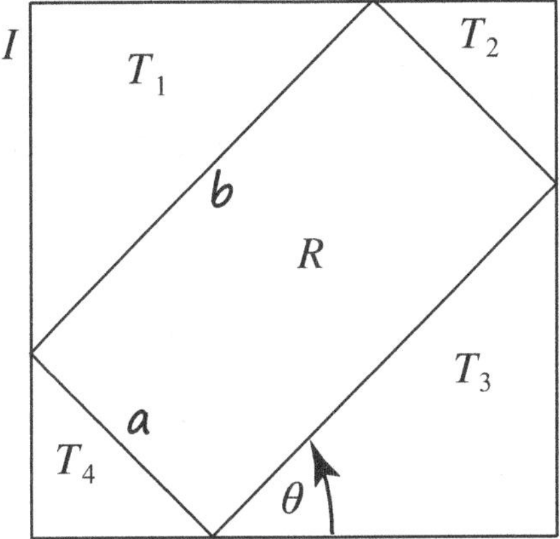

Figure 0.7 Measure of a rotated rectangle $|R| = |I| - |T_1| - |T_2| - |T_3| - |T_4|$.

circumscribed interval I minus the measures of the right triangles R_1, R_2, R_3, and interval I_1, that is $|T| = |I| - |R_1| - |R_2| - |R_3| - |I_1|$.

Euclidean measure of a rotated figure: We can similarly measure a bounded rectangle R in $\mathbb{R}^2$ with sides a and b obtained by rotating an interval I an angle of θ as shown in Figure 0.7. Elementary geometry on the circumscribed rectangle

$$|R| = |I| - |T_1| - |T_2| - |T_3| - |T_4|$$

will result in the expected Euclidean measure $|R| = a \cdot b$.

Euclidean measure of a polygon: Finally, any two-dimensional polygon P is a nonoverlapping union of triangles, so the Euclidean measure $|P|$ is the sum of the measures of these triangles.

Euclidean measure of a polytope:[4] Although more difficult to visualize, the above steps can be performed for n-dimensional polytopes, as well. The Euclidean measure of polytopes behave as we expect:

- The definition guarantees translation invariance. That is, $|P + x| = |P|$ for all n-polytopes and all $x \in \mathbb{R}^n$.
- Euclidean measure is invariant under rotation.
- Theorem 0.35 leads to the result that the measure of a subset cannot be greater than that of the entire set.
- Theorem 0.33 results in finite additivity of the Euclidean measure.

However, Euclidean measure will not take us much beyond the measure of polytopes. A sphere $S(x, r)$ in $\mathbb{R}^2$ is not a polygon. Archimedes went beyond Euclidean measure and used **countable additivity**, as defined below, to show that the area of a sphere $S(x, r)$ in $\mathbb{R}^2$ is equivalent to that of a right triangle with altitude equal to the radius r and base equal to the circumference c; that is, Area $= \frac{1}{2}rc$. The formula $c = 2\pi r$ gives us the familiar formula Area $= \pi r^2$.

Any theory of measure, such as that of Archimedes, that goes beyond Euclidean measure should include the usual Euclidean measure of n-dimensional intervals I (and finite union of intervals J and n-polytopes P), as well as the property of countable additivity. Namely a theory of measure $m(A)$ of sets $A \subset \mathbb{R}^n$, should have the following properties:

- For each n-dimensional open interval I, we have $m(I) = |I|$.
- **Countable additivity**: If sets $A_1, A_2, A_3, \ldots$ have measures $m(A_1)$, $m(A_2), m(A_3), \ldots$, then there is a measure $m(A)$ of the countable union $A = \bigcup_{k=1}^{\infty} A_k$. Further, if the sets A_k are disjoint, then $m(A) = \sum_{k=1}^{\infty} m(A_k)$.

In Chapter 1 on **Lebesgue measure** we do precisely that. We first extend the notion of measure from intervals I and finite union of intervals, to countable unions of intervals J, then open and closed sets G and F, and finally an entire class of sets called **Lebesgue measurable sets** $\mathcal{L}$. Although the class $\mathcal{L}$ is large, it **cannot** be extended to include **all** of the subsets of $\mathbb{R}^n$.

In the next section we consider an important special case, the sets of **Lebesgue measure zero**.

[4] A polytope or n-polytope is an n-dimensional generalization of a polygon. A two-dimensional polytope is a polygon. A three-dimensional polytope is a polyhedron.

Exercises for Section 0.9

0.9.1 Consider the x-axis as the closed two-dimensional interval

$$I = \left\{(x, y) \mid x = 0, \; -\infty < y < \infty\right\}.$$

Find its Euclidean measure $|I|$ in $\mathbb{R}^2$.

0.9.2 Find the Euclidean measure of the x-axis when considered as a one-dimensional open interval in $\mathbb{R}$. Also find the Euclidean measure of the x-axis when considered as a subset of the two-dimensional $\mathbb{R}^2$.

0.9.3 Prove Theorem 0.35.

0.9.4 Let I be an interval in $\mathbb{R}^n$ and let J be a finite union of intervals in $\mathbb{R}^n$. Prove $|J| = |J \cap I| + |J - I|$.

0.9.5 Let J and J' be finite unions of n-intervals. Show that

$$|J \cup J'| + |J \cap J'| = |J| + |J'|.$$

0.10 Sets of Measure Zero

Definition 0.20 Recall the definition of an n-dimensional interval as given by Equation (0.18). A **countable open interval cover** of a set $A \subset \mathbb{R}^n$ consists of a countable collection of n-dimensional open intervals $I_1, I_2, \ldots$ such that

$$A \subset L, \quad \text{where} \quad L = \bigcup_i I_i.$$

The cover of open intervals I_i need not be disjoint, nor nonoverlapping.

Definition 0.21 A subset A of $\mathbb{R}^n$ has **measure zero** if, for any given $\epsilon > 0$, the set A has a countable open interval cover $L = \bigcup_i I_i$ with

$$\sum |I_i| < \epsilon.$$

We write $m(A) = \inf\left\{ \sum_{i=1}^{\infty} |I_i| \;\middle|\; A \subset L = \bigcup_{i=1}^{\infty} I_i \right\} = 0$.

Example 0.2 A bounded line segment of the x-axis in $\mathbb{R}^2$ has measure zero in $\mathbb{R}^2$ as illustrated in Figure 0.8.

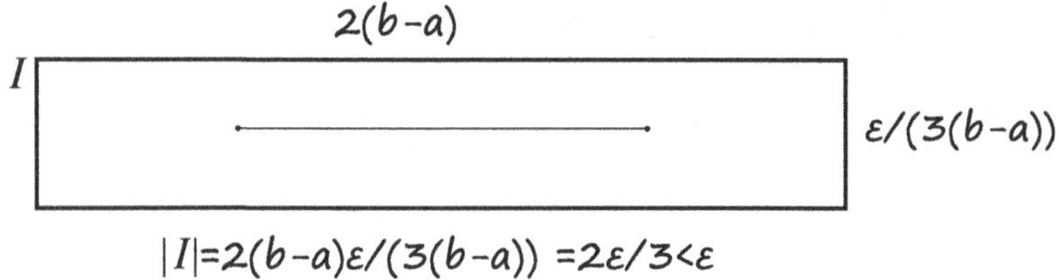

Figure 0.8 A line segment $A = [a, b]$ on the x-axis of $\mathbb{R}^2$ has $\mathbb{R}^2$-measure zero.

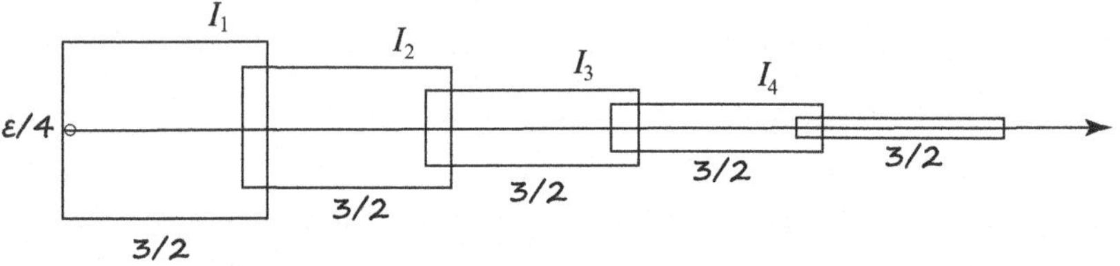

Figure 0.9 The positive x-axis of $\mathbb{R}^2$ has $\mathbb{R}^2$-measure zero.

Example 0.3 In the Euclidean space $\mathbb{R}^2$, the set $A = \{(x, y) \mid x > 0,\ y = 0\}$, consisting of the positive x-axis, has measure zero.[5] For any given $\epsilon > 0$ we can cover the positive x-axis with two-dimensional intervals $A \subset L = \bigcup_{k=1}^{\infty} I_k$ as follows:

$$I_1 = \left\{(x, y) \in \mathbb{R}^2 \mid \quad 0 < x < 1 + \frac{1}{2}, \quad -\frac{\epsilon}{8} < y < \frac{\epsilon}{8}\right\}$$

$$I_2 = \left\{(x, y) \in \mathbb{R}^2 \mid \quad 1 < x < 2 + \frac{1}{2}, \quad -\frac{\epsilon}{16} < y < \frac{\epsilon}{16}\right\}$$

$$\vdots$$

$$I_i = \left\{(x, y) \in \mathbb{R}^2 \mid i - 1 < x < i + \frac{1}{2}, \ -\frac{\epsilon}{2^{i+2}} < y < \frac{\epsilon}{2^{i+2}}\right\}$$

$$\vdots$$

For this open interval cover, and as illustrated in Figure 0.9, we have $\sum_{i=1}^{\infty} |I_i| = \sum_{i=1}^{\infty} \frac{3}{2} \cdot \frac{\epsilon}{2^{i+1}} = \frac{3}{4}\epsilon < \epsilon$.

The above argument can be modified to obtain the following theorem.

[5] Here we are considering the two-dimensional measure of the x-axis. The one-dimensional Euclidean measure (length) of the x-axis, being unbounded, is defined to be infinite, as noted in Section 0.9.

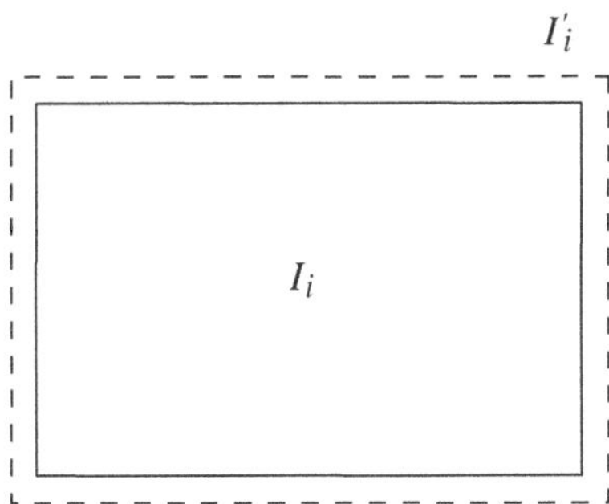

Figure 0.10 An interval I_i covered by a slightly larger *open* interval I'_i.

Theorem 0.36 *In* $\mathbb{R}^n$, *any degenerate interval has measure zero.*

Theorem 0.37 *Any countable set* $A = \{x^1, x^2, \ldots\}$ *in* $\mathbb{R}^n$ *has measure zero.*

Proof Let $\epsilon > 0$ be given. For each $x^i = (x^i_1, \ldots, x^i_n) \in A$, let I_i be the open interval $I_i = \{(x_1, \ldots, x_n) \in \mathbb{R}^n \mid a_k < x_k < b_k$, for $k = 1, 2, \ldots, n\}$ with $a_1 = x^i_1 - \frac{\epsilon}{2^{i+2}}$, $b_1 = x^i_1 + \frac{\epsilon}{2^{i+2}}$ and $a_k = x^i_k - \frac{1}{2}$, $b_k = x^i_k + \frac{1}{2}$ for $k = 2, \ldots, n$. Clearly $|I_i| = (b_1 - a_1)(b_2 - a_2) \cdots (b_n - a_n) = \left(\frac{\epsilon}{2^{i+1}}\right)(1)(1) \cdots (1) = \frac{\epsilon}{2^{i+1}}$. The open intervals I_i cover A and $\sum_i |I_i| = \sum_i \frac{\epsilon}{2^{i+1}} = \frac{\epsilon}{2} < \epsilon$. ∎

Theorem 0.38 *The countable union* $A = \bigcup A_i$ *of sets of measure zero still has measure zero.*

The proof uses the same ideas as in Theorem 0.37 by covering each set A_i with open intervals whose total measure is less than $\frac{\epsilon}{2^{i+1}}$. The details are left for Exercise 0.10.4.

We show next that we need not be restricted to *open* covering intervals in the definition of a set of measure zero.

Theorem 0.39 *A set A is of measure zero if and only if, for every* $\epsilon > 0$, *there exist intervals* $I_1, I_2, \ldots$ *covering* $A \subset \bigcup I_i$ *for which* $\sum |I_i| < \epsilon$. *The covering intervals need not be open.*

Proof **($\Rightarrow$):** Is obviously true. **($\Leftarrow$):** Let $\epsilon > 0$ be given, and suppose we have covering intervals $A \subset \bigcup I_i$ for which $\sum |I_i| < \epsilon/2$. It is clear that for each interval I_i, we can find a slightly larger *open* interval I'_i such that

$$I_i \subset I'_i \quad \text{and} \quad |I'_i| < |I_i| + \frac{\epsilon}{2^{i+1}}.$$

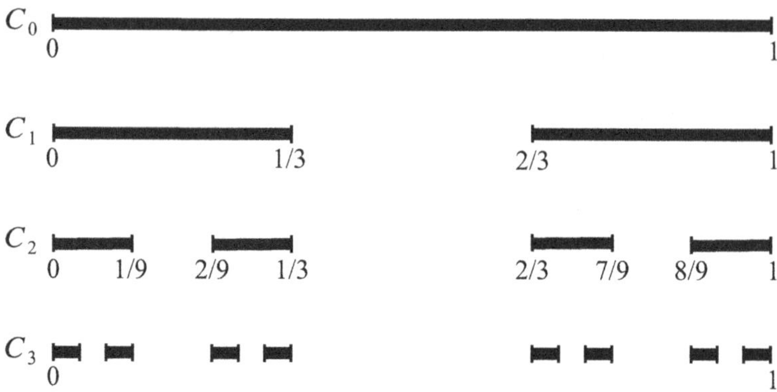

Figure 0.11 The sets C_0, C_1, C_2, and C_3.

The intervals I_i' form an open interval cover of $A \subset \bigcup I_i \subset \bigcup I_i'$ and

$$\sum_i |I_i'| \leq \sum_i \left(|I_i| + \frac{\epsilon}{2^{i+1}}\right) \leq \sum_i |I_i| + \sum_i \frac{\epsilon}{2^{i+1}} < \frac{\epsilon}{2} + \frac{\epsilon}{2} = \epsilon. \qquad \blacksquare$$

Although every countable set in $\mathbb{R}$ has measure zero, some *uncountable* sets in $\mathbb{R}$ also have measure zero. The famous **Cantor set** is an example.

Definition 0.22 The Cantor set: Let $C_0 = [0,1] \subset \mathbb{R}$. We have $|C_0| = 1$. From C_0 remove the middle third $\left(\frac{1}{3}, \frac{2}{3}\right)$ to obtain the union of two disjoint closed intervals $C_1 = \left[0, \frac{1}{3}\right] \cup \left[\frac{2}{3}, 1\right]$. We have $|C_1| = 1/3 + 1/3 = 2/3$. From C_1 remove the middle thirds $\left(\frac{1}{9}, \frac{2}{9}\right)$ and $\left(\frac{7}{9}, \frac{8}{9}\right)$ of the two disjoint closed intervals to obtain the union of four disjoint closed intervals $C_2 = \left[0, \frac{1}{9}\right] \cup \left[\frac{2}{9}, \frac{3}{9}\right] \cup \left[\frac{6}{9}, \frac{7}{9}\right] \cup \left[\frac{8}{9}, 1\right]$. We have $|C_2| = (2/3)^2$. Then remove the middle thirds of the four closed intervals of C_2, to obtain eight intervals in C_3 with $|C_3| = (2/3)^3$. This approach is illustrated in Figure 0.11.

And so on. For each $i = 0, 1, 2, \ldots$, the set C_i is the union of 2^i disjoint closed intervals. The **Cantor set** is the intersection of all of these sets

$$C = \bigcap_{i=0}^{\infty} C_i.$$

Theorem 0.40 *The Cantor set C has measure zero.*

Proof Let $\epsilon > 0$ be given. Since $C_0 = [0,1]$, we have $|C_0| = 1$. The set C_1 is obtained from C_0 by removing the middle third $\left(\frac{1}{3}, \frac{2}{3}\right)$. It is a union of two closed intervals and we have $|C_1| = 2/3$. For each i, the set C_i is obtained from C_{i-1} by removing the middle one third of each subinterval. This makes

C_i a finite union of 2^i closed intervals in $\mathbb{R}$ with a Euclidean measure of $|C_i| = (2/3)^i$. For each i, C_i is a closed interval cover of $C \subset C_i$, and for sufficiently large i, we have $|C_i| = (2/3)^i < \epsilon$. ∎

Theorem 0.41 *The Cantor set C is uncountable. It is actually conumerous with $\mathbb{R}$.*

Proof Represent the numbers x in $[0, 1]$ in a base 3 (ternary) expansion:

$$x = 0.a_1a_2a_3\ldots = \frac{a_1}{3} + \frac{a_2}{3^2} + \frac{a_3}{3^3} + \cdots.$$

Replace any terminating ternary digit 1 with the equivalent ternary digits ending with repeated 2s. For example, instead of $\frac{1}{3} = 0.1$, we write $\frac{1}{3} = 0.0222\ldots$. Similarly, $\frac{4}{9} = 0.11 = 0.10222\ldots$ and $\frac{10}{27} = 0.101 = 0.100222\ldots$ but $\frac{2}{3} = 0.1222\ldots = 0.2$.

With this rule, $C_1 = \left[0, \frac{1}{3}\right] \cup \left[\frac{2}{3}, 1\right]$, consists of all numbers $0 \leq x \leq 1$ with a_1 being 0 or 2. In the same way, $C_2 = \left[0, \frac{1}{9}\right] \cup \left[\frac{2}{9}, \frac{3}{9}\right] \cup \left[\frac{6}{9}, \frac{7}{9}\right] \cup \left[\frac{8}{9}, 1\right]$ consists of all numbers $x \in [0, 1]$ with a_1 and a_2 being 0 or 2. Since $C = \bigcap_{i=1}^{\infty} C_i$, the numbers $x \in C$ have ternary expansions with only 0s and 2s.

Using Cantor's diagonal argument, as in the proof of Theorem 0.2, we can show that C is uncountable. Furthermore, it is clear that there are as many ternary expansions with only 0s and 2s as there are binary expansions (of 0s and 1s). Thus there exists a one-to-one and onto mapping between C and all binary expansions. Since we can create a one-to-one mapping between all binary expansions and all of $\mathbb{R}$, it shows that C is conumerous with $\mathbb{R}$. ∎

Definition 0.23 If a specific property happens everywhere except in a set of measure zero, it is said to happen **almost everywhere** or happen **a.e.** The French abbreviation **p.p.** (**presque partout**) is also commonly used.

Example 0.4 Consider the function f defined by

$$f(x) = \begin{cases} 1 & \text{if } x = 1/2 \\ 0 & \text{if } x \neq 1/2. \end{cases}$$

This function is zero almost everywhere.

Example 0.5 Dirichlet function: Consider the indicator function of the set of rational numbers $\mathbb{Q}$ defined by

$$f_{\mathbb{Q}}(x) = \begin{cases} 1 & \text{if } x \in \mathbb{Q} \\ 0 & \text{if } x \notin \mathbb{Q}. \end{cases}$$

Since $\mathbb{Q}$ is countable, this function is zero almost everywhere. We write $f_{\mathbb{Q}}(x) = 0$ a.e.

Remark John C. Wenger has pointed out (in private conversation) that one can prove that any real interval $[a,b]$, with $b > a$, is uncountable without using the Cantor diagonal argument. If $[a,b]$ were countable, then the measure would be zero by Theorem 0.37). Yet we can see that the total measure of intervals covering $[a,b]$ cannot be made less than ϵ if $0 < \epsilon < b - a$.

Exercises for Section 0.10

0.10.1 Let $A \subset \mathbb{R}^n$. Suppose that for each $\epsilon > 0$, there exists a finite union of n-intervals $J = \bigcup_{k=1}^{N} I_k$ such that $A \subset J$ and $|J| < \epsilon$. Show that $m(A) = 0$.

0.10.2 How does the condition described in Exercise 0.10.1 differ from the definition of measure zero? Give an example of a set of measure zero which does not satisfy this condition.

0.10.3 Suppose $A \subset \mathbb{R}$ is a set of measure zero. Show that the set $A^2 = \{x^2 \in \mathbb{R} \mid x \in A\}$ is also a set of measure zero.

0.10.4 Prove Theorem 0.38.

0.10.5 Show that in $\mathbb{R}^2$ the entire x-axis has $\mathbb{R}^2$-measure zero.

0.10.6 Show that in $\mathbb{R}^2$ the line segment joining the origin $(0,0)$ with the point $(1,1)$ has $\mathbb{R}^2$-measure zero.

0.10.7 Show that in $\mathbb{R}$ the set of irrational numbers is *not* a set of measure zero.

0.10.8 Show that every open interval $(a,b) \subset \mathbb{R}$ is a countable union of closed intervals. Also show that every closed interval $[a,b] \subset \mathbb{R}$ is a countable intersection of open intervals.

0.10.9 Prove Exercise 0.10.8, for n-dimensional open intervals $I \subset \mathbb{R}^n$.

0.10.10 Show that an open set G in $\mathbb{R}^n$ cannot be a set of measure zero.

0.10.11 Suppose $A \subset \mathbb{R}^n$ is a set of measure zero. Show that there exist countable open sets G_k whose intersection $B = \bigcap_{k=1}^{\infty} G_k$ is of measure zero and $A \subset B$.

0.10.12 Show that the Cantor set C is a compact subset of $\mathbb{R}$.

0.10.13 If $A, B \subset \mathbb{R}$, define $A + B = \{x + y \mid x \in A,\ y \in B\}$. Show that, for the Cantor set C, we have $C + C = [0,2]$. Hint: Consider the ternary expansion of each $0 \leq z \leq 1$ and write $z = x + y$ in such a way that the expansions of x and y contain only 0s and 1s. This will show $[0,1] \subset \frac{1}{2}C + \frac{1}{2}C$, where we define $\frac{1}{2}C = \{\frac{x}{2} \in \mathbb{R} \mid x \in C\}$.

0.10.14 If $A, B \subset \mathbb{R}$, define $A - B = \{x - y \mid x \in A,\ y \in B\}$. Note that this set is *not* the usual difference of sets. Here $A - B$ is the set of differences in $\mathbb{R}$. Show that, for the Cantor set C, we have $C - C = [-1,1]$.

$*$0.11 Overview of Integration

You are familiar with the Riemann integral from your courses in calculus and real analysis. The Lebesgue integral, which was introduced in 1902 in the PhD thesis[6] of Henri Lebesgue, is more general than the Riemann integral. Chapter 2 formally deals with the Lebesgue integral. In this section we give an overview of some of these results for Riemann and Lebesgue integration on $\mathbb{R}^n$. This overview will permit you to bypass Part I on measure and integration and start with functional analysis in Part II.

$*$0.11.1 Overview of Riemann Integration

Definition 0.24 A **partition** of an n-dimensional interval I is a finite collection of nonoverlapping **subintervals** $\mathcal{P} = \{I_1, I_2, \ldots, I_N\}$, as illustrated in Figure 0.12, whose union is I

$$I = \bigcup_{k=1}^{N} I_k.$$

Definition 0.25 A **step function** $s\colon I \longrightarrow \mathbb{R}$ defined on a *bounded* n-dimensional interval I is one for which there exists a partition $\mathcal{P} = \{I_1, I_2, \ldots, I_N\}$ such that s is constant on the interiors of the subintervals I_k:

$$s(x) = \alpha_k \in \mathbb{R} \quad \text{whenever } x \text{ is in the interior of } I_k.$$

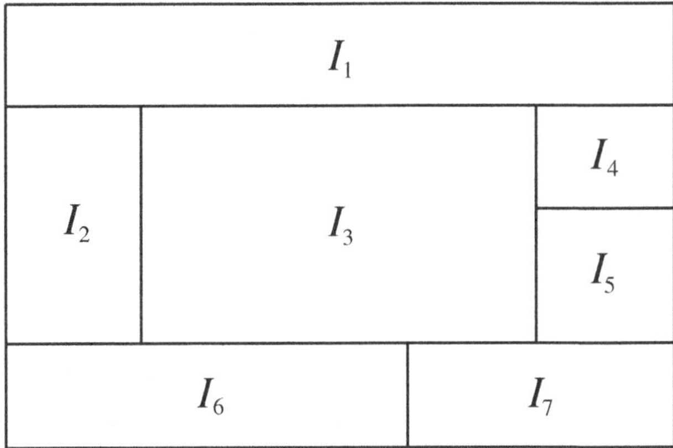

Figure 0.12 A partition of a two-dimensional *bounded interval*.

[6] Henri Léon Lebesgue, "Intégrale, Longuere, Aire," Ph.D. thesis, Univ. of Nancy, 1902.

The values of s on the boundaries of I_k do not matter since the measures are not affected.

Definition 0.26 The **Riemann integral** of such a step function s is defined to be

$$\int_I s(x)\,dx = \sum_{k=1}^{N} \alpha_k |I_k|.$$

Here $|I_k|$ is the Euclidean measure of the n-dimensional interval I_k.

Definition 0.27 Let f be a real-valued *bounded* function defined on an n-dimensional interval I. The **lower Riemann integral** of f is defined to be

$$\underline{\int_I} f(x)\,dx = \sup\left\{ \int_I s(x)\,dx \,\middle|\, s \text{ is a step function defined on } I \text{ and } s \le f \right\}.$$

The **upper Riemann integral** of f is defined to be

$$\overline{\int_I} f(x)\,dx = \inf\left\{ \int_I s(x)\,dx \,\middle|\, s \text{ is a step function defined on } I \text{ and } f \le s \right\}.$$

We say that f is **Riemann integrable** on I if the upper and lower integrals are equal. Then the **Riemann integral** of f is the common value

$$\int_I f(x)\,dx = \underline{\int_I} f(x)\,dx = \overline{\int_I} f(x)\,dx.$$

Suppose $f(t) = u(t) + iv(t)$ is a bounded *complex*-valued function defined on an n-dimensional interval I. Here $u = u(t)$ and $v = v(t)$ are bounded *real*-valued functions defined on I. Then the Riemann integral is defined to be

$$\int_I f(t)\,dt = \int_I u(t)\,dt + i \int_I v(t)\,dt. \tag{0.20}$$

A formal treatment of the Riemann integral is given later in Section 3.2.The definition of the Riemann integral above was given by Jean Gaston Darboux in 1875. The original version, involving limits of Riemann sums, was given by Bernhard Riemann in 1854. Here we summarize some of the results for Riemann integration.

R1. A Riemann integrable function is necessarily bounded on I (Darboux's definition requires boundedness. Riemann's definition does not, but it is still a consequence as shown in Theorem 3.4).

R2. The set of Riemann integrable functions on a fixed interval I is a linear space (defined on p. 9) of functions. This space includes all of the continuous functions on I.

R3. The Riemann integral is a linear functional on this space (Exercise 3.2.2).

One of the first important results of Lebesgue's work was the following characterization of functions that are Riemann integrable.

R4. The Lebesgue–Vitali theorem (Theorem 3.8): Let f be defined on a bounded n-dimensional closed interval I. Let D be the set of points in I where f is not continuous. Then f is Riemann integrable on I if and only if f is bounded on I and D is of measure zero.

As a consequence of the Lebesgue–Vitali theorem, the Dirichlet function $f_{\mathbb{Q}}$ of $\mathbb{Q}$, as given in Example 0.5, restricted to $I = [0, 1]$ is not Riemann integrable since the function is discontinuous everywhere.[7]

*0.11.2 Overview of Lebesgue Integration

Definition 0.28 The **indicator function** or **characteristic function** of a set $A \subset \mathbb{R}^n$ is

$$f_A(x) = \begin{cases} 1 & \text{if } x \in A \\ 0 & \text{if } x \notin A. \end{cases}$$

Definition 0.29 A real-valued **simple function** defined on an n-dimensional interval I is any function which assumes only a finite number of distinct values. A simple function s can be written as a (finite) linear combination of indicator functions

$$s(x) = \sum_{k=1}^{N} \alpha_k f_{A_k}(x), \text{ with disjoint sets } A_k \text{ and distinct } \alpha_k \in \mathbb{R}. \tag{0.21}$$

If the sets A_k are subintervals $A_k = I_k$ of a partition of a bounded interval I, then the simple function s is a **step function** (see Definition 0.25).

If sets A_k are Lebesgue measurable (defined in Section 0.9 and Chapter 1) with Lebesgue measures $m(A_k)$, then we say that $s = \sum_{k=1}^{N} \alpha_k f_{A_k}$ is a **measurable simple function** and we define the **Lebesgue integral** of such a simple function to be

$$\int_I s\, dm = \sum_{k=1}^{N} \alpha_k m(A_k).$$

We use the usual convention that $0 \cdot m(A_k) = 0$, even if $m(A_k) = \infty$.

[7] Actually one does not need the powerful Lebesgue–Vitali theorem to see this. One can easily show from basic principles that the lower and upper Riemann integrals are unequal:

$$\underline{\int_I} f_{\mathbb{Q}}(x)\, dx = 0 < \overline{\int_I} f_{\mathbb{Q}}(x)\, dx = 1.$$

Let f be a *nonnegative* real-valued function defined on a n-dimensional interval I. The nonnegative function f is **Lebesgue measurable**[8] on I if there exists an increasing sequence $0 \le s_1 \le s_2 \le s_3 \cdots$ of measurable simple functions which converge to f on I. We then define the **Lebesgue integral** to such a nonnegative function f to be

$$\int_I f\,dm = \lim_{k\to\infty} \int_I s_k\,dm.$$

A nonnegative function f is said to be **Lebesgue integrable** on I if

$$\int_I f\,dm < \infty.$$

In order to define Lebesgue integrability for a function $f\colon I \longrightarrow \mathbb{R}$ which is not necessarily nonnegative, we first define

$$f^+(x) = \sup\{f(x), 0\} \qquad f^-(x) = \sup\{-f(x), 0\}.$$

Note that $f^+ \ge 0,\ f^- \ge 0,\ f = f^+ - f^-$, and $|f| = f^+ + f^-$.

In this general case, f is said to be **Lebesgue integrable** if both nonnegative functions f^+ and f^- are Lebesgue integrable, and then we define the Lebesgue integral to be

$$\int_I f\,dm = \int_I f^+\,dm - \int_I f^-\,dm.$$

Here are the basic results of Lebesgue integration.

L1. The collection of functions that are Lebesgue integrable includes all Riemann integrable functions. If the Riemann integral exists, then the Lebesgue integral exists and the values of the two integrals are the same.

L2. Although Riemann integrable functions must be bounded, Lebesgue integrable functions need not be. They are even allowed to have ∞ and $-\infty$ as values. Furthermore, Lebesgue integrable functions may be defined on an unbounded interval I, which may even be all of $\mathbb{R}^n$.

L3. The collection of Lebesgue integrable functions on $\mathbb{R}^n$ is denoted by $L(\mathbb{R}^n)$, or L^1, or simply L. The integral of a function f defined on $\mathbb{R}^n$ is denoted by

$$\int f\,dm.$$

To restrict a Lebesgue integral to an interval I, use the indicator function f_I of I. Then

[8] Although this is not the formal definition of a Lebesgue measurable function, Theorem 2.16 shows that this is an equivalent formulation.

$$f_I \cdot f(x) = \begin{cases} f(x) & \text{if } x \in I \\ 0 & \text{if } x \notin I \end{cases} \quad \text{and define } \int_I f\,dm = \int f_I \cdot f\,dm.$$

Because of Property **L1** above, we may use the same notation for Lebesgue and Riemann integrals. So for $x = (x_1, \ldots, x_n)$, we use

$$\int f(x)\,dx \quad \text{or} \quad \int_{-\infty}^{\infty} \cdots \int_{-\infty}^{\infty} f(x_1, \ldots, x_n)\,dx_1 \cdots dx_n.$$

L4. Unlike the Lebesgue–Vitali characterization of Riemann integrable functions given above, there is no easy characterization of Lebesgue integrable functions. However, Lebesgue integration applies to most functions of interest in analysis and its applications.

L5. If the functions f, g are equal almost everywhere and f is Lebesgue integrable on an interval I, then so is g with the same value of the integral. For example, the Dirichlet function $f_{\mathbb{Q}}$ of $\mathbb{Q}$ as given in Example 0.5 is almost everywhere equal to 0, so it is Lebesgue integrable with integral 0. As already noted, the Lebesgue–Vitali theorem shows that this function is not Riemann integrable.

L6. **Fubini's theorem** (Theorem 3.2): Let $f : \mathbb{R}^2 \longrightarrow \mathbb{R}$ be Lebesgue integrable. Then for almost all y, the y-sections $f_y : \mathbb{R} \longrightarrow \mathbb{R}$, defined by $f_y(x) = f(x, y)$, are Lebesgue integrable. Also the function

$$F(y) = \int f_y(x)\,dx = \int f(x, y)\,dx$$

is Lebesgue integrable and

$$\int f\,dm = \int F(y)\,dy.$$

This may also be written as

$$\int f\,dm = \int \left\{ \int f(x, y)\,dx \right\} dy.$$

The main advantages of Lebesgue integration are certain limit theorems which do not hold for Riemann integration. Here are the two main limit theorems.

L7. **Lebesgue's monotone convergence theorem** (Theorem 2.38): Suppose that $f_1, f_2, f_3, \ldots$ is a sequence of Lebesgue integrable functions on $\mathbb{R}^n$ such that

$$0 \le f_1 \le f_2 \le f_3 \le \cdots.$$

Then

$$\lim_{k\to\infty}\int f_k\,dm = \int\Big(\lim_{k\to\infty} f_k\Big)\,dm.$$

We are permitting $\lim_{k\to\infty} f_k(x)$ to be infinite. The integral may also turn out to be infinite.

L8. Lebesgue's dominated convergence theorem (Theorem 2.41): Suppose that $f_1, f_2, f_3, \ldots$ is a convergent sequence of Lebesgue integrable functions and suppose that there exists a Lebesgue integrable functions g such that $|f_k(x)| \le g(x)$ for all $x \in \mathbb{R}$. Then $\lim_{k\to\infty} f_k$ is Lebesgue integrable and

$$\lim_{k\to\infty}\int f_k\,dm = \int\Big(\lim_{k\to\infty} f_k\Big)\,dm.$$

*0.12 Functions of Bounded Variation

Let $f\colon [a,b] \longrightarrow \mathbb{R}$. If f is discontinuous at an interior point $a < c < b$ yet both one-sided limits $f(c+) = \lim_{x\to c+} f(x)$ and $f(c-) = \lim_{x\to c-} f(x)$ exist, then f is said to have a **discontinuity of the first kind** at c. At the endpoints of $[a,b]$, if $f(a) \neq f(a+)$ or $f(b) \neq f(b-)$, it is also said there is a discontinuity of the first kind. Any other discontinuity is called a **discontinuity of the second kind**. With a discontinuity of the second kind, the function may be **unbounded** or **oscillate** too much. In this section we study the degree to which a function can oscillate in an interval.

A **partition** of a closed and bounded interval $[a,b]$ in $\mathbb{R}$ is a strictly increasing finite sequence $\mathcal{P} = \{x_0, x_1, \ldots, x_N\}$ of points such that

$$a = x_0 < x_1 < \cdots < x_N = b.$$

For any $f\colon [a,b] \longrightarrow \mathbb{R}$, we find the following sum associated with $\mathcal{P}$

$$V_f(\mathcal{P}) = \sum_{k=1}^{N} |f(x_k) - f(x_{k-1})|.$$

This number is called the **variation of f on $[a,b]$ for the partition $\mathcal{P}$.** Note that N depends on $\mathcal{P}$. The supremum of $V_f(\mathcal{P})$ for all partitions of $[a,b]$ is called the **total variation of f on $[a,b]$** and denoted by

$$V = V_f = V_f[a,b] = \sup_{\mathcal{P}} V_f(\mathcal{P}).$$

For any specific partition $\mathcal{P}$, we have $0 \le V_f(\mathcal{P}) < \infty$.

So for the supremum, we have $0 \leq V_f \leq \infty$.

If $V_f < \infty$, then f is said to be of **bounded variation**[9] on $[a,b]$.

Example 0.6 Monotonic functions: Let $f\colon [a,b] \longrightarrow \mathbb{R}$ be monotonically increasing. Then, for every partition $\mathcal{P}$, we have $V_f(\mathcal{P}) = f(b) - f(a)$. Thus f is of bounded variation with $V_f = f(b) - f(a) = |f(b) - f(a)|$. Similarly, for a monotonically decreasing function, we have $V = f(a) - f(b) = |f(b) - f(a)|$. Combining the two cases, we have $V_f = |f(b) - f(a)|$ for any monotonic function f on $[a,b]$.

Example 0.7 Dirichlet function: Consider the indicator function $f_{\mathbb{Q}}$ of the set of rational numbers

$$f_{\mathbb{Q}}(x) = \begin{cases} 1 & \text{if } x \in \mathbb{Q} \\ 0 & \text{if } x \notin \mathbb{Q}. \end{cases}$$

For any interval $[a,b]$ with $a < b$ we have $V[a,b] = \infty$.

Example 0.8 A continuous function need not be of bounded variation: Consider the following function f on the interval $[0,1]$:

$$f(x) = \begin{cases} x \sin \frac{\pi}{2x} & \text{if } x \neq 0 \\ 0 & \text{if } x = 0. \end{cases}$$

At all points $x \neq 0$ the function is clearly continuous. At $x = 0$ we can show $\lim_{x\to 0} x \sin \frac{\pi}{2x} = 0$; hence f is also continuous at $x = 0$. Actually f is uniformly continuous on the interval $[0,1]$ For the partition $\mathcal{P}_N = \{0, \frac{1}{2N}, \frac{1}{2N-1}, \ldots, \frac{1}{3}, \frac{1}{2}, \frac{1}{1}\}$ we have $f(1) = 1, f(1/2) = 0, f(1/3) = -1/3$, $f(1/4) = 0, f(1/5) = 1/5, \ldots, f(\frac{1}{2N}) = 0, f(0) = 0$. Thus

$$\begin{aligned} V(\mathcal{P}_N) &= 0 + \frac{1}{2N-1} + \cdots + \frac{1}{5} + \frac{1}{5} + \frac{1}{3} + \frac{1}{3} + 1 \\ &> \frac{1}{2N} + \cdots + \frac{1}{6} + \frac{1}{5} + \frac{1}{4} + \frac{1}{3} + \frac{1}{2}. \end{aligned}$$

Since the harmonic series $\sum \frac{1}{k}$ diverges, we have $V_f = \sup_N V(\mathcal{P}_N) = \infty$.

[9] Bounded variation can be defined for functions of the form $f\colon [a,b] \longrightarrow \mathbb{R}^n$ by using the norm instead of the absolute value. In this case, we define

$$V_f(\mathcal{P}) = \sum_{k=1}^{N} \|f(x_k) - f(x_{k-1})\|.$$

Many of the results of this section hold in the n-dimensional case.

Thus continuity on a bounded interval $I = [a,b]$ (and hence, uniform continuity on I) is not sufficient for bounded variation. We now define a stronger condition which is sufficient for bounded variation.

Definition 0.30 Lipschitz continuity: A function f defined on $[a,b]$ is said to be Lipschitz continuous if it satisfies the **Lipschitz condition of order 1**,[10] wherein there exists a constant $M > 0$ such that $|f(x) - f(y)| \le M|x - y|$ for all $x, y \in [a,b]$. If $M < 1$, then f is a contraction function as defined in Exercise 0.8.9.

Theorem 0.42 *Every Lipschitz continuous function f on a bounded interval $[a,b]$ is of bounded variation on $[a,b]$.*

Proof Let $\mathcal{P} = \{x_0, \dots, x_N\}$ be a partition of $[a,b]$. If $|f(x) - f(y)| \le M|x - y|$ for all $x, y \in [a,b]$, then

$$V_f(\mathcal{P}) = \sum_{k=1}^{N} \left| f(x_k) - f(x_{k-1}) \right| \le \sum_{k=1}^{N} M(x_k - x_{k-1}) = M(b-a).$$

Thus $V_f \le M(b-a) < \infty$. ∎

Corollary 0.43 *If $f\colon [a,b] \longrightarrow \mathbb{R}$ has a bounded derivative, then f is Lipschitz continuous and hence of bounded variation on $[a,b]$.*

Corollary 0.44 *The space $\mathcal{C}^1[a,b]$, of all functions with a continuous derivative is included in the collection of functions of bounded variation on $[a,b]$.*

Theorem 0.45 *For any two functions f, g defined on $[a,b]$, we have*

$$V_{f+g}[a,b] \le V_f[a,b] + V_g[a,b].$$

The proof is left as Exercise 0.12.12.

Although the sum and product of monotonic functions need not be monotonic, the collection of functions of bounded variation is closed with respect to addition and multiplication. This makes it a vector space and even an associative algebra.[11] Theorem 0.46 summarizes some of the **basic properties of functions of bounded variation**. All proofs are left as exercises.

[10] The Lipschitz condition of order $\alpha > 0$ ($\alpha \in \mathbb{R}$) requires the inequality $|f(x) - f(y)| \le M|x - y|^{\alpha}$ for all $x, y \in [a,b]$.

[11] An **algebra** is a vector space that also has an operation of multiplication, satisfying the axiom $\alpha fg = (\alpha f)g = f(\alpha g)$ for all scalars α. In this case, function multiplication is also commutative and associative.

Theorem 0.46 *Functions of bounded variation have the following properties:*

(bv1) *If f is of bounded variation on $[a,b]$, then so is αf for any $\alpha \in \mathbb{R}$.*
(bv2) *If f and g are of bounded variation on $[a,b]$, then so are $f+g, f-g$.*
(bv3) *If f and g are of bounded variation on $[a,b]$, then so is fg.*
(bv4) *If $[c,d] \subset [a,b]$, then $V_f[c,d] \leq V_f[a,b]$.*
(bv5) *If $a < c < b$, then $V_f[a,b] = V_f[a,c] + V_f[c,b]$.*

Corollary 0.47 *If f and g are both monotonically increasing on $[a,b]$, then $f-g$ is of bounded variation on $[a,b]$.*

Theorem 0.48 shows that *all* functions of bounded variation can be so characterized as a difference of two monotonically increasing functions.

Definition 0.31 Let f be a function of bounded variation on $[a,b]$ and define the **variation function** of f by

$$F(x) = V_f[a,x] \quad \text{for} \quad 0 \leq x \leq b.$$

By Property **(bv4)**, the variation function is monotonically increasing. It starts with $F(a) = 0$ and ends with $F(b) = V_f[a,b]$. Furthermore, by **(bv5)**

$$F(y) - F(x) = V_f[x,y] \quad \text{for any} \quad a \leq x < y \leq b. \tag{0.22}$$

Theorem 0.48 *A function f is of bounded variation on $[a,b]$ if and only if it is the difference $F-G$ of two monotonically increasing functions on $[a,b]$.*

Proof ($\Leftarrow$): This follows from Property **(bv2)**.

($\Rightarrow$): Let f be of bounded variation on $[a,b]$ and let $F(x) = V_f[a,x]$ be the variation function of f. Let $G = F - f$. Clearly F is monotonically increasing. It remains to be shown that $G = F - f$ is also monotonically increasing. If $a \leq x < y \leq b$, we have by the definition of G and Equation (0.22),

$$\begin{aligned} G(y) - G(x) &= F(y) - F(x) - [f(y) - f(x)] \\ &= V_f[x,y] - [f(y) - f(x)] \geq 0. \end{aligned}$$

The final inequality follows from $V_f[x,y] \geq V_f(\mathcal{P})$ for any partition $\mathcal{P}$. So taking the trivial partition $\mathcal{P} = \{x,y\}$, we have $V_f[x,y] \geq V_f(\mathcal{P}) = |f(y) - f(x)|$. Thus G is also monotonically increasing and $f = F - G$. ∎

Corollary 0.49 is an easy consequence of Theorem 0.48 and Exercises 0.12.4 and 0.12.5.

Corollary 0.49 *Suppose that f is of bounded variation on $[a,b]$. Then f has only a countable number of discontinuities, all of them of the first kind.*

Theorem 0.50 *Let f be of bounded variation on $[a,b]$ with decomposition $f = F - G$ into the difference of monotonic functions as in Theorem 0.48. The function f is right-continuous (or left-continuous, or continuous) at $c \in [a,b]$ if and only if both $F(x) = V_f[a,b]$ and $G(x) = F(x) - f(x)$ are right-continuous (or left-continuous, or continuous) at $c \in [a,b]$.*

Proof Suppose that F is right continuous at c. For the partition $\mathcal{P} = \{c, c+h\}$ of the interval $[c, c+h]$, we have $V_f[c,c+h] \geq V_f(\mathcal{P}) = |f(c+h) - f(c)|$. By Equation (0.22) above we have $F(c+h) - F(c) \geq |f(c+h) - f(c)| \geq 0$. Letting $h \to 0+$ we have $F(c+) - F(c) \geq |f(c+) - f(c)| \geq 0$. So right continuity $F(c+) = F(c)$ of F implies $f(c+) = f(c)$. The argument for left-continuity (and hence continuity) is similar.

For the converse, suppose that f is right-continuous at $c \in [a,b]$. Let $\epsilon > 0$ be given. Then there exists $\delta > 0$ such that $0 \leq |f(x) - f(c)| < \epsilon/2$ whenever $c < x < c+\delta$. By definition of $V_f[c,b]$, there exists a partition $\mathcal{P}' = \{c < x_1 < \cdots < x_N = b\}$ of the interval $[c,b]$ such that

$$V_f[c,b] < V_f(\mathcal{P}') + \frac{\epsilon}{2} = \sum_{k=1}^{N} |f(x_k) - f(x_{k-1})| + \frac{\epsilon}{2}.$$

Adding points to $\mathcal{P}'$ will only increase $V_f(\mathcal{P}')$. So we may add a point to $\mathcal{P}'$, if necessary, to ensure that $c < x_1 < c + \delta$. Then we have

$$\begin{aligned}
F(x_1) - F(c) &= V_f[c,x_1] = V_f[c,b] - V_f[x_1,b] \\
&< \sum_{k=1}^{n} |f(x_k) - f(x_{k-1})| + \frac{\epsilon}{2} - V_f[x_1,b] \\
&= |f(x_1) - f(c)| + \sum_{k=2}^{N} |f(x_k) - f(x_{k-1})| + \frac{\epsilon}{2} - V_f[x_1,b] \\
&= |f(x_1) - f(c)| + \frac{\epsilon}{2} - \left(V_f[x_1,b] - \sum_{k=2}^{N} |f(x_k) - f(x_{k-1})| \right) \\
&< \frac{\epsilon}{2} + \frac{\epsilon}{2} - \left(V_f[x_1,b] - \sum_{k=2}^{N} |f(x_k) - f(x_{k-1})| \right) \leq \epsilon.
\end{aligned}$$

The last inequality follows from the fact that $\mathcal{P}'' = \{x_1 < x_2 < \cdots < x_N\}$ is a partition of $[x_1,b]$; so that $V_f(\mathcal{P}'') = \sum_{k=2}^{N} |f(x_k) - f(x_{k-1})| \leq V_f[x_1,b]$. Finally, since the variation function is monotonically increasing, we have

$$0 \leq F(x) - F(c) \leq F(x_1) - F(c) < \epsilon \text{ whenever } c < x < c + \delta',$$

where $\delta' = x_1 - c > 0$. So F is right-continuous. The same holds for $G = F - f$. Proof for left-continuity (and hence continuity) is similar. ∎

The decomposition of a function of bounded variation as a difference of two monotonically increasing functions is *never* unique. If $f = F - G$ is such an expression, and g is monotonically increasing, then $f = (F+g) - (G+g)$ is another. However, the decomposition $f = F - G$, where $F(x) = V_f[a,x]$, is unique in the following sense.

Theorem 0.51 *If f is a function of bounded variation on $[a,b]$ and $f = F_1 - G_1$ is a decomposition into a difference of two monotonically increasing functions. Then*

$$F_1(x) \geq F(x) = V_f[a,x] \text{ and } G_1(x) \geq G(x) = V_f[a,x] - f(x).$$

The proof is left as Exercise 0.12.23.

Exercises for Section 0.12

0.12.1 Let $f(x) = \begin{cases} x+1 & \text{if } -2 \leq x < -1 \\ -x-1 & \text{if } -1 \leq x < 0 \\ x+1 & \text{if } 0 \leq x \leq 1. \end{cases}$

[a] Find all discontinuities and identify the kinds.
[b] Find the total variation $V_f[-2,1]$.
[c] Find the variation function $F(x)$ on $[-2,1]$ and $G(x) = F(x) - f(x)$.

0.12.2 Let $f(x) = \begin{cases} 0 & \text{if } x < 0 \\ x+1 & \text{if } 0 \leq x < 3 \\ 5 & \text{if } x = 3 \\ x^2 & \text{if } x > 3. \end{cases}$

[a] Find all discontinuities and identify the kinds.
[b] Find the total variation $V_f[-1,4]$.
[c] Find the variation function $F(x)$ on $[-1,4]$ and $G(x) = F(x) - f(x)$.

0.12.3 Let $f(x) = \begin{cases} \sin\frac{\pi}{2x} & \text{if } x \neq 0 \\ 0 & \text{if } x = 0. \end{cases}$

[a] Find all discontinuities and identify the kinds.
[b] Find the total variation $V_f[0,1]$.

0.12.4 Show that a monotonic function only has discontinuities of the first kind.

0.12.5 Use Exercise 0.12.4 to show that a monotonic function can only have a countable number of discontinuities.

0.12.6 Show that the Dirichlet function given in Example 0.7 has only discontinuities of the second kind.

0.12.7 Consider the function given in Example 0.8. Prove the assertion that f is continuous at 0. Namely, show $\lim_{x\to 0} f(x) = f(0) = 0$.

0.12.8 Show that the function $f(x) = x^2$ satisfies the Lipschitz condition of order 1 on the interval $I = [-1,+1]$ but not on all of $\mathbb{R} = (-\infty,\infty)$.

0.12.9 Prove Corollary 0.43.

0.12.10 Consider the following function f on the interval $[-1,1]$.

$$f(x) = \begin{cases} x^2 \sin \frac{\pi}{2x} & \text{if } x \neq 0 \\ 0 & \text{if } x = 0. \end{cases}$$

Show that this function f is Lipschitz continuous $[-1,1]$, and has a derivative on $[-1,1]$ but that the derivative is not continuous at $x=0$.

0.12.11 Show that Lipschitz continuity does not imply everywhere differentiability. Hint: consider $f(x) = |x|$ on $[-1,1]$.

0.12.12 Prove Theorem 0.45.

0.12.13 Show that if f is of bounded variation on $[a,b]$, then f is bounded on $[a,b]$.

0.12.14 Show that for two functions f and g defined on $[a,b]$ with bounds $M = \sup_{a\le x\le b} |f(x)|$ and $N = \sup_{a\le x\le b} |g(x)|$, we have

$$V_{fg}[a,b] \le N \cdot V_f[a,b] + M \cdot V_g[a,b].$$

0.12.15 Prove Property **(bv1)** of Theorem 0.46.

0.12.16 Prove Property **(bv2)** of Theorem 0.46. Hint: Use Exercise 0.12.12.

0.12.17 Prove Property **(bv3)** of Theorem 0.46. Use Exercises 0.12.14 and 0.12.13.

0.12.18 Prove Property **(bv4)** of Theorem 0.46.

0.12.19 Prove Property **(bv5)** of Theorem 0.46.

0.12.20 Consider the function $f(x) = \sqrt{x}$ on $[0,1]$. Show that f is uniformly continuous and of bounded variation but not Lipschitz continuous on $[0,1]$.

0.12.21 Show that if f is of bounded variation and $1/f$ is bounded on $[a,b]$, then $1/f$ is of bounded variation on $[a,b]$. Hint: Use

$$\sum_{k=1}^{N} \left| \frac{1}{f(x_k)} - \frac{1}{f(x_{k-1})} \right| = \sum_{k=1}^{N} \left| \frac{f(x_{k-1}) - f(x_k)}{f(x_k) f(x_{k-1})} \right|.$$

0.12.22 Show that the function $f(x) = \begin{cases} 0 & \text{if } x = 0 \\ \frac{1}{x} & \text{if } x \neq 0 \end{cases}$ is of bounded variation on every closed subinterval of $(0, 1)$ yet is not of bounded variation on $[0, 1]$.

0.12.23 Prove Theorem 0.51.

*0.13 Inequalities

Here we present several useful inequalities which occur in this book.

Theorem 0.52 Triangle Inequality: *For any $a, b \in \mathbb{C}$, we have*

$$|a + b| \leq |a| + |b|.$$

Proof For any $z \in \mathbb{C}$, we have $|z|^2 = z\bar{z}$ and $2\text{Re}(z) = z + \bar{z} \leq 2|z|$. By Equation (0.3) we have

$$\begin{aligned} |a+b|^2 &= (a+b)\overline{(a+b)} = |a|^2 + 2\text{Re}(a\bar{b}) + |b|^2 \\ &\leq |a|^2 + 2|a\bar{b}| + |b|^2 = |a|^2 + 2|a||b| + |b|^2 \\ &= (|a| + |b|)^2. \end{aligned}$$

■

Definition 0.32 Let $p, q \in \mathbb{R}$. If $p, q > 1$ and $\frac{1}{p} + \frac{1}{q} = 1$, we say that p and q are **Hölder conjugates**. Note: $1 < p < 2$ if and only if $2 < q < \infty$. Also $p = 2$ is conjugate to $q = 2$. The Hölder conjugate of $p = 1$ is defined to be $q = \infty$ and the conjugate of $p = \infty$ is $q = 1$.

Theorem 0.53 W.H. Young's inequality: *Suppose $p, q > 1$ are Hölder conjugates. Then for any nonnegative real numbers a, b, we have*

$$ab \leq \frac{a^p}{p} + \frac{b^q}{q}.$$

We have equality if and only if $a^p = b^q$.

Proof Fix a and let b be variable. Consider the function $f(b) = ab - \frac{b^q}{q}$ with derivative $f'(b) = a - b^{q-1}$. Elementary calculus shows that f attains its maximum only at $b = a^{\frac{1}{q-1}} = a^{\frac{p}{q}}$. So for nonnegative b we have

$$ab - \frac{b^q}{q} = f(b) \leq f\left(a^{\frac{p}{q}}\right) = a^{\frac{p}{q}+1} - \frac{a^p}{q} = a^p - \frac{a^p}{q} = \frac{a^p}{p}.$$

Equality holds if and only if $b = a^{\frac{p}{q}}$. ■

Definition 0.33 Let $p \in \mathbb{R}$ with $p \geq 1$ and $x = (x_1, \ldots, x_n) \in \mathbb{R}^n$ (or $\mathbb{C}^n$). We define the p**-norm** of x to be

$$\|x\|_p = \left(\sum_{k=1}^{n} |x_k|^p \right)^{1/p}.$$

Note that in the case $p = 2$, this is the **Euclidean norm** (defined in Section 0.5).

Theorem 0.54 Hölder's inequality: *Suppose $p > 1$, $q > 1$ are Hölder conjugates. Then for any $x, y \in \mathbb{R}^n$ (or $\mathbb{C}^n$), we have*

$$\sum_{k=1}^{n} |x_k y_k| \leq \left(\sum_{k=1}^{n} |x_k|^p \right)^{1/p} \left(\sum_{k=1}^{n} |y_k|^q \right)^{1/q} = \|x\|_p \|y\|_q. \tag{0.23}$$

If we define the product sequence $xy = (x_1 y_1, x_2 y_2, \ldots, x_n y_n)$, the inequality is equivalent to

$$\|xy\|_1 \leq \|x\|_p \|y\|_q.$$

The special case $p = q = 2$ is know as **Cauchy's inequality (E3)** (Theorem 0.6).

Proof If $x = (0, \ldots, 0)$ or $y = (0, \ldots, 0)$, the result is clear. Otherwise, $\|x\|_p \neq 0$ and $\|y\|_q \neq 0$. By Young's inequality, for each $k = 1, \ldots, n$, with $a = \frac{x_k}{\|x\|_p}$ and $b = \frac{y_k}{\|y\|_q}$, we have

$$\frac{|x_k y_k|}{\|x\|_p \|y\|_q} \leq \frac{|x_k|^p}{p \cdot (\|x\|_p)^p} + \frac{|y_k|^q}{q \cdot (\|y\|_q)^q}. \tag{0.24}$$

Summing both sides gives us

$$\frac{\sum |x_k y_k|}{\|x\|_p \|y\|_q} \leq \frac{\sum |x_k|^p}{p \cdot (\|x\|_p)^p} + \frac{\sum |y_k|^q}{q \cdot (\|y\|_q)^q} = \frac{1}{p} + \frac{1}{q} = 1,$$

which is the desired result. ∎

Theorem 0.55 Hölder's inequality for ω: *Suppose $p > 1$, $q > 1$ are Hölder conjugates. Then for any $x, y \in \omega$, we have*

$$\|xy\|_1 = \sum_{k=1}^{\infty} |x_k y_k| \leq \left(\sum_{k=1}^{\infty} |x_k|^p \right)^{1/p} \left(\sum_{k=1}^{\infty} |y_k|^q \right)^{1/q} = \|x\|_p \|y\|_q. \tag{0.25}$$

These sums may be infinite.

Proof In Inequality (0.23), let n tend to infinity, first on the right side to obtain an upper bound, and then on the left side to obtain the desired result. ∎

Theorem 0.56 Minkowski's inequality: *Suppose $p \geq 1$. Then for any x, $y \in \mathbb{C}^n$, we have*

$$\left(\sum_{k=1}^{n} |x_k + y_k|^p\right)^{1/p} \leq \left(\sum_{k=1}^{n} |x_k|^p\right)^{1/p} + \left(\sum_{k=1}^{n} |y_k|^p\right)^{1/p}. \tag{0.26}$$

The inequality is the following p-norm triangle inequality:

$$\|x + y\|_p \leq \|x\|_p + \|y\|_p.$$

Proof

$$\begin{aligned}\sum_{k=1}^{n} |x_k + y_k|^p &= \sum_{k=1}^{n} |x_k + y_k||x_k + y_k|^{p-1} \leq \sum_{k=1}^{n} (|x_k| + |y_k|)|x_k + y_k|^{p-1} \\ &= \sum_{k=1}^{n} |x_k||x_k + y_k|^{p-1} + \sum_{k=1}^{n} |y_k||x_k + y_k|^{p-1}.\end{aligned}$$

Applying Hölder's inequality to each of the last sums, we obtain

$$\begin{aligned}\sum_{k=1}^{n} |x_k + y_k|^p \leq &\left(\sum_{k=1}^{n} |x_k|^p\right)^{1/p} \left(\sum_{k=1}^{n} |x_k + y_k|^{(p-1)q}\right)^{1/q} \\ &+ \left(\sum_{k=1}^{n} |y_k|^p\right)^{1/p} \left(\sum_{k=1}^{n} |x_k + y_k|^{(p-1)q}\right)^{1/q}\end{aligned}$$

with $\frac{1}{p} + \frac{1}{q} = 1$. Since $(p-1)q = p$, we have

$$\sum_{k=1}^{n} |x_k + y_k|^p \leq \left(\|x\|_p + \|y\|_p\right) \left(\sum_{k=1}^{n} |x_k + y_k|^p\right)^{1/q}.$$

Dividing by $\left(\sum |x_k + y_k|^p\right)^{1/q}$, leads to

$$\left(\sum_{k=1}^{n} |x_k + y_k|^p\right)^{1-\frac{1}{q}} = \left(\sum_{k=1}^{n} |x_k + y_k|^p\right)^{1/p} \leq \|x\|_p + \|y\|_p. \quad \blacksquare$$

Theorem 0.57 Minkowski's inequality for ω: *Suppose $p \geq 1$. For any x, $y \in \omega$, we have*

$$\left(\sum_{k=1}^{\infty} |x_k + y_k|^p\right)^{1/p} \leq \left(\sum_{k=1}^{\infty} |x_k|^p\right)^{1/p} + \left(\sum_{k=1}^{\infty} |y_k|^p\right)^{1/p}.$$

In particular, for any $p \geq 1$, the space ℓ_p is closed under addition, and hence a vector space. The inequality does not hold for $0 < p < 1$.

Proof In Minkowski's inequality (Theorem 0.56), let n tend to infinity, first on the right side to obtain an upper bound, and then on the left side to obtain the desired result. That this fails for $0 < p < 1$ is left as Exercise 0.13.3. ∎

Theorem 0.58 Jensen's inequality: *If* $0 < p < q$ *and* $x \in \mathbb{C}^n$, *then*

$$\|x\|_q = \left(\sum_{k=1}^{n} |x_k|^q\right)^{1/q} \leq \left(\sum_{k=1}^{n} |x_k|^p\right)^{1/p} = \|x\|_p. \tag{0.27}$$

That is, $\|x\|_p$ *is a decreasing function of* p *for* $p > 0$.

Proof The case $x = 0$ is clear. Otherwise, we will show that $\frac{\|x\|_p}{\|x\|_q} \geq 1$ for $0 < p < q$. Let $A = \|x\|_q > 0$. Then for each k, we have $\frac{|x_k|}{A} \leq 1$. Since $0 < p < q$ we have $\left(\frac{|x_k|}{A}\right)^p \geq \left(\frac{|x_k|}{A}\right)^q$. It follows that

$$\frac{\|x\|_p}{\|x\|_q} = \frac{\|x\|_p}{A} = \left(\sum_{k=1}^{n}\left(\frac{|x_k|}{A}\right)^p\right)^{1/p}$$
$$\geq \left(\sum_{k=1}^{n}\left(\frac{|x_k|}{A}\right)^q\right)^{1/p} = \left(\frac{\sum_{k=1}^{n}|x_k|^q}{A^q}\right)^{1/p}.$$

The last expression becomes $\left(\frac{\|x\|_q}{A}\right)^{q/p} = 1^{q/p} = 1$. Thus $\frac{\|x\|_p}{\|x\|_q} \geq 1$. ∎

Theorem 0.59 Jensen's inequality for ω: *If* $0 < p < q$ *and* $x \in \omega$, *then*

$$\left(\sum_{k=1}^{\infty} |x_k|^q\right)^{1/q} \leq \left(\sum_{k=1}^{\infty} |x_k|^p\right)^{1/p}.$$

In particular, it follows that, if $1 \leq p \leq q$, *then* $\ell_p \subset \ell_q$.

Proof In Jensen's inequality (0.27), let n tend to infinity, first on the right side to obtain an upper bound, and then on the left to obtain the desired result. ∎

Definition 0.34 For $x \in \mathbb{R}^n$, $\mathbb{C}^n$, and ω, we define $\|x\|_\infty = \sup_k |x_k|$.

This definition is inspired by the following two results.

Theorem 0.60 Supremum norm for $\mathbb{R}^n$ and $\mathbb{C}^n$: *For each $x \in \mathbb{R}^n$ and $x \in \mathbb{C}^n$, we have*

$$\lim_{p\to\infty} \|x\|_p = \sup_k |x_k| = \|x\|_\infty.$$

Proof Since $1 \le k \le n$, the supremum $\sup_k |x_k|$ is really a maximum, which means that there exists a j such that $|x_j| = \sup_k |x_k|$. Let $1 < p < \infty$. We have $|x_j|^p \le \sum_{k=1}^n |x_k|^p \le n|x_j|^p$ and thus

$$\|x\|_p = \left(\sum_{k=1}^n |x_k|^p\right)^{1/p} \ge \left(|x_j|^p\right)^{1/p} = |x_j| = \sup_k |x_k|.$$

For the inequality in the opposite direction, we see that

$$\|x\|_p = \left(\sum_{k=1}^n |x_k|^p\right)^{1/p} \le \left(n|x_j|^p\right)^{1/p} = n^{1/p}|x_j| = n^{1/p} \sup_k |x_k|,$$

which tends to $\sup\limits_k |x_k|$ as $p \to \infty$. ∎

Theorem 0.61 Supremum norm for ω: *If, for some $1 \le p < \infty$ and $x \in \omega$, we have $\|x\|_p < \infty$, then*

$$\lim_{q\to\infty} \|x\|_q = \sup_k |x_k| = \|x\|_\infty.$$

Proof If $\|x\|_p < \infty$ for some $1 \le p < \infty$, by Jensen's inequality we have $\|x\|_q < \infty$ for all $q > p$. By the theory of convergent series, $x_k \to 0$, so just as in the proof of Theorem 0.60, there exists a j for which $|x_j| = \sup_k |x_k|$ and $\|x\|_q \ge |x_j| = \sup_k |x_k| = \|x\|_\infty$. So the decreasing $\|x\|_q$ is bounded below and $\lim\limits_{q\to\infty} \|x\|_q \ge \|x\|_\infty$.

For the inequality in the opposite direction, given $\epsilon > 0$, choose $N > 0$ such that $\left(\sum_{k=N+1}^\infty |x_k|^p\right)^{1/p} < \epsilon$. Let $u_k = x_k$ for $1 \le k \le N$ and $u_k = 0$ for $k > N$, and let $v_k = x_k - u_k$. Minkowski's and Jensen's inequalities on $x_k = u_k + v_k$ show that for $q > p$,

$$\begin{aligned}\|x\|_q &= \left(\sum_{k=1}^\infty |u_k + v_k|^q\right)^{1/q} \le \left(\sum_{k=1}^N |x_k|^q\right)^{1/q} + \left(\sum_{k=N+1}^\infty |x_k|^q\right)^{1/q}\\ &\le N^{1/q} \sup_k |x_k| + \epsilon.\end{aligned}$$

As $q \to \infty$, the term $N^{1/q} \sup_k |x_k|$ tends to $\sup_k |x_k|$. ∎

Exercises for Section 0.13

0.13.1 Let $x = (2, 1, 2)$, $y = (3, 4, 3)$ in $\mathbb{R}^3$. Check Hölder's inequality for $p = 3/2$.

0.13.2 Prove Hölder's inequality for ω in the case $p = 1$ with conjugate $q = \infty$.

0.13.3 Give an example for $0 < p < 1$ where Minkowski's inequality does not hold.

0.13.4 Let $0 < p < 1$ and $x \in \mathbb{R}^n$. Show that $\sum_{k=1}^{n} |x_k| \leq \left(\sum_{k=1}^{n} |x_k|^p\right)^{1/p}$.

0.13.5 Suppose $1 < p, q, r < \infty$ and $\dfrac{1}{p} + \dfrac{1}{q} + \dfrac{1}{r} = 1$ and $x, y, w \in \mathbb{R}^n$. Prove

$$\sum_{k=1}^{n} |x_k y_k w_k| \leq \|x\|_p \|y\|_q \|w\|_r.$$

0.13.6 Suppose $1 < p, q, r < \infty$ and $\frac{1}{r} = \frac{1}{p} + \frac{1}{q}$ and $x, y \in \mathbb{R}^n$. Extend Hölder's inequality by proving

$$\|xy\|_r \leq \|x\|_p \|y\|_q.$$

*0.14 The Axiom of Choice

Axiom of Choice The axiom of choice says that for any nonempty family of nonempty sets $\mathcal{A}$, there exists a function f such that for any set $A \in \mathcal{A}$, we have $f(A) \in A$. Such a function is called a **choice function**. It assigns to each nonempty set A, an element of A. The axiom of choice was first stated by Ernst Zermelo (1871–1953) in E. Zermelo, Beweis dass jede Menge wohlgeordnet werden kann. Mathematische Annalen. 59(4) (1904), 514–16.

It is known that the axiom of choice for uncountable sets cannot be proven from the other eight axioms of set theory (known as the **Zermelo–Fraenkel axioms**, or **ZF axioms**). When the axiom of choice is adjoined to the ZF axioms, they are known as the **ZFC axioms**.

In 1938, Kurt Gödel showed that the axiom of choice is consistent with the ZF axioms. That is, if there is no contradiction within the ZF axioms, then no contradiction can arise if the axiom of choice is assumed. In 1963, Paul Cohen proved that the axiom of choice could be false without contradicting the ZF axioms. As a consequence of the results of Gödel and Cohen, one is free to either assume or not assume the axiom of choice.

There are various statements, often known as "lemmas," "principles," or "theorems," that are actually equivalent to the axiom of choice.[12]

There are important results in set theory, algebra, and analysis that require the axiom of choice. In this book we use it only in four proofs:

(1) Every vector space has an algebraic basis[13] (Theorem 0.63).
(2) Sequential compactness in $\mathbb{R}^n$ implies compactness (Theorem 5.40).
(3) There exist nonmeasurable sets (Theorem 1.44).
(4) The Hahn–Banach extension principle (Theorem 6.19).

We now give a well-known statement, Zorn's lemma, which is equivalent to the axiom of choice. In many situations, Zorn's lemma is more easily applicable than the axiom of choice. First, some definitions.

Definitions Suppose $\mathcal{A}$ is a collection of sets. Any set A is said to be a **maximal set** in $\mathcal{A}$ if for all $B \in \mathcal{A}$, we have $A \subset B \Rightarrow A = B$.

A subcollection $\mathcal{C}$ of $\mathcal{A}$ is said to be a **chain** if for all $A, B \in \mathcal{C}$, either $A \subset B$ or $B \subset A$.

Theorem 0.62 Zorn's lemma: *Suppose $\mathcal{A}$ is a nonempty collection of sets. If, for any chain $\mathcal{C}$ in $\mathcal{A}$, the set $\bigcup_{C \in \mathcal{C}} C$ is in $\mathcal{A}$, then $\mathcal{A}$ contains a maximal set.*

The lemma was stated and proved (using the axiom of choice) in 1935 by Max Zorn.[14] We will not prove it; we simply accept it as an axiom.

Here is a result, item **(1)** above, whose proof uses Zorn's lemma.

Theorem 0.63 *Every linear space V has an algebraic basis (as defined in Section 0.3).*

Proof Let $\mathcal{A}$ be the collection of all linearly independent subsets of V. We show that for every chain $\mathcal{C}$ in $\mathcal{A}$, the union $U_{\mathcal{C}} = \bigcup_{C \in \mathcal{C}} C$ is a linearly independent set (that is, $U_{\mathcal{C}} \in \mathcal{A}$).

[12] Examples of such statements are Tuckey's lemma, Zorn's lemma, the Hausdorff maximality principle, and the well-ordering theorem.

[13] Algebraic basis is defined in Section 0.3.

[14] Zorn's lemma is generally stated in terms of a partially ordered set instead of subsets of $\mathcal{A}$. It states that any nonempty partially ordered set in which each chain has an upper bound has a maximal element. For a definition of partially ordered sets, further discussion, and proof of its equivalence to the axiom of choice, see any standard text on logic.

Consider a *finite* set of elements $x^1, \dots, x^n$ in $U_{\mathcal{C}}$, with

$$\alpha_1 x^1 + \cdots + \alpha_n x^n = 0.$$

For each $k = 1, \dots, n$, there exists $C_k \in \mathcal{C}$ such that $x^k \in C_k$. Since $\mathcal{C}$ is a chain, there must exist j with $1 \le j \le n$ such that C_j contains all of $x^1, \dots, x^n$. Since C_j is a linearly independent set, we have

$$\alpha_1 x^1 + \cdots + \alpha_n x^n = 0 \implies \alpha_1 = \cdots = \alpha_n = 0.$$

Thus $U_{\mathcal{C}}$ is linearly independent.

By Zorn's lemma, there exists a maximal linearly independent set B in $\mathcal{A}$. We show that B is a basis of V. Since $B \in \mathcal{A}$ the set B is linearly independent. It is thus sufficient to show that $\operatorname{span}(B) = V$. Assume not. Then there exists $x \in V - \operatorname{span}(B)$. But then $B \cup \{x\}$ would be a linearly independent set strictly containing B. This contradicts the maximality of B. Thus B is an algebraic basis. ∎

This theorem is actually equivalent to Zorn's lemma. Hence, it is equivalent to the axiom of choice. In functional analysis, other types of bases, such as orthonormal bases (Definition 7.4) are more useful than algebraic bases.

PART I

Measure and Integration

1
Lebesgue Measure

At this point you should have read Section 0.9 about Euclidean measure on $\mathbb{R}^n$ and Section 0.10 about sets of measure zero. In this chapter we extend measure theory on $\mathbb{R}^n$ beyond Euclidean measure and sets of measure zero.

It is not possible to extend the notion of measure to *all* subsets of $\mathbb{R}^n$ but to at most a proper family $\mathcal{L}$ of subsets of $\mathbb{R}^n$, called the Lebesgue measurable sets. Once we have defined $\mathcal{L}$ and the Lebesgue measure m of its sets, our goal in this chapter is to prove the basic properties set out in Theorem 1.1.

Theorem 1.1 *The Lebesgue measure $m(A)$ of a Lebesgue measurable set $A \in \mathcal{L}$ in $\mathbb{R}^n$ has the following properties:*

(m1) *The empty set $\emptyset$ is measurable ($\emptyset \in \mathcal{L}$), and $m(\emptyset) = 0$.*
(m2) *If A, B are measurable and $A \subset B$, then $0 \leq m(A) \leq m(B)$.*
(m3) *If A, B are measurable, so are the following sets:*

- **(m3a)** A^c *(and B^c).*
- **(m3b)** $A - B$ *(and $B - A$).*
- **(m3c)** $A \cap B$.

(m4) *If $A_1, A_2, \ldots$ are measurable, so is the union $A = \bigcup_{k=1}^{\infty} A_k$ and we have*

- **(m4a)** $m(A) \leq \sum_{k=1}^{\infty} m(A_k)$.
- **(m4b)** *Further, if the sets A_k are disjoint, we have $m(A) = \sum_{k=1}^{\infty} m(A_k)$.*

Here is the organization of this chapter. After the definition of Lebesgue measure in Section 1.1, we prove the various properties **(m1)** through **(m4)** given above for this measure in Theorems 1.12, 1.14, 1.15, 1.33, and 1.36. To achieve this, we progress through the following special cases.

- First for finite union of intervals J in $\mathbb{R}^n$ (Section 1.2).
- Then for countable unions of intervals L in $\mathbb{R}^n$ (Section 1.4).
- Then open and closed sets G and F in $\mathbb{R}^n$ (Section 1.5).
- And finally, the maximal class of measurable sets $\mathcal{L}$ (Section 1.7).

1.1 Outer and Inner Measure of Sets

Definition 1.1 Consider any set $A \subset \mathbb{R}^n$. Cover it with countably many n-dimensional open intervals $A \subset L = \bigcup_{k=1}^{\infty} I_k$, as illustrated in Figure 1.1. This is always possible since $\mathbb{R}^n$ is itself an open interval. Then take the infimum of the sums of Euclidean measures $\sum_{k=1}^{\infty} |I_k|$ for all such countable covers. We call this infimum the **outer measure** (or for emphasis, **Lebesgue outer measure**) $m^*(A)$ of A,

$$m^*(A) = \inf \left\{ \sum_{k=1}^{\infty} |I_k| \,\middle|\, A \subset L = \bigcup_{k=1}^{\infty} I_k \right\}.$$

The symbol $|I_k|$ denotes the Euclidean measure of the open interval I_k, as discussed in Section 0.9. A cover could be finite $A \subset \bigcup_{k=1}^{N} I_k$ or countably infinite $A \subset \bigcup_{k=1}^{\infty} I_k$, and the intervals need not be disjoint. But we can limit ourselves to covers where all intervals I_k are bounded. This is because every unbounded interval I_k is a countable union $I_k = \bigcup_j I_{kj}$ of nonoverlapping bounded intervals with the same Euclidean measure $|I_k| = \sum_j |I_{kj}|$.

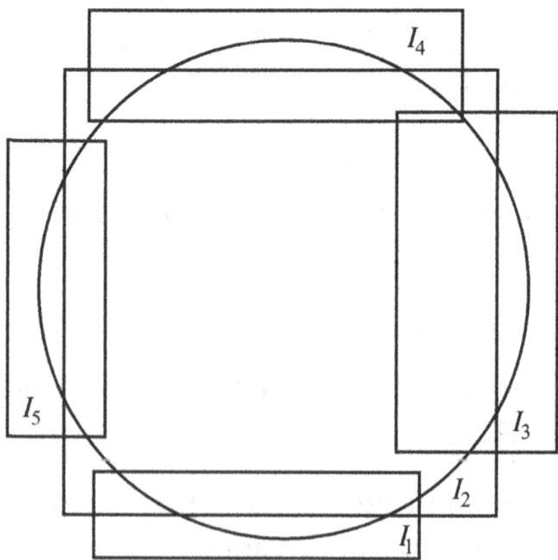

Figure 1.1 The cover of a two-dimensional sphere by intervals $I_1, \ldots, I_5$.

Although every subset A of $\mathbb{R}^n$ clearly has an outer measure $m^*(A)$, this cannot be used as the definition of Lebesgue measure $m(A)$ because the equality in additivity property **(m4b)** does not hold in general for outer measure. Indeed, we will show in Section 1.9 (see Exercise 1.9.9) that there exist disjoint sets A_1 and A_2, for which additivity fails. Namely,

$$A_1 \cap A_2 = \emptyset \quad \text{but} \quad m^*(A_1 \cup A_2) < m^*(A_1) + m^*(A_2).$$

Failure to obtain equality for all disjoint sets is a disappointment. In his 1902 Ph.D. thesis, Lebesgue chose to restrict the concept of measure to a class of sets that would ensure all of the properties **(m1)** through **(m4)**. This was done by way of the **inner measure** $m_*(A)$ which we define later in this section. Among other things, for a set $A \subset \mathbb{R}^n$ to be **Lebesgue measurable** ($A \in \mathcal{L}$) we will need the inner and outer measures of A to be equal.

Example 1.1 • If a set Z has measure zero, as defined in Section 0.10, it has outer measure zero $m^*(Z) = 0$, and conversely. That is, the notions of outer measure zero and measure zero are the same.

- A bounded set $A \subset \mathbb{R}^n$ will always have a finite outer measure.
- An unbounded set may have either finite or infinite outer measure.
- The positive x-axis of $\mathbb{R}^2$, as shown in Example 0.3 (Section 0.10), is an unbounded set in $\mathbb{R}^2$ which has (2-dimensional) outer measure zero.

Theorem 1.2 *The outer measure of any n-dimensional interval I is its Euclidean measure*

$$m^*(I) = |I|.$$

Proof If I is a degenerate interval, Theorem 0.36 shows that it has measure zero, hence I also has outer measure zero.

Next we consider a bounded nondegenerate n-dimensional interval I. If I is closed, it is of the form

$$I = \{(x_1, \ldots, x_n) \in \mathbb{R}^n \mid a_k \le x_k \le b_k, \text{ for } k = 1, 2, \ldots, n\}.$$

If not closed, then some of the above $\le$ signs are replaced by $<$ signs. Given $\epsilon > 0$, we can find a slightly larger *open* interval,

$$I' = \{(x_1, \ldots, x_n) \in \mathbb{R}^n \mid c_k < x_k < d_k, \text{ for } k = 1, 2, \ldots, n\},$$

with $c_k \le a_k < b_k \le d_k$ such that $|I| \le |I'| \le |I| + \epsilon$ (as illustrated in Figure 0.10; Theorem 0.39). Since I' is an open interval cover of I, it is thus clear that $m^*(I) \le |I'| \le |I| + \epsilon$. Since this is true for all $\epsilon > 0$, we have $m^*(I) \le |I|$.

To obtain the opposite inequality, consider a slightly smaller *closed* interval I'' such that $|I''| \le |I| \le |I''| + \epsilon$. Any open interval cover $I_1, I_2, \ldots$

of I is also an open cover of I''. Since I'' is closed and bounded, it is compact by the Heine–Borel theorem (Theorem 0.23). Hence, the open cover of I'' has a finite subcover, say $I_{i_1}, I_{i_2}, \ldots, I_{i_N}$. Then $|I''| \le \sum_{k=1}^{N} |I_{i_k}| \le \sum_{i=1}^{\infty} |I_i|$. Taking the infimum with respect to all open interval covers of I, we obtain $|I| \le |I''| + \epsilon \le m^*(I) + \epsilon$. Since this is true for all $\epsilon > 0$, we have $|I| \le m^*(I)$.

Finally, the case of an unbounded nondegenerate interval I in $\mathbb{R}^n$ is left as Exercise 1.1.2. ∎

Remark As a consequence of the above theorem we see that, in the definition of outer measure, the interval covers of a set A need not be restricted to *open* intervals.

The following Corollary is an easy consequence of Theorem 1.2 and Theorem 0.32.

Corollary 1.3 *For all finite unions $J = \bigcup_{k=1}^{N} I_k$ of nonoverlapping intervals, we have*

$$m^*(J) = |J| = \sum_{k=1}^{N} |I_k|.$$

Theorem 1.4 *The outer measure has the following three properties:*

(m*1) $m^*(\emptyset) = 0$.
(m*2) *If $A \subset B$, then $0 \le m^*(A) \le m^*(B)$.*
(m*3) *If $A = \bigcup_{k=1}^{\infty} A_k$, then $m^*(A) \le \sum_{k=1}^{\infty} m^*(A_k)$.*

Proof The proof of **(m*1)** is immediate. To prove **(m*2)**, note that every open interval cover of B is automatically an open interval cover of A. Details of the proof are left as Exercise 1.1.1.

Property **(m*3)**: We assume $m^*(A_k) < \infty$ for all k (otherwise, the proof is trivial). Let $\epsilon > 0$ be given. For each set A_k, let $L_k = \bigcup_{i=1}^{\infty} I_{ki}$ be an intervals cover of A_k such that

$$m^*(A_k) \le \sum_i |I_{ki}| < m^*(A_k) + \frac{\epsilon}{2^k}.$$

The union of the sets L_k will then cover A,

$$A = \bigcup_{k=1}^{\infty} A_k \subset \bigcup_{k=1}^{\infty} L_k = \bigcup_{k=1}^{\infty} \left(\bigcup_{i=1}^{\infty} I_{ki} \right).$$

Thus

$$m^*(A) \le \sum_{k=1}^{\infty}\left(\sum_{i=1}^{\infty}|I_{ki}|\right) \le \sum_{k=1}^{\infty}\left(m^*(A_k)+\frac{\epsilon}{2^k}\right) = \left[\sum_{k=1}^{\infty} m^*(A_k)\right]+\epsilon.$$

Since this inequality is true for all $\epsilon > 0$, it is true for the infimum of all $\epsilon > 0$, which gives us the Property **(m*3)**. ∎

Property **(m*3)**, which is the inequality $m^*(\bigcup_{k=1}^{\infty} A_k) \le \sum_{k=1}^{\infty} m^*(A_k)$, cannot in general be change to equality, even if the sets are disjoint. As mentioned before, an example with strict inequality will be given in Section 1.9. However, there are special cases in which equality holds. One such case is when the sets A_k are separated by nonoverlapping intervals.

Theorem 1.5 *If sets $A_1, A_2, \ldots$ are contained in nonoverlapping intervals $I_1, I_2, \ldots$, respectively ($A_1 \subset I_1$, $A_2 \subset I_2, \ldots$), then we have the following equality:*

$$\sum_{k=1}^{\infty} m^*(A_k) = m^*\left(\bigcup_{k=1}^{\infty} A_k\right). \tag{1.1}$$

Proof Suppose $A_1 \subset I_1, A_2 \subset I_2, \ldots$ for nonoverlapping $I_1, I_2, \ldots$. Let $A = \bigcup_{k=1}^{\infty} A_k$. Because of Property **(m*3)**, it is sufficient to show the inequality in the direction $\sum_{k=1}^{\infty} m^*(A_k) \le m^*(A)$. For each $N = 1, 2, \ldots,$ let $B_N = \bigcup_{k=1}^{N} A_k$. We first show that $\sum_{k=1}^{N} m^*(A_k) \le m^*(B_N)$. Consider any interval cover $\bigcup_j I'_j$ of B_N. For each j and $k = 1, 2, \ldots, N$, let $I'_{jk} = I'_j \cap I_k$. Then for each j, we have $\bigcup_{k=1}^{N} I'_{jk} \subset I'_j$ and $m^*(\bigcup_{k=1}^{N} I'_{jk}) \le m^*(I'_j) = |I'_j|$. Since the intervals I'_{jk} are nonoverlapping we also have, by Corollary 1.3,

$$m^*\left(\bigcup_k I'_{jk}\right) = \sum_{k=1}^{N} m^*(I_{kj}) = \sum_{k=1}^{N} |I_{kj}| \le |I'_j|.$$

For each $k = 1, 2, \ldots, N$, the intervals I'_{jk} cover A_k. So $m^*(A_k) \le \sum_j |I'_{jk}|$.

Then

$$\sum_{k=1}^{N} m^*(A_k) \le \sum_{k=1}^{N}\sum_j |I'_{jk}| = \sum_j \sum_{k=1}^{N} |I'_{jk}| \le \sum_j |I'_j|.$$

This holds for every interval cover $\bigcup_j I'_j$ of B_N, so $\sum_{k=1}^{N} m^*(A_k) \le m^*(B_N)$. To complete the proof, note that $B_N \subset A$. Thus $\sum_{k=1}^{N} m^*(A_k) \le m^*(A)$. Letting $N \to \infty$, we obtain the desired inequality $\sum_{k=1}^{\infty} m^*(A_k) \le m^*(A)$. ∎

Corollary 1.6 *Consider intervals $I_1 \subset I_2$ with $A_1 \subset I_1$ and $A_2 \subset I_2 - I_1$. Then*

$$m^*(A_1 \cup A_2) = m^*(A_1) + m^*(A_2).$$

This can be proven by partitioning $I_2 - I_1$ into nonoverlapping intervals. The partitioning splits A_2 into parts separated by these nonoverlapping intervals. Then apply Theorem 1.5 to these parts. Details are left as Exercise 1.1.3.

We can extend this result to more increasing nonoverlapping intervals $I_1 \subset I_2 \subset I_3 \subset \cdots \subset I_N$ with $A_1 \subset I_1, A_2 \subset I_2 - I_1, \ldots, A_N \subset I_N - I_{N-1}$. Then let $N \to \infty$ to prove the following corollary.

Corollary 1.7 *Consider n-dimensional intervals $I_1 \subset I_2 \subset I_3 \subset \cdots$ with*

$$A_1 \subset I_1,\ A_2 \subset I_2 - I_1,\ \cdots,\ A_k \subset I_k - I_{k-1},\ \cdots.$$

Then for $N = 1, 2, 3, \ldots$ we have $m^*\left(\bigcup_{k=1}^N A_k\right) = \sum_{k=1}^N m^*(A_k)$ and

$$m^*\left(\bigcup_{k=1}^{\infty} A_k\right) = \sum_{k=1}^{\infty} m^*(A_k).$$

Corollary 1.8 *If a finite union J of nonoverlapping intervals is disjoint from a set A, then*

$$m^*(A \cup J) = m^*(A) + m^*(J) = m^*(A) + |J|.$$

No matter how we choose to define the Lebesgue measure m of sets, the additivity property **(m4)** is vitally important. This will be attained by restricting ourselves to "Lebesgue measurable" set. In the case of *two disjoint* "measurable" sets A_1 and A_2, the desired property is

$$m(A_1 \cup A_2) = m(A_1) + m(A_2). \tag{1.2}$$

We begin the definition with the case where $A_1 = A$ is a *bounded* set (say bounded by a cube $C_M \subset \mathbb{R}^n$ centered at the origin of side $M > 0$) and $A_2 = C_M - A_1$ is its complement with respect to C_M. Such a cube of side $M > 0$ and centered at the origin, we call an ***M*-cube** (see Figure 1.2).

The sets $A_1 = A$ and $A_2 = C_M - A$ are disjoint and bounded. So the desired Equation (1.2) is, in this case,

$$m(C_M) = m(A_1 \cup A_2) = m(A_1) + m(A_2) = m(A) + m(C_M - A)$$

$$\text{or} \qquad m(A) = m(C_M) - m(C_M - A). \tag{1.3}$$

Note that $m^*(C_M) = |C_M| = M^n$. Since the Equality (1.2) (and hence (1.3)) is not guaranteed for the outer measure, we use this desired (1.3) to motivate the definition of the **inner measure** $m_*(A)$.

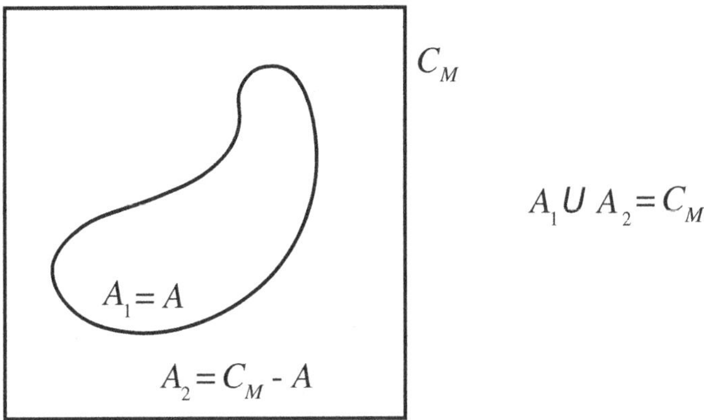

Figure 1.2 An M-cube C_M covering a set A.

Definition 1.2 In the bounded case $A \subset C_M \subset \mathbb{R}^n$, the **inner measure** (for emphasis **Lebesgue inner measure**) of A is defined to be

$$m_*(A) = m^*(C_M) - m^*(C_M - A) = M^n - m^*(A^c \cap C_M) \geq 0. \qquad (1.4)$$

Claim: This definition is well defined because the value of the inner measure does not depend on $M > 0$ as long as $A \subset C_M$.

Proof of Claim: Given $A \subset C_{M_1} \subset C_{M_2}$, let $J = C_{M_2} - C_{M_1}$. Since J is a finite union of nonoverlapping intervals and it is disjoint from C_{M_1}, by Corollary 1.3, we have $|C_{M_2}| = |C_{M_1}| + |J|$ or $|J| = M_2^n - M_1^n$. Since J is disjoint from A we have by Corollary 1.8,

$$\begin{aligned} m^*(A^c \cap C_{M_2}) &= m^*(A^c \cap (C_{M_1} \cup J)) \\ &= m^*((A^c \cap C_{M_1}) \cup J) \\ &= m^*(A^c \cap C_{M_1}) + |J| \\ &= m^*(A^c \cap C_{M_1}) + M_2^n - M_1^n. \end{aligned}$$

So $$m_*(A) = M_2^n - m^*(A^c \cap C_{M_2}) = M_1^n - m^*(A^c \cap C_{M_1}).$$ ∎

Lemma 1.9 *For any set A and M-cube C_M, we have $m_*(A \cap C_M) + m^*(C_M - A) = |C_M|$.*

Now in the *unbounded* case $A \subset \mathbb{R}^n$, we use the bounded sets $A \cap C_M$ to define the **Lebesgue inner measure** as follows:

$$m_*(A) = \lim_{M \to \infty} m_*(A \cap C_M) = \sup_M m_*(A \cap C_M).$$

Of course the inner measure of an unbounded set may be infinite.

Definition 1.3 A *bounded* set A is said to be **measurable** (for emphasis **Lebesgue measurable**) if

$$m^*(A) = m_*(A), \tag{1.5}$$

and we define the **Lebesgue measure** $m(A)$ to be this common value

$$m(A) = m^*(A) = m_*(A). \tag{1.6}$$

An *unbounded* set A is said to be **Lebesgue measurable** if

$$m(A \cap C_M) = m^*(A \cap C_M) = m_*(A \cap C_M) \text{ for every } M\text{-cube } C_M. \tag{1.7}$$

Actually, in the bounded case, $A = A \cap C_M$ for some $M > 0$, so Equations (1.7) can be used as the definition of Lebesgue measure in *both* the bounded and unbounded cases. The **Lebesgue measure** for a measurable set A is then defined to be

$$m(A) = \sup_M m(A \cap C_M). \tag{1.8}$$

Since $m(A \cap C_M)$ increases with M, we also have the equivalent

$$m(A) = \lim_{M \to \infty} m(A \cap C_M). \tag{1.9}$$

The definition of the inner measure may seem awkward. But we can look forward in Theorem 1.28 (after we show that every closed set is measurable) to the cleaner result

$$m_*(A) = \sup\{m(F) \mid F \text{ closed and } F \subset A\}.$$

Theorem 1.10 *For any* $A \subset \mathbb{R}^n$, *we have* $0 \le m_*(A) \le m^*(A) \le \infty$.

Proof In the bounded case, say $A \subset C_M$, Equation (1.4) and the inequality $m^*(A) + m^*(C_M - A) \ge m^*(A \cup (C_M - A))$ (from Property **(m*3)**) show

$$\begin{aligned} m^*(A) - m_*(A) &= m^*(A) - \{m^*(C_M) - m^*(C_M - A)\} \\ &= m^*(A) + m^*(C_M - A) - m^*(C_M) \\ &\ge m^*\big(A \cup (C_M - A)\big) - m^*(C_M) \\ &= m^*(C_M) - m^*(C_M) = 0. \end{aligned}$$

In the unbounded case, we use the result for the bounded case along with **(m*2)**, to obtain

$$m_*(A) = \lim_{M\to\infty} m_*(A \cap C_M) \le \lim_{M\to\infty} m^*(A \cap C_M) \le \lim_{M\to\infty} m^*(A) = m^*(A).$$

This completes the proof. ∎

Corollary 1.11 *Sets of measure zero are measurable.*

Proof A set Z of measure zero has $m^*(Z) = 0$. The inequality of Theorem 1.10 shows that $m_*(Z) = 0$. Hence, for every M-cube C_M, we have $m^*(Z \cap C_M) = m_*(Z \cap C_M) = m(Z \cap C_M) = 0$. ∎

This completes the proof of Property **(m1)**.

Theorem 1.12 Property (m1): *The empty set $\emptyset$ is measurable, and $m(\emptyset) = 0$.*

Definition 1.4 The collection of all measurable subsets of $\mathbb{R}^n$ is denoted by $\mathcal{L}(\mathbb{R}^n)$ or simply $\mathcal{L}$.

Theorem 1.13 *If $A \in \mathcal{L}$, then $m(A) = m_*(A) = m^*(A)$.*

Proof Let $A \in \mathcal{L}$. By the definitions of inner measure and of measurability

$$m(A) = \sup_M m_*(A \cap C_M) = \sup_M m^*(A \cap C_M) = m_*(A).$$

It is sufficient to prove $m^*(A) = \sup_M m^*(A \cap C_M)$. Since $m^*(A \cap C_M)$ increases with $M > 0$, it is sufficient to prove

$$m^*(A) = \sup_N m^*(A \cap C_N) \quad \text{for} \quad N = 1, 2, 3, \ldots. \tag{1.10}$$

Let $A_1 = A \cap C_1, A_2 = (A \cap C_2) - C_1, \ldots, A_k = (A \cap C_k) - C_{k-1}, \ldots$. Note $\bigcup_{k=1}^N A_k = A \cap C_N$ and $\bigcup_{k=1}^\infty A_k = A$. Then by Corollary 1.7:

$$m^*(A \cap C_N) = m^*\left(\bigcup_{k=1}^N A_k\right) = \sum_{k=1}^N m^*(A_k).$$

Finally, we obtain (1.10): $\sup_N m^*(A \cap C_N) = \sum_{k=1}^\infty m^*(A_k) = m^*(A)$. ∎

Warning We *cannot* simply define sets A to be measurable whenever $m^*(A) = m_*(A)$. If it were so, then in view of Theorem 1.10, all sets with $m_*(A) = \infty$ would be measurable. That would lead to *every* set in $\mathbb{R}^n$ being measurable (By Exercise 1.1.6, every set in $\mathbb{R}^n$ can be written as an intersection of two sets with infinite inner measure). But this cannot be because nonmeasurable sets *do* exist, as demonstrated later in Section 1.9.

Theorem 1.14 Property (m2): *If A, B are measurable and $A \subset B$ then $m(A) \leq m(B)$.*

This follows from $m^*(A) \leq m^*(B)$ (Theorems 1.4) Equations (1.7) and (1.8).

Theorem 1.15 Property (m3a): *If a set A is measurable, so is its complement A^c.*

Proof Consider an M-cube C_M. From Lemma 1.9 and $(A^c)^c = A$, we have

$$m_*(A \cap C_M) = |C_M| - m^*(A^c \cap C_M) \quad \text{and}$$
$$m_*(A^c \cap C_M) = |C_M| - m^*(A \cap C_M).$$

If A is measurable, then $m^*(A \cap C_M) = m_*(A \cap C_M)$, which leads to

$$m_*(A^c \cap C_M) = |C_M| - m^*(A \cap C_M)$$
$$= |C_M| - m_*(A \cap C_M) = m^*(A^c \cap C_M).$$

The set A^c thus satisfies the definition of Lebesgue measurability. ∎

Deriving the remaining properties of Theorem 1.1, namely **(m3b)**, **(m3c)**, and **(m4)**, is tricky. We will first show that the class $\mathcal{L}$ of measurable sets includes all open and closed sets, and then obtain an approximation theorem that will take us to the goal (Theorem 1.36).

Remark How many measurable sets are there? The Cantor set C is conumerous with $\mathbb{R}$; that is, its cardinality is c. Since C has measure zero, every subset of C has measure zero. Hence, all of these subset are measurable. Since C is conumerous with $\mathbb{R}$, there are as many subsets of C as there are subsets of $\mathbb{R}$. This cardinality is defined to be 2^c. So the cardinality of $\mathcal{L}(\mathbb{R})$ is 2^c. The same goes for the cardinality of $\mathcal{L}(\mathbb{R}^n)$.

Exercises for Section 1.1

1.1.1 Prove Property **(m*2)** of Theorem 1.4.

1.1.2 Finish the proof of Theorem 1.2, which states that $m^*(I) = |I|$, by considering the case of an unbounded nondegenerate interval I in $\mathbb{R}^n$. Since $|I| = \infty$, it remains to be shown that $m^*(I) = \infty$.

1.1.3 Prove Corollary 1.6 without the containment condition $I_1 \subset I_2$.

1.1.4 Prove Corollary 1.7.

1.1.5 Prove Corollary 1.8.

1.1.6 Show that *every* set E in $\mathbb{R}^n$ can be written as an intersection of two sets with infinite inner measure. Hint: Find two disjoint subsets of $\mathbb{R}^n$ of infinite inner measure. Then adjoin E to both sets.

1.1.7 Show that for $A \subset B$, we have $m_*(A) \leq m_*(B)$.

1.1.8 Show that every nonempty open subset G of $\mathbb{R}^n$ has nonzero inner measure.

1.2 Finite Union of Intervals

Although the definition of Euclidean measure is very different from that of Lebesgue measure, in this section we show that every finite union of n-intervals J is Lebesgue measurable ($J \in \mathcal{L}$) and the measures have the same value; that is, $m(J) = |J|$. First, the case where J is an n-interval I.

Theorem 1.16 *Every n-dimensional interval I is Lebesgue measurable and $m(I) = |I|$.*

Proof To show measurability of I, it is sufficient to show that

$$m^*(I \cap C_M) = m_*(I \cap C_M) = |C_M| - m^*(C_M - I) \quad \text{for all } C_M. \tag{1.11}$$

The set $I \cap C_M$ is an interval. Thus the left side of (1.11) is $m^*(I \cap C_M) = |I \cap C_M|$ by Theorem 1.2.

The set $C_M - I$ need not be an interval but it can be partitioned by extending all of the boundaries of I within C_M to obtain a finite union of nonoverlapping intervals, say $C_M - I = \bigcup_{k=1}^{N} I_k$ (see Figure 1.3).

Then $m^*(C_M - I) = |C_M - I| = \sum_{k-1}^{N} |I_k|$ by Corollary 1.3. The cube C_M is a nonoverlapping union of intervals $I \cap C_M, I_1, I_2, \ldots, I_N$. Equation (1.11) follows because by Theorem 0.32 we have

$$|C_M| = |(I \cap C_M) \cup (C_M - I)| = |I \cap C_M| + \sum_{k=1}^{N} |I_k|.$$

It remains to be shown that $m(I) = |I|$. In the bounded case, Equation (1.6) and Corollary 1.3 show $m(I) = m^*(I) = |I|$. In the unbounded case, clearly $|I|$, $m^*(I)$, $m_*(I)$, hence $m(I)$, are all infinite. ∎

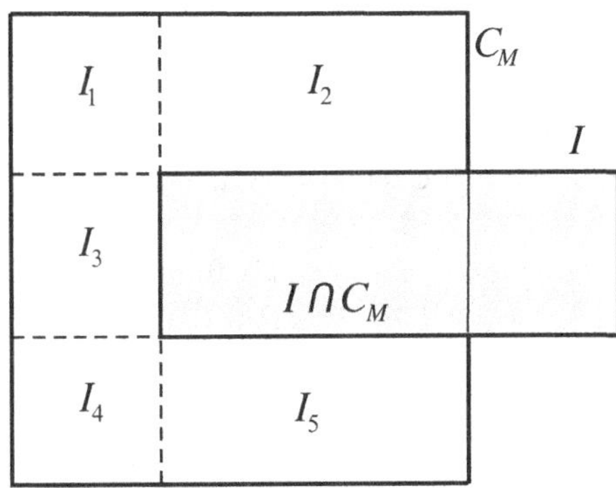

Figure 1.3 A bounded 2-dimensional interval I is measurable.

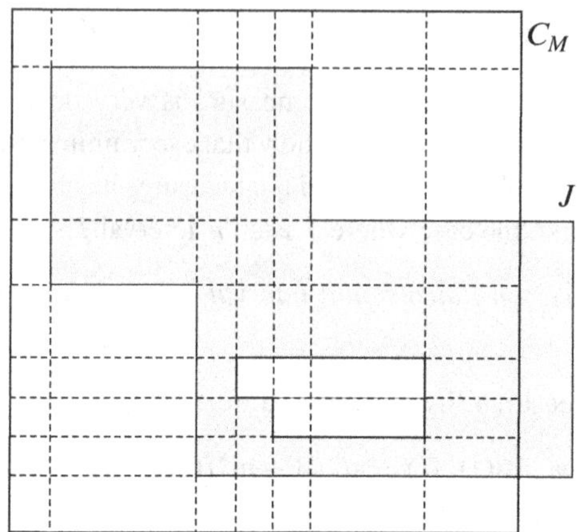

Figure 1.4 C_M is the union of intervals forming $J \cap C_M$ and $C_M - J$.

Theorem 1.17 *A finite union $J = \bigcup_{k=1}^{N} I_k$ of nonoverlapping intervals is Lebesgue measurable and $m(J) = \sum_{k=1}^{N} |I_k|$.*

Proof The proof follows that of the preceding Theorem 1.16. Suppose $J = \bigcup_{k=1}^{N} I_k$ is a finite union of nonoverlapping intervals $I_1, I_2, \dots I_N$. For measurability of J, we need the equality

$$m_*(J \cap C_M) = m^*(J \cap C_M) \quad \text{for all } M\text{-cubes } C_M.$$

By definition of inner measure, $m_*(J \cap C_M) = |C_M| - m^*(C_M - J)$. So it is sufficient to show that for every C_M we have

$$|C_M| = m^*(J \cap C_M) + m^*(C_M - J). \tag{1.12}$$

But $J \cap C_M$ is the finite union of the intervals $I_k \cap C_M$ for $k = 1, 2, \dots, N$. So $m^*(J \cap C_M) = |J \cap C_M| = \sum_{k=1}^{N} |I_k \cap C_M|$. By extending all boundaries of the intervals I_k within C_M, we see that $C_M - J$ is also a finite union of nonoverlapping intervals, say $C_M - J = \bigcup_{l=1}^{L} I'_l$.

By Corollary 1.3, $m^*(J \cap C_M) + m^*(C_M - J) = \sum_{k=1}^{N} |I_k \cap C_M| + \sum_{l=1}^{L} |I'_l|$.

Further, C_M is the union of all intervals forming $J \cap C_M$ and $C_M - J$ (illustrated in Figure 1.4), and by Theorem 0.32, $|C_M|$ is the sum of the measures of all of these intervals. This results in (1.12), which proves of the measurability of J.

Finally, $m(J) = m^*(J) = \sum_{k=1}^{N} |I_k|$ by Corollary 1.3. ∎

Corollary 1.18 *Any finite union of intervals J, whether overlapping or not, is Lebesgue measurable and $m(J) = |J|$.*

The proof follows that of Theorem 0.34 for Euclidean measure.

*1.3 Carathéodory Characterization

By Lemma 1.9, if a set A is measurable, then for every M-cube C_M

$$m^*(A \cap C_M) + m^*(C_M - A) = |C_M|. \tag{1.13}$$

In light of the results of the previous section, the measures of finite unions of intervals behave as expected. Equation (1.13) can easily be extended from M-cubes to intervals I as follows. If a set A is measurable, then for every interval I we have

$$m^*(A \cap I) + m^*(I - A) = |I|. \tag{1.14}$$

This is a basis of the following characterization of measurability. It is often used as a definition of measurability because this approach does not require the use of the inner measure.

Theorem 1.19 Carathéodory's theorem: *A set A is measurable if and only if for every "test" set $T \subset \mathbb{R}^n$, we have*

$$m^*(A \cap T) + m^*(T - A) = m^*(T).$$

Proof ($\Leftarrow$): Take the M-cube $T = C_M$. Then $A \cap T = A \cap C_M \subset C_M$, and by Equations (1.4), we have

$$m_*(A \cap C_M) = |C_M| - m^*(C_M - A) = m^*(T) - m^*(T - A).$$

This is $m^*(A \cap C_M)$, by hypothesis. Thus $m_*(A \cap C_M) = m^*(A \cap C_M)$ for every M-cube C_M, which means A is measurable.

($\Rightarrow$): Suppose A is measurable and consider a countable open interval cover $I_1, I_2, \ldots$ of T. Then

$$A \cap T \subset \bigcup_k (A \cap I_k) \quad \text{and} \quad T - A = A^c \cap T \subset \bigcup_k (A^c \cap I_k) = \bigcup_k (I_k - A).$$

Then by **(m*2)** and **(m*3)**, we have

$$\begin{aligned} m^*(A \cap T) + m^*(T - A) &\le \sum_k m^*(A \cap I_k) + \sum_k m^*(A^c \cap I_k) \\ &= \sum_k \{m^*(A \cap I_k) + m^*(I_k - A)\} = \sum_k |I_k|. \end{aligned}$$

The last equality follows from (1.14) applied to the intervals I_k. Taking the infimum with respect to all open interval covers of T, we have

$$m^*(A \cap T) + m^*(T - A) \le m^*(T).$$

Since $(A \cap T) \cup (T - A) = T$ the opposite inequality follows from Property **(m*3)**. This completes the proof of the theorem. ∎

1.4 Countable Union of Intervals

Theorem 1.20 *Any countable union of intervals* $L = \bigcup_{k=1}^{\infty} I_k$ *is a countable union of* nonoverlapping *intervals.*

Proof As in the proof of Theorem 0.34, let

$$J_1 = I_1,\ J_2 = I_2 - J_1,\ J_3 = I_3 - (J_1 \cup J_2),\ J_4 = I_4 - (J_1 \cup J_2 \cup J_3), \ldots .$$

Note that the sets J_k are finite unions of intervals, that they are nonoverlapping, and that $L = \bigcup_{k=1}^{\infty} J_k$. ∎

The decomposition of a set L into a countable union of nonoverlapping intervals is not unique. Fortunately, the next theorem shows that the measure does not depend on the decomposition.

Theorem 1.21 *If set L is written as a countable union of nonoverlapping intervals in two ways,* $L = \bigcup_k I_k = \bigcup_i I'_i$, *then*

$$m^*(L) = \sum_{k=1}^{\infty} |I_k| = \sum_{i=1}^{\infty} |I'_i|. \tag{1.15}$$

Proof Degenerate intervals may be ignored since their measures are zero. If any of the intervals I_k and I'_i are unbounded and nondegenerate, then both sums in (1.15) are infinite, hence equal. So we may suppose that all of the intervals are bounded.

Assume one sum is less than the other, say $\sum |I'_i| < \sum |I_k|$. Then there exists $N > 0$ and $\epsilon > 0$ such that

$$\sum_{i=1}^{\infty} |I'_i| + \epsilon = \sum_{k=1}^{N} |I_k|. \tag{1.16}$$

For each I'_i, choose a slightly larger open interval $G_i \supset I'_i$ such that

$$|I'_i| \le |G_i| < |I'_i| + \frac{\epsilon}{2^{i+1}}.$$

For each I_k, choose a slightly smaller closed interval $F_k \subset I_k$ such that

$$|F_k| \le |I_k| < |F_k| + \frac{\epsilon}{2^{k+1}} \quad \text{for} \quad k = 1, \ldots, N.$$

Let $K = \bigcup_{k=1}^{N} F_k$. This set K, being a finite union of closed and bounded intervals, is closed and bounded. By the Heine–Borel theorem (Theorem 0.23),

K is compact. The open sets $G_1, G_2, \ldots$ cover L, and hence also K. By compactness, there is a finite subcover of K, say $G_1, G_2, \ldots, G_M$. Then

$$\sum_{k=1}^{N} |I_k| < \sum_{k=1}^{N} \left(|F_k| + \frac{\epsilon}{2^{k+1}}\right) \leq \sum_{i=1}^{M} |G_i| + \frac{\epsilon}{2} \leq \sum_{i=1}^{\infty} |G_i| + \frac{\epsilon}{2} \leq \sum_{i=1}^{\infty} |I_i'| + \epsilon.$$

This contradicts (1.16). Consequently, the sums in (1.15) are equal. That this sum is the outer measure is clear from Property **(m*2)**. ∎

Theorem 1.22 *If L is a countable union of intervals, $L = \bigcup_k I_k$, then $L \in \mathcal{L}$. If the intervals are nonoverlapping, then $m(L) = m^*(L) = \sum_{k=1}^{\infty} |I_k|$.*

Proof We may assume that the intervals are nonoverlapping (see Theorem 1.20). By (1.15), $m^*(L) = \sum_{k=1}^{\infty} |I_k|$. First, we consider the case of bounded L. Then $m^*(L) < \infty$. Let $\epsilon > 0$ be given. Then there exists $N > 0$ such that $\sum_{k=1}^{N} |I_k| > m^*(L) - \epsilon$. By Theorem 1.17,

$$m_*(L) \geq m_*\left(\bigcup_{k=1}^{N} I_k\right) = m^*\left(\bigcup_{k=1}^{N} I_k\right) = \sum_{k=1}^{N} |I_k| > m^*(L) - \epsilon.$$

Since $\epsilon > 0$ is arbitrary, and since $m_*(L) \leq m^*(L)$ (Theorem 1.10), we have equality $m_*(L) = m^*(L) = \sum_{k=1}^{\infty} |I_k|$. The bounded case is done.

For the unbounded case, repeat the above argument for $L \cap C_M$. Since $L \cap C_M$ is a countable union of intervals, we obtain for all M-cubes C_M,

$$m(L \cap C_M) = m_*(L \cap C_M) = m^*(L \cap C_M).$$

Hence, by definition, $L \in \mathcal{L}$. The proof is complete. ∎

Corollary 1.23 *If L and L' are each countable unions of intervals, then $L \cup L' \in \mathcal{L}$.*

If further, L and L' are disjoint, then

$$m(L \cup L') = m(L) + m(L').$$

The proof of this corollary is left as Exercise 1.4.2.

Remark If L and L' are *finite* unions of intervals, so are $L \cup L'$, $L \cap L'$, $L' - L$, and L^c. If L and L' are *countable* unions of intervals, so are $L \cup L'$ and $L \cap L'$, but we cannot assume that for $L' - L$ and L^c. For example, the Cantor set (Definition 0.22) is the difference $C = L' - L$ of the closed interval $L' = [0, 1]$ and the countable union of open intervals $L = \left(\frac{1}{3}, \frac{2}{3}\right) \cup \left(\frac{1}{9}, \frac{2}{9}\right) \cup \left(\frac{7}{9}, \frac{8}{9}\right) \cup \left(\frac{1}{27}, \frac{2}{27}\right) \cup \cdots$. Yet C contains no intervals (except degenerate ones of the form $[a,a]$, but then it is not a countable union of such intervals). Similarly, L^c is not a countable union of intervals.

Exercises for Section 1.4

1.4.1 Use the Cantor set to show that the family of countable unions of intervals is not closed under countable intersections.

1.4.2 Prove Corollary 1.23.

1.5 Open and Closed Sets

Theorem 1.24 *Every open set G in $\mathbb{R}^n$ can be written as a countable union of nonoverlapping closed cubes.*

Proof Let G be an open set in $\mathbb{R}^n$. Tile $\mathbb{R}^n$ with nonoverlapping cubes of side 1. There are countably many such cubes. Select those cubes that are entirely contained in G. Bisect each side of the remaining cubes to obtain subcubes of side $\frac{1}{2}$. In $\mathbb{R}^n$, each cube has 2^n such subcubes. Select those subcubes that are entirely contained in G. Continue bisecting and selecting this way. Since G is open and the cubes become arbitrarily small, each point of G will eventually be caught in some selected cube of side $1/2^N$, for some N. The countable union of the selected cubes will thus be all of G. ∎

Corollary 1.25 *All open sets and all closed sets are measurable.*

Proof All cubes are intervals and thus measurable. For open sets, the result follows from Theorem 1.24. For closed sets, it follows because they are complements of open sets and because of Property **(m3a)** (Theorem 1.15). ∎

The following result follows from Corollary 1.23.

Corollary 1.26 *If G_1 and G_2 are open and disjoint, then $m(G_1 \cup G_2) = m(G_1) + m(G_2)$.*

Theorem 1.27 *For any set A, we have*

$$m^*(A) = \inf\{m(G) \mid G \text{ is open and } A \subset G\}.$$

Proof If G is open and $A \subset G$, then by **(m*2)** we have $m^*(A) \leq m^*(G) = m(G)$. So $m^*(A) \leq \inf\{m(G) \mid G \text{ is open and } A \subset G\}$. The opposite inequality follows from the definition of $m^*(A)$ as the infimum of the measure of all open *interval* covers. ∎

Theorem 1.28 *For any set A, we have*

$$m_*(A) = \sup\{m(K) \mid K \text{ compact and } K \subset A\}$$
$$= \sup\{m(F) \mid F \text{ closed and } F \subset A\}.$$

Proof If F is closed and $F \subset A$, then $m(F) = m_*(F) \leq m_*(A)$. Since every compact set is closed, $m_*(A) \geq \sup\{m(F) \mid F \text{ closed and } F \subset A\} \geq \sup\{m(K) \mid K \text{ compact and } K \subset A\}$. For the other direction, first consider the bounded case, say $A \subset C_M$. By definition,

$$m_*(A) = |C_M| - m^*(C_M - A)$$

and by Theorem 1.27 we have

$$m^*(C_M - A) = \inf\{m(G) \mid G \text{ open and } C_M - A \subset G\}.$$

Given $\epsilon > 0$, there exists an open set G such that $C_M - A \subset G$ and

$$m(G) - \epsilon < m^*(C_M - A) \leq m(G).$$

Then $m_*(A) = |C_M| - m^*(C_M - A) < |C_M| - m(G) + \epsilon = m_*(C_M - G) + \epsilon$. Since $C_M - G$ is closed and bounded, it is compact. Additionally, since $C_M - G \subset A$, we have $m_*(A) < \sup\{m(K) \mid K \text{ compact and } K \subset A\} + \epsilon$.

Taking the infimum with respect to all $\epsilon > 0$, we obtain

$$m_*(A) \leq \sup\{m(K) \mid K \text{ compact and } K \subset A\}.$$

Now for the unbounded case. The set $A \cap C_M$ is bounded for any C_M. By the above argument,

$$m_*(A \cap C_M) \leq \sup\{m(K) \mid K \text{ compact and } K \subset A \cap C_M\}.$$

This quantity is bounded above by $\sup\{m(K) \mid K \text{ compact and } K \subset A\}$. So $m_*(A) = \lim_{M\to\infty} m_*(A \cap C_M) \leq \sup\{m(K) \mid K \text{ compact and } K \subset A\}$. ∎

Theorem 1.29 *If F_1 and F_2 are closed and disjoint, then $m(F_1 \cup F_2) = m(F_1) + m(F_2)$.*

Proof First consider the bounded case. In this case, F_1 and F_2 are actually compact. Let G be an open set that covers $F_1 \cup F_2$; that is, $F_1 \cup F_2 \subset G$. By Theorem 0.29, there exists $r > 0$ such that $d(x, y) > r$ for all $x \in F_1$ and $y \in F_2$. Then the open sets

$$G_1 = G \cap \bigcup_{x\in F_1} S(x, r/2) \quad \text{and} \quad G_2 = G \cap \bigcup_{x\in F_2} S(x, r/2)$$

are disjoint and cover F_1 and F_2, respectively. We then have

$$m(F_1)+m(F_2) \le m(G_1)+m(G_2) = m(G_1 \cup G_2) \le m(G).$$

By Theorem 1.27 above we have $m(F_1)+m(F_2) \le m(F_1 \cup F_2)$. The opposite inequality follows from Property **(m*3)**.

In the unbounded case, since the equality follows for bounded $F_1 \cap C_M$ and $F_2 \cap C_M$, it still holds when taking the limit $M \to \infty$. ∎

Theorem 1.30 *If $A_1, A_2, \ldots$ are disjoint sets in $\mathbb{R}^n$, then $m_*\left(\bigcup_{k=1}^{\infty} A_k\right) \ge \sum_{k=1}^{\infty} m_*(A_k)$.*

Notice that this inequality is the reverse of that for outer measures.

Proof If $F_1, F_2, \ldots, F_N$ are closed subsets of $A_1, A_2, \ldots, A_N$, respectively, they are also disjoint. Then by Theorem 1.29, we have

$$m_*\left(\bigcup_{k=1}^{\infty} A_k\right) \ge m_*\left(\bigcup_{k=1}^{N} F_k\right) = \sum_{k=1}^{N} m(F_k).$$

Take the supremum with respect to all closed $F_k \subset A_k$. By Theorem 1.28,

$$m_*\left(\bigcup_{k=1}^{\infty} A_k\right) \ge \sum_{k=1}^{N} m_*(A_k).$$

Finally, take the supremum with respect to N to complete the proof. ∎

Corollary 1.31 *If $F \subset G$, where F is closed and G is open, then $G - F$ is measurable and*

$$m(G - F) = m(G) - m(F).$$

Proof Clearly $G - F$ is measurable, since it is open. By Theorem 1.30,

$$m(G) = m((G-F) \cup F) \ge m_*(G-F) + m_*(F) = m(G-F) + m(F).$$

The opposite inequality follows from Property **(m*3)**. ∎

Exercises for Section 1.5

1.5.1 Let $\epsilon > 0$ be given. Show that there exists an open set G in $\mathbb{R}$ which contains all of the rational numbers but $m(G) < \epsilon$.

1.5.2 Open sets on the real line have an interesting property that does not hold for higher dimensions. Show that every open set G in $\mathbb{R}$ can be written as a countable union of *disjoint* open intervals. Hint: For each $x \in G$, Let I_x be the largest open interval containing x which is in G. Show that these intervals are disjoint.

1.5.3 Exercise 1.5.2 *cannot* be extended to higher dimensions.[1] In particular, show that in $\mathbb{R}^2$, the open sphere $S(0,1)$ cannot be written as a countable union of disjoint open intervals.

1.5.4 Show that the Cantor set, which is closed, cannot be written as a countable union of nonoverlapping intervals.

1.6 The Approximation Theorem

Theorem 1.32 Approximation theorem: *A set A is measurable if and only if for each $\epsilon > 0$ there exists a closed set F and an open set G such that*

$$F \subset A \subset G \quad \text{and} \quad m(G - F) < \epsilon. \tag{1.17}$$

Proof ($\Leftarrow$): Suppose that for each $\epsilon > 0$, there exists a closed F and an open G such that (1.17) holds. Then

$$m^*(A) \le m(G) = m(F) + m(G - F) < m(F) + \epsilon \le m_*(A) + \epsilon.$$

Since $\epsilon > 0$ is arbitrary, we have $m^*(A) \le m_*(A)$. Since $m_*(A) \le m^*(A)$ always holds we have equality. In the bounded case, we are done.

For the unbounded case, let F and G be as above. Also let C_M be a *closed* M-cube and let $C_{M'}$ be a slightly larger *open* cube such that $m(C_{M'} - C_M) < \epsilon$. Then $F \cap C_M$ is closed and $G \cap C_{M'}$ is open with $F \cap C_M \subset A \cap C_M \subset G \cap C_{M'}$. So

$$\begin{aligned} m^*(A \cap C_M) &\le m(G \cap C_{M'}) \\ &\le m(F \cap C_M) + m\big((G \cap C_{M'}) - (F \cap C_M)\big) \\ &= m(F \cap C_M) + m(G - F) + m(C_{M'} - C_M) \\ &\le m_*(A \cap C_M) + \epsilon + \epsilon. \end{aligned}$$

Since $\epsilon > 0$ is arbitrary, we conclude $m^*(A \cap C_M) = m_*(A \cap C_M)$ for every M-cube. By definition, we have $A \in \mathcal{L}$.

($\Rightarrow$): First consider the case of a measurable A with $m(A) < \infty$. By Theorem 1.27, there exists an open set $G \supset A$ such that

$$m^*(A) \le m(G) < m^*(A) + \epsilon/2,$$

[1] This fact makes the theory of Lebesgue measure more difficult on $\mathbb{R}^n$ for $n > 1$.

and by Theorem 1.28 there exists a closed set F such that $F \subset A$ and

$$m_*(A) - \epsilon/2 < m(F) \le m_*(A).$$

Combining the two and using Corollary 1.31, we have

$$m(G - F) = m(G) - m(F) < m^*(A) - m_*(A) + \epsilon.$$

Since A is measurable, this results in $m(G - F) < \epsilon$.

In the case of a measurable set A with $m(A) = \infty$, for each $M = 1, 2, \ldots$, the set $A \cap C_M$ is measurable with finite measure. By the above case, we can find closed F_M and open G_M such that

$$F_M \subset A \cap C_M \subset G_M \quad \text{and} \quad m(G_M - F_M) < \frac{\epsilon}{2^M}. \quad \text{Let}$$

$$F = \bigcup_{M=1}^{\infty} F_M \quad \text{and} \quad G = \bigcup_{M=1}^{\infty} G_M.$$

It is clear that $F \subset A \subset G$ and that G is open.

A countable union of closed sets need not be closed. However, in this case, F *is* closed. To show this, consider a boundary point x of F. We show that x must belong to F. Since $\bigcup C_M = \mathbb{R}^n$, x is an interior point of some C_M. Every open sphere $S(x,r)$ contains some, perhaps smaller, open sphere $S(x,r') \subset C_M$. The open sphere $S(x,r')$ must intersect both F and F^c, and hence also $F_M = F \cap C_M$ and $F^c \subset F_M^c$. This means that x is a boundary point of F_M. Since F_M is closed, we have $x \in F_M \subset F$. This shows that F is closed. Finally,

$$m(G - F) \le m\left(\bigcup_{M=1}^{\infty} (G_M - F_M)\right) \le \sum_{M=1}^{\infty} m(G_M - F_M) < \sum_{M=1}^{\infty} \frac{\epsilon}{2^M} = \epsilon. \quad \blacksquare$$

1.7 General Measurable Sets

We have already shown in Theorem 1.15 that if A is measurable, then A^c is measurable. To complete the proof of Property **(m3)**, we need only show the following.

Theorem 1.33 Remainder of Property (m3): *If A and B are measurable, then so are*

$$A - B \quad \textit{and} \quad A \cap B.$$

Proof We use the approximation theorem to obtain the measurability of $A-B$. Given $\epsilon > 0$, find $F_1 \subset A \subset G_1$ and $F_2 \subset A \subset G_2$ such that $m(G_1 - F_1) < \epsilon/2$ and $m(G_2 - F_2) < \epsilon/2$. Let $F = F_1 - G_2$ and $G = G_1 - F_2$. Clearly F is closed and G is open and $F \subset A - B \subset G$. Then $G - F \subset (G_1 - F_1) \cup (G_2 - F_2)$ and so

$$m(G - F) \leq m\big((G_1 - F_1) \cup (G_2 - F_2)\big) \leq m(G_1 - F_1) + m(G_2 - F_2) < \epsilon.$$

That $A \cap B$ is measurable follows from the above argument and

$$A \cap B = A - (A - B). \qquad \blacksquare$$

Corollary 1.34 *If A and B are measurable, then so is $A \cup B$.*

Proof We only need to use Property **(m3)** along with the De Morgan law

$$A \cup B = \left(A^c \cap B^c\right)^c. \qquad \blacksquare$$

Corollary 1.35 *Suppose sets A and B differ by a set of measure zero (that is, $A - B$ and $B - A$ are both sets of measure zero). If one is measurable, so is the other.*

We leave the proof to Exercise 1.7.1. Finally we prove Property **(m4)**.

Theorem 1.36 Property (m4): *If $A_1, A_2, \ldots$ are measurable, then so is $A = \bigcup_{k=1}^{\infty} A_k$ and we have* **(m4a)** $m(A) \leq \sum_{k=1}^{\infty} m(A_k)$.

Further, if the A_k are disjoint, we have equality **(m4b)** $m(A) = \sum_{k=1}^{\infty} m(A_k)$.

Proof We prove the disjoint case first. By Theorems 1.13, 1.30, and Property **(m*3)**, we have,

$$m_*(A) = m_*\left(\bigcup_{k=1}^{\infty} A_k\right) \geq \sum_{k=1}^{\infty} m_*(A_k) = \sum_{k=1}^{\infty} m^*(A_k) \geq m^*\left(\bigcup_{k=1}^{\infty} A_k\right) = m^*(A).$$

Additionally, since the inner measure cannot exceed the outer measure we have equalities. In particular, we have

$$m^*(A \cap C_M) = m_*(A \cap C_M) = m(A \cap C_M) \text{ for every } M\text{-cube } C_M.$$

So in the disjoint case, A is measurable and we have the equality **(m4b)**.

In the general case, we can write A as a disjoint union as follows.

$$B_1 = A_1,\ B_2 = A_2 - B_1,\ \ldots,\ B_k = A_k - (B_1 \cup B_2 \cup \cdots \cup B_{k-1}),\ \ldots.$$

Clearly the sets B_k are disjoint. Inductively, we can use Theorem 1.33 and the disjoint case of this proof to show that the sets B_k are measurable. Since

$A = \bigcup_{k=1}^{\infty} B_k$ is a disjoint union, A is measurable. It follows that $m(A) = \sum_{k=1}^{\infty} m(B_k)$. Finally, since $B_k \subset A_k$, we have $m(A) = \sum_{k=1}^{\infty} m(B_k) \leq \sum_{k=1}^{\infty} m(A_k)$, which completes the proof of **(m4)**. ∎

Corollary 1.37 *If $A_1, A_2, \ldots$ are measurable, then so is the intersection $\bigcap_{k=1}^{\infty} A_k$.*

The proof follows from Properties **(m3a)**, **(m4)**, and the De Morgan law

$$\bigcap A_k = \left(\bigcup A_k^c\right)^c.$$

Corollary 1.38 *If $A_1 \subset A_2 \subset A_3 \subset \cdots$, with each A_k measurable, then*

$$m\left(\bigcup_{k=1}^{\infty} A_k\right) = \lim_{k\to\infty} m(A_k).$$

Proof Let $B_1 = A_1$ and $B_k = A_k - A_{k-1}$ for $k > 1$. Then the sets B_k are disjoint and measurable and

$$\bigcup_{k=1}^{\infty} B_k = \bigcup_{k=1}^{\infty} A_k \quad \text{and} \quad A_N = \bigcup_{k=1}^{N} B_k \quad \text{for all } N.$$

By Property **(m4)** we have

$$m\left(\bigcup_{k=1}^{\infty} A_k\right) = \sum_{k=1}^{\infty} m(B_k) = \lim_{N\to\infty} \sum_{k=1}^{N} m(B_k) = \lim_{N\to\infty} m(A_N).$$ ∎

Corollary 1.39 *All countable unions of closed sets are measurable. They are called **F_σ sets**. All countable intersections of open sets are measurable, and are called **G_δ sets**.*

The first statement follows immediately from Property **(m4)** once we have the fact that all closed sets are measurable (Theorem 1.25). Similarly, the second uses Corollary 1.37.

Exercises for Section 1.7

1.7.1 Prove Corollary 1.35.

1.7.2 Show that if $A_1 \supset A_2 \supset A_3 \supset \cdots$ are measurable with $m(A_1) < \infty$, then

$$m\left(\bigcap_{k=1}^{\infty} A_k\right) = \lim_{k\to\infty} m(A_k).$$

1.7.3 Give an example where the conclusion in Exercise 1.7.2 does not hold for $m(A_1) = \infty$.

1.7.4 Consider the set $A = \{(x, y) \in \mathbb{R}^2 \mid 0 \le x \le 1, 0 \le y \le x^2\}$. Show $m(A) = 1/3$ by building closed sets F_N and F'_N with $F_N \subset A \subset F'_N$ such that $\lim_{N\to\infty} m(F_N) = \lim_{N\to\infty} m(F'_N) = 1$.

1.7.5 Consider the set $A = \{(x, y) \in \mathbb{R}^2 \mid 0 \le x < \infty, 0 \le y \le e^{-x}\}$. Show $m(A) = 1$ by building F_σ sets P_N and Q_N with $P_N \subset A \subset Q_N$ such that $\lim_{N\to\infty} m(P_N) = \lim_{N\to\infty} m(Q_N) = 1$.

1.8 Borel Sets

Definition 1.5 A collection of subsets of $\mathbb{R}^n$ which contains the empty set $\emptyset$ and which is algebraically closed under complements and countable unions is called a **σ-algebra**. By De Morgan's laws, it is necessarily also closed under countable intersections. An example of a σ-algebra is the power set of $\mathbb{R}^n$ consisting of *all* subsets of $\mathbb{R}^n$. Another example is the family $\mathcal{B}$ of **Borel**[2] **sets**, defined to be the σ-algebra generated by the collection in $\mathbb{R}^n$ of all open sets.[3]

This Borel family of sets $\mathcal{B}$ contains all countable intersections of open sets (the **G_δ sets**) and all countable unions of closed sets (the **F_σ sets**). However, $\mathcal{B}$ contains much more than just the G_δ and F_σ sets. It is a very large family. Clearly $\mathcal{B} \subset \mathcal{L}$. Yet it is still possible to find Lebesgue measurable sets which are not Borel sets.

Theorem 1.40 *Most sets of measure zero fail to be Borel sets.*

Proof We use a cardinality argument. First consider the case $\mathbb{R}$. Every open set G in $\mathbb{R}$ is a countable union of disjoint open intervals (Exercise 1.5.2). Each such open interval is the countable union of open intervals with rational endpoints. Thus every open set G in $\mathbb{R}$ can be associated with a *sequence* of rational numbers (the sequence of endpoints). The cardinality of such

[2] These sets are named in honor of Émile Borel (1871–1956), a French mathematician. Many results are named after him, including the Heine–Borel theorem (Theorem 0.23 and Theorem 5.45). At age 22, he was already the head of the Mathematics Department of the University of Lille. Later, when he became head of the department at the *École Normale Supérieure* in Paris, Lebesgue was an undergraduate.

[3] It is easy to see that the intersection of any family of σ-algebras is a σ-algebra (Exercise 1.8.4). Thus $\mathcal{B}$ is the intersection of all σ-algebras containing all the open sets of $\mathbb{R}^n$.

a collection of sequences is c, the same as the cardinality of $\mathbb{R}$ (See Cantor's theorem [Theorem 0.2]). Since $\mathcal{B}$ is the smallest collection containing the open sets which is also algebraically closed under the operations of complements, countable unions, and countable intersections, it can be shown that the cardinality of $\mathcal{B}$ is c as well. Recall the Remark at the end of Section 1.1 where it was shown that there are 2^c sets of measure zero. Since the cardinality of $\mathcal{B}$ is merely c, there must be (many) sets of measure zero that are not Borel sets.

For the general case $\mathbb{R}^n$, we can use the fact that the intersection of a Borel set with the x-axis will be a Borel set on the x-axis. The above case shows there are sets of measure zero on the x-axis which are not Borel. ∎

This cardinality argument shows the existence of measurable sets which are not Borel. In Section 1.10 we will give an explicit example.

Corollary 1.41 *The family $\mathcal{B}$ is a proper subset of $\mathcal{L}$.*

Proof The family $\mathcal{L}$ is algebraically closed under complements, and countable unions and intersections. It also contains all open sets. So $\mathcal{L}$ contains $\mathcal{B}$. Yet $\mathcal{L}$ also contains all of the sets of measure zero, which $\mathcal{B}$ does not. ∎

Even though most measurable sets are not Borel sets, we can show that the difference between $\mathcal{B}$ and $\mathcal{L}$ can be explained merely by the sets of measure zero.

Theorem 1.42 *For every measurable set $A \subset \mathbb{R}^n$, there exists an F_σ set $B = \bigcup F_k$ such that $B \subset A$ and $m(A - B) = 0$. Thus $A = B \cup Z$ is the disjoint union of a Borel set B and a set $Z = A - B$ of measure zero.*

Similarly, for every measurable set A, there exists a G_δ set $C = \bigcap G_k$ such that $A \subset C$ and $m(C - A) = 0$.

The proof is Exercise 1.8.8.

Corollary 1.43 *$\mathcal{L}$ is the σ-algebra generated by $\mathcal{B}$ and the sets of measure zero.*

Remark If all we wanted were the basic properties **(m1)** through **(m4)** of measure (as given in Theorem 1.1), then the family of Borel sets $\mathcal{B}$ would do it. Actually, this is the smallest family of sets containing the open intervals and satisfying **(m1)**–**(m4)**. However, it does not contain all of the sets of measure zero. The larger Lebesgue family $\mathcal{L}$, which is defined in terms of equality of inner and outer measures, maximizes the sets in $\mathbb{R}^n$ that have the usual measure of intervals and the properties **(m1)**–**(m4)**. So any method of measuring sets in $\mathbb{R}^n$ having these properties cannot be larger than $\mathcal{L}$. In the next section we will show that $\mathcal{L}$ cannot include *all* subsets of $\mathbb{R}^n$.

Definition 1.6 A σ-algebra $\mathcal{M}$ is said to be **complete** if it contains all subsets of measure zero; that is, $A \in \mathcal{M}$ whenever $A \subset Z \in \mathcal{M}$ and $m(Z) = 0$. The fact that the σ-algebra of Borel sets $\mathcal{B}$ is not complete creates some theoretical difficulties. By Corollary 1.43, $\mathcal{L}$ is the σ-algebra generated by $\mathcal{B}$ and the sets of measure zero. We say that $\mathcal{L}$ is the **completion** of the σ-algebra $\mathcal{B}$. We will discuss the completion more in Section 4.1.

Exercises for Section 1.8

1.8.1 Show that $\mathcal{L}$ and $\mathcal{B}$ are σ-algebras.

1.8.2 Show that any half-open interval $(a,b]$ is a Borel set in $\mathbb{R}$.

1.8.3 Show that for any two Borel sets A, B, the difference $A - B$ is a Borel set.

1.8.4 Show that the intersection of any collection of σ-algebras is a σ-algebra.

1.8.5 Let $\mathcal{F}$ be any fixed collection of subsets of $\mathbb{R}^n$. Show that the intersection of all σ-algebras containing $\mathcal{F}$ is itself a σ-algebra containing $\mathcal{F}$. This is called the σ-algebra generated by the collection $\mathcal{F}$.

1.8.6 See Exercise 1.8.5. Let $\mathcal{F}$ be a collection of all n-dimensional intervals in $\mathbb{R}^n$. Show that the σ-algebra generated by the collection $\mathcal{F}$ is the family of Borel sets $\mathcal{B}$.

1.8.7 Show that for every measurable set A, there exists an F_σ set B and a G_δ set C such that $B \subset A \subset C$ and $m(C - B) = 0$. For each $k = 1,2,\ldots$ use the approximation theorem (Theorem 1.32) with $\epsilon = 1/k$ to find an open set $A \subset G_k$ and a closed set $F_k \subset A$.

1.8.8 Prove Theorem 1.42, using Exercise 1.8.7.

1.8.9 Show that every set of measure zero is a subset of a Borel set of measure zero.

*1.9 Nonmeasurable Sets

We show in this section that not every subset of $\mathbb{R}^n$ is Lebesgue measurable. This is *not* due to a fault in the theory of Lebesgue measure. It turns out that, if we want a method of measuring sets in $\mathbb{R}^n$ to have the following three properties,[4] then not every subset of $\mathbb{R}^n$ can be measurable.

[4] Indeed, a measure having these three properties on a family of subsets of $\mathbb{R}^n$ must be the Lebesgue measure on that family. This was shown by the Hungarian mathematician Alfréd Haar (1885–1933). See Exercises 1.9.2–1.9.4.

(h1) The countable additivity property **(m4)** (given in Theorem 1.1).
(h2) The measure of the unit cube is 1, $m(C_1) = 1$.
(h3) The measure is translation invariant. That is, for each measurable set $A \subset \mathbb{R}^n$ and $x \in \mathbb{R}^n$, we have $m(A + x) = m(A)$.

Below we give an example of a nonmeasurable set given in 1905 by the Italian mathematician Giuseppe Vitali (1875–1932).

Theorem 1.44 Vitali's 1905 Example: *There exists a nonmeasurable set in* $\mathbb{R}$.

Proof Consider the equivalence relation $x \sim y$ on $\mathbb{R}$ to mean $x - y$ is rational. The verification that $\sim$ is an equivalence relation is left as Exercise 1.9.5. The equivalence classes are sets of the form

$$[x] = x + \mathbb{Q} = \{x + r \mid r \in \mathbb{Q}\}.$$

For example, $[0] = \mathbb{Q}$ and $[\sqrt{2}] = \left\{\sqrt{2}, \sqrt{2} + 1, \sqrt{2} - 1, \dots, \sqrt{2} + \frac{27}{17}, \dots\right\}$. Equivalence classes partition $\mathbb{R}$ in the sense that

$$\text{for any } x, y \in \mathbb{R}, \text{ we have either } [x] = [y] \quad \text{or} \quad [x] \cap [y] = \emptyset. \tag{1.18}$$

Since $\mathbb{Q}$ is dense in $\mathbb{R}$, each equivalence class $[x]$ is dense in $\mathbb{R}$. In particular, each equivalence class intersects the interval $[-1, 1]$. From each equivalence class, we choose one element that lies in the interval $[-1, 1]$. Denote this set by $\mathcal{V}$. This is accomplished by the use of the axiom of choice[5] (see Section 0.14). This set $\mathcal{V}$ has the property that for each $x \in \mathbb{R}$, there is exactly one in $v \in \mathcal{V}$ such that $r = x - v \in \mathbb{Q}$. Such a set is called a **Vitali set**. Here $\mathcal{V} \subset [-1, 1]$. We will show that $\mathcal{V}$ is not measurable.

By (1.18) above we have a disjoint union

$$\mathbb{R} = \bigcup_{v \in \mathcal{V}} [v].$$

Note that for each rational r, the set

$$r + \mathcal{V} = \{r + v \mid v \in \mathcal{V}\}$$

also consists of exactly one element from each equivalence class $[x]$.

For each $x \in [-1, 1]$ let v be the uniquely chosen element of $\mathcal{V}$ in the equivalence class $[x]$. Then $r = x - v$ is rational and $-2 \le r \le 2$. That is, every $x \in [-1, 1]$ belongs to $r + \mathcal{V}$ for some rational $r \in [-2, 2]$.

[5] Actually the axiom of choice was discovered as a result of examining the proof of Vitali's example.

Let $r_1, r_2, \ldots$ be an enumeration of the rationals in $[-2,2]$. Then

$$[-1,1] \subset \bigcup_{k=1}^{\infty} \left(r_k + \mathcal{V}\right) \subset [-3,3]. \tag{1.19}$$

It is easy to see that $m^*(r_k + \mathcal{V}) = m^*(\mathcal{V})$ for each k. If the outer measure $m^*(\mathcal{V})$ were zero, then $m^*(r_k + \mathcal{V}) = 0$ for all k, and then by Property **(m*3)**,

$$2 = m^*([-1,1]) \leq \sum_{k=1}^{\infty} m^*(r_k + \mathcal{V}) = 0,$$

which is false. Thus the outer measure of $\mathcal{V}$ is positive: $m^*(\mathcal{V}) > 0$.

Now we look at the inner measure. By Theorem 1.30 we have

$$\sum_{k=1}^{\infty} m_*(\mathcal{V}) = \sum_{k=1}^{\infty} m_*(r_k + \mathcal{V}) \leq m_* \left(\bigcup_{k=1}^{\infty} (r_k + \mathcal{V}) \right)$$
$$\leq m_*([-3,3]) = 6 < \infty.$$

This can happen only if $m_*(\mathcal{V}) = 0$. Thus $m_*(\mathcal{V}) = 0 < m^*(\mathcal{V})$, so $\mathcal{V}$ is not measurable. ∎

Vitali's example can be expanded to $\mathbb{R}^n$. See Exercise 1.9.8.

If a set Z has zero measure, then all subsets have zero measure and are thus measurable. The following corollary shows that all *other* sets have nonmeasurable subsets.

Corollary 1.45 *If a set $A \subset \mathbb{R}$ has positive outer measure $m^*(A) > 0$, then there exists a nonmeasurable subset $\mathcal{V}_A$ of A.*

Proof If A is nonmeasurable, then A itself is a nonmeasurable subset of A. If $A \subset \mathbb{R}$ is measurable with $m(A) > 0$, there exists an $N > 0$ such that $A_N = A \cap [-N, N]$ has positive measure. If we let $r_1, r_2, \ldots$ be an enumeration of the rationals in $[-N-1, N+1]$, the inclusion (1.19) becomes

$$[-N,N] \subset \bigcup_{k=1}^{\infty} \left(r_k + \mathcal{V}\right) \subset [-N-2, N+2]. \tag{1.20}$$

Intersecting with the set A, we obtain

$$A_N = A \cap [-N,N] \subset \bigcup_{k=1}^{\infty} \left(\left(r_k + \mathcal{V}\right) \cap A \right) \subset A \cap [-N-2, N+2].$$

Taking the outer measure, $0 < m^*(A_N) \leq \sum_{k=1}^{\infty} m^*\left((r_k + \mathcal{V}) \cap A\right) \leq 2(N+2)$. The sum is positive, so there must be a k for which $m^*\big((r_k + \mathcal{V}) \cap A\big) > 0$.

Denote this set by $\mathcal{V}_A = (r_k + \mathcal{V}) \cap A$. We have $\mathcal{V}_A \subset A$ and $m^*(\mathcal{V}_A) > 0$. Looking at the inner measure, we obtain

$$0 \le m_*(\mathcal{V}_A) = m_*\big((r_k + \mathcal{V}) \cap A\big) \le m_*(r_k + \mathcal{V}) = m_*(\mathcal{V}) = 0.$$

So $0 = m_*(\mathcal{V}_A) < m^*(\mathcal{V}_A)$; that is, $\mathcal{V}_A$ is a nonmeasurable subset of A. ∎

Exercises for Section 1.9

1.9.1 Show that the Lebesgue measure satisfies Property **(h3)** given in Section 1.9.

1.9.2 Suppose that a method of measuring sets in $\mathbb{R}^2$ has the Properties **(h1)**, **(h2)**, and **(h3)** as given in this section. If we bisect the sides of the unit cube into 4 subcubes, we can easily see that the subcubes will all have equal measure, hence measure $\frac{1}{4}$. Instead of bisecting, we could divide into more parts. Show that a 2-dimensional interval with rational sides r and q will have the usual measure rq.

1.9.3 Continue with Exercise 1.9.2 by extending it to intervals with real sides. This will give us the Euclidean measure of all 2-dimensional intervals.

1.9.4 Extend the above two exercises to $\mathbb{R}^n$. This will give us the Euclidean measure of n-dimensional intervals, which is where this chapter started.

1.9.5 Show that the relation $\sim$ defined in the proof of Theorem 1.44 is an equivalence relation. That is, show that the following properties hold for all $x, y, z \in \mathbb{R}$:

(EQ1) $x \sim x$ (reflexivity);
(EQ2) If $x \sim y$, then $y \sim x$ (symmetry);
(EQ3) If $x \sim y$ and $y \sim z$, then $x \sim z$ (transitivity).

1.9.6 Prove statement (1.18) in the proof of Theorem 1.44.

1.9.7 From the proof of Theorem 1.44 show that for each rational r, the set $r + \mathcal{V} = \{r + v \mid v \in \mathcal{V}\}$ consists of exactly one element from each equivalence class $[x]$.

1.9.8 Prove Theorem 1.44 for general dimension n. Use the set $\mathbb{Q}^n$ of points in $\mathbb{R}^n$ whose coordinates are rational.

1.9.9 Use Corollary 1.45 to show that there exist (nonmeasurable) disjoint sets A_1 and A_2, for which $m^*(A_1 \cup A_2) < m^*(A_1) + m^*(A_2)$.

1.9.10 Give a proof of Corollary 1.45 for the case $\mathbb{R}^n$ with $n > 1$.

*1.10 A Measurable Set Which Is Not Borel

In Section 1.8, it was shown that there exist sets of measure zero which are not Borel. This section constructs an example. Recall that the Cantor set C (Definition 0.22) is a set of measure zero. Hence, it and all of its subsets are Lebesgue measurable. Our strategy is to construct a subset of the Cantor set C which is not a Borel set. We start by constructing a continuous function that maps the Cantor set onto a set of positive measure.

Recall that the Cantor set is the intersection $C = \bigcap_{k=0}^{\infty} C_k$, where

$C_0 = [0, 1]$.
C_1 is obtained by removing the middle third $\left(\frac{1}{3}, \frac{2}{3}\right)$ of C_0.
C_2 is obtained by removing the middle thirds $\left(\frac{1}{9}, \frac{2}{9}\right)$ and $\left(\frac{7}{9}, \frac{8}{9}\right)$ of C_1.
And so on. For each $k = 1, 2, \ldots$, the set C_k is obtained by removing the middle thirds of the 2^{k-1} subintervals of C_{k-1}.
In all there are $1 + 2^1 + 2^2 + \cdots + 2^{k-1} = 2^k - 1$ subintervals removed from C_0 to obtain C_k.

Denote these removed intervals by $D_{k,1}, D_{k,2}, D_{k,3}, \ldots, D_{k,2^k-1}$, arranged in increasing order. For example, for $k = 3$ we have the $2^3 - 1 = 7$ intervals $D_{2,1}, D_{2,2}, \ldots, D_{2,7} =$

$$\left(\frac{1}{27}, \frac{2}{27}\right), \left(\frac{1}{9}, \frac{2}{9}\right), \left(\frac{7}{27}, \frac{8}{27}\right), \left(\frac{1}{3}, \frac{2}{3}\right), \left(\frac{19}{27}, \frac{20}{27}\right), \left(\frac{7}{9}, \frac{8}{9}\right), \left(\frac{25}{27}, \frac{26}{27}\right).$$

Define, for $k = 1, 2, \ldots$, the function

$$F_k(x) = \begin{cases} 0 & \text{if } x = 0 \\ \dfrac{j}{2^k} & \text{if } x \in D_{k,j}, \text{ for } j = 1, 2, \ldots, 2^k - 1 \\ 1 & \text{if } x = 1. \end{cases}$$

For the remaining parts of $[0, 1]$ (that is, for $0 < x < 1$ with $x \in C_k$), make F_k linear on each subinterval of C_k (see Figure 1.5).

It is clear that $|F_1(x) - F_0(x)| < 1/2$, $|F_2(x) - F_1(x)| < 1/2^2$ and, in general, $|F_k(x) - F_{k-1}(x)| < 1/2^k$. Thus for $n < m$ we have

$$\begin{aligned} |F_m(x) - F_n(x)| &\le |F_m(x) - F_{m-1}(x)| + \cdots + |F_{n+1}(x) - F_n(x)| \\ &\le \frac{1}{2^m} + \cdots + \frac{1}{2^{n+1}} < \frac{1}{2^n}. \end{aligned}$$

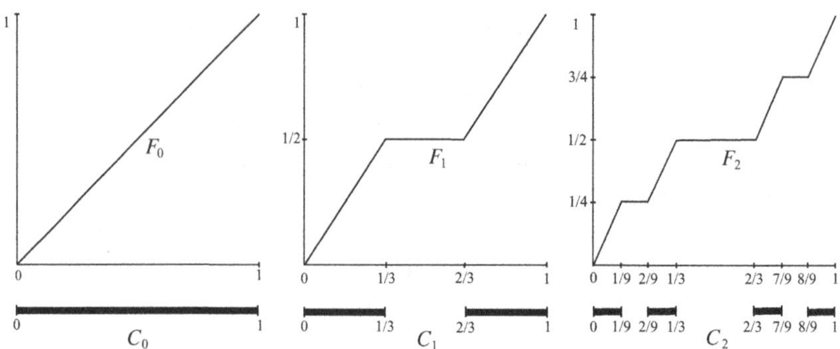

Figure 1.5 The function F_0, F_1, F_2 with sets C_0, C_1, C_2 underneath.

Since, for each x, $F_k(x)$ is Cauchy, the functions F_k converge to a limit F. This function F is called the **Cantor ternary function.**[6] It is sometimes called the **Devil's staircase** or the **Cantor–Lebesgue function**.

Theorem 1.46 *The Cantor ternary function F is continuous and monotonically increasing on [0, 1]. It is differentiable almost everywhere on* $[0,1]$ *with* $F'(x) = 0$ *almost everywhere.*

Proof Each of the functions F_k is continuous. Furthermore $|F_m(x) - F_n(x)| < \frac{1}{2^n}$ for $n < m$. This shows that convergence is uniform on $[0,1]$; that is, it does not depend on x. It is a standard result of real analysis that the uniform limit F of continuous functions F_k is continuous. Thus F is continuous.

For each F_k we have $F_k(x) \leq F_k(y)$ whenever $x < y$. Letting $k \to \infty$, we have $F(x) \leq F(y)$ for $x < y$. Thus F is monotonically increasing.

Notice that the Cantor ternary function is constant on all open intervals $D_{k,j}$, whose union is $[0,1] - C$. Thus $F'(x) = 0$ on $[0,1] - C$. Since C is of measure zero, we have $F'(x) = 0$ almost everywhere. ∎

[6] There is an explicit formulation of the Cantor ternary function. Represent the numbers $x \in [0,1]$ in base 3 (ternary) expansion

$$x = 0.a_1a_2a_3\ldots = \frac{a_1}{3} + \frac{a_2}{3^2} + \frac{a_3}{3^3} + \cdots.$$

To obtain uniqueness of the representation, replace any terminating ternary digit 1 with the equivalent ternary digits ending with repeated 2s. It was shown in Theorem 0.41 that the elements of C are those $x \in [0,1]$ for which none of the a_k are 1. In the case $x \in C$, let $b_k = a_k/2$ for $k = 1,2,3,\ldots$ and let $F(x) = \sum_{k=1}^{\infty} \frac{b_k}{2^k}$. For $x \notin C$ let N be the smallest value of k for which $a_k = 1$, and then let $b_k = a_k/2$ for $k = 1,2,3,\ldots,N-1$ and $b_N = a_N = 1$. Then let $F(x) = \sum_{k=1}^{N} \frac{b_k}{2^k}$.

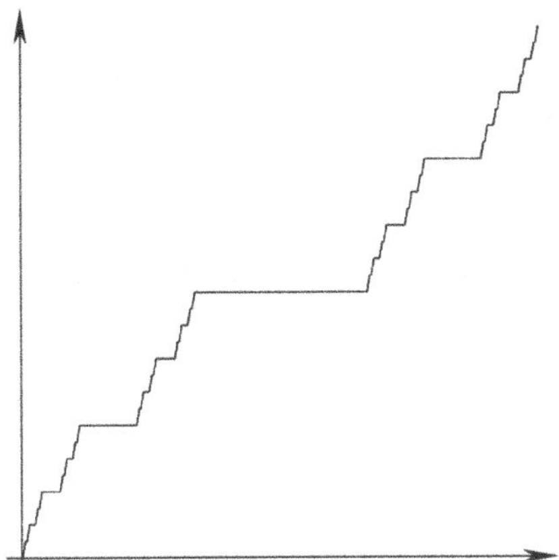

Figure 1.6 The Cantor ternary function F or Devil's staircase.[7]

Theorem 1.47 *The function $h(x) = F(x) + x$ is a strictly increasing and continuous function that maps $[0, 1]$ onto $[0, 2]$ and maps C onto a set of Lebesgue measure 1. That is, $m\big(h(C)\big) = 1$. The inverse $h^{-1}\colon [0, 2] \longrightarrow [0, 1]$ is also continuous and increasing.*

Proof Since $F\colon [0, 1] \longrightarrow [0, 1]$ is monotonically increasing and continuous, h will be strictly increasing and continuous. We have $h(0) = 0$ and $h(1) = 2$. By the intermediate value theorem, h maps $[0, 1]$ onto $[0, 2]$.

Since the measure of C_k is $\left(\frac{2}{3}\right)^k$, the measure of the complement $A_k = D_{k,1} \cup D_{k,2} \cup \cdots \cup D_{k,2^k-1}$ is $1 - \left(\frac{2}{3}\right)^k$. Since F is constant on all the sets $D_{k,j}$ we have that the measure of $h(A_k)$ is $1 - \left(\frac{2}{3}\right)^k$. Hence, the measure of $h(C_k)$ must be $2 - \left(1 - \left(\frac{2}{3}\right)^k\right) = 1 + \left(\frac{2}{3}\right)^k$. Letting $k \to \infty$, we obtain the measure $m\big(h(C)\big) = 1$. That h^{-1} is continuous and strictly increasing, is left as Exercise 1.10.1. ∎

Theorem 1.48 *There is a measurable subset of C which is not a Borel set.*

Proof We can now use a Vitali set given by Corollary 1.45. Consider the function h given in Theorem 1.47. Since $m\big(h(C)\big) = 1 > 0$, there is a nonmeasurable Vitali set $\mathcal{V} \subset h(C) \subset [0, 2]$. Consider the inverse image $h^{-1}(\mathcal{V}) \subset C$. Since the set $\mathcal{V}$ is nonmeasurable, it cannot be a Borel set. For if $h^{-1}(\mathcal{V})$ were a Borel set, then $\mathcal{V} = h(h^{-1}(\mathcal{V}))$ would also be a Borel set. Thus $h^{-1}(\mathcal{V})$

[7]

is not Borel set. On the other hand, $h^{-1}(\mathcal{V})$ is a subset of C, therefore it has measure zero. Further, the set $h^{-1}(\mathcal{V})$ is measurable because every set of measure zero is measurable. ∎

Exercises for Section 1.10

1.10.1 Show that the inverse function h^{-1} of Theorem 1.47 is increasing and continuous.

1.10.2 Show that the function h of Theorem 1.47 and its inverse h^{-1} map Borel sets to Borel sets.

2
Lebesgue Integral

In this chapter we consider Lebesgue integration of functions from $\mathbb{R}^n$ to $\mathbb{R}$. At this point you should have read Chapter 1 as well as the earlier Sections 0.9 and 0.10.

2.1 Measurable Functions

Definition 2.1 A function $f\colon \mathbb{R}^n \longrightarrow \mathbb{R}$ is said to be **measurable** (for emphasis **Lebesgue measurable**) if for every open interval $I = (a,b) \subset \mathbb{R}$, the inverse image $f^{-1}(I)$ is a measurable set in $\mathbb{R}^n$.

Theorem 2.1 *Let $f\colon \mathbb{R}^n \longrightarrow \mathbb{R}$. The following statements are equivalent:*

(a) *f is measurable; that is, for every open interval $I = (a,b) \subset \mathbb{R}$, the inverse image $f^{-1}(I) \subset \mathbb{R}^n$ is measurable;*
(b) *for every closed interval $I' = [a,b] \subset \mathbb{R}$, the inverse image $f^{-1}(I') \subset \mathbb{R}^n$ is measurable;*
(c) *for every interval $A = (a,\infty)$, the inverse image $f^{-1}(A)$ is measurable;*
(d) *for every interval $A' = [a,\infty)$, the inverse image $f^{-1}(A')$ is measurable;*
(e) *for every interval $B = (-\infty,b)$, the inverse image $f^{-1}(B)$ is measurable;*
(f) *for every interval $B' = (-\infty,b]$, the inverse image $f^{-1}(B')$ is measurable.*

Proof The equivalences are not difficult to prove. The equivalences $(c) \Leftrightarrow (f)$ and $(d) \Leftrightarrow (e)$ are proven by simply taking complements.

To prove $(d) \Rightarrow (c)$, note that the inverse image of $A = (a, \infty)$ is the countable union of inverse images of sets of the form $[r, \infty)$ for r rational and $r > a$. The fact that a countable union of measurable sets is measurable will yield this result.

To prove $(c) \Rightarrow (d)$, note that the inverse image of $A' = [a, \infty)$ is the countable intersection of inverse images of sets of the form (r, ∞) for r rational and $r < a$. The fact that a countable intersection of measurable sets is measurable will yield this result.

To prove $(a) \Rightarrow (c)$, note that the inverse image of $A = (a, \infty)$ is the countable union of inverse images of sets of the form $(a + k, a + k + 2)$ for $k = 0, 1, 2, \ldots$. A similar argument can be used to prove $(b) \Rightarrow (d)$.

And so on, for the other implications. In particular we can show that **(b)** follows from **(c)** and **(e)**. Also, **(a)** follows from **(d)** and **(f)**. Both proofs are left as Exercise 2.1.2. ∎

Remark For measurability of functions, we may also consider extended real valued functions to those having values ∞ and $-\infty$ as well as $\mathbb{R}$. In the extended real case, we have to include inverse images of intervals such as $[a, \infty]$, $(a, \infty]$, $[-\infty, b]$, and $[-\infty, b)$.

Example 2.1 It is *not* sufficient to consider inverse images only of singletons $\{x\} \subset \mathbb{R}$, as this example shows.

Let A be a nonmeasurable subset of $[0, 1]$, such as Vitali's example $A = \mathcal{V}$ (Theorem 1.44). Define the function $f : \mathbb{R} \longrightarrow \mathbb{R}$ as follows

$$f(x) = \begin{cases} e^x & \text{if } x \in A \\ -e^x & \text{if } x \notin A. \end{cases}$$

The function is not measurable since the inverse image of the open interval $G = (0, 3)$ is $f^{-1}(G) = A$, which is nonmeasurable. Since the function is one-to-one, the inverse image of each single point $\{x\}$ is either a single point or is empty. All such inverse images are of measure zero, hence measurable.

Theorem 2.2 *A function $f : \mathbb{R}^n \longrightarrow \mathbb{R}$ is measurable if and only if, for every open set G in $\mathbb{R}$, the inverse image of $f^{-1}(G)$ is a measurable subset of $\mathbb{R}^n$.*

Proof The proof of $(\Leftarrow)$ is clear by letting $G = (a, b)$. Conversely, let f be measurable and let G be an open subset of $\mathbb{R}$. Every open subset of $\mathbb{R}$ is a countable union of disjoint open intervals (see Exercise 1.5.2), say $G = \bigcup_k I_k$

for some $I_k = (a_k, b_k)$. Since for each k, the set $f^{-1}(I_k)$ is measurable by definition, and $f^{-1}(G) = \bigcup_k f^{-1}(I_k)$, the result follows. ∎

Recall from Section 0.8 (in particular, Theorem 0.24) that a function $f\colon \mathbb{R}^n \longrightarrow \mathbb{R}$ is continuous if and only if the inverse images of open sets are open. This can be compared to the above result stating that a function is measurable if and only if the inverse image of an open set is measurable. Since every open set is measurable, we have the following.

Corollary 2.3 *Every continuous function $f\colon \mathbb{R}^n \longrightarrow \mathbb{R}$ is measurable.*

Corollary 2.4 *Every monotonic function $f\colon \mathbb{R} \longrightarrow \mathbb{R}$ is measurable.*

Proof Suppose that f is monotonically increasing. For every open interval (a,b), the inverse image is the open interval $f^{-1}(a,b) = \{x \in \mathbb{R} \mid a < f(x) < b\}$, which is an interval I from $f^{-1}(a)$ to $f^{-1}(b)$. If f is strictly increasing then I is open but in general it may be closed or half-open. In any case, I is measurable. Similarly, if f is monotonically decreasing, then $f^{-1}(a,b)$ is an interval from $f^{-1}(b)$ to $f^{-1}(a)$. ∎

Theorem 2.5 *If f is measurable and $f(x) = g(x)$ almost everywhere, then g is measurable.*

Proof Let $Z = \{x \in \mathbb{R}^n \mid f(x) \neq g(x)\}$ with $m(Z) = 0$. Let $I = (a,b) \subset \mathbb{R}$. Since f is measurable, the inverse image $f^{-1}(I)$ is a measurable subset of $\mathbb{R}^n$. The inverse image $g^{-1}(I)$ differs from the measurable set $f^{-1}(I)$ by a subset of Z. By Corollary 1.35 (Section 1.7), the set $g^{-1}(I)$ is also measurable. ∎

Theorem 2.6 *Let α be a real constant. If f is measurable, then so are $f + \alpha$ and αf.*

Proof Suppose $\alpha \in \mathbb{R}$, f is measurable, and $g = f + \alpha$. The inverse image of any open interval (a,b) under g is measurable since it is the inverse image of $(a - \alpha, b - \alpha)$ under f. The measurability of αf is left for Exercise 2.1.9. ∎

Lemma 2.7 *Suppose functions f and g are measurable. Then the set $A = \{x \in \mathbb{R}^n \mid f(x) < g(x)\}$ is a measurable subset of $\mathbb{R}^n$.*

Proof For each rational $r \in \mathbb{Q}$, define the set $A_r = \{x \in \mathbb{R}^n \mid f(x) < r < g(x)\}$. The set A_r is measurable since it is the intersection of the measurable sets $\{x \in \mathbb{R}^n \mid f(x) < r\}$ and $\{x \in \mathbb{R}^n \mid r < g(x)\}$. Finally, the set A is measurable since it is the countable union of the measurable sets A_r. ∎

Theorem 2.8 *If functions f and g are measurable, then so is $f+g$.*

Proof Consider any open interval $I=(a,b)\subset\mathbb{R}$. Suppose f and g are measurable and let $h=f+g$. The inverse image $h^{-1}(I)$ is the intersection

$$h^{-1}(I)=\Big\{x\in\mathbb{R}^n \mid f(x)>-g(x)+a\}\bigcap\{x\mid -g(x)>f(x)-b\Big\},$$

which is measurable by Theorem 2.6 and Lemma 2.7. ∎

Lemma 2.9 *If $f:\mathbb{R}^n\longrightarrow\mathbb{R}$ is measurable, then so is the function f^2.*

Proof Let $h=f^2$. We use Theorem 2.1(c). If $a<0$, then $h^{-1}(a,\infty)=\mathbb{R}^n$. Otherwise, it is the union

$$\begin{aligned}h^{-1}(a,\infty)&=\{x\in\mathbb{R}^n \mid f^2(x)>a\}\\&=\Big\{x\mid f(x)>\sqrt{a}\}\bigcup\{x\mid f(x)<-\sqrt{a}\Big\},\end{aligned}$$

which is measurable. This proves the lemma. ∎

Theorem 2.10 *If functions f and g are measurable, then so is fg.*

Proof Observing that $fg=\dfrac{1}{4}(f+g)^2-\dfrac{1}{4}(f-g)^2$ and using the previous results, we have the measurability of fg. ∎

Definition 2.2 For any function $f:\mathbb{R}^n\longrightarrow\mathbb{R}$, define the **positive part** of f to be

$$f^+(x)=\begin{cases}f(x) & \text{if } f(x)\geq 0\\ 0 & \text{if } f(x)<0,\end{cases}$$

and the **negative part** of f to be

$$f^-(x)=\begin{cases}0 & \text{if } f(x)\geq 0\\ -f(x) & \text{if } f(x)<0.\end{cases}$$

Note that, for each function $f:\mathbb{R}^n\longrightarrow\mathbb{R}$, we have

$$f=f^+-f^-\quad\text{and}\quad |f|=f^++f^-. \tag{2.1}$$

Theorem 2.11 *If a function f is measurable, then so are f^+, f^-, and $|f|$.*

Proof Let $I=(a,b)$ and suppose f is measurable. The inverse image of I under f^+ is $f^{-1}\big(I\cap[0,\infty)\big)$, which is measurable. This shows that f^+ is measurable. Similarly, f^- is measurable. Since $|f|=f^++f^-$, we have its measurability by Theorem 2.8. ∎

Definition 2.3 Let $f_1, f_2, \ldots$ be a sequence of functions from $\mathbb{R}^n$ to $\mathbb{R}$. For all $x \in \mathbb{R}^n$ and $j = 1, 2, \ldots$, let $g_j(x) = \sup_{k \geq j} f_k(x)$ and $h_j(x) = \inf_{k \geq j} f_k(x)$. Note that these sequences are monotonic and

$$-\infty \leq h_j(x) \leq h_{j+1}(x) \leq g_{j+1}(x) \leq g_j(x) \leq \infty.$$

Define the functions $\limsup\limits_{k\to\infty} f_k$ and $\liminf\limits_{k\to\infty} f_k$ as follows:

$$\limsup_{k\to\infty} f_k(x) = \lim_{j\to\infty} g_j(x) = \inf_{j\geq 1}\Big(\sup_{k\geq j} f_k(x)\Big), \tag{2.2}$$

$$\liminf_{k\to\infty} f_k(x) = \lim_{j\to\infty} h_j(x) = \sup_{j\geq 1}\Big(\inf_{k\geq j} f_k(x)\Big). \tag{2.3}$$

Theorem 2.12 *If $f_1, f_2, \ldots$ are measurable, then so are the following functions.*

$$\sup_k f_k, \quad \inf_k f_k, \quad \limsup_{k\to\infty} f_k, \quad \liminf_{k\to\infty} f_k.$$

Proof Let $h = \sup_k f_k$. Observe the inverse image

$$h^{-1}(-\infty, b) = \left\{x \;\middle|\; \sup_k f_k(x) \leq b\right\} = \bigcap_k \{x \mid f_k(x) \leq b\}.$$

This shows that $\sup_k f_k$ is measurable. Similarly, $\inf_k f_k$ is measurable. Applying these two facts to the definitions of $\limsup_k f_k$ and $\liminf_k f_k$ in (2.2) and (2.3) above, the remainder of the results follow. ∎

If a sequence of functions converges, then the limit is equal to the lim sup, which is also equal to the lim inf. Hence, the following corollary.

Corollary 2.13 *The limit of a convergent sequence of measurable functions is measurable.*

Definition 2.4 Let $A \subset \mathbb{R}^n$. The **indicator function** or **characteristic function** of A is

$$f_A(x) = \begin{cases} 1 & \text{if } x \in A \\ 0 & \text{if } x \notin A. \end{cases}$$

The symbol χ_A is also commonly used instead of f_A.

Definition 2.5 A **simple function** is any function $s : \mathbb{R}^n \longrightarrow \mathbb{R}$ which assumes only a finite number of distinct values. A simple function s can be written as a linear combination of indicator functions

$$s = \sum_{k=1}^{N} \alpha_k f_{A_k}, \text{ with disjoint sets } A_k \text{ and distinct } \alpha_k \in \mathbb{R}. \tag{2.4}$$

If the sets A_k are n-dimensional *intervals*, then s is called a **step function**.

Theorem 2.14 *A simple function expressed in the form* (2.4) *is measurable if and only if all of the sets A_k are measurable.*

This is easy to prove and left as Exercise 2.1.6.

Theorem 2.15 *If s and t are simple functions, then so are*

$$\alpha s + \beta t, \ \ s^+, \ \ s^-, \ \ |s|, \ \textit{and} \ \ st \ \textit{for any} \ \alpha, \beta \in \mathbb{R}.$$

Further, if t is never zero, then $1/t$ and s/t are also simple. The never zero *restriction can be dropped in the case of extended real valued functions as described in the Remark near the beginning of this chapter.*

The proofs are routine and omitted.

Theorem 2.16 *A* nonnegative *function $f\colon \mathbb{R}^n \longrightarrow \mathbb{R}$ is measurable if and only if there exists an increasing sequence of measurable simple functions*

$$0 \leq s_1 \leq s_2 \leq s_3 \leq \cdots \leq f$$

that converge everywhere to f.

Proof Suppose $f\colon \mathbb{R}^n \longrightarrow \mathbb{R}$ is measurable. For each $N = 1, 2, \ldots$ define the simple function s_N as follows. For $k = 0, 1, \ldots, N2^N - 1$, let

$$A_k = \left\{ x \in \mathbb{R}^n \;\middle|\; \frac{k}{2^N} \leq f(x) < \frac{k+1}{2^N} \right\},$$

and for $k = N2^N$, let $A_k = \{x \in \mathbb{R}^n \mid f(x) \geq N\}$.

Clearly the sets A_k for $k = 0, 1, 2, \ldots, N2^N$ are disjoint and measurable. Define s_N as the following linear combination of indicator functions:

$$s_N(x) = \sum_{k=0}^{N2^N} \frac{k}{2^N} f_{A_k}(x).$$

Clearly $s_1 \leq s_2 \leq s_3 \leq \cdots \leq f$. If $0 \leq f(x) < N$, then $|s_N(x) - f(x)| < 1/2^N$. And if $f(x) \geq N$ for all N, then $f(x) = \infty$ and $s_N(x) = N \to f(x)$ as $N \to \infty$. The converse follows from Corollary 2.13. ∎

This theorem may be extended to general measurable functions $f = f^+ - f^-$ by applying the theorem to the nonnegative functions f^+ and f^-.

Corollary 2.17 Approximation by simple functions: *A function $f : \mathbb{R}^n \longrightarrow \mathbb{R}$ is measurable if and only if there exists a sequence of measurable simple functions which converge everywhere to f.*

Exercises for Section 2.1

2.1.1 Prove implication $(b) \Rightarrow (d)$ of Theorem 2.1. See the proof of $(a) \Rightarrow (c)$.

2.1.2 Finish the proof of Theorem 2.1 by showing that statement **(b)** follows from **(c)** and **(e)**, and that statement **(a)** follows from **(d)** and **(f)**.

2.1.3 Show that if f is continuous on $\mathbb{R}$ and $f = 0$ a.e., then $f = 0$.

2.1.4 Show that if f and g are continuous on $\mathbb{R}$ and $f = g$ a.e., then $f = g$.

2.1.5 Let $f_1, f_2, \ldots$ be a sequence of functions from $\mathbb{R}^n$ to $\mathbb{R}$. Show that

$$-\infty \leq \liminf_{k\to\infty} f_k \leq \limsup_{k\to\infty} f_k \leq \infty.$$

2.1.6 Prove Theorem 2.14.

2.1.7 Give a detailed proof of Corollary 2.17.

2.1.8 Continuing with Exercise 2.1.7, show that the convergence by simple functions is uniform.

2.1.9 Finish proving Theorem 2.6 by showing that for measurable f and constant α, the function αf is measurable. Consider $\alpha = 0$, $\alpha > 0$, $\alpha < 0$ separately.

2.2 Egorov's Theorem

Egorov's theorem is important and useful in dealing with almost everywhere convergent sequences of measurable functions. It shows that, although almost everywhere convergence is a weaker property than uniform convergence, it is in some sense not much weaker. The proof is rather delicate because it involves two independent epsilons.

Theorem 2.18 Egorov's theorem: *Suppose $f_1, f_2, \ldots$ is a sequence of measurable functions such that $f_k \to f$ a.e. on a measurable set $A \subset \mathbb{R}^n$ with $m(A) < \infty$. Then for every $\epsilon > 0$, there exists a measurable subset B of A with $m(A - B) < \epsilon$, such that the convergence $f_k \to f$ is uniform on the set B.*

Proof To say $f_k \to f$ a.e., means there exists a subset $Z \subset A$ of measure zero such that $f_k(x) \to f(x)$ for all $x \in A - Z$. Let $\epsilon > 0$ be given.

We start with $\epsilon_1 = \epsilon/2$. Then for $k = 1, 2, \ldots$, let

$$A_k = \{x \in A - Z \mid |f(x) - f_j(x)| < \epsilon_1 \quad \text{for all} \quad j \geq k\}.$$

The A_k are clearly measurable subsets of $A - Z$ and

$$A_1 \subset A_2 \subset A_3 \subset \cdots.$$

Since $f_k \to f$ on $A - Z$, each $x \in A - Z$ belongs to some A_k. Hence

$$\bigcup_{k=1}^{\infty} A_k = A - Z.$$

By Corollary 1.38 (Section 1.7), $\lim_{k\to\infty} m(A_k) = m(A - Z) = m(A)$. Since $m(A_k) \uparrow m(A) < \infty$, there exists $k_1 > 0$ such that $m(A) - \epsilon_1 < m(A_{k_1}) \leq m(A)$. Let $B_1 = A_{k_1}$.

In summary, for $\epsilon_1 = \epsilon/2$ we have found

(1) $0 < k_1$, and

(2) a set $B_1 \subset A$ such that $m(A - B_1) < \epsilon/2$, and

(3) $|f(x) - f_j(x)| < \epsilon/2$ for all $x \in B_1$ and $j \geq k_1$.

Repeating this argument for $\epsilon_i = \epsilon/2^i$, we can find

(1) $0 < k_1 < k_2 < \cdots < k_i$, and

(2) set $B_i \subset A$ such that $m(A - B_i) < \epsilon/2^i$, and

(3) $|f(x) - f_j(x)| < \epsilon/2^i$ for all $x \in B_i$ and $j \geq k_i$.

Let $B = \bigcap_{i=1}^{\infty} B_i$. Then $A - B = \bigcup_{i=1}^{\infty}(A - B_i)$. So

$$m(B - A) \leq \sum_{i=1}^{\infty} m(A - B_i) \leq \sum_{i=1}^{\infty} \frac{\epsilon}{2^i} = \epsilon.$$

The sequence f_k converges uniformly to f on B because, given any $\epsilon' > 0$, we can find $\epsilon_i = \epsilon/2^i < \epsilon'$; and then $|f(x) - f_j(x)| < \epsilon/2^i < \epsilon'$ for all $x \in B \subset B_i$ and $j \geq k_i$. ∎

Exercises for Section 2.2

2.2.1 Consider the sequence of functions $f_k(x) = x^k$ for $k = 1, 2, \ldots$ and $f(x) = 0$ on the interval $A = [0, 1] \subset \mathbb{R}$. Demonstrate Egorov's theorem in this case by finding the set $Z \subset A$ and the set B for each $\epsilon > 0$.

2.2.2 Give an example which shows that the condition $m(A) < \infty$ is necessary for Egorov's theorem.

2.2.3 Show that Egorov's theorem is no longer true if the condition that the functions f_k be measurable is dropped.

2.2.4 A sequence of measurable functions f_k is said to **converge in measure** to function f if, given $\epsilon > 0$, the measure of the set $A_k = \{x \in \mathbb{R}^n \mid |f_k(x) - f(x)| \geq \epsilon\}$ converges to zero as $k \to \infty$. Show that if a sequence of measurable f_k convergence to f almost everywhere on a set A with $\mu(A) < \infty$, then it converges in measure on A.

2.3 The Lebesgue Integral

We define the Lebesgue integral of measurable functions in stages. First, we do it for simple functions in Subsection 2.3.1. Then we define it for nonnegative functions in Subsection 2.3.2. Finally we consider the general case in Subsection 2.3.3.

2.3.1 Measurable Simple Functions

Definition 2.6 If s is a measurable simple function in the form $s = \sum_{k=1}^{N} \alpha_k f_{A_k}$, with disjoint sets A_k, the **Lebesgue integral** of s is defined to be

$$\int s\, dm = \sum_{k=1}^{N} \alpha_k m(A_k).$$

The integral may be infinite. We use the convention that, if $\alpha_k = 0$, then $\alpha_k m(A_k) = 0$, even if $m(A_k) = \infty$. In particular, if s is zero on all of $\mathbb{R}^n$, then $\int s\, dm = 0$.

We may always assume that $\bigcup_{k=1}^{N} A_k = \mathbb{R}^n$. For if not, we can adjoin the term $\alpha_{N+1} f_{A_{N+1}}$ to the representation of s, where $\alpha_{N+1} = 0$ and $A_{N+1} = \mathbb{R}^n - \bigcup_{k=1}^{N} A_k$.

We can restrict the Lebesgue integral to any n-dimensional interval I as follows. Multiply s by the indicator function f_I and write

$$\int_I s\, dm = \int s \cdot f_I\, dm = \sum_{k=1}^{N} \alpha_k m(A_k \cap I).$$

Theorem 2.19 *If s is a measurable simple function with $s \geq 0$, then $\int s\, dm \geq 0$.*

The proof is easy and omitted.

Although we have not yet shown that two representations of s, say

$$s = \sum_{k=1}^{N} \alpha_k f_{A_k} = \sum_{j=1}^{M} \alpha'_j f_{A'_j},$$

will result in the same value of the integral $\sum_{k=1}^{N} \alpha_k m(A_k) = \sum_{j=1}^{M} \alpha'_j m(A'_j)$, Theorem 2.19 holds, regardless of the representation of s. That the Lebesgue integral is well defined for simple functions will be deferred until Theorem 2.23. Momentarily, we show the truth of the following Theorems 2.20, 2.21, and 2.22 for *particular representations* of s and t.

Theorem 2.20 *If s and t are measurable simple functions, then*

$$\int (s+t)\, dm = \int s\, dm + \int t\, dm.$$

Proof Suppose $s = \sum_{k=1}^{N} \alpha_k f_{A_k}$ and $t = \sum_{j=1}^{M} \beta_j f_{B_j}$. Here $\int s\, dm = \sum_{k=1}^{N} \alpha_k m(A_k)$ and $\int t\, dm = \sum_{j=1}^{M} \beta_j m(B_j)$. We assume $\bigcup_{k=1}^{N} A_k = \mathbb{R}^n$ and $\bigcup_{j=1}^{M} B_j = \mathbb{R}^n$. Note $A_k = \bigcup_{j=1}^{M} (A_k \cap B_j)$ and $B_j = \bigcup_{k=1}^{N} (A_k \cap B_j)$. Then we have the following representations of s and t:

$$s = \sum_{k=1}^{N} \alpha_k f_{A_k} = \sum_{k=1}^{N} \sum_{j=1}^{M} \alpha_k f_{A_k \cap B_j},$$

$$t = \sum_{j=1}^{M} \beta_j f_{B_j} = \sum_{j=1}^{M} \sum_{k=1}^{N} \beta_j f_{A_k \cap B_j} = \sum_{k=1}^{N} \sum_{j=1}^{M} \beta_j f_{A_k \cap B_j}.$$

Adding, we obtain

$$s + t = \sum_{k=1}^{N} \alpha_k f_{A_k} + \sum_{j=1}^{M} \beta_j f_{B_j} = \sum_{k=1}^{N} \sum_{j=1}^{M} (\alpha_k + \beta_j) f_{A_k \cap B_j},$$

and so $\int s\, dm + \int t\, dm = \int (s+t)\, dm$. ∎

Theorem 2.21 *If s is a measurable simple function with $\alpha \in \mathbb{R}$, then*

$$\int \alpha s\, dm = \alpha \int s\, dm.$$

The proof is easy and omitted.

Theorem 2.22 *If s and t are measurable simple functions with $t \le s$, then*

$$\int t\,dm \le \int s\,dm.$$

Proof Let $h = s - t \ge 0$. Theorem 2.19 shows $\int (s-t)\,dm \ge 0$. Applying Theorems 2.20 and 2.21 to the simple functions s and $-t$, the result follows. ∎

Actually, this is sufficient to show that the integral of a measurable simple function is well defined.

Theorem 2.23 *The definition of $\int s\,dm$ is well defined for simple functions. That is,*

if $\sum_{k=1}^{N} \alpha_k f_{A_k} = \sum_{j=1}^{M} \alpha'_j f_{A'_j}$, then $\sum_{k=1}^{N} \alpha_k m(A_k) = \sum_{j=1}^{M} \alpha'_j m(A'_j)$.

Proof If $s = t$ but there are two representations of the same function,

$$s = \sum_{k=1}^{N} \alpha_k f_{A_k} \quad \text{and} \quad t = \sum_{j=1}^{M} \alpha'_j f_{A'_j},$$

then Theorem 2.22 shows (since $s \le t$),

$$\int s\,dm = \sum_{k=1}^{N} \alpha_k m(A_k) \le \sum_{j=1}^{M} \alpha'_j m(A'_j) = \int t\,dm.$$

The reverse inequality holds as well. Thus $\int s\,dm = \int t\,dm$. ∎

Theorem 2.24 *If s and t are measurable simple functions and $s = t$ almost everywhere, then*

$$\int s\,dm = \int t\,dm.$$

The result follows easily from the definition.

Theorem 2.25 *If $0 \le s_1 \le s_2 \le s_3 \le \cdots$ is a nondecreasing sequence of measurable simple functions converging to a measurable simple function s, then*

$$\lim_{k\to\infty} \int s_k\,dm = \int s\,dm. \tag{2.5}$$

Proof Suppose s is a measurable simple function. Then s is a linear combination of indicator functions, say $s = \sum_{i=1}^{N} \alpha_i f_{A_i}$ where the A_i are measurable and disjoint, and the α_i are positive. Let $A = \bigcup_{i=1}^{N} A_i = \{x \in \mathbb{R}^n \mid s(x) > 0\}$.

If $m(A) = 0$, then Equality (2.5) holds since all of the integrals are zero.

Next we consider the case $0 < m(A) < \infty$. Let $\alpha = \max_i \alpha_i = \sup_x s(x)$. We use Egorov's theorem (Theorem 2.18). Given $\epsilon > 0$, there exists a subset $C \subset A$ with $m(C) < \epsilon/\alpha$ such that $s_k \to s$ uniformly on the set $B = A - C$. Therefore, there exists $N > 0$ such that $0 \leq s - s_k < \epsilon/m(A)$ on $B = A - C$ for all $k > N$. Note $s - s_k = 0$ on the complement of A. So for $k > N$, we have

$$0 \leq s - s_k = (s - s_k) f_C + (s - s_k) f_{A-C} < \alpha f_C + \frac{\epsilon}{m(A)} f_{A-C}.$$

Hence $0 \leq \int s\, dm - \int s_k\, dm \leq \alpha m(C) + \frac{\epsilon}{m(A)} m(A) < \epsilon + \epsilon$.

Letting k tend to infinity, it follows that $\int s\, dm = \lim_k \int s_k\, dm$.

Finally, the case $m(A) = \infty$. Here $\int s\, dm = \sum_{i=1}^{N} \alpha_i m(A_i) = \infty$. If we restrict the functions to an M-cube $C_M \subset \mathbb{R}^n$ of sides $M > 0$ centered at the origin, then by the above case we have

$$\int_{C_M} s\, dm = \lim_{k\to\infty} \int_{C_M} s_k\, dm \leq \lim_{k\to\infty} \int s_k\, dm.$$

But $\int_{C_M} s\, dm = \sum_{i=1}^{N} \alpha_i m(A_i \cap C_M)$. As $M \to \infty$, this quantity tends to $\int s\, dm = \sum_{i=1}^{N} \alpha_i m(A_i) = \infty$. Hence, both sides of (2.5) are infinite. ∎

2.3.2 Nonnegative Measurable Functions

Now that we have the Lebesgue integral for measurable simple functions, we next define it for nonnegative measurable functions.

Definition 2.7 Suppose $f\colon \mathbb{R}^n \longrightarrow \mathbb{R}$ is nonnegative and measurable. By Theorem 2.16, there is an increasing sequence of measurable simple functions

$$0 \leq s_1 \leq s_2 \leq s_3 \leq \cdots \leq f$$

converging everywhere to f. We define the **Lebesgue integral** of f to be

$$\int f\, dm = \lim_{k\to\infty} \int s_k\, dm.$$

In light of Theorem 2.22, the sequence of integrals is also increasing.

The following theorem shows that this definition is well defined.

Theorem 2.26 *Let $f\colon \mathbb{R}^n \longrightarrow \mathbb{R}$ be nonnegative and measurable. If $0 \leq s_1 \leq s_2 \leq \cdots \leq f$ and $0 \leq t_1 \leq t_2 \leq \cdots \leq f$ are increasing sequences of measurable simple functions both converging to f, then*

$$\lim_{k\to\infty} \int s_k\, dm = \lim_{k\to\infty} \int t_k\, dm. \tag{2.6}$$

Proof Fix N and let $h_k = \min\{s_N, t_k\}$. It is clear that $0 \le h_1 \le h_2 \le \cdots$ are measurable simple functions that converge to s_N. By Theorem 2.25 we have

$$\int s_N\,dm = \lim_{k\to\infty} \int h_k\,dm \le \lim_{k\to\infty} \int t_k\,dm.$$

Letting N tend to infinity, we have $\lim_{N\to\infty} \int s_N\,dm \le \lim_{k\to\infty} \int t_k\,dm$.

By symmetry, the inequality also holds in the other direction. ∎

Theorem 2.27 *Suppose $f : \mathbb{R}^n \longrightarrow \mathbb{R}$ is nonnegative and measurable. Then $\int f\,dm = 0$ if and only if $f = 0$ a.e.*

Proof ($\Rightarrow$): For $k = 1, 2, \ldots$, let $A_k = \{x \in \mathbb{R}^n \mid f(x) \ge 1/k\}$. Then

$$m(A_k) = \int f_{A_k}\,dm \le \int kf\,dm = k\int f\,dm = 0.$$

It follows that $\{x \in \mathbb{R}^n \mid f(x) > 0\} = \bigcup_k A_k$ must be of measure zero.

($\Leftarrow$): This is left as Exercise 2.3.1. ∎

Corollary 2.28 *A measurable function f is zero a.e. if and only if $\int |f|\,dm = 0$.*

Theorem 2.29 *If $\alpha > 0$, and f and g are nonnegative measurable functions, then so are αf and $f + g$. In this case,*

$$\int \alpha f\,dm = \alpha \int f\,dm \quad \textit{and} \quad \int (f+g)\,dm = \int f\,dm + \int g\,dm.$$

The proof is straightforward from Theorems 2.6 and 2.8 and the definition of the Lebesgue integral of nonnegative measurable functions. Another straightforward result is the following.

Theorem 2.30 *If f and g are measurable and $0 \le f \le g$, then $0 \le \int f\,dm \le \int g\,dm$.*

An alternative definition of the Lebesgue integral of a nonnegative measurable function is that given by Equation (2.7).

Theorem 2.31 *Suppose $f : \mathbb{R}^n \longrightarrow \mathbb{R}$ is nonnegative and measurable. Then*

$$\int f\,dm = \sup\left\{\int s\,dm \mid s \text{ is measurable, simple, and } 0 \le s \le f\right\}. \tag{2.7}$$

Proof Since a limit cannot exceed a supremum, it is clear that the integral does exceed the supremum in (2.7). For the other direction, consider any measurable simple function s satisfying $0 \le s \le f$. Then $0 \le f - s \le f$ is a measurable function. By Theorem 2.16, there exists an increasing sequence $0 \leq t_1 \leq t_2 \leq \cdots$ of measure simple functions which converges to $f - s$. Adding s to both sides, we get the increasing sequence of measurable simple functions $0 \leq 0 + s \leq t_1 + s \leq t_2 + s \leq \cdots$ which converges to $(f - s) + s = f$. By Theorem 2.22, $\int s\, dm \leq \lim_{k\to\infty} \int (t_k + s)\, dm = \int f\, dm$. Taking the supremum with respect to s gives us the inequality in the other direction. Hence, (2.7) holds. ∎

2.3.3 The General Case

Definition 2.8 If $f\colon \mathbb{R}^n \longrightarrow \mathbb{R}$ is an arbitrary (not necessarily nonnegative) measurable function, then the nonnegative functions f^+ and f^- are also measurable (Theorem 2.11). Each of them has a Lebesgue integral. We say that f is **Lebesgue integrable** if *both* $\int f^+\, dm < \infty$ and $\int f^-\, dm < \infty$. In this case we write $f \in L$ or $f \in L^1$ or $f \in L^1(\mathbb{R}^n)$ and we define the Lebesgue integral of f to be

$$\int f\, dm = \int f^+\, dm - \int f^-\, dm.$$

Note that an integrable function f need not be bounded but we must have both $\int f^+\, dm$ and $\int f^-\, dm$ finite. Without this double finiteness condition, $L^1(\mathbb{R}^n)$ would fail to be a linear space. Also an integrable function is permitted to have values $+\infty$ and $-\infty$, but only on sets of measure zero.

If A is a measurable subset of $\mathbb{R}^n$, such as an n-dimensional interval $A = I$, we define the **integral over the set A** by multiplying f with the indicator function f_A as follows

$$\int_A f\, dm = \int f \cdot f_A\, dm.$$

In the one-dimensional case with $A = I = [a,b]$, we write

$$\int_I f\, dm = \int_a^b f\, dm = \int_a^b f(x)\, dx.$$

Theorem 2.32 *If f is Lebesgue integrable and $f = g$ almost everywhere, then g is also integrable and*

$$\int f dm = \int g\, dm.$$

The result follows easily from the Theorem 2.5 and the definitions in that section. Consequently, even if a measurable function is undefined in a set of measure zero, the Lebesgue integral of the function makes sense.

Example 2.2 The indicator function of the rational numbers $f_{\mathbb{Q}} : \mathbb{R} \longrightarrow \mathbb{R}$ is equal to the zero function almost everywhere. Thus it is Lebesgue integrable and $\int f_{\mathbb{Q}}\, dm = 0$. This function is *not* Riemann integrable over any interval $[a, b]$ (see Exercise 2.3.2).

Theorem 2.33 *Suppose* $f : \mathbb{R}^n \longrightarrow \mathbb{R}$ *is measurable. The following are equivalent:*

(a) $f \in L^1$,
(b) $|f| \in L^1$,
(c) $\int |f|\, dm < \infty$.

In this case, $|\int f\, dm| \leq \int |f|\, dm$.

The proof is left as Exercise 2.3.3.

Corollary 2.34 *If* f *is measurable and* g *is integrable and* $|f| \leq |g|$, *then* f *is integrable and*

$$\int |f|\, dm \leq \int |g|\, dm.$$

Example 2.3 Let $I = [0, 1] \subset \mathbb{R}$ and $f(x) = \sin \dfrac{1}{x}$ on I. Even though $f(0)$ is undefined, the function f is integrable on I since $|f| \leq f_I$ and the indicator function f_I is integrable.

Now a housekeeping result.

Theorem 2.35 *If* $f, g \in L^1$ *and* $\alpha \in \mathbb{R}$, *then* $\alpha f \in L^1$ *(and* $\alpha g \in L^1$*) and* $f + g \in L^1$. *Then*

$$\int \alpha f\, dm = \alpha \int f\, dm \quad \text{and} \quad \int f + g\, dm = \int f\, dm + \int g\, dm.$$

Proof The integrability of αf is left as Exercise 2.3.4.

If $f, g \in L^1$, then $|f|, |g| \in L^1$ (Theorem 2.33). Since $|f + g| \leq |f| + |g|$, by Corollary 2.34 we have $f + g \in L^1$. Note that

$$f + g = (f^+ + g^+) - (f^- + g^-).$$

By definition

$$\int f + g\, dm = \int (f^+ + g^+)\, dm - \int (f^- + g^-)\, dm. \tag{2.8}$$

Rearranging terms, Equation (2.8) becomes

$$\int f^+\, dm - \int f^-\, dm + \int g^+\, dm - \int g^-\, dm = \int f\, dm + \int g\, dm. \quad \blacksquare$$

Corollary 2.36 *L^1 is a linear space of functions and the Lebesgue integral is a linear functional on this space.*

Theorem 2.37 Integral approximation by simple functions: *Let f be an integrable function. Given any $\epsilon > 0$, we can find a measurable simple function s such that*

$$\int |f - s|\, dm < \epsilon.$$

Proof Write f as a difference of nonnegative functions $f = f^+ - f^-$. By Theorem 2.31, there exist measurable simple functions t and u such that

$$0 \le \int f^+\, dm - \int t\, dm < \frac{\epsilon}{2} \quad \text{and} \quad 0 \le \int f^-\, dm - \int u\, dm < \frac{\epsilon}{2}.$$

Let $s = t - u$. Then $f - s = (f^+ - t) + (u - f^-)$ and

$$\int |f - s|\, dm \le \int |f^+ - t|\, dm + \int |f^- - u|\, dm < \frac{\epsilon}{2} + \frac{\epsilon}{2} = \epsilon. \qquad \blacksquare$$

Exercises for Section 2.3

2.3.1 Prove the ($\Leftarrow$) part of Theorem 2.27.

2.3.2 Prove that the indicator function of the rationals $f_{\mathbb{Q}} \colon \mathbb{R} \longrightarrow \mathbb{R}$ is not Riemann integrable on any interval $[a,b]$ by showing that the lower Riemann integral is always 0 and the upper Riemann integral is always $b - a$.

2.3.3 Prove Theorem 2.33.

2.3.4 Prove the part of Theorem 2.35 which states that a constant multiple αf of an integrable function f is integrable. Note: You have to consider the cases $\alpha = 0$, $\alpha > 0$, and $\alpha < 0$ separately.

2.3.5 Prove Corollary 2.36.

2.3.6 Show that $\int |f - g|\, dm = 0$ if and only if $f = g$ almost everywhere.

2.3.7 Give an example of a function $f \colon \mathbb{R} \longrightarrow \mathbb{R}$ for which $|f| \in L^1$ but $f \notin L^1$.

2.3.8 Give an example of a function $f \colon \mathbb{R} \longrightarrow \mathbb{R}$ for which $f \in L^1$ but $f^2 \notin L^1$. Note that if f is Riemann integrable on an interval $[a,b]$, then so is f^2.

2.3.9 Show that if $f \in L^1$, then f is finite almost everywhere.

2.4 Limit Theorems for the Lebesgue Integral

Theorem 2.38 Lebesgue's monotone convergence: *Suppose that $f_1, f_2, f_3, \ldots$ is an increasing sequence of nonnegative measurable functions on $\mathbb{R}^n$:*

$$0 \le f_1 \le f_2 \le f_3 \le \cdots .$$

Let $f(x) = \lim_{k\to\infty} f_k(x)$, where $f(x)$ is permitted to be infinite. Then f is measurable and

$$\lim_{k\to\infty} \int f_k \, dm = \int f \, dm.$$

Proof The function f is measurable by Corollary 2.13. Since

$$0 \le f_1 \le f_2 \le f_3 \le \cdots \le f, \quad \text{we have}$$

$$0 \le \int f_1 \, dm \le \int f_2 \, dm \le \int f_3 \, dm \le \cdots \le \int f \, dm,$$

and thus

$$\lim_{k\to\infty} \int f_k \, dm \le \int f \, dm.$$

This may be infinite.

It remains to prove the opposite inequality. Let s be any measurable simple function satisfying $0 \le s \le f$ and let $0 < \delta < 1$. Define

$$A_k = \{x \in \mathbb{R}^n \mid \delta s(x) \le f_k(x)\}.$$

Clearly, $A_1 \subset A_2 \subset A_3 \subset \cdots$ and

$$\delta \int_{A_k} s \, dm = \int_{A_k} \delta s \, dm \le \int_{A_k} f_k \, dm \le \int f_k \, dm. \tag{2.9}$$

Since $f_k \uparrow f$, we have $\bigcup_{k=1}^{\infty} A_k = \mathbb{R}^n$. For each k, the product sf_{A_k} is a simple function and $sf_{A_k} \uparrow s$. Thus by the definition of the Lebesgue integral (Definition 2.7) (or Theorem 2.26),

$$\lim_{k\to\infty} \int_{A_k} s \, dm = \lim_{k\to\infty} \int s f_{A_k} \, dm = \int s \, dm.$$

Combining this with (2.9), we have

$$\delta \int s \, dm \le \lim_{k\to\infty} \int f_k \, dm.$$

Since this holds for all $0 < \delta < 1$, we have

$$\int s \, dm \le \lim_{k\to\infty} \int f_k \, dm.$$

Since s is an arbitrary measurable simple function satisfying $0 \leq s \leq f$, we have by the definition of the integral of the Lebesgue integral (or Theorem 2.31), that

$$\int f\,dm \leq \lim_{k\to\infty} \int f_k\,dm. \qquad \blacksquare$$

Corollary 2.39 *Suppose that $f_1, f_2, f_3, \ldots$ is a sequence of nonnegative measurable functions on $\mathbb{R}^n$. Then their sum is integrable and*

$$\int \left(\sum_{k=1}^{\infty} f_k\right) dm = \sum_{k=1}^{\infty} \int f_k\,dm.$$

Theorem 2.40 Fatou's lemma: *Suppose $f_1, f_2, \ldots$ is a sequence of nonnegative measurable functions. Then $\liminf_{k\to\infty} f_k$ is measurable and*

$$\int \left(\liminf_{k\to\infty} f_k\right) dm \leq \liminf_{k\to\infty} \int f_k\,dm.$$

Proof If we define $g_k = \inf\{f_k, f_{k+1}, f_{k+2}, \ldots\}$, then $0 \leq g_1 \leq g_2 \leq \cdots$. Recall the meaning of lim inf:

$$\liminf_{k\to\infty} f_k = \lim_{k\to\infty} g_k.$$

By Lebesgue's monotone convergence theorem above, we have

$$\int \left(\liminf_{k\to\infty} f_k\right) dm = \int \lim_{k\to\infty} g_k\,dm = \lim_{k\to\infty} \int g_k\,dm = \liminf_{k\to\infty} \int g_k\,dm.$$

Finally, since $g_k \leq f_k$, we have $\liminf_{k\to\infty} \int g_k\,dm \leq \liminf_{k\to\infty} \int f_k\,dm$. $\blacksquare$

The most useful of the convergence theorems is the following which was proved by Henri Lebesgue in 1904.

Theorem 2.41 Lebesgue's dominated convergence: *Suppose that $f_1, f_2, f_3, \ldots$ are measurable functions that converge to f and suppose that g is a nonnegative integrable function with $|f_k(x)| \leq g(x)$ for all $x \in \mathbb{R}^n$. Then $f = \lim_{k\to\infty} f_k$ is integrable and*

$$\int f\,dm = \int \left(\lim_{k\to\infty} f_k\right) dm = \lim_{k\to\infty} \int f_k\,dm.$$

Proof Since $|f_k| \leq g$ for all k, by Corollary 2.34, f_k is integrable for all k. The function f is measurable because it is a limit of measurable functions. And it is also integrable since $|f| = \left|\lim_{k\to\infty} f_k\right| \le g$. Since $g + f_k \ge 0$ we can apply Fatou's lemma to obtain

$$\begin{aligned}\int g\,dm + \int f\,dm &= \int (g+f)\,dm \le \liminf_{k\to\infty} \int (g+f_k)\,dm \\ &= \int g\,dm + \liminf_{k\to\infty} \int f_k\,dm.\end{aligned}$$

Subtracting the finite $\int g\,dm$, we have

$$\int f\,dm \le \liminf_{k\to\infty} \int f_k\,dm. \tag{2.10}$$

Since $g - f_k \ge 0$, we can again apply Fatou's lemma to obtain

$$\begin{aligned}\int g\,dm - \int f\,dm &= \int (g-f)\,dm \le \liminf_{k\to\infty} \int (g-f_k)\,dm \\ &= \int g\,dm + \liminf_{k\to\infty} \int (-f_k)\,dm.\end{aligned}$$

Similarly, since

$$\liminf_{k\to\infty} \int (-f_k)\,dm = \liminf_{k\to\infty}\left(-\int f_k\,dm\right) = -\limsup_{k\to\infty} \int f_k\,dm,$$

we obtain $\limsup_{k\to\infty} \int f_k\,dm \le \int f\,dm$. Combined with (2.10), we have the desired result $\int f\,dm = \lim_{k\to\infty} \int f_k\,dm$. ∎

Corollary 2.34 can also be considered a corollary of Lebesgue's dominated convergence theorem.

Corollary 2.42 Lebesgue's bounded convergence: *Suppose that $f_1, f_2, f_3, \ldots$ is a convergent sequence of measurable functions defined on a bounded measurable set A and suppose that $|f_k| \le M$ for some constant M and all k. Then $\lim_{k\to\infty} f_k$ is integrable on A and*

$$\lim_{k\to\infty} \int_A f_k\,dm = \int_A \left(\lim_{k\to\infty} f_k\right) dm.$$

To prove this, use the integrable function $g = Mf_A$.

Corollary 2.43 *Suppose $f\colon \mathbb{R}^n \longrightarrow \mathbb{R}$ is measurable. If $\int_I f\,dm = 0$ for all n-dimensional intervals I, then $f = 0$ almost everywhere.*

Proof Assume $\int_I f\,dm = 0$ for all n-dimensional intervals I but f is *not* 0 a.e. By definition, both $\int_I f^+\,dm$ and $\int_I f^+\,dm$ must be finite. Then $\int_I |f|\,dm$ is also finite. We first consider the case where $f > 0$ on a set A with $m(A) > 0$. We may restrict A to a subset of some n-dimensional bounded open interval I'. Then there exists a closed set $F \subset A$ with $m(F) > 0$. So $\int_F f\,dm > 0$. Since $I' - F$ is open, it is the countable union of nonoverlapping closed cubes (Theorem 1.24), say $I' - F = \bigcup_{k=1}^{\infty} C_{M_k}$. By assumption,

$$\int_{I'-F} f\,dm = \sum_{k=1}^{\infty} \int_{C_{M_k}} f\,dm = 0 \text{ and } \int_{I'} f\,dm = 0.$$

Then

$$0 = \int_{I'} f\,dm = \int_{I'-F} f\,dm + \int_F f\,dm = 0 + \int_F f\,dm.$$

This contradicts $\int_F f\,dm > 0$. Similarly, it can be shown that $f < 0$ cannot happen on a set of positive measure. So $f = 0$, almost everywhere. ∎

Example 2.4 Lebesgue's convergence theorems do not hold for the Riemann integral. Let $r_1, r_2, \ldots$ be an enumeration of the rational numbers between 0 and 1. Define the functions

$$f_k(x) = \begin{cases} 1 & \text{if } x \in \{r_1, \ldots, r_k\} \\ 0 & \text{otherwise.} \end{cases}$$

Each f_k is Riemann integrable (with Riemann integral 0) and

$$0 \le f_1 \le f_2 \le f_3 \le \cdots.$$

The limit $f = \lim_{k\to\infty} f_k$ is the indicator function of the set of rational numbers between 0 and 1. The conditions of Lebesgue's monotone convergence theorem are satisfied. The function f is Lebesgue integrable but we know that it is not Riemann integrable.

We finish the chapter with two outstanding results regarding Riemann integration. They will be proven in Section 3.2. We state them here for the one-dimensional case $f: \mathbb{R} \longrightarrow \mathbb{R}$, although they also hold for higher dimensions $f: \mathbb{R}^n \longrightarrow \mathbb{R}$.

Theorem 3.7 *If $f: [a,b] \longrightarrow \mathbb{R}$ is Riemann integrable, then it is Lebesgue integrable. In that case, the Riemann integral is equal to the Lebesgue integral.*

Theorem 3.8 Lebesgue–Vitali: *Let $f: \mathbb{R} \longrightarrow \mathbb{R}$ be defined on interval $[a,b]$ and let D be the points on $[a,b]$ where f is discontinuous. Then f is Riemann integrable on $[a,b]$ if and only if f is bounded on $[a,b]$ and $m(D) = 0$. That is, f is Riemann integrable on $[a,b]$ if and only if f is bounded on $[a,b]$ and continuous a.e.*

Exercises for Section 2.4

2.4.1 Suppose $f : \mathbb{R}^n \longrightarrow \mathbb{R}$ is integrable. For each $k = 1, 2, 3, \ldots$, define

$$f_k(x) = \begin{cases} f(x) & \text{if } |f(x)| \le k \text{ and } \|x\|_2 \le k \\ 0 & \text{otherwise.} \end{cases}$$

Here the inequality $\|x\|_2 \le k$ defines the closed unit ball about the origin of radius k. Show that

$$\lim_{k\to\infty} \int f_k \, dm = \int f \, dm.$$

2.4.2 Suppose $f : \mathbb{R} \longrightarrow \mathbb{R}$ is integrable. For each $k = 1, 2, 3, \ldots$, define

$$f_k(x) = \begin{cases} f(x)\left(1 - \frac{|x|}{k}\right) & \text{if } |x| \le k \\ 0 & \text{otherwise.} \end{cases}$$

Show that

$$\lim_{k\to\infty} \int f_k \, dm = \int f \, dm.$$

2.4.3 Suppose $f : \mathbb{R} \longrightarrow \mathbb{R}$ is integrable. For each $k = 1, 2, 3, \ldots$, define

$$f_k(x) = \begin{cases} f(x)e^{-\frac{|x|}{k}} & \text{if } |x| \le k \\ 0 & \text{otherwise.} \end{cases}$$

Show that

$$\lim_{k\to\infty} \int f_k \, dm = \int f \, dm.$$

2.4.4 Show that the conditions of Lebesgue's bounded convergence theorem (Corollary 2.42) can be relaxed to hold if the measurable functions f_k are convergent and uniformly bounded by M *almost everywhere* on the bounded measurable set A.

3
Some Calculus

In the previous two chapters we defined Lebesgue measure and integration and gave their basic properties. They include the limit theorems that are fundamental to mathematical analysis. In the present chapter we consider topics that are traditionally known as part of calculus.

*3.1 Fubini's Theorem

Fubini's theorem permits integration over $\mathbb{R}^n$ to be performed as n integrations over $\mathbb{R}$. We state it for the case $f : \mathbb{R}^2 \longrightarrow \mathbb{R}$ but it holds more generally for $f : \mathbb{R}^n \longrightarrow \mathbb{R}$. First, a definition and a preliminary result.

Definition 3.1 If $A \subset \mathbb{R}^2$, then for each $y \in \mathbb{R}$, the **y-section of A** is the set (illustrated in Figure 3.1)

$$A_y = \{x \in \mathbb{R} \mid (x, y) \in A\}.$$

Theorem 3.1 Cavalieri's principle: *Let A be a measurable subset of* $\mathbb{R}^2$ *with* $m(A) < \infty$. *Then for almost all y, the y-sections* A_y *are measurable sets in* $\mathbb{R}$. *Furthermore the function* $F(y) = m(A_y)$ *is a measurable function of y, defined for almost all y, and*

$$\int F(y)\,dy = \int m(A_y)\,dy = m(A). \tag{3.1}$$

Proof

Claim 1: The result holds for 2-dimensional intervals $A = I = [a,b] \times [c,d]$. In this case, $A_y = I_y = [a,b]$ for $c \le y \le d$ and $A_y = \emptyset$ otherwise. The integral (3.1) becomes $\int m(A_y)\,dy = \int_c^d (b-a)\,dy = (b-a)(d-c) = m(A)$.

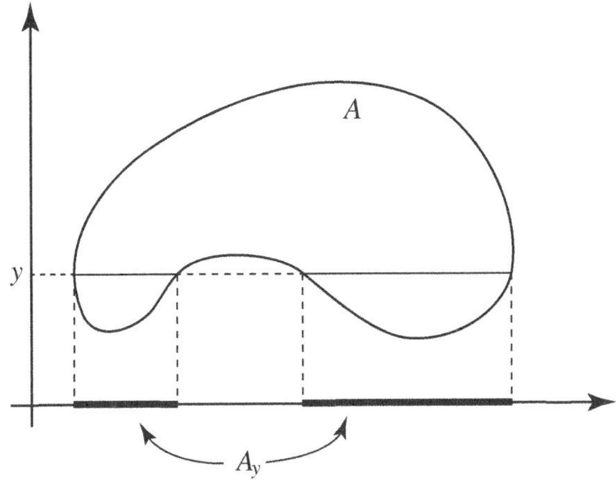

Figure 3.1 The y-section A_y of a set $A \subset \mathbb{R}^2$.

Claim 2: The result holds for countable unions of *nonoverlapping* intervals $A = \bigcup I_k$. Here $m(A) = \sum |I_k|$. By Claim 1, the y-sections $(I_k)_y$ are measurable and $F_k(y) = m\big((I_k)_y\big)$ are measurable functions of y and

$$\int F_k(y)\,dy = \int m\big((I_k)_y\big)\,dy = m(I_k) = |I_k|.$$

Since the I_k are nonoverlapping and $A_y = (\bigcup I_k)_y = \bigcup (I_k)_y$, we have $m(A_y) = \sum m\big((I_k)_y\big)$. That is, $F = \sum F_k$. By Corollary 2.39 (Section 2.4) of Lebesgue's monotone convergence theorem, F is measurable and

$$\int m(A_y)\,dy = \int \sum_{k=1}^{\infty} F_k(y)\,dy = \sum_{k=1}^{\infty} \int F_k(y)\,dy = \sum_{k=1}^{\infty} |I_k| = m(A).$$

Claim 3: The result holds for open sets $A = G$. Since every open set is a countable union of nonoverlapping closed intervals (Theorem 1.24), this claim follows from Claim 2.

Claim 4: The result holds for countable intersections of open sets of finite measure. Suppose $A = \bigcap G_k$ where each G_k is open. We may assume $G_{k+1} \subset G_k$ for all k and $m(G_1) < \infty$. The functions $F_k(y) = m\big((G_k)_y\big)$ are measurable by Claim 3. Since $F = \lim_{k\to\infty} F_k$, with $0 \le F_k \le F_1$, and F_1 integrable, we have by Lebesgue's dominated convergence theorem,

$$\int m(A_y)\,dy = \lim_{k\to\infty} \int m\big((G_k)_y\big)\,dy = \lim_{k\to\infty} m(G_k) = m(A).$$

Claim 5: The result holds for $m(A) = 0$. From the definition of measure zero, we can find open interval covers G_k of A with $m(G_k) < 1/2^k$. Let $C = \bigcap G_k$. Then $m(C) = 0$ and, by Claim 4, we have $\int m(C_y)\,dy = m(C) = 0$. So by Theorem 2.27 (Subsection 2.3.2), $m(C_y) = 0$ for almost all y. Since $A \subset C$, we have $m(A_y) = 0$ for almost all y. Hence,

$$\int F(y)\,dy = \int m(A_y)\,dy = 0 = m(A).$$

Claim 6: The result holds for measurable A with $m(A) < \infty$. By Theorem 1.42 (Section 1.8), there is a countable intersection of open sets C such that $A \subset C$ and $m(C - A) = 0$. Here $m(C) = m(C - A) + m(A) = m(A) < \infty$. By Claim 4, the y-sections C_y are all measurable. By Claim 5, the y-sections of $C - A$ are measurable for almost all y. Since $A = C - (C - A)$, it follows from Property **(m3b)** that the y-sections A_y are measurable for almost all y. Furthermore, A_y and C_y differ on a set of measure zero, so $F(y) = m(A_y) = m(C_y)$. Thus

$$\int F(y)\,dy = \int m(A_y)\,dy = \int m(C_y)\,dy = m(C) = m(A). \qquad \blacksquare$$

Definition 3.2 If $f : \mathbb{R}^2 \longrightarrow \mathbb{R}$, then for each $y \in \mathbb{R}$, the **y-section of f** is the function

$$f_y(x) = f(x, y).$$

Theorem 3.2 Fubini's theorem: *Let $f : \mathbb{R}^2 \longrightarrow \mathbb{R}$ be Lebesgue integrable. Then the y-sections $f_y : \mathbb{R} \longrightarrow \mathbb{R}$ are integrable for almost all y. Also*

$$F(y) = \int f_y(x)\,dx \text{ is integrable and } \int f\,dm = \int F(y)\,dy.$$

This may also be written as an ***iterated integral*** *and the roles of x and y may be interchanged*

$$\int f\,dm = \int \left\{ \int f(x, y)\,dx \right\} dy = \int \left\{ \int f(x, y)\,dy \right\} dx.$$

Proof

Claim 1: The result holds for any integrable indicator function $f = f_A$. Here $f(x, y) = f_A(x, y) = 1$ if $(x, y) \in A$, and $f(x, y) - 0$ otherwise. Clearly the y-sections f_y of f are the indicator functions of the y-sections A_y. Integrability of f means that A is measurable and $m(A) < \infty$. By definition of the Lebesgue integral, we have $\int f\,dm = \int f_A\,dm = m(A) < \infty$. Then by Cavalieri's principle $F(y) = m(A_y)$ and $\int F(y)\,dy = m(A)$.

Claim 2: The result holds for integrable simple functions. This follows from Claim 1 since an integrable simple function is a linear combination of integrable indicator functions.

Claim 3: The result holds for nonnegative integrable f. If f is nonnegative, then by Theorem 2.16 (Section 2.1) there is an increasing sequence of simple functions

$$0 \le s_1 \le s_2 \le s_3 \le \cdots$$

that converges to f. Clearly, $0 \le (s_1)_y \le (s_2)_y \le (s_3)_y \le \cdots$ is an increasing sequence that converges to f_y. Hence, f_y is measurable. For each k, let $F_k(y) = \int (s_k)_y(x)\,dx$. By Claim 1, each F_k is integrable and by Lebesgue's monotone convergence theorem, we have

$$F(y) = \int f_y(x)\,dx = \lim_{k\to\infty} F_k(y) = \lim_{k\to\infty} \int s_k(x, y)\,dx.$$

The function F, being a limit of measurable functions, is measurable. Using Lebesgue's monotone convergence theorem with Claim 2, and then Lebesgue's monotone convergence again, we have

$$\begin{aligned}
\int F(y)\,dy &= \int \lim_{k\to\infty} \left\{ \int s_k(x, y)\,dx \right\} dy \\
&= \lim_{k\to\infty} \int \left\{ \int s_k(x, y)\,dx \right\} dy \\
&= \lim_{k\to\infty} \int s_k(x, y)\,dm = \int f\,dm.
\end{aligned}$$

Claim 4: The result holds for integrable f. This follows since f is a difference of two nonnegative functions $f = f^+ - f^-$. Applying Claim 3 to f^+ and f^- will prove this general case. ∎

Example 3.1 It should be noted that the existence of the iterated integrals

$$\int \left\{ \int f(x, y)\,dx \right\} dy \quad \text{and} \quad \int \left\{ \int f(x, y)\,dy \right\} dx$$

does *not* guarantee integrability of f. For example, define

$$f(x, y) = \frac{x^2 - y^2}{(x^2 + y^2)^2}$$

on the 2-dimensional interval $X = [0, 1] \times [0, 1]$. The function fails to be integrable yet both iterated integrals exist. The iterated integrals fail to be equal. We have

$$\int_0^1 \left\{ \int_0^1 f(x,y)\, dx \right\} dy = -\frac{\pi}{4} \quad \text{and} \quad \int_0^1 \left\{ \int_0^1 f(x,y)\, dy \right\} dx = \frac{\pi}{4}.$$

However, there is a useful converse of Fubini's theorem, known as Tonelli's theorem, which we state without proof.

Theorem 3.3 Tonelli's theorem: *Let $f : \mathbb{R}^2 \longrightarrow \mathbb{R}$ be Lebesgue measurable. If one of the iterated integrals*

$$\int \left\{ \int |f(x,y)|\, dx \right\} dy \quad \text{or} \quad \int \left\{ \int |f(x,y)|\, dy \right\} dx$$

exists, then so does the other, and they are equal. Also the function f is Lebesgue integrable. So Fubini's theorem applies.

Exercises for Section 3.1

3.1.1 Evaluate the two iterated integrals given in Example 3.1.

3.1.2 Show $\int_0^\infty e^{-x^2}\, dx = \frac{\sqrt{\pi}}{2}$. Use Fubini's theorem on $f(x,y) = e^{-(x^2+y^2)}$ and convert to polar coordinates to evaluate $\int_0^\infty \int_0^\infty f(x,y)\, dx\, dy = \pi/4$.

*3.2 The Riemann Integral

Now we consider the Riemann integral for functions $f : \mathbb{R} \longrightarrow \mathbb{R}$, but the results are easily extended to functions $f : \mathbb{R}^n \longrightarrow \mathbb{R}$. We will define and state some basic properties of the Riemann integral. Although the standard integral in mathematical analysis is the more general Lebesgue integral, the Riemann integral is still useful because of its simplicity.

Definition 3.3 Let $[a,b]$ be a closed and bounded interval in $\mathbb{R}$. A **partition** of $[a,b]$ is a strictly increasing finite sequence $\mathcal{P} = \{x_0, x_1, \ldots, x_N\}$ of points such that

$$a = x_0 < x_1 < \cdots < x_N = b.$$

The partition divides the interval I into N **subintervals**

$$[x_0, x_1], [x_1, x_2], \ldots, [x_{N-1}, x_N].$$

The **mesh** of the partition is defined to be

$$\|\mathcal{P}\| = \max_{1 \le k \le N} (x_k - x_{k-1}).$$

Definition 3.4 Let f be a function defined on $I = [a,b]$. A **Riemann sum** for f over a partition $\mathcal{P}$ is *any* sum of the form

$$S(f,\mathcal{P}) = \sum_{k=1}^{N} f(x_k')(x_k - x_{k-1}) \text{ where } x_{k-1} \le x_k' \le x_k.$$

Definition 3.5 Let $f\colon [a,b] \longrightarrow \mathbb{R}$ and let $A \in \mathbb{R}$. We say that A is the **Riemann integral** of f on $[a,b]$ if for every $\epsilon > 0$, there exists a $\delta = \delta(\epsilon) > 0$ such that

$$|S(f,\mathcal{P}) - A| < \epsilon \text{ whenever } \|\mathcal{P}\| < \delta.$$

If there exists such an $A \in \mathbb{R}$ we say that f is **Riemann integrable** on $[a,b]$ and we write

$$A = \int_a^b f(x)\,dx.$$

The number A is uniquely determined (Exercise 3.2.1). The set of Riemann integrable functions on an interval $[a,b]$ is a linear space and the Riemann integral is a linear functional on this space (Exercise 3.2.2).

Theorem 3.4 *A function which is Riemann integrable on* $[a,b]$ *must be bounded on* $[a,b]$.

Proof Assume that f is unbounded on $[a,b]$ yet f is Riemann integrable with integral A. For $\epsilon = 1$, there exists $\delta > 0$ such that

$$|S(f,\mathcal{P}) - A| < \epsilon = 1 \text{ whenever } \|\mathcal{P}\| < \delta.$$

Let $\mathcal{P} = \{x_0, x_1, \ldots, x_N\}$ be any partition with $\|\mathcal{P}\| < \delta$. Then for any choice of $x_{k-1} \le x_k' \le x_k$, we have

$$\left| \sum_{k=1}^{N} f(x_k')(x_k - x_{k-1}) - A \right| < \epsilon = 1.$$

In particular, we may let $x_k' = x_k$ for all $1 \le k \le N$ to obtain

$$\left| \sum_{k=1}^{N} f(x_k)(x_k - x_{k-1}) - A \right| < \epsilon = 1.$$

Since f is unbounded on $[a,b]$, it must be unbounded on at least one of the subintervals $[x_{j-1}, x_j]$ for $1 \le j \le N$. In such a subinterval $[x_{j-1}, x_j]$, choose $x_{j-1} \le x_j'' \le x_j$ such that $|f(x_j'')| > \frac{2}{x_j - x_{j-1}} + |f(x_j)|$. Then

$$|f(x_j'') - f(x_j)|(x_j - x_{j-1}) \ge \big(|f(x_j'')| - |f(x_j)|\big)(x_j - x_{j-1}) > 2. \quad (3.2)$$

For all other $k \neq j$, choose $x_k'' = x_k$. Then

$$|f(x_j'')-f(x_j)|(x_j-x_{j-1}) \leq \Big|\sum_{k=1}^{N} f(x_k'')(x_k-x_{k-1})-\sum_{k=1}^{N} f(x_k)(x_k-x_{k-1})\Big|$$

$$\leq \Big|\sum_{k=1}^{N} f(x_k'')(x_k-x_{k-1})-A\Big|+\Big|\sum_{k=1}^{N} f(x_k')(x_k-x_{k-1})-A\Big| < 2\epsilon = 2.$$

This contradicts (3.2), so our assumption that f is unbounded is false. ∎

Definition 3.6 Let $\mathcal{P} = \{x_0, x_1, \ldots, x_N\}$ be a partition of the closed and bounded interval $[a,b]$. A **step function** is one that is constant on the open subintervals (x_{k-1}, x_k) for $k = 1, 2, \ldots, N$.

Definition 3.7 Let f be a bounded function on the closed and bounded interval $[a,b]$ and let $\mathcal{P} = \{x_0, x_1, \ldots, x_N\}$ be a partition of $[a,b]$.

The **lower step function** f_L of f with respect to $\mathcal{P}$ is the step function with values $m_k = \inf_{x_{k-1} \leq x \leq x_k} f(x)$ on the open subintervals (x_{k-1}, x_k).

The **upper step function** f_U of f with respect to $\mathcal{P}$ is the step function with values $M_k = \sup_{x_{k-1} \leq x \leq x_k} f(x)$ on the open subintervals (x_{k-1}, x_k).

The values at the n partition points $\{x_0, x_1, \ldots, x_N\}$ do not matter.

Definition 3.8 Let f be a bounded function on a closed and bounded interval $[a,b]$. The **lower Riemann sum** of f with respect to $\mathcal{P}$ is

$$L(f,\mathcal{P}) = \sum_{k=1}^{N} m_k(x_k - x_{k-1}) = \int_a^b f_L(x)\,dx$$

and the **upper Riemann sum** of f with respect to $\mathcal{P}$ is

$$U(f,\mathcal{P}) = \sum_{k=1}^{N} M_k(x_k - x_{k-1}) = \int_a^b f_U(x)\,dx.$$

Lemma 3.5 *If a partition $\mathcal{P}$ is finer than another partition $\mathcal{Q}$ (that is, $\mathcal{Q} \subset \mathcal{P}$), then*

$$L(f,\mathcal{Q}) \leq L(f,\mathcal{P}) \leq U(f,\mathcal{P}) \leq U(f,\mathcal{Q}). \tag{3.3}$$

Furthermore, if $\mathcal{P}$ is obtained by adjoining j points to $\mathcal{Q}$, then

$$\begin{aligned} &0 \leq U(f,\mathcal{Q}) - U(f,\mathcal{P}) \leq 2jM\|\mathcal{Q}\| \text{ and} \\ &0 \leq L(f,\mathcal{P}) - L(f,\mathcal{Q}) \leq 2jM\|\mathcal{Q}\|, \end{aligned} \tag{3.4}$$

where $M = \sup_{a \leq x \leq b} |f(x)|$ and $\|\mathcal{Q}\|$ is the mesh of $\mathcal{Q}$.

Proof First, consider the case where $\mathcal{P}$ is obtained by adjoining a single point to the partition $\mathcal{Q}$, say a point t between x_{k-1} and x_k. Let M' and M'' be the suprema of f in the new subintervals $[x_{k-1},t]$ and $[t,x_k]$, respectively. Note $M' \le M_k \le M$ and $M'' \le M_k \le M$. Then,

$$\begin{aligned} U(f,\mathcal{Q}) - U(f,\mathcal{P}) &= M_k(x_k - x_{k-1}) - M'(t - x_{k-1}) - M''(x_k - t) \\ &= (M_k - M')(t - x_{k-1}) + (M_k - M'')(x_k - t) \ge 0 \end{aligned}$$

because the other terms of $U(f,\mathcal{P})$ and $U(f,\mathcal{Q})$ cancel. Furthermore,

$$(M_k - M')(t - x_{k-1}) + (M_k - M'')(x_k - t) \le 2M(x_k - x_{k-1}) \le 2M\|\mathcal{Q}\|.$$

A similar argument establishes $0 \le L(f,\mathcal{P}) - L(f,\mathcal{Q}) \le 2M\|\mathcal{Q}\|$. Repeating the argument for j added points, results in the inequalities of (3.4). The middle inequality of (3.3) is obvious from the definition. ∎

Theorem 3.6 *Consider a bounded function f on a closed and bounded interval $[a,b]$. The following are equivalent.*

(a) *The function f is Riemann integrable on $[a,b]$.*
(b) *For each $\epsilon > 0$, there exists a partition $\mathcal{P}$ such that $U(f,\mathcal{P}) - L(f,\mathcal{P}) < \epsilon$.*
(c) *For each $\epsilon > 0$, there exist step functions s and t such that*

$$s(x) \le f(x) \le t(x) \ \text{ on } \ [a,b] \ \text{ and } \ \int_a^b \big(t(x) - s(x)\big)\,dx < \epsilon.$$

Proof ***(a)*** $\Rightarrow$ ***(b)*:** Given $\epsilon > 0$, suppose there exists a $\delta = \delta(\epsilon) > 0$ such that

$$|S(f,\mathcal{P}) - A| < \epsilon \ \text{ whenever } \ \|\mathcal{P}\| < \delta. \qquad \text{Let } \ \|\mathcal{P}\| < \delta.$$

Then $A - \epsilon < \sum_{k=1}^N f(x_k')(x_k - x_{k-1}) < A + \epsilon$ for all choices of $x_{k-1} \le x_k' \le x_k$. Since $A - \epsilon$ is a lower bound for all $S(f,\mathcal{P}) = \sum_{k=1}^N f(x_k')(x_k - x_{k-1})$, as each x_k' ranges over the subinterval $[x_{k-1},x_k]$, we may take the infimum in each subinterval $m_k = \inf\{f(x') \mid x_{k-1} \le x_k' \le x_k\}$ to obtain the inequality

$$A - \epsilon \le \sum_{k=1}^N m_k(x_k - x_{k-1}).$$

Similarly for the supremum, we have

$$\begin{aligned} A - \epsilon \le L(f,\mathcal{P}) &= \sum_{k=1}^N m_k(x_k - x_{k-1}) \le U(f,\mathcal{P}) \\ &= \sum_{k=1}^N M_k(x_k - x_{k-1}) \le A + \epsilon. \end{aligned}$$

(b) $\Rightarrow$ ***(c)*:** This is clear if we let $s = f_L$ and $t = f_U$.

(*c*) ⇒ (*b*): Let $\mathcal{P}'$ be the partition that defines the step function s. Then for the partition $\mathcal{P}'$, the lower sum function f_L' satisfies $s \le f_L' \le f$. Similarly, if $\mathcal{P}''$ is the partition defining the step function t, then the upper sum function f_U'' satisfies $f \le f_U'' \le t$. Let $\mathcal{P}$ be the union (refinement) of the partitions $\mathcal{P}'$ and $\mathcal{P}''$ with corresponding lower and upper functions f_L, f_U. Refinements make lower sums larger and upper sums smaller:

$$s \le f_L' \le f_L \le f \le f_U \le f_U'' \le t.$$

Then
$$U(f,\mathcal{P}) - L(f,\mathcal{P}) \le \int_a^b \big(t(x) - s(x)\big)\, dx < \epsilon.$$

(*b*) ⇒ (*a*): For any partition $\mathcal{P}$ and any $x_{k-1} \le x_k' \le x_k$, we clearly have

$$\sum_{k=1}^{N} m_k(x_k - x_{k-1}) \le \sum_{k=1}^{N} f(x_k')(x_k - x_{k-1}) \le \sum_{k=1}^{N} M_k(x_k - x_{k-1}).$$

Thus, for any Riemann sum $S(f,\mathcal{P})$, we have $L(f,\mathcal{P}) \le S(f,\mathcal{P}) \le U(f,\mathcal{P})$.

Also **(b)**, along with the inequalities of (3.3), imply that there exists only one number A such that

$$L(f,\mathcal{P}) \le A \le U(f,\mathcal{P}).$$

Let $\epsilon > 0$ be given, and suppose (b). There exists a partition $\mathcal{P}'$ such that

$$L(f,P') \le U(f,\mathcal{P}') < L(f,\mathcal{P}') + \epsilon/2.$$

Suppose that the partition $\mathcal{P}'$ contains j points. Let $\delta = \epsilon/4Mj$, where M is the supremum of f on $[a,b]$. Let $\mathcal{P}$ be any partition with $\|\mathcal{P}\| < \delta$ and let $\mathcal{P}'' = \mathcal{P} \cup \mathcal{P}'$. By the inequalities of (3.4), we have

$$L(f,\mathcal{P}') \le L(f,\mathcal{P}'') \le U(f,\mathcal{P}'') \le U(f,\mathcal{P}) \le U(f,\mathcal{P}'') + 2jM\|\mathcal{P}\| \quad \text{and}$$
$$U(f,\mathcal{P}'') + 2jM\|\mathcal{P}\| < U(f,\mathcal{P}'') + \epsilon/2 \le U(f,\mathcal{P}') + \epsilon/2 < L(f,\mathcal{P}') + \epsilon.$$

Since A and $S(f,\mathcal{P})$ are both between $L(f,\mathcal{P}')$ and $L(f,\mathcal{P}') + \epsilon$, we have $|S(f,\mathcal{P}) - A| < \epsilon$ whenever $\|\mathcal{P}\| < \delta$. ∎

Theorem 3.7 *If $f \colon [a,b] \longrightarrow \mathbb{R}$ is Riemann integrable, then it is Lebesgue integrable. In that case the Riemann integral is equal to the Lebesgue integral.*

Proof For $\epsilon = 1/k$, there exist step functions s_k', and t_k' such that

$$s_k' \le f \le t_k' \quad \text{on} \quad [a,b] \quad \text{and} \quad \int \big(t_k' - s_k'\big)\, dm < \frac{1}{k}.$$

Let $s_k = \max\{s'_1, s'_2, \dots, s'_k\}$ and $t_k = \min\{t'_1, t'_2, \dots, t'_k\}$. The sequence of functions s_k is nondecreasing and bounded. It thus converges to some function g. Similarly, the nonincreasing sequence t_k converges to some function h. By Lebesgue's monotone convergence theorem, both h and g are Lebesgue integrable and $g \le f \le h$. Then $h - g \ge 0$ and the Lebesgue integral vanishes $\int_{[a,b]} h - g = 0$. This shows that $h = f = g$ almost everywhere, and hence f is also Lebesgue integrable.

Finally, by Lebesgue's monotone convergence theorem, the Lebesgue integral is

$$\int_{[a,b]} f = \lim_{k\to\infty} \int s_k \, dm \le \sup_{s \le f} \int s \, dm = \int_a^b f(x)\, dx$$

and similarly

$$\int_{[a,b]} f = \lim_{k\to\infty} \int t_k \, dm \ge \inf_{f \le t} \int t \, dm = \int_a^b f(x)\, dx.$$

Thus the Riemann integral equals the Lebesgue integral. ∎

Theorem 3.8 Lebesgue–Vitali: *Let $f\colon [a,b] \longrightarrow \mathbb{R}$ and let D be the points of $[a,b]$ where f is discontinuous. Then f is Riemann integrable on $[a,b]$ if and only if f is bounded on $[a,b]$ and $m(D) = 0$. That is, f is Riemann integrable on $[a,b]$ if and only if f is bounded on $[a,b]$ and f is continuous almost everywhere.*

Proof Suppose f is Riemann integrable on $[a,b]$. The proof of Theorem 3.7 has a nondecreasing sequence of step functions s_k and a nonincreasing sequence of step functions t_k, both converging almost everywhere to f, say everywhere except in a set B of measure zero. For each k, let $\mathcal{P}_k$ be the finite number of endpoints of the step functions s_k and t_k. Include the points $\mathcal{P}_k$ to define $Z = B \cup \left(\bigcup_{k=1}^{\infty} \mathcal{P}_k\right)$, which is of measure zero.

Claim: $D \subset Z$; that is, f is continuous on $[a,b] - Z$. Let $\epsilon > 0$ and $x \in [a,b] - Z$ be given, and let the functions $s_k \le f \le t_k$ be as in the proof of Theorem 3.7. Choose k such that

$$t_k(x) - \epsilon < f(x) < s_k(x) + \epsilon.$$

For the partition $\mathcal{P}_k$, the point x belongs to some subinterval (x_{i-1}, x_i). The step functions s_k and t_k are constant on (x_{i-1}, x_i). So if $y \in (x_{i-1}, x_i)$, then

$$t_k(x) - \epsilon < f(y) < s_k(x) + \epsilon.$$

Thus

$$-\epsilon < s_k(x) - f(x) \le f(y) - f(x) \le t_k(x) - f(x) < \epsilon.$$

For y sufficiently close to x, we have $|f(y) - f(x)| < \epsilon$. We conclude that f is continuous on $[a,b] - Z$.

To prove the converse, assume f is continuous almost everywhere and let x be a point of continuity of f. Let $\epsilon > 0$ be given. There exists $\delta > 0$ such that

$$f(x) - \epsilon < f(y) < f(x) + \epsilon \tag{3.5}$$

whenever $|y - x| < \delta$. Let $\mathcal{Q}_k$ be the partition of $[a,b]$ into 2^k subintervals of equal length. For sufficiently large k, the partition $\mathcal{Q}_k$ has mesh $\|\mathcal{Q}_k\| < \delta$. Then x belongs to some subinterval $[x_{i-1}, x_i]$ of $\mathcal{Q}_k$. We have $|y - x| < \delta$ for all $x, y \in [x_{i-1}, x_i]$. Taking the infimum of $f(y)$ we see from (3.5) that the lower step function s_k of the partition $\mathcal{Q}_k$ satisfies

$$f(x) - \epsilon \le s_k(x) < f(x) + \epsilon.$$

Similarly, for the upper step function t_k we have

$$f(x) - \epsilon < t_k(x) \le f(x) + \epsilon.$$

This shows that at every point of continuity, $s_k(x) \uparrow f(x)$ and $t_k(x) \downarrow f(x)$. Lebesgue's monotone convergence theorem shows that $f \in \mathcal{L}$ and both

$$L(f, \mathcal{Q}_k) \uparrow \int f\,dm \quad \text{and} \quad U(f, \mathcal{Q}_k) \downarrow \int f\,dm.$$

By part (c) of Theorem 3.6, f is Riemann integrable. ∎

Example 3.2 Let C be the Cantor set (Definition 0.22 in Section 0.10). The indicator function $f_C(x)$ is Riemann integrable on $[0,1]$.

Example 3.3 The indicator function $f_{\mathbb{Q}}$ of the rational numbers (the Dirichlet function) is *not* Riemann integrable on $[0,1]$.

Corollary 3.9 **(a)** *All functions continuous on $I = [a,b]$ are Riemann integrable on I.*

(b) *All monotonic functions on I are Riemann integrable on I.*

(c) *All functions of bounded variation on I are Riemann integrable on I.*

Exercises for Section 3.2

3.2.1 Show that the Riemann integral $A = \int_a^b f(x)\,dx$ is unique. That is, if there are two real numbers A and A' satisfying the ϵ, δ conditions of the definition, it can be shown that $|A - A'| < 2\epsilon$.

3.2.2 Let $\alpha, \beta \in \mathbb{R}$. Show that if f, g are Riemann integrable on $[a,b]$, then so is $\alpha f + \beta g$. Further, show that the Riemann integral is a linear functional. That is, show

$$\int_a^b \alpha f(x) + \beta g(x)\,dx = \alpha \int_a^b f(x)\,dx + \beta \int_a^b g(x)\,dx.$$

3.2.3 Let f be a nonnegative Riemann integrable function on $[a,b]$. Show

$$\int_a^b f\,dx = \sup\left\{\int_a^b s(x)\,dx \mid s \text{ is a step function and } 0 \le s \le f\right\}.$$

3.2.4 Let f be a Riemann integrable function on $[a,b]$. Prove that for any $\epsilon > 0$ we can find a step function s such that

$$\int_a^b |f(x) - s(x)|\,dx < \epsilon.$$

∗3.3 Fundamental Theorem of Calculus

Let $f : \mathbb{R} \longrightarrow \mathbb{R}$ be integrable on an interval $[a,b]$. Here we consider the integral function

$$F(x) = \int_a^x f(t)\,dt \quad \text{for} \quad a \le x \le b.$$

First, we show that this integral operation smooths the function:

(a) If f is integrable and bounded, then F is continuous.
(b) If f is continuous, then F is differentiable and

$$F'(x) = f(x) \quad \text{for} \quad a \le x \le b.$$

Then we give the fundamental theorem of calculus for the Riemann integral. Later in this section we will see how the theorem fares for the Lebesgue integral.

Theorem 3.10 *Suppose that $f : \mathbb{R} \longrightarrow \mathbb{R}$ is Lebesgue integrable and bounded on an interval $[a,b]$. Then*

$$F(x) = \int_a^x f(t)\,dt \quad \textit{for} \quad a \leq x \leq b$$

is (uniformly) continuous on $[a,b]$.

Proof Suppose that f is bounded by M. That is, $|f(t)| < M$ for all $a \leq t \leq b$. Given $\epsilon > 0$, let $\delta = \epsilon/M$. If $x, y \in [a,b]$ with $|y-x| < \delta$ (say $x \leq y$), then

$$|F(y) - F(x)| = \left| \int_x^y f(t)\,dt \right| \leq \int_x^y |f(t)|\,dt \leq M|y-x| < M\delta = \epsilon. \quad \blacksquare$$

Since Riemann integrable functions are necessarily bounded (Theorem 3.4), we have the following corollary.

Corollary 3.11 *Suppose that $f : \mathbb{R} \longrightarrow \mathbb{R}$ is Riemann integrable on an interval $[a,b]$. Then*

$$F(x) = \int_a^x f(t)\,dt \quad \textit{for} \quad a \leq x \leq b$$

is (uniformly) continuous on $[a,b]$.

Later (Theorem 3.19) we will strengthen the result to **absolute continuity**, a form of continuity stronger than uniform continuity.

Theorem 3.12 *Suppose that $f : \mathbb{R} \longrightarrow \mathbb{R}$ is Lebesgue integrable on $[a,b]$ and continuous at some interior point $a < c < b$. Then*

$$F(x) = \int_a^x f(t)\,dt \textit{ is differentiable at } c \textit{ and } F'(c) = f(c).$$

Proof Let $\epsilon > 0$. If f is continuous at c, then there exists $\delta > 0$ such that $|f(t) - f(c)| < \epsilon$ whenever $|t - c| < \delta$. Then for $0 < h < \delta$, we have

$$\begin{aligned} \left| \frac{F(c+h) - F(c)}{h} - f(c) \right| &= \left| \frac{1}{h} \int_c^{c+h} \{f(t) - f(c)\}\,dt \right| \\ &\leq \frac{1}{|h|} \int_c^{c+h} \big| f(t) - f(c) \big|\,dt \leq \epsilon. \end{aligned}$$

For $-\delta < h < 0$, we have the same bound ϵ. Thus $F'(c) = f(c)$. $\blacksquare$

Theorem 3.13 **Fundamental theorem of calculus:** *Suppose $f : \mathbb{R} \longrightarrow \mathbb{R}$ is Riemann integrable on $[a,b]$ and F is an antiderivative of f on $[a,b]$; that is, $F'(x) = f(x)$ for all $x \in [a,b]$. Then*

$$\int_a^b f(t)\,dt = F(b) - F(a).$$

Proof Let $\mathcal{P} = \{x_0, x_1, \ldots, x_N\}$ be a partition of $[a,b]$. On each subinterval $[x_{k-1}, x_k]$, the mean value theorem for F states that there exists some $x_k' \in (x_{k-1}, x_k)$ such that

$$F(x_k) - F(x_{k-1}) = F'(x_k')(x_k - x_{k-1}) = f(x_k')(x_k - x_{k-1}).$$

Summing over k we have

$$F(b) - F(a) = \sum_{k=1}^{N} \{F(x_k) - F(x_{k-1})\} = \sum_{k=1}^{N} f(x_k')(x_k - x_{k-1}).$$

This is a Riemann sum of f on $[a,b]$. Letting the mesh of $\mathcal{P}$ tend to 0, we conclude that

$$F(b) - F(a) = \int_a^b f(t)\,dt. \qquad \blacksquare$$

Corollary 3.14 *If f is differentiable on $[a,b]$ and the derivative f' is Riemann integrable on $[a,b]$, then*

$$\int_a^b f'(x)\,dx = f(b) - f(a).$$

Example 3.4 CAUTION: Being differentiable almost everywhere is not enough. The Cantor–Lebesgue ternary function F as described on p. 94 has derivative $F'(x) = 0$ almost everywhere on $[0,1]$. So $f = F'$ is Riemann integrable and $\int_0^1 f(x)\,dx = 0$. Yet $F(1) - F(0) = 1$. Below is a simpler example.

Example 3.5 The form of the fundamental theorem of calculus as given by the corollary above does not hold for functions f that are merely Riemann integrable and differentiable almost everywhere. For example, let f be the indicator function of $[\frac{1}{2}, 1]$. Then $f'(x) = 0$ almost everywhere, hence it is Riemann integrable. Yet

$$\int_0^1 f'(x)\,dx = 0 \neq f(1) - f(0) = 1.$$

Theorem 3.15 Integration by parts: *If f and g are differentiable on $[a,b]$ and f' and g' are integrable on $[a,b]$ then*

$$\int_a^b fg' = \{f(b)g(b) - f(a)g(a)\} - \int_a^b f'g.$$

The proof is an easy consequence of the product formula $(fg)' = fg' + f'g$ and the above corollary of the fundamental theorem of calculus.

Theorem 3.16 Mean value theorem for integrals: *Let f be continuous on $[a,b]$. Then there exists $a < c < b$ such that*

$$f(c) = \frac{1}{b-a}\int_a^b f(x)\,dx.$$

Proof This is an easy consequence of the maximum value and intermediate value theorems for continuous functions. If $m = \min_{a\le x\le b} f(x) = f(x')$ and $M = \max_{a\le x\le b} f(x) = f(x'')$, then $f(x') \le f(x) \le f(x'')$ for all $a \le x \le b$. Thus

$$m = f(x') \le \frac{1}{b-a}\int_a^b f(x)\,dx \le f(x'') = M.$$

By the intermediate value theorem, there exists c between x' and x'' such that $f(c) = \frac{1}{b-a}\int_a^b f(x)\,dx$. ∎

Now we strengthen Theorem 3.10 in two ways: by dropping the condition of boundedness for f and obtaining absolute continuity for F, which is stronger than uniform continuity.

Definition 3.9 A function f is **absolutely continuous** on $[a,b]$ if, given $\epsilon > 0$, there exists $\delta > 0$ such that for any finite set of pairwise disjoint open subintervals $\{(a_1,b_1),(a_2,b_2),\ldots,(a_N,b_N)\}$ for which $\sum_{k=1}^N (b_k - a_k) < \delta$, we have

$$\sum_{k=1}^{N} |f(b_k) - f(a_k)| < \epsilon.$$

Note that here $a \le a_1 < b_1 \le a_2 < b_2 \le \cdots \le a_N < b_N \le b$. Limiting to $N = 1$ gives us uniform continuity.

Before proceeding to the remainder of this section, it is a good idea to review the topic of functions of bounded variation (Section 0.12).

Theorem 3.17 *If $f\colon [a,b] \longrightarrow \mathbb{R}$ is absolutely continuous on $[a,b]$, then f is of bounded variation on $[a,b]$.*

Proof Suppose f is absolutely continuous on $[a,b]$. For $\epsilon = 1$, let $\delta > 0$ be specified by the definition of absolute continuity. A consequence of the definition is that, for $a < x < y < b$, we have $V_f[x,y] < 1$ whenever $y - x < \delta$. Find a whole number N such that $\frac{b-a}{\delta} < N$ and break the interval $[a,b]$ into N subintervals $[x_{k-1},x_k]$, where $x_k = a + k\frac{b-a}{N}$. Here $x_0 = a$ and $x_N = b$. Each subinterval has length $\frac{b-a}{N} < \delta$. Thus, using Property **(bv5)** of Theorem 0.46 (Section 0.12), we have

$$V_f[a,b] = V_f[x_0,x_1] + \cdots + V_f[x_{N-1},x_N] < 1 + \cdots + 1 = N < \infty. \quad \blacksquare$$

The converse is not true because functions of bounded variation need not even be continuous.

Corollary 3.18 *If f is absolutely continuous on $[a,b]$, then it is the difference $F - G$ of two monotonically increasing continuous functions on $[a,b]$.*

Proof This follows from Theorem 0.48 along with Theorem 0.50. ∎

Theorem 3.19 *Suppose that $f\colon \mathbb{R} \longrightarrow \mathbb{R}$ is Lebesgue integrable on $[a,b]$. Then the function*

$$F(x) = \int_a^x f(t)\,dt \quad \textit{for} \quad a \le x \le b$$

is absolutely continuous on $[a,b]$.

Proof First, we consider that case where f is bounded. Suppose $|f(x)| < M$ for all $a \le x \le b$. Given $\epsilon > 0$, let $\delta = \epsilon/M$. Then for $a \le a_k < b_k \le b$, we have

$$|F(b_k) - F(a_k)| \le \int_{a_k}^{b_k} |f(t)|\,dt \le M(b_k - a_k).$$

If $\sum_{k=1}^N (b_k - a_k) < \delta = \epsilon/M$, then $\sum_{k=1}^N |F(b_k) - F(a_k)| < \epsilon$.

Now the general case where f is Lebesgue integrable. Given $\epsilon > 0$, we can find a measurable simple function s such that $\int_a^b |f(x) - s(x)|\,dx < \frac{\epsilon}{2}$ (Theorem 2.37 in Subsection 2.3.3). Since s is bounded, we can use the bounded case on s to find $\delta > 0$ such that $\sum_{k=1}^N \int_{a_k}^{b_k} |s(x)|\,dx < \epsilon/2$. Then

$$\begin{aligned}
\sum_{k=1}^N |F(b_k) - F(a_k)| &\le \sum_{k=1}^N \int_{a_k}^{b_k} |f(x)|\,dx \\
&\le \sum_{k=1}^N \int_{a_k}^{b_k} |f(x) - s(x)|\,dx + \sum_{k=1}^N \int_{a_k}^{b_k} |s(x)|\,dx \\
&\le \int_a^b |f(x) - s(x)|\,dx + \sum_{k=1}^N \int_{a_k}^{b_k} |s(x)|\,dx < \frac{\epsilon}{2} + \frac{\epsilon}{2}.
\end{aligned}$$

∎

The converse that every absolutely continuous function is an integral function is also true. It is the fundamental theorem of calculus for Lebesgue integration. First, we need to formally define an integral function.

Definition 3.10 A function F defined on an interval $[a,b]$ is called an **indefinite integral** if there exists an integrable function f on $[a,b]$ and a constant α such that

$$F(x) = \alpha + \int_a^x f(t)\,dt.$$

Example 3.6 An indefinite integral is not in general a function of the form

$$F(x) = \int_c^x f(t)\,dt \quad \text{for some } c.$$

For example, consider the constant function $F(x) = 1$.

Example 3.7 The condition $F'(x) = f(x)$ almost everywhere is not sufficient to ensure that F is an indefinite integral. For example, let F be the indicator function of the interval $\left[\frac{1}{2}, 1\right]$. Then $F'(x) = 0$ almost everywhere on $[0,1]$ but F is not an indefinite integral.

Finally, the fundamental theorem of calculus for Lebesgue integration. It is of limited applicability so we state it without proof.

Theorem 3.20 Fundamental theorem of calculus for Lebesgue integration: *A function F defined on an interval $[a,b]$ is an indefinite integral if and only if it is absolutely continuous. If F is an indefinite integral with*
$F(x) = \alpha + \int_a^x f(x)\,dx$, then $F'(x) = f(x)$ almost everywhere.

Exercises for Section 3.3

3.3.1 Prove the integration by parts formula, Theorem 3.15.

3.3.2 Show that $f(x) = \sqrt{x}$ is absolutely continuous on $[0,1]$.

3.3.3 A function f defined on $[a,b]$ is said to satisfy the **Lipschitz condition of order 1** if there exists a constant $M > 0$ such that $|f(x) - f(y)| \leq M|x - y|$ for all $x, y \in [a,b]$. Show that if a function satisfies this condition, then it is absolutely continuous on $[a,b]$.

3.3.4 Show that a function which has a continuous derivative on $[a,b]$ satisfies the Lipschitz condition of order 1. Use the mean value theorem for differentiable functions that you learned in calculus and/or real analysis.

3.3.5 Consider the function $f\colon [0,1] \longrightarrow \mathbb{R}$ defined by $f(0) = 0$ and $f(x) = x \sin 1/x$ for $x \neq 0$. Show that f is continuous on $[0,1]$ but that f is not absolutely continuous on $[0,1]$.

3.3.6 Suppose f and g are absolutely continuous on $[a,b]$ and let $\alpha, \beta \in \mathbb{R}$. Show that $\alpha f + \beta g$ is also absolutely continuous on $]a,b]$.

3.3.7 Suppose f and g are absolutely continuous on $[a,b]$. Show that the product fg is also absolutely continuous on $[a,b]$. Hint: First show that

$$\sum_{k=1}^{N} |f(b_k)g(b_k) - f(a_k)g(a_k)|$$
$$\leq M_g \sum_{k=1}^{N} |f(b_k) - f(a_k)| + M_f \sum_{k=1}^{N} |g(b_k) - g(a_k)|.$$

3.3.8 Suppose that $f\colon [a,b] \longrightarrow \mathbb{R}$ is absolutely continuous on $[a,b]$ with $f(x) \neq 0$ for all $a \leq x \leq b$. Show that the reciprocal f^{-1} is absolutely continuous.

3.3.9 If a bounded function $f\colon [0,1] \longrightarrow \mathbb{R}$ is absolutely continuous on the interval $[\epsilon, 1]$ for all $\epsilon > 0$, is it necessarily true that f is absolutely continuous on $[0,1]$? Either prove it or find a counterexample.

3.3.10 Show that the Cantor ternary function of Section 1.10 is absolutely continuous.

4
Abstract Measures

4.1 Abstract Measure Spaces

Lebesgue integration, as developed in the previous chapters, is based on the measure $m(A)$ of measurable sets A in Euclidean space $\mathbb{R}^n$, which in turn is based on the Euclidean measure $|I|$ of n-dimensional intervals I. Here, we abstract the concept of measure to other spaces, called **abstract measure spaces**. This approach starts with a set X and a collection of "measurable" subsets and a nonnegative "measure" on these subsets. The advantage of this approach is that it applies to many nongeometric situations where measures are useful. An example of this is in probability theory, where "events" are measured by their probabilities. This makes probability theory a form of measure theory.

We start with some basic definitions.

Definition 4.1 Let X be any set. An **algebra of subsets** is a collection $\mathcal{M}$ of subsets of X satisfying the following properties:

(A1) $\emptyset \in \mathcal{M}$ and $X \in \mathcal{M}$.
(A2) If $A, B \in \mathcal{M}$, then $A \cup B \in \mathcal{M}$.
(A3) If $A \in \mathcal{M}$, then $A^c \in \mathcal{M}$.

Using De Morgan's laws, it is easy to see that if $A, B \in \mathcal{M}$, then we have also $A \cap B \in \mathcal{M}$ and $A - B \in \mathcal{M}$. That is, $\mathcal{M}$ is closed under the operations of union, intersection, difference, and complement of sets. The algebraic closure of these operations justifies the term **algebra**.

Definition 4.2 Let X be any set. A **σ-algebra** is an algebra $\mathcal{M}$ of subsets of X that is also closed under countable unions.

(A4) If $A_k \in \mathcal{M}$ for $k = 1, 2, \ldots$, then $\bigcup_{k=1}^{\infty} A_k \in \mathcal{M}$.

By De Morgan's laws, it follows that a σ-algebra is also closed under countable intersections.

Example 4.1 Let X be any set. The entire power set $\mathcal{M} = 2^X$ is a σ-algebra. At the other extreme, the set $\mathcal{M} = \{\emptyset, X\}$ is a σ-algebra. Actually, for any σ-algebra $\mathcal{M}$, we have $\{\emptyset, X\} \subset \mathcal{M} \subset 2^X$; that is, $\{\emptyset, X\}$ is the smallest possible σ-algebra and 2^X is the largest possible σ-algebra of X.

Example 4.2 Consider any nonempty set X and any collection of subsets $\mathcal{F}$ of X. The intersection of all σ-algebras that contain $\mathcal{F}$ (note that 2^X is such a σ-algebra) is also a σ-algebra (see Exercise 4.1.3). This is the *smallest* σ-algebra containing $\mathcal{F}$, and it is called the **σ-algebra generated** by $\mathcal{F}$.

Example 4.3 Recall the definition of the family $\mathcal{B}$ of Borel sets (Definition 1.5 in Section 1.8). It is the σ-algebra generated by the collection of all open sets in $\mathbb{R}^n$.

Definition 4.3 Let X be a nonempty set, and let $\mathcal{M}$ be a σ-algebra of subsets of X. A real-valued function μ defined on $\mathcal{M}$ is called a **measure on $\mathcal{M}$** if it satisfies the following conditions:

(AM1) $0 \le \mu(A) \le \infty$ for all $A \in \mathcal{M}$.
(AM2) $\mu(\emptyset) = 0$.
(AM3) If $A_1, A_2, \ldots$ are disjoint sets in $\mathcal{M}$, then $\mu\left(\bigcup_{k=1}^{\infty} A_k\right) = \sum_{k=1}^{\infty} \mu(A_k)$.

The family $\mathcal{M}$ is then called the collection of **μ-measurable sets**, and the triplet $(X, \mathcal{M}, \mu)$ is called an (abstract) **measure space**.

A measure μ for which $\mu(X) = 1$ is called a **probability measure**, and, in this case, the triplet $(X, \mathcal{M}, \mu)$ is called a **probability space**.

Example 4.4 Lebesgue measure: The triplet $(\mathbb{R}^n, \mathcal{L}, m)$ is a measure space. That is, the family of Lebesgue measurable subsets of $\mathbb{R}^n$ with the Lebesgue measure m is a measure space. Indeed, the notion of measure space is an abstraction of this example.

Example 4.5 Borel measure: The triplet $(\mathbb{R}^n, \mathcal{B}, m)$ is a measure space. That is, the family of Borel sets with Lebesgue measure m, is a measure space.

Example 4.6 Zero measure: Denote the family of sets of measure zero in $\mathbb{R}^n$ by $\mathcal{Z}$. Then, $(\mathbb{R}^n, \mathcal{Z}, m)$ is a measure space.

Example 4.7 Mass: For any physical object A in $\mathbb{R}^3$, let $\mu(A)$ be its mass. It is easy to see that the three properties of a measure are satisfied. If the mass distribution of the objects is not uniform, then mass is not simply proportional to the Lebesgue measure. In physics, measure spaces such as this one are frequently encountered, not just with mass but with other physical measurements of objects. On the other hand, the electric charge of an object can be both positive and negative, so it is not a measure space in this sense. It requires a more general "signed" measure, discussed later in Section 4.4.

Example 4.8 Probability: Consider an experiment whose outcome depends on chance. Each possible outcome of the experiment is called a **sample point**, and the collection of all sample points is called the **sample space** S. Let $\mathcal{M}$ be a σ-algebra of subsets of S. These sets in $\mathcal{M}$ are called **events**. Note that if $S = \mathbb{R}^n$, not all subsets of S can be events since we know that $\mathbb{R}^n$ has nonmeasurable sets. However, in the case of a countable sample space, $\mathcal{M}$ may consist of the entire power set 2^S, as shown in Exercise 4.1.11 below. A measure Pr defined on $\mathcal{M}$ for which $Pr(S) = 1$ is a **probability measure**, and the triplet $(S, \mathcal{M}, Pr)$ is a **probability space**. Probability measures are discussed further in Section 4.3.

It is important in measure theory for a σ-algebra to contain *all* subsets of any set of measure zero. That is, it is important to be able to make the following argument:

For any $A \in \mathcal{M}$, if $\mu(A) = 0$ and $B \subset A$, then $B \in \mathcal{M}$ and $\mu(B) = 0$.

This is not a guaranteed result for σ-algebras. Hence, we need to formally define the **completion** of a σ-algebra.

Definition 4.4 Let $(X, \mathcal{M}, \mu)$ be a measure space. We define its **completion** $(X, \overline{\mathcal{M}}, \overline{\mu})$ as follows:

(CAM1) $A \in \overline{\mathcal{M}} \iff \exists\, B, C \in \mathcal{M}$ with $B \subset A \subset C$ and $\mu(C - B) = 0$.
(CAM2) $\overline{\mu}(A) = \mu(B) = \mu(C)$.

For the rest of this section, we will presume that $(X, \mathcal{M}, \mu)$ is a fixed *complete* measure space.

We now introduce the notion of μ-integrable functions. As before, functions may take values $\pm\infty$. Proofs are omitted since they follow exactly the corresponding ones in Chapter 2 if we replace $\mathbb{R}^n$ by X, $\mathcal{L}$ by $\mathcal{M}$, and m by μ.

Definition 4.5 A function $f\colon X \longrightarrow [-\infty, +\infty]$ is **μ-measurable** if for all open intervals G of the form $G = (a,b)$, with $-\infty \le a < b \le +\infty$, or $G = (a, +\infty]$, or $G = [-\infty, b)$, we have $f^{-1}(G) \in \mathcal{M}$.

Theorem 4.1 *Let $f\colon X \longrightarrow [-\infty, +\infty]$ and $-\infty \le a < b \le +\infty$. The following statements are equivalent.*

(a) *f is μ-measurable.*
(b) *For every $C = [a,b]$, the inverse image $f^{-1}(C)$ is μ-measurable.*
(c) *For every $A = (a,\infty)$, the inverse image $f^{-1}(A)$ is μ-measurable.*
(d) *For every $A' = [a,\infty)$, the inverse image $f^{-1}(A')$ is μ-measurable.*
(e) *For every $B = (-\infty,b)$, the inverse image $f^{-1}(B)$ is μ-measurable.*
(f) *For every $B' = (-\infty,b]$, the inverse image $f^{-1}(B')$ is μ-measurable.*

The proof follows that of Theorem 2.1.

Theorem 4.2 *A function $f\colon X \longrightarrow [-\infty, +\infty]$ is μ-measurable if and only if for every open set G in $\mathbb{R}$, the inverse image of $f^{-1}(G)$ is μ-measurable.*

The proof follows that of Theorem 2.2.

Definition 4.6 A relation on X is said to hold ***μ*-almost everywhere** (μ-a.e.) if the set of points Z for which it fails is μ-measurable with $\mu(Z) = 0$.

Theorem 4.3 *If f is μ-measurable and $f(x) = g(x)$ μ-a.e., then g is μ-measurable.*

The proof follows that of Theorem 2.5.

Recall that for any function $f\colon X \longrightarrow [-\infty, +\infty]$, we define the **positive part** of f to be

$$f^+(x) = \begin{cases} f(x) & \text{if } f(x) \ge 0 \\ 0 & \text{if } f(x) < 0, \end{cases}$$

and the **negative part** of f to be

$$f^-(x) = \begin{cases} 0 & \text{if } f(x) \ge 0 \\ -f(x) & \text{if } f(x) < 0. \end{cases}$$

Theorem 4.4 *Let α be a real constant. If functions f and g are μ-measurable, then so are*

$$f^+, \quad f^-, \quad f+\alpha, \quad \alpha f, \quad f+g, \quad \textit{and} \quad fg.$$

The proofs follow that of Theorems 2.6, 2.8, 2.10, and 2.11.

Theorem 4.5 *The limit of a convergent sequence of μ-measurable functions is μ-measurable.*

The proof follows that of Theorem 2.12 and its Corollary.

Definition 4.7 A **simple** function $s\colon X \longrightarrow [-\infty, +\infty]$ is one that assumes only a finite number of distinct values.

Theorem 4.6 *A nonnegative function* $f\colon X \longrightarrow [0, +\infty]$ *is* μ*-measurable if and only if there exists an increasing sequence of* μ*-measurable simple functions*

$$0 \le s_1 \le s_2 \le s_3 \le \cdots \le f \tag{4.1}$$

that converge to f *everywhere on* X.

The proof follows that of Theorem 2.16.

A μ-measurable simple function s can be written as a linear combination $s = \sum_{k=1}^{N} \alpha_k f_{A_k}$ of indicator functions f_{A_k} of disjoint μ-measurable sets A_k and distinct α_k, $(-\infty \le \alpha_k \le +\infty)$. We may also assume $X = \bigcup A_k$.

Definition 4.8 For any μ-measurable simple function $s = \sum_{k=1}^{N} \alpha_k f_{A_k}$, we define

$$\int_X s\, d\mu = \sum_{k=1}^{N} \alpha_k \mu(A_k).$$

If a nonnegative function $f\colon X \longrightarrow [0, +\infty]$ is μ-measurable, we define

$$\int_X f\, d\mu = \lim_{k\to\infty} \int_X s_k\, d\mu,$$

where s_k is any increasing sequence of μ-measurable simple functions converging to f, as in Equation (4.1). As was shown for the Lebesgue measure in Section 2.3, this definition is well defined.

A function $f\colon X \longrightarrow [-\infty, +\infty]$ is said to be ***μ*-integrable** if $\int f^+\, d\mu < \infty$ and $\int f^-\, d\mu < \infty$. In this case, we write $f \in \mathcal{L}(\mu)$ and

$$\int f\, d\mu = \int f^+\, d\mu - \int f^-\, d\mu.$$

Theorem 4.7 *Let* $f\colon X \longrightarrow \mathbb{R}$ *be* μ*-measurable. Then* $f \in \mathcal{L}(\mu)$ *if and only if* $|f| \in \mathcal{L}(\mu)$.

In this case, $\left|\int f\, d\mu\right| \le \int |f|\, d\mu$.

With these definitions, all of the integration theorems of Chapter 2 hold for a measure space $(X, \mathcal{M}, \mu)$. If we replace $\mathbb{R}^n$ by X, and $\mathcal{L}$ by $\mathcal{M}$, and m by μ, the proofs of these theorems go through. The following are two examples.

Theorem 4.8 Lebesgue's monotone convergence theorem: *Suppose that* f_1, $f_2, f_3, \ldots$ *is a sequence of* μ*-measurable functions on* X *such that*

$$0 \le f_1 \le f_2 \le f_3 \le \cdots,$$

and let $f(x) = \lim_{k\to\infty} f_k(x)$. *We are permitting* $f(x)$ *to be infinite. Then* f *is* μ*-measurable and*

$$\lim_{k\to\infty} \int f_k \, d\mu = \int f \, d\mu.$$

The proof follows that of Theorem 2.38.

Theorem 4.9 Lebesgue's dominated convergence theorem: *Suppose that* $f_1, f_2, f_3, \ldots$ *is a convergent sequence of* μ*-measurable functions and suppose that* g *is a nonnegative* μ*-integrable function with* $|f_k(x)| \le g(x)$ *for all* $x \in X$. *Then* $\lim_{k\to\infty} f_k \in \mathcal{L}(\mu)$ *and*

$$\lim_{k\to\infty} \int f_k \, d\mu = \int \left(\lim_{k\to\infty} f_k \right) d\mu.$$

The proof follows that of Theorem 2.41.

Exercises for Section 4.1

4.1.1 Show that, for $X = \mathbb{R}^n$, the collection $\mathcal{L}$ of Lebesgue measurable sets is a σ-algebra.

4.1.2 Every countable set in $\mathbb{R}^n$ has Lebesgue measure zero. Give an example of an abstract measure space $(X, \mathcal{M}, \mu)$ for which some countable sets have infinite measure.

4.1.3 Prove the claim of Example 4.2 that for any collection of subsets $\mathcal{F}$ of X, the intersection of all σ-algebras that contain $\mathcal{F}$ is a σ-algebra.

4.1.4 Consider any infinite set X, and let $\mathcal{M}$ consist of all sets A for which either A is finite or A^c is finite. Show that $\mathcal{M}$ is an algebra but is not a σ-algebra.

4.1.5 Consider any infinite set X, and let $\mathcal{M}$ consist of all sets A for which either A is countable or A^c is countable. Show that $\mathcal{M}$ is a σ-algebra.

4.1.6 Consider an uncountable set X with the σ-algebra of Exercise 4.1.5. Define the function μ as follows. If A is a countable subset of X, let $\mu(A) = 0$. If A is uncountable, let $\mu(A) = \infty$. Show μ is a measure on X.

4.1.7 Let X be any nonempty set, and let $\mathcal{M} = 2^X$. Define the **counting measure** μ as follows. If A is a finite subset of X, let $\mu(A)$ be the number of elements in A. If A is an infinite subset of X, let $\mu(A) = \infty$. Show that the counting measure is a measure on X.

4.1.8 Consider $\mathbb{N}$ with the counting measure and $X = \mathbb{N} \times \mathbb{N}$, also with the counting measure. Give an interpretation of Fubini's theorem (Theorem 3.2) in this case.

4.1.9 Let $X = \{a_1, a_2, a_3, \ldots\}$ be a countable set, and let $\mathcal{M} = 2^X$. A **discrete measure** is defined as follows. Let $\{p_1, p_2, p_3, \ldots\}$ be a sequence of nonnegative numbers. If $A = \{a_{k_1}, a_{k_2}, a_{k_3}, \ldots\}$ is a subset of X, let $\mu(A) = \sum_j p_{k_j} = p_{k_1} + p_{k_2} + p_{k_3} + \cdots$. Show that this is a measure on X.

4.1.10 Consider a σ-algebra $\mathcal{M}$ on a nonempty set X. Show that a (finite) linear combination of measures on $\mathcal{M}$ is a measure on $\mathcal{M}$.

4.1.11 This is an example of a probability space with a discrete measure as described in Exercise 4.1.9. Consider a random experiment with only countably many sample points $S = \{a_1, a_2, a_3, \ldots\}$. For each $k = 1, 2, \ldots$, let p_k be the probability of the sample point a_k. Note that $p_1 + p_2 + p_3 + \cdots = 1$. For each subset A of X, the probability $Pr(A)$ of A is the sum of the probabilities of the sample points in A. Show that $(S, 2^S, Pr)$ forms a probability space. (For example, count the number of tosses of a fair coin until heads comes up for the first time. In this example the sample points are $S = \{1, 2, 3, \ldots\}$ and $p_k = 2^{-k}$.)

4.1.12 Let X be any nonempty set, and let $\mathcal{M} = 2^X$. Fix an element $a \in X$, and define the **Dirac**[1] **measure** $\delta_a : \mathcal{M} \longrightarrow \mathbb{R}$ as follows. If $a \in A$, let $\delta_a(A) = 1$, and if $a \notin A$, let $\delta_a(A) = 0$. The measure δ_a is commonly called the Dirac delta "function." Show that the Dirac measure at a is a measure on X.

4.1.13 Show that the completion $\overline{\mathcal{M}}$ of a σ-algebra $\mathcal{M}$ is a σ-algebra.

4.1.14 Show that for any measure μ on X, the completion $\overline{\mu}$ is well defined and is a measure on X.

4.1.15 Show that $\overline{\mu}(A) = 0$ if and only if there exists $Z \in \mathcal{M}$ with $A \subset Z$ such that $\mu(Z) = 0$.

4.1.16 Is the Borel measure space $(\mathbb{R}^n, \mathcal{B}, m)$ complete? Explain.

[1] Paul Adrien Maurice Dirac (1902–1984) was an English theoretical physicist who won the Nobel Prize at the age of 31 for his work in the field of quantum theory. The following is a quote of Dirac criticizing poetry: The aim of science is to make difficult things understandable in a simpler way; the aim of poetry is to state simple things in an incomprehensible way. The two are incompatible. (H. Kragh, *Dirac: A scientific biography*, Cambridge University Press (1990), p. 258.)

4.2 Lebesgue–Stieltjes Measure

Consider a monotonically increasing function $F\colon \mathbb{R} \longrightarrow \mathbb{R}$. For such a function, both one-sided limits exist at each point $a \in \mathbb{R}$. That is, the right-hand limit of F at a is

$$F(a+) = \lim_{x\to a+} F(x) = \lim_{h\to 0+} F(a+h),$$

and the left-hand limit is

$$F(a-) = \lim_{x\to a-} F(x) = \lim_{h\to 0+} F(a-h).$$

Clearly, F is continuous at $a \in \mathbb{R}$ if and only if $F(a+) = F(a-)$.

Definition 4.9 Let F be a monotonically increasing function on $\mathbb{R}$, and let I be an interval in $\mathbb{R}$ with endpoints $-\infty < a \le b < \infty$. The ***F*-measure** (or for emphasis **Lebesgue–Stieltjes**[2] ***F*-measure**) of I, denoted by $\mu_F(I)$, is the change in the function F over the interval I. More formally,

$$\begin{aligned}
\mu_F([a,b]) &= F(b+) - F(a-),\\
\mu_F([a,b)) &= F(b-) - F(a-),\\
\mu_F((a,b]) &= F(b+) - F(a+),\\
\mu_F((a,b)) &= F(b-) - F(a+) \quad \text{for} \quad a < b,\\
\mu_F((a,a)) &= \mu_F(\emptyset) = 0.
\end{aligned}$$

The rule is that if an endpoint is included, take the limit from the outside, and if an endpoint is not included, take the limit from the inside. The following properties are consequences of this definition:

$$\mu_F(I') \le \mu_F(I'') \text{ for any intervals } I' \subset I''. \tag{4.2}$$

Let I' and I'' be disjoint intervals in $\mathbb{R}$, and $I = I' \cup I''$. Then

$$\mu_F(I) = \mu_F(I') + \mu_F(I''). \tag{4.3}$$

If the function F is continuous at the endpoints of I, then

$$\mu_F(I) = F(b) - F(a).$$

[2] Thomas Jan Stieltjes (1856–1894) was a Dutch mathematician. He is known as the father of the theory of continued fractions. He was passed over for a Chair at Groningen University because he lacked a doctorate degree. A plan was devised to award him an honorary doctorate at Leiden University, but, because of a communication failure, he missed the ceremony. Instead, he took a position at the University of Toulouse in France.

In particular, if $F(x) = x$, then μ_F gives the usual Euclidean measure of intervals $\mu_F(I) = |I| = b - a$.

Example 4.9 Although the *Lebesgue* measures of the intervals $[a,b], (a,b]$, $[a,b), (a,b)$ are all the same, this is not always the case for the *Lebesgue–Stieltjes* measures. Furthermore, the Lebesgue–Stieltjes measure of a single point may be nonzero. For example, let

$$F(x) = \begin{cases} 0 & \text{if } x < 1 \\ 1 & \text{if } x \geq 1. \end{cases}$$

Then

$$\begin{aligned}
\mu_F((0,1)) &= F(1-) - F(0+) = F(1-) - F(0) = 0 - 0 = 0, \\
\mu_F((0,1]) &= F(1+) - F(0+) = F(1+) - F(0) = 1 - 0 = 1, \\
\mu_F([1,1]) &= F(1+) - F(1-) = 1 - 0 = 1, \\
\mu_F((1,2)) &= F(2-) - F(1+) = F(2) - F(1+) = 1 - 1 = 0, \\
\mu_F([1,2)) &= F(2-) - F(1-) = F(2) - F(1-) = 1 - 0 = 1.
\end{aligned}$$

Example 4.10 Let F be a right-continuous monotonically increasing step function with jumps $j_k > 0$ at the points a_k, for $k = 1, 2, \ldots$. That is,

$$j_k = F(a_k) - F(a_k-) \quad \text{for} \quad k = 1, 2, \ldots.$$

Then for any interval $I = (a,b)$ with $a < b$, we have $\mu_F(I) = \sum_{a<a_k<b} j_k$.

We may extend the disjoint union property (4.3) above to all countable unions of intervals as follows.

Theorem 4.10 *Consider the interval bounded interval set $I = [a,b] \subset \mathbb{R}$. If I is a disjoint countable union of intervals $I = \bigcup_k I_k$, then*

$$\mu_F(I) = \sum_{k=1}^{\infty} \mu_F(I_k). \tag{4.4}$$

Proof From properties (4.2) and (4.3) we have, for any finite disjoint union of intervals, $\mu_F(I) \geq \sum_{k=1}^{N} \mu_F(I_k)$. Letting $N \to \infty$ results in the inequality

$$\mu_F(I) \geq \sum_{k=1}^{\infty} \mu_F(I_k).$$

It remains to show the reverse inequality. Let $\epsilon > 0$ be given. If I is not already closed, replace it with a slightly smaller closed interval J such that $\mu_F(J) \leq \mu_F(I) < \mu_F(J) + \epsilon$. Then for each k, if I_k is not already open, replace it with a slightly larger open interval I'_k such that $\mu_F(I_k) \le \mu_F(I'_k) < \mu_F(I_k) + \frac{\epsilon}{2^{k+1}}$. The open intervals I'_k form an open cover of I, and hence they form an open cover of $J \subset I$. Since J is closed and bounded, there is a finite subcover of J (by Heine–Borel; Theorem 0.23), say $J \subset \bigcup_{k=1}^{N} I'_k$. Then

$$\mu_F(J) \le \sum_{k=1}^{N} \mu_F(I'_k) < \sum_{k=1}^{N} \left(\mu_F(I_k) + \frac{\epsilon}{2^{k+1}}\right) < \sum_{k=1}^{N} \mu_F(I_k) + \epsilon.$$

Thus $\mu_F(I) < \mu_F(J) + \epsilon < \sum_{k=1}^{N} \mu_F(I_k) + 2\epsilon \le \sum_{k=1}^{\infty} \mu_F(I_k) + 2\epsilon$. Taking the infimum with respect to all $\epsilon > 0$, the proof is complete. ∎

This theorem shows that any two decompositions of an interval into a disjoint union of intervals will yield the same F-measure. More generally, Corollary 4.11 shows that any decomposition of an open set $G \subset \mathbb{R}$ into a disjoint union of intervals, will yield the same F-measure.

Corollary 4.11 *If an open set $G \subset \mathbb{R}$ can be written as a countable union of disjoint bounded intervals in two ways, $G = \bigcup_i I_i = \bigcup_j J_k$, then*

$$\sum_{i=1}^{\infty} \mu_F(I_i) = \sum_{j=1}^{\infty} \mu_F(J_j). \tag{4.5}$$

Proof Let $I_{ij} = I_i \cap J_j$. Then $\mu_F(I_i) = \sum_j \mu_F(I_{ij})$ and $\mu_F(J_j) = \sum_i \mu_F(I_{ij})$. So $\sum_{i=1}^{\infty} \mu_F(I_i) = \sum_{ij} \mu_F(I_{ij}) = \sum_{j=1}^{\infty} \mu_F(J_j)$. ∎

That every open set can be written as a union of disjoint intervals was proven in Theorem 1.24. Thus (4.5) can be used as the definition of the F-measure $\mu_F(G)$ of any open set G in $\mathbb{R}$. This definition can include the F-measures of unbounded intervals

$$\mu_F((a,\infty)) = F(\infty) - F(a+),$$
$$\mu_F((-\infty,b)) = F(b-) - F(-\infty), \text{ and}$$
$$\mu_F(\mathbb{R}) = \mu_F((-\infty,\infty)) = F(\infty) - F(-\infty).$$

We can then proceed to define Lebesgue–Stieltjes F-measurable sets just as we did Lebesgue measurable sets. They form a σ-algebra. By the preceding theorem this σ-algebra includes all open sets, and hence all Borel sets.

Definition 4.10 The integral with respect to an F-measure is called the **μ_F-Lebesgue–Stieltjes integral**. For a μ_F-integrable function $g : \mathbb{R} \longrightarrow \mathbb{R}$, we write

$$\int g \, d\mu_F = \int g(x) \, dF(x).$$

The function F is called the **distribution function** of the integral.

Example 4.11 Lebesgue Integral: If $F(x) = x$, then the Lebesgue–Stieltjes integral becomes the Lebesgue integral.

$$\int g \, d\mu_F = \int g(x) \, dx = \int g \, dm.$$

Example 4.12 Series: For the right-continuous function given in Example 4.10 above and a function $g : \mathbb{R} \longrightarrow \mathbb{R}$, we have

$$\int g(x) dF(x) = \sum_k g(a_k) j_k.$$

In particular, in the case $a_k = k$ and $j_k = 1$ for all k, then

$$\int g(x) dF(x) = \sum_k g(k).$$

This shows that all series of real numbers are Lebesgue–Stieltjes integrals.

These two examples show us that integrals and series can be considered as part of the same theory, instead of separate theories.

We now consider the **Riemann–Stieltjes integral**, which will be defined in terms of **Stieltjes sums** just as the Riemann integral was defined in terms of Riemann sums. The Riemann–Stieltjes integrable functions form a subclass of the Lebesgue–Stieltjes integrable functions. The ability to write limits of Stieltjes sums as integrals is useful in many applications in physics.

Definition 4.11 Suppose that g is a *bounded* function defined on the interval $I = [a,b]$. Let $\mathcal{P} = \{x_0, x_1, \ldots, x_N\}$ be a partition of I, and let F be a monotonically increasing function on I. A **Stieltjes sum** is any sum of the form

$$S(g, \mathcal{P}, F) = \sum_{k=1}^{N} g(x_k')\big(F(x_k) - F(x_{k-1})\big), \text{ where } x_{k-1} \leq x_k' \leq x_k.$$

Definition 4.12 Let g be a *bounded* function defined on the interval $I = [a,b]$ and let $A \in \mathbb{R}$. We say that A is the **Riemann–Stieltjes integral** of g on $[a,b]$ with respect to the distribution F if for every $\epsilon > 0$, there exists a $\delta = \delta(\epsilon) > 0$ such that

$$|S(g, \mathcal{P}, F) - A| < \epsilon \quad \text{whenever} \quad \|\mathcal{P}\| < \delta.$$

If there exists such an $A \in \mathbb{R}$, we write $A = \int_a^b g(x) \, dF(x)$.

Just as we defined the lower and upper Riemann sums in Section 3.2, we can use the lower step function g_L and upper step function g_U to define the **lower Stieltjes sum** and **upper Stieltjes sum** of g with respect to $\mathcal{P}$ as follows.

$$L(g,\mathcal{P},F) = \int_a^b g_L(x)\,dF(x) = \sum_{k=1}^{N} m_k\big(F(x_k) - F(x_{k-1})\big),$$

$$U(g,\mathcal{P},F) = \int_a^b g_U(x)\,dF(x) = \sum_{k=1}^{N} M_k\big(F(x_k) - F(x_{k-1})\big).$$

The basic equivalence result for Riemann integrals (Theorem 3.6) can be extended to Riemann–Stieltjes integration.

Theorem 4.12 *Consider a bounded function g on a closed interval $I = [a,b]$, where $-\infty < a < b < \infty$, and a monotonically increasing distribution function F. The following are equivalent.*

(a) *The function g is Riemann–Stieltjes integrable on I with respect to F.*
(b) *For each $\epsilon > 0$, there exists a partition $\mathcal{P}$ such that*

$$U(g,\mathcal{P},F) - L(g,\mathcal{P},F) < \epsilon.$$

(c) *For each $\epsilon > 0$, there exist step functions s and t such that*

$$s(x) \le g(x) \le t(x) \;\; on \;\; [a,b] \;\; and \quad \int_a^b \Big(t(x) - s(x)\Big)\,dF(x) < \epsilon.$$

The proof follows that of Theorem 3.6.

Theorem 4.13 *If $g\colon [a,b] \longrightarrow \mathbb{R}$ is Riemann–Stieltjes integrable with respect to F, then it is Lebesgue–Stieltjes integrable with respect to F. In this case the Riemann–Stieltjes integral is equal to the Lebesgue–Stieltjes integral.*

The proof follows that of Theorem 3.7.

Theorem 4.14 *If g is continuous and F is monotonically increasing on $I = [a,b]$, then g is Riemann–Stieltjes integrable with respect to F on $[a,b]$.*

Proof We show that condition **(b)** of Theorem 4.12 above is satisfied. Let $\epsilon > 0$ be given. Since g is continuous, it is uniformly continuous on $[a,b]$. Thus there exists $\delta > 0$ such that, for $x', x'' \in I$,

$$|g(x') - g(x'')| < \frac{\epsilon}{F(b) - F(a)} \text{ whenever } |x' - x''| < \delta.$$

Let $\mathcal{P}$ be a partition with mesh $\|\mathcal{P}\| < \delta$. Then $M_k - m_k < \frac{\epsilon}{F(b)-F(a)}$ for each $k = 1, 2, \ldots, N$, and

$$\begin{aligned} U(g,\mathcal{P},F) - L(g,\mathcal{P},F) &= \sum_{k=1}^{N}(M_k - m_k)\Big(F(x_k) - F(x_{k-1})\Big) \\ &< \frac{\epsilon}{F(b)-F(a)}\sum_{k=1}^{N}\Big(F(x_k) - F(x_{k-1})\Big) \\ &= \frac{\epsilon}{F(b)-F(a)}\Big(F(b) - F(a)\Big) = \epsilon. \end{aligned}$$ ∎

Theorem 4.15 *If g is a function of bounded variation on $I = [a,b]$ and F is continuous (and monotonically increasing) on I, then g is Riemann–Stieltjes integrable with respect to F on I.*

Proof We show that condition **(b)** of Theorem 4.12 above is satisfied. Let $\epsilon > 0$ be given. The case $F(a) = F(b)$ is trivial, so we assume $F(a) < F(b)$. For any integer N, we form a partition $\mathcal{P} = \{x_0, x_1, x_2, \ldots, x_N\}$ as follows. By the intermediate value theorem for continuous functions, for each c with $F(a) < c < F(b)$, we can find $a < x < b$ such that $F(x) = c$. So for an even spacing $c_0 = F(a) < c_1 < c_2 < \cdots < c_{N-1} < F(b) = c_N$ with $c_k - c_{k-1} = [F(b) - F(a)]/N$, we can find $a = x_0 \le x_1 \le x_2 \le \cdots \le x_N = b$ such that $F(x_k) = c_k$ for $k = 0, 1, 2, \ldots, N$.

First, consider the case when g is monotonically increasing. Then we have $m_k = g(x_{k-1})$ and $M_k = g(x_k)$ for $k = 0, 1, 2, \ldots, N$. And thus

$$\begin{aligned} U(g,\mathcal{P},F) - L(g,\mathcal{P},F) &= \sum_{k=1}^{N}(M_k - m_k)\big(F(x_k) - F(x_{k-1})\big) \\ &= \frac{1}{N}\sum_{k=1}^{N}\big(g(x_k) - g(x_{k-1})\big)\big(F(b) - F(a)\big) \\ &= \frac{1}{N}\big(g(b) - g(a)\big)\big(F(b) - F(a)\big) \end{aligned}$$

which, for sufficiently large N, is less than ϵ. The same argument works when g is monotonically decreasing, except that $m_k = g(x_k)$ and $M_k = g(x_{k-1})$.

Theorem 0.48 states that every function of bounded variation can be decomposed into a difference of two monotonically increasing functions. This fact can be used to obtain the result when g is a function of bounded variation. ∎

Theorem 4.16 *If a monotonically increasing function* $F\colon \mathbb{R} \longrightarrow \mathbb{R}$ *is differentiable and if both* g *and* F' *are Riemann integrable on* $[a,b]$, *then* g *is Riemann–Stieltjes integrable on* $[a,b]$ *with respect to* F *and*

$$\int g(x)\,dF(x) = \int g(x)F'(x)\,dx.$$

Proof We show that condition **(b)** of Theorem 4.12 above is satisfied. Let $\epsilon > 0$. For any partition $\mathcal{P} = \{x_0, x_1, x_2, \ldots, x_N\}$, by the mean value theorem for differentiable F, we have $F(x_k) - F(x_{k-1}) = F'(x_k')(x_k - x_{k-1})$ for some $x_{k-1} \le x_k' \le x_k$ and all $k = 1, 2, \ldots, N$. Since F is monotonically increasing, and $F' \ge 0$, and since F' is Riemann integrable, it is bounded. Say $0 \le F'(x) \le M$ for all $x \in [a,b]$. Then

$$\begin{aligned} U(g,\mathcal{P},F) - L(g,\mathcal{P},F) &= \sum_{k=1}^{N}(M_k - m_k)\Big(F(x_k) - F(x_{k-1})\Big) \\ &\le M\sum_{k=1}^{N}(M_k - m_k)\big(x_k - x_{k-1}\big) \\ &= M\Big(U(g,\mathcal{P}) - L(g,\mathcal{P})\Big). \end{aligned}$$

Finally, since g is Riemann integrable, by Property **(b)** of Theorem 3.6, there exists a partition such that $U(g,\mathcal{P}) - L(g,\mathcal{P}) < \epsilon/M$. This shows that $U(g,\mathcal{P},F) - L(g,\mathcal{P},F) < \epsilon$. ∎

Exercises for Section 4.2

4.2.1 Suppose that $F\colon \mathbb{R} \longrightarrow \mathbb{R}$ is monotonically increasing. Show that for all $a \in \mathbb{R}$, we have $F(a-) \le F(a) \le F(a+)$.

4.2.2 Suppose that $F\colon \mathbb{R} \longrightarrow \mathbb{R}$ is monotonically increasing and $a < b$. Show that $F(a+) \le F(b-)$.

4.2.3 Use the above two exercises to confirm property (4.2).

4.2.4 Prove property (4.3).

4.2.5 Let $F(x) = \begin{cases} 0 & \text{if } x < 0 \\ x+1 & \text{if } 0 \le x < 3 \\ 5 & \text{if } x = 3 \\ x^2 & \text{if } x > 3. \end{cases}$

Sketch a graph of F and find (a) $\mu_F([-1,1])$, (b) $\mu_F([0,3])$ (c) $\mu_F([0,3))$, (d) $\mu_F([-1,3])$, (e) $\mu_F([-1,3))$, (f) $\mu_F(\{0\})$, (g) $\mu_F(\{2\})$, and (h) $\mu_F(\{3\})$.

4.2.6 Let $g(x) = 1$ on $I = [0,1]$, and let $F(x) = \begin{cases} 0 & \text{if } 0 \le x \le \frac{1}{2} \\ 1 & \text{if } \frac{1}{2} < x \le 1 \end{cases}$.

$$\text{Find} \quad \int_0^1 g(x)dF(x).$$

4.2.7 Find an example of functions g and F defined on the interval $I = [0,1]$ such that g is *Lebesgue–Stieltjes* integrable with respect to F but g is *not Riemann–Stieltjes* integrable with respect to F.

4.2.8 Suppose the function g has bounds $A \le g(x) \le B$ on $[a,b]$. If g is Riemann–Stieltjes integrable with respect to a monotonically increasing function F, show that $A\big(F(b) - F(b)\big) \le \int_a^b g\,dF \le B\big(F(b) - F(a)\big)$.

4.3 Probability Measure

Consider an experiment whose outcome depends on chance. Each possible outcome of the experiment is called a **sample point**, and the collection S of all sample points is called the **sample space**. Let $\mathcal{M}$ be a σ-algebra of subsets of the sample space S. The sets in $\mathcal{M}$ are called **events**. A measure Pr on $\mathcal{M}$ for which $Pr(S) = 1$ is called a **probability measure** and the triplet $(S, \mathcal{M}, Pr)$ is called a **probability space**.

A probability measure Pr satisfies the following axioms:[3]

(Pr1) For any event $A \in \mathcal{M}$, we have $0 \le Pr(A) \le 1$.
(Pr2) $Pr(\emptyset) = 0$ and $Pr(S) = 1$.
(Pr3) If $A_1, A_2, \ldots \in \mathcal{M}$ are pairwise disjoint events, then

$$Pr\Big(\bigcup_k A_k\Big) = \sum_k Pr(A_k).$$

Definition 4.13 A measurable function $X\colon S \longrightarrow \mathbb{R}$ is called a **random variable**. Traditionally, random variables are denoted by upper case letters, such as X and Y. The **probability distribution** (sometimes called the **cumulative probability distribution**) of a random variable X is the function $F = F_X$ from $\mathbb{R}$ to the interval $I = [0,1]$ given by

$$F(x) = Pr\big(\{y \mid X(y) \le x\}\big) \quad \text{for all} \quad x \in \mathbb{R}.$$

[3] These axioms are due to Andrey Kolmogorov. In 1933, he showed in his publication *Grundberiffe der Wahrscheinlichkeitsrechnung* (translated as: A. Kolmogorov, *Foundations of probability theory* (2nd ed.), Chelsea Publishing Company, New York (1956)) that all properties of probability theory can be derived from these three axioms.

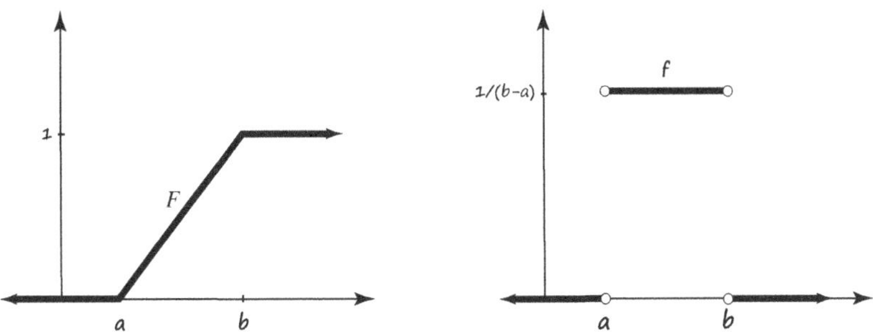

Figure 4.1 A uniform distribution function F, and its density function f.

A probability distribution F has the following properties:

(PD1) F is a monotonically increasing function on $\mathbb{R}$.
(PD2) F is right-continuous.
(PD3) $F(-\infty) = \lim_{x \to -\infty} F(x) = 0$ and $F(\infty) = \lim_{x \to \infty} F(x) = 1$.

Random variables associate events in $\mathcal{M}$ with subsets of $\mathbb{R}$. It is then possible to use the probability distribution F to define the Lebesgue–Stieltjes F-measure. The definitions of F and μ_F give us

$$Pr\Big(\{y \mid a < X(y) \leq b\}\Big) = F(b+) - F(a+) = F(b) - F(a) = \mu_F\big((a,b]\big).$$

In general, for any F-measurable subset B of $\mathbb{R}$, we have,

$$Pr\Big(\{y \mid X(y) \in B\}\Big) = \mu_F(B).$$

If F is continuous, we say that the probability measure Pr has a **continuous distribution**. If F is differentiable, the derivative $f = F'$ is called the **probability density function** for the probability measure Pr (see Figure 4.1). If f is Riemann integrable, then the fundamental theorem of calculus (Theorem 3.13) gives us

$$F(a) = Pr(-\infty < x \leq a) = \int_{-\infty}^{a} f(x)\,dx \quad \text{and} \quad Pr(A) = \int_A f(x)\,dx.$$

Example 4.13 Uniform distribution: Let $-\infty < a < b < \infty$. Define the uniform distribution function F on $I = [a,b]$ by

$$F(x) = \begin{cases} 0 & \text{if } x < a \\ \frac{x-a}{b-a} & \text{if } a \leq x < b \\ 1 & \text{if } b \leq x. \end{cases}$$

This distribution F is continuous and it is differentiable everywhere except at a and b. Here, the probability density function is

$$f(x) = F'(x) = \begin{cases} 0 & \text{if } x < a \\ \frac{1}{b-a} & \text{if } a < x < b \\ 0 & \text{if } x > b. \end{cases}$$

For any event $A \subset I$, the probability is

$$Pr(A) = \int_A d\mu_F = \frac{1}{b-a}\int_a^b f_A\, dm = \frac{m(A)}{b-a},$$

where $m(A)$ is the Lebesgue measure of the set A.

Example 4.14 Discrete distribution: Let F be a right-continuous monotonically increasing step function with jumps $p_k > 0$ at the points a_k, with $8a_1 < a_2 < \cdots$ and $\sum_k p_k = 1$. Then

$$F(x) = \begin{cases} 0 & \text{if } x < a_1 \\ p_1 + p_2 + \cdots + p_k & \text{if } a_k \le x < a_{k+1}. \end{cases}$$

If there are only a finite number of points $a_1 < \cdots < a_N$, then, additionally,

$$F(x) = p_1 + \cdots + p_N = 1 \qquad \text{if } \; x \ge a_N.$$

An immediate consequence of this probability measure is that all subsets A of $\mathbb{R}$ are events and $Pr(A) = \sum_{a_k \in A} p_k$.

Example 4.15 Continuing the previous example, let us consider the experiment of tossing a fair coin until it comes up "heads" for the first time. Let p_k denote the probability that the first "heads" comes up on the kth toss; that is, "tails" comes up for $k-1$ times followed by "heads." Then $a_k = k$ and $p_k = (1/2)^k$ for $k = 1, 2, \ldots$. Note that $\sum_{k=1}^{\infty} p_k = \sum_{k=1}^{\infty} (1/2)^k = 1$.

Let us compute the probability that the first "heads" comes up on an odd toss. Here the event is $A = \{1, 3, 5, \ldots\}$. We have

$$Pr(A) = \sum_{a_k \in A} p_k = p_1 + p_3 + p_5 + \cdots = \frac{1}{2}\sum_{k=0}^{\infty}\left(\frac{1}{4}\right)^k = \frac{1}{2}\frac{1}{1-\frac{1}{4}} = \frac{2}{3}.$$

Exercises for Section 4.3

4.3.1 Suppose a random experiment takes only values 0 and 1 with probabilities q and p, respectively. Note $p + q = 1$. Find the probability distribution function.

4.3.2 In rolling a fair die, the number of dots showing on the top face can be 1, 2, 3, 4, 5, or 6, with equal probabilities. Find and sketch a graph of the probability distribution function F.

4.3.3 In a certain population of people, suppose that 5% test positive for COVID-19. If *two* people are randomly selected from that population, let X be the number who test positive. Find (a) the sample space S, (b) all values $p_k = Pr(\{k\})$ for $k \in S$, and (c) the probability distribution function F.

4.3.4 Consider the probability density function

$$f(x) = \begin{cases} 0 & \text{if } x < 0 \\ \lambda e^{-\lambda x} & \text{if } x > 0, \end{cases}$$

for some constant $\lambda > 0$. Find the probability distribution function F. This is called an exponential distribution.

4.3.5 Find the density function of the following probability distribution function:

$$F(x) = \begin{cases} 0 & \text{if } \quad x < 0 \\ \frac{x^2}{4} & \text{if } 0 \le x < 2 \\ 1 & \text{if } \quad x \ge 2. \end{cases}$$

4.4 Signed Measure

Let $(X, \mathcal{M}, \mu)$ be a measure space. So far, we have only considered measures μ with the property

$$\mu(A) \ge 0 \quad \text{for every set } A \in \mathcal{M}.$$

A measure μ satisfying this condition is called a **positive measure**. Furthermore, if μ satisfies

$$0 \le \mu(A) < \infty \quad \text{for every set } A \in \mathcal{M},$$

it is called a **finite positive measure**. In this section we look at what happens if we allow a measure to take both positive and negative values.

Example 4.16 Let f be an integrable function on a measure space $(X, \mathcal{M}, \mu)$. For each measurable A, define $\mu_f(A) = \int_A f \, d\mu$. If f is nonnegative, then $(X, \mathcal{M}, \mu_f)$ satisfies all three conditions **(AM1)**, **(AM2)**, and **(AM3)** of an abstract measure on $\mathcal{M}$ as given in Definition 4.3. Otherwise, if f is *not* nonnegative, then condition **(AM1)** fails, but conditions **(AM2)** and **(AM3)**

are still satisfied. This is an example of a signed measure on $\mathcal{M}$. We will return to measures of this kind μ_f for $f \in \mathcal{L}(\mu)$ later in Section 4.5.

Example 4.17 Let μ_1 and μ_2 be positive measures on X with the same σ-algebra $\mathcal{M}$, and let $\nu = \mu_1 - \mu_2$. To avoid the situation $\nu(A) = \infty - \infty$, at least one of μ_1, μ_2 must be a *finite* positive measure. Both μ_1 and μ_2 satisfy **(AM1)**, **(AM2)**, and **(AM3)**. However, ν might satisfy only **(AM2)** and **(AM3)**.

Definition 4.14 Let X be any nonempty set, and let $\mathcal{M}$ be a σ-algebra of subsets of X. A real valued function ν defined on $\mathcal{M}$ is called a **signed measure on** $\mathcal{M}$ if it satisfies the following conditions:

(SM1) $\nu(\emptyset) = 0$.
(SM2) If $A_1, A_2, \ldots$ are disjoint sets in $\mathcal{M}$, then $\nu\left(\bigcup_{k=1}^{\infty} A_k\right) = \sum_{k=1}^{\infty} \nu(A_k)$.
(SM3) ν does not take both values $+\infty$ and $-\infty$.

If we restrict ourselves to *finite* measures, it is clear that the sum of two such signed measures is a finite signed measure. Additionally, the product of a finite signed measure with a real number is also a finite signed measure. Thus the finite signed measures form a vector space. The positive measures do *not* form a vector space.

Example 4.18 Signed Lebesgue–Stieltjes measure: Recall that f is of bounded variation on $[a,b]$ if and only if it is the difference the of two monotonically increasing functions $f = F - G$ (Theorem 0.48). This leads to the **signed Lebesgue–Stieltjes measure**

$$\mu_f = \mu_F - \mu_G,$$

where both μ_F and μ_G are positive Lebesgue–Stieltjes measures. We conclude that every Lebesgue–Stieltjes measure can be expressed as a difference of two positive Lebesgue–Stieltjes measures.

The corresponding **signed Lebesgue–Stieltjes integral** is

$$\begin{aligned}\int_a^b g(x)\,df(x) &= \int_a^b g(x)\,d\mu_F - \int_a^b g(x)\,d\mu_G \\ &= \int_a^b g(x)\,dF(x) - \int_a^b g(x)\,dG(x).\end{aligned}$$

The formula for integration by parts has a new look in the context of Lebesgue–Stieltjes integration.

Theorem 4.17 Integration by parts: *Suppose that f and g are continuous functions of bounded variation on the interval $[a,b]$. Then*

$$\int_a^b f(x)\,dg(x) + \int_a^b g(x)\,df(x) = \mu_{fg}[a,b] = f(b)g(b) - f(a)g(a).$$

We now show that *every* signed measure can be expressed as a difference of two positive measures. We consider two ways of doing this: the **Jordan decomposition** and the **Hahn decomposition**.

First we define the total variation $|\nu|$ of a signed measure.

Definition 4.15 Let ν be a signed measure on a measure space $(X,\mathcal{M},\nu)$ and let $A \in \mathcal{M}$. Define its **total variation** $|\nu|$ as follows.

$$|\nu|(A) = \sup \sum |\nu(A_k)|,$$

where the supremum is over all finite disjoint unions $A = \bigcup_{k=1}^N A_k$ with $A_k \in \mathcal{M}$. Note that $|\nu(A)| \le |\nu|(A)$ for every measurable A.

Theorem 4.18 *For any signed measure ν on a measure space $(X,\mathcal{M},\nu)$, its total variation $|\nu|$ is a positive measure.*

Proof Conditions **(AM1)** and **(AM2)** for an abstract measure (Definition 4.3) are clearly satisfied by $|\nu|$. It remains to prove the countable additivity condition **(AM3)**. Suppose $A = \bigcup_{k=1}^\infty A_k$ is a countable disjoint union of ν-measurable sets. We are done if we can show

$$|\nu|(A) = \sum_{k=1}^{\infty} |\nu|(A_k). \tag{4.6}$$

Proof of $\ge$: Taking a partial sum of the right side, we have

$$\sum_{k=1}^{M} |\nu|(A_k)| = \sum_{k=1}^{M} \sup \sum_{j=1}^{N} |\nu(B_{kj})| = \sup \sum_{k=1}^{M}\sum_{j=1}^{N} |\nu(B_{kj})|, \tag{4.7}$$

where the supremum is taken of all finite disjoint unions $A_k = \bigcup_{j=1}^N B_{kj}$. Since $\bigcup_{k=1}^M \bigcup_{j=1}^N B_{kj}$ forms a possible finite disjoint union of $\bigcup_{k=1}^M A_k$, the quantity (4.7) is less than or equal to $|\nu|\left(\bigcup_{k=1}^M A_k\right) \le |\nu|(A)$. Letting M tend to ∞, completes the proof.

Proof of $\le$: As before, suppose $A = \bigcup_{k=1}^\infty A_k$ is a countable disjoint union. If the series in (4.6) diverges, both sides are infinite. Otherwise, given $\epsilon > 0$, we can find a finite disjoint decomposition $A = \bigcup_{j=1}^{M} B_j$ such that $|\nu|(A) < \sum_{j=1}^{M} |\nu(B_j)| + \epsilon$. Let $B_{kj} = A_k \cap B_j$ for $k = 1, 2, 3, \dots$ and $j = 1, 2, \dots, M$. By countable additivity of ν, we have $\nu(B_j) = \sum_{k=1}^{\infty} \nu(B_{kj})$. Also, since $A_k = \bigcup_{j=1}^{M} B_{kj}$, we have $\sum_{j=1}^{M} |\nu(B_{kj})| \leq |\nu|(A_k)$. So

$$|\nu|(A) < \sum_{j=1}^{M} |\nu(B_j)| + \epsilon \leq \sum_{j=1}^{M} \left| \sum_{k=1}^{\infty} \nu(B_{kj}) \right| + \epsilon \leq \sum_{j=1}^{M} \sum_{k=1}^{\infty} |\nu(B_{kj})| + \epsilon$$

$$= \sum_{k=1}^{\infty} \sum_{j=1}^{M} |\nu(B_{kj})| + \epsilon \leq \sum_{k=1}^{\infty} |\nu|(A_k) + \epsilon.$$

The infimum over all $\epsilon > 0$ proves the $\leq$ inequality of (4.6). ∎

Definition 4.16 Let ν be a signed measure and let $|\nu|$ be its total variation measure. Define the **positive variation** measure (or **upper variation**) and **negative variation** measure (or **lower variation**) of ν to be, respectively,

$$\nu^+ = \frac{|\nu| + \nu}{2} \quad \text{and} \quad \nu^- = \frac{|\nu| - \nu}{2}.$$

It is clear that both ν^+ and ν^- are positive measures. Furthermore,

$$\nu = \nu^+ - \nu^- \quad \text{and} \quad |\nu| = \nu^+ + \nu^-.$$

The expression of the signed measure ν as a difference of two positive measures $\nu = \nu^+ - \nu^-$ is known as the **Jordan decomposition** of ν. A decomposition of ν as a difference of positive measures is *never* unique because for any positive measure μ on the same measure space, we can also write, for $\nu_P = \nu^+ + \mu$ and $\nu_N = \nu^- + \mu$, to obtain

$$\nu = \nu_P - \nu_N = (\nu^+ + \mu) - (\nu^- + \mu) = \nu^+ - \nu^-.$$

However, the Jordan decomposition is unique in the following sense.

Theorem 4.19 *Let ν be a signed measure. In any expression $\nu = \nu_P - \nu_N$ as a difference of positive measures, the Jordan decomposition satisfies the following minimal condition:*

$$\nu^+ \leq \nu_P \quad \textit{and} \quad \nu^- \leq \nu_N. \tag{4.8}$$

Proof Let A be ν-measurable. We first show that $|\nu|(A) \leq \nu_P(A) + \nu_N(A)$. Let $A = \bigcup_{k=1}^{N} A_k$ be any finite disjoint union. Then for each k,

$$|\nu(A_k)| = |\nu_P(A_k) + \nu_N(A_k)| \leq |\nu_P(A_k)| + |\nu_N(A_k)| = \nu_P(A_k) + \nu_N(A_k).$$

So, for the supremum with respect to all finite disjoint unions, we have

$$|\nu|(A) = \sup \sum |\nu(A_k)| \leq \sup \sum \left\{ \nu_P(A_k) + \nu_N(A_k) \right\} = \nu_P(A) + \nu_N(A).$$

Now *adding* the expressions $\nu^+ + \nu^- = |\nu| \le \nu_P + \nu_N$ and $\nu = \nu^+ - \nu^- = \nu_P - \nu_N$, we obtain $2\nu^+ \le 2\nu_P$, and thus the first inequality of (4.8). Subtracting, instead of adding, gives us the other inequality. ∎

Another way of expressing a signed measure ν on a measure space $(X, \mathcal{M}, \nu)$ as a difference of two positive measures is by means of the **Hahn decomposition** of X into a **positive set** P and a **negative set** N.

Definition 4.17 A measurable set P is called a **positive set** with respect to a measure space $(X, \mathcal{M}, \nu)$ if $\nu(A) \ge 0$ for every measurable subset $A \subset P$. Similarly N is called a **negative set** if $\nu(A) \le 0$ for every measurable subset $A \subset N$.

A set which is both a positive set and a negative set with respect to ν is called a **null set**. A ν-measurable set is a null set if and only if *every measurable subset* has ν-measure zero. So every null set must itself have ν-measure zero. However, the converse is not true. A set of measure zero may well be a union of two sets whose measures are not zero but negatives of each other. Similarly, a ν-measurable set A with positive ν-measure $\nu(A) > 0$ may fail to be a positive set.

Lemma 4.20 *Suppose P is a positive set and N is a negative set with respect to a measure space $(X, \mathcal{M}, \nu)$. Define, for every measurable set A, the following:*

$$\nu_P(A) = \nu(A \cap P) \quad \textit{and} \quad \nu_N(A) = -\nu(A \cap N).$$

Then ν_P and ν_N are both positive measures on X.

The proof is left as Exercise 4.4.13.

Theorem 4.21 Hahn decomposition: *Consider a measure space $(X, \mathcal{M}, \nu)$ with a signed measure ν. Then there is a ν-measurable positive set P and ν-measurable negative set N, such that $P \cup N = X$ and $P \cap N = \emptyset$.*

Proof

Claim 1: There exists a "largest" positive set P and a *disjoint* negative set N in X whose union is all of X. Since v cannot take both values $+\infty$ and $-\infty$ (condition **(SM3)** in Definition 4.14), let us assume $\infty \le \nu(A) < \infty$ for all measurable A (if otherwise, use $-\nu$). If there is no measurable positive set in X, we take $P = \emptyset$ and $N = X$, and we are done. Since $\nu(A) < \infty$ for all A, ν is bounded above. Thus, there exists $\alpha \in \mathbb{R}$ such that $\sup \nu(A) < \alpha$, where the supremum ranges over all measurable positive sets A. We can then find positive sets $P_1, P_2, P_3, \ldots$, such that $\lim \nu(P_k) = \alpha$. If we let $P = \bigcup P_k$, then $\nu(P) = \alpha$.

Claim 2: P is a measurable positive set. Clearly a countable *disjoint* union of measurable positive sets is a positive set. Also, every measurable subset of a positive set is a positive set. Since the union $P = \bigcup P_k$ can be written as a disjoint union $P = \bigcup P'_k$, of the measurable positive sets with $P'_k = P_k - \{P_1 \cup P_2 \cup \cdots \cup P_{k-1}\}$, this proves the claim.

Claim 3: $N = X - P$ is a negative set. Assume not. That is, we assume that it is not the case that $\nu(A) \leq 0$ for all measurable subsets of N. Then there has to exist a measurable subset $P_0 \subset N$, such that $\nu(P_0) > 0$. But by disjointness of P and N, we have $\nu(P \cup P_0) = \nu(P) + \nu(P_0) = \alpha + \nu(P_0) > \alpha$. This contradicts the maximality condition $\alpha = \sup \nu(A)$. ∎

For the Hahn decomposition of X into P and N, the **Hahn positive part** ν_P and the **Hahn negative part** ν_N of the signed measure ν are defined as follows. For any ν-measurable set A, we define

$$\nu_P(A) = \nu(P \cap A) \geq 0 \quad \text{and} \quad \nu_N(A) = -\nu(N \cap A) \geq 0.$$

We then obtain the Hahn decomposition of ν into a difference of positive measures, as follows.

$$\nu = \nu_P - \nu_N.$$

Although the Jordan decomposition (Definition 4.16) is unique, the Hahn decomposition is unique only to within a null set.

Exercises for Section 4.4

4.4.1 Let f be an integrable function on a measure space $(X, \mathcal{M}, \mu)$. Show that the measure μ_f of Example 4.16 is a signed measure. That is, show that μ_f satisfies conditions **(AM2)** and **(AM3)**.

4.4.2 Let ν_1 and ν_2 be finite signed measures and $\alpha, \beta \in \mathbb{R}$. Show that $\alpha\nu_1 + \beta\nu_2$ is a finite signed measure.

4.4.3 Show that $|\alpha\nu| = |\alpha||\nu|$ for any finite signed measure ν and any $\alpha \in \mathbb{R}$.

4.4.4 For any finite signed measure ν and every measurable set A, prove

$$-\nu^-(A) \leq \nu(A) \leq \nu^+(A) \quad \text{and} \quad |\nu(A)| \leq |\nu|(A).$$

4.4.5 If ν_1 and ν_2 are two finite signed measures, show that $|\nu_1 + \nu_2| \leq |\nu_1| + |\nu_2|$.

4.4.6 Suppose $\nu = \mu_1 - \mu_2$ is a signed measure expressed as a difference of positive measures. Show that there exists a positive measure μ such that $\mu_1 = \nu^+ + \mu$ and $\mu_2 = \nu^- + \mu$. Consequently, $\nu = \mu_1 - \mu_2 = (\nu^+ + \mu) - (\nu^- + \mu)$.

4.4.7 Suppose that a function g is Riemann–Stieltjes integrable on $[a,b]$ with respect to a function of bounded variation f. And suppose $|g(x)| \leq M$ on $[a,b]$. Show that $\left| \int_a^b g(x)\, df(x) \right| \leq M \cdot V_f[a,b]$.

4.4.8 Let ν be a signed measure on X. Show that any countable union of measurable positive sets is a positive set.

4.4.9 Give an example of a signed measure space and a set Z of measure zero which is not a null set.

4.4.10 Give an example of a signed measure ν and a measurable set A with $\nu(A) > 0$ which has a subset $B \subset A$ with $\nu(B) < 0$.

4.4.11 Let ν be a signed measure on X. Show that a measurable set N_0 is a negative set whenever $\nu^+(N_0) = 0$.

4.4.12 Let ν be a signed measure on X. Show that a measurable set A is a null set whenever $|\nu|(A) = 0$.

4.4.13 Prove the Hahn decomposition lemma (Lemma 4.20) above.

4.4.14 Let $X = P \cup N$ be a Hahn decomposition of $(X, \mathcal{M}, \nu)$ and let $\nu = \nu^+ + \nu^-$ be the Jordan decomposition. Show that $\nu^+(N) = 0$ and $\nu^-(P) = 0$.

4.4.15 Suppose that $X = P \cup N$ and $X = P' \cup N'$ are two Hahn decompositions of X. Show that $P \cap N'$ and $N \cap P'$ are both null sets.

4.4.16 Give an example of two Hahn decompositions that differ by a nonempty null set.

∗4.5 Absolute Continuity

In this section we work with a fixed positive measure space $(X, \mathcal{M}, \mu)$.

Theorem 4.22 *Let μ be a positive measure on X and let $f \in \mathcal{L}(\mu)$. Recall the measure μ_f given in Example 4.16 defined by*

$$\mu_f(A) = \int_A f\, d\mu \quad \text{for all} \quad A \in \mathcal{M}.$$

Given any $\epsilon > 0$, there exists $\delta > 0$ such that $\mu_f(A) < \epsilon$ whenever $\mu(A) < \delta$. In this case, we say $\mu_f(A) \to 0$ as $\mu(A) \to 0$.

Proof Because $|\int f\,d\mu| \leq \int |f|\,d\mu$ (Theorem 4.7), it is sufficient to consider the case of nonnegative f. Let $\epsilon > 0$ be given. By the definition of μ-integrability (Definition 4.8), we can find a simple function s such that $0 \leq s \leq f$ and

$$0 \leq \int f\,d\mu - \int s\,d\mu < \epsilon/2.$$

Let A be any measurable set with $\mu(A) > 0$. Then

$$0 \leq \int_A f\,d\mu < \int_A s\,d\mu + \epsilon/2.$$

The simple function s takes only a finite number of distinct values, so it has some bound $0 \leq s(x) \leq M$ for all $x \in X$. Then

$$0 \leq \int_A f\,d\mu < M\mu(A) + \epsilon/2 < \epsilon \quad \text{whenever} \quad \mu(A) < \epsilon/2M.$$

The proof is complete if we let $\delta = \epsilon/2M$. ∎

This theorem leads to the following definition.

Definition 4.18 A positive measure ν is said to be **absolutely continuous** with respect to a positive measure μ if, given any $\epsilon > 0$, there exists $\delta > 0$ such that $\nu(A) < \epsilon$ whenever $\mu(A) < \delta$. In that case, we write $\nu \ll \mu$. If ν is absolutely continuous with respect to the Lebesgue measure m, we simply say that ν is absolutely continuous. For *signed* measures ν and μ, we say ν is absolutely continuous with respect to μ whenever $|\nu| \ll |\mu|$. Here we also write $\nu \ll \mu$.

Theorem 4.22 above shows $\mu_f \ll \mu$ whenever $f \in \mathcal{L}(\mu)$. The converse will be proved in the next section. Now we give an important characterization of absolute continuity.

Theorem 4.23 *Suppose ν and μ are finite measures on X with the same σ-algebra $\mathcal{M}$. Then $\nu \ll \mu$ if and only if, for all $A \in \mathcal{M}$, $\nu(A) = 0$ whenever $\mu(A) = 0$.*

Proof Suppose that $\nu \ll \mu$ and $\mu(A) = 0$. Given $\epsilon > 0$, there exists $\delta > 0$ such that $\nu(A) < \epsilon$ whenever $\mu(A) < \delta$. Since $\mu(A) = 0$, we have $\mu(A) < \delta$ for *all* $\epsilon > 0$. Thus $\nu(A) < \epsilon$ for all $\epsilon > 0$, or $\nu(A) = 0$.

Conversely, suppose that $\nu(A) = 0$ whenever $\mu(A) = 0$. But assume it is not the case that $\nu \ll \mu$. Then for some $\epsilon > 0$, there exist measurable sets

$A_1, A_2, \ldots$ such that $\mu(A_k) < 1/2^k$ yet $\nu(A_k) \geq \epsilon$. For each k, let $B_k = A_k \cup A_{k+1} \cup \cdots$ and let $A = \bigcap_k B_k$. Then we have

$$\mu(A) \leq \mu(B_k) \leq \sum_{j=k}^{\infty} \mu(A_j) \leq \frac{1}{2^{k-1}} \quad \text{for every } k = 1, 2, \ldots.$$

This implies $\mu(A) = 0$. But, since $B_1 \supset B_2 \supset \cdots$ and $A = \bigcap_k B_k$, we have

$$\nu(A) = \lim_k \nu(B_k) \geq \liminf_k \nu(A_k) \geq \epsilon.$$

Under the assumption that ν is not absolutely continuous with respect to μ we have found a set A with $\mu(A) = 0$ but $\nu(A) \geq \epsilon > 0$. This contradiction proves the theorem. ∎

Theorem 4.24 *If ν and μ are finite measures on X with $\nu \ll \mu$ and $\nu \neq 0$, there exists $\epsilon > 0$ and a measurable set $B \in \mathcal{M}$ of positive ν-measure such that $\nu(A) \geq \epsilon\mu(A)$ for all $A \subset B$. That is, B is a positive set for $\nu - \epsilon\mu$.*

Proof For each $k = 1, 2, 3, \ldots$, let $X = P_k \cup N_k$ be the Hahn decomposition of X with respect to the (signed) measure $\nu - \frac{1}{k}\mu$. Let $N = \bigcap N_k$ and $P = \bigcup P_k$. Then for all k, we have

$$\nu(N) \leq \nu(N_k) \leq \frac{1}{k}\mu(N_k) \leq \frac{1}{k}\mu(X).$$

So $\nu(N) = 0$. Since $X = P \cup N$ and $\nu \neq 0$, we have $\nu(P) > 0$. And since $\nu \ll \mu$ it follows that $\mu(P) > 0$. Hence, $\mu(P_k) > 0$ for some k. Since P_k is a positive set for $\nu - \frac{1}{k}\mu$, we have $\nu(A) - \frac{1}{k}\mu(A) \geq 0$ for all subsets of P_k. For this k, if we let $B = P_k$ with $\epsilon = \frac{1}{k}$, we are done. ∎

Exercises for Section 4.5

4.5.1 Suppose $(X, \mathcal{M}, \mu)$ is measure space, $f \in \mathcal{L}(\mu)$, and $\mu_f(A) = \int_A f\, d\mu$ for every μ-measurable A. Show that

$$P = \{x \in X \mid f(x) > 0\} \quad \text{and} \quad N = \{x \in X \mid f(x) \leq 0\}$$

is a Hahn decomposition of X with respect to the measure μ.

4.5.2 Show that if $\nu_1 \ll \mu$ and $\nu_2 \ll \mu$, then $\alpha_1\nu_1 + \alpha_2\nu_2 \ll \mu$ for α_1, $\alpha_2 \in \mathbb{R}$.

4.5.3 Show that if $\nu \ll \mu$ and $\mu \ll \lambda$, then $\nu \ll \lambda$.

4.5.4 Suppose ν and μ are signed measures. Show that $\nu \ll \mu$ if and only if $\nu^+ \ll \mu$ and $\nu^- \ll \mu$.

4.5.5 Suppose ν and μ are measures on X with the same σ-algebra $\mathcal{M}$. Show that $\nu \ll \nu + \mu$.

*4.6 The Radon–Nikodym Theorem

Our goal is to show that for a positive finite measure μ, we have $\nu \ll \mu$ if and only if there exists $f \in \mathcal{L}(\mu)$ such that

$$\nu(A) = \mu_f(A) = \int_A f \, d\mu \quad \text{for every } \mu\text{-measurable set } A.$$

One direction we already have from Theorem 4.22. The converse is the following theorem.

Theorem 4.25 Radon–Nikodym: *Let $(X, \mathcal{M}, \mu)$ be a finite measure space and ν, μ be positive with $\nu \ll \mu$. Then there exists a unique (μ-a.e.) nonnegative $f \in \mathcal{L}(\mu)$ such that*

$$\nu(A) = \mu_f(A) = \int_A f \, d\mu \quad \textit{for every } \mu\textit{-measurable set } A. \tag{4.9}$$

Proof Let $\mathcal{N}$ be the set of all nonnegative measurable functions h on X such that for every $A \in \mathcal{M}$, we have

$$\int_A h \, d\mu \leq \nu(A).$$

Since ν is positive, the set $\mathcal{N}$ contains at least the zero function.

Claim: $\mathcal{N}$ is closed under finite suprema. For $h_1, h_2 \in \mathcal{N}$, define $h = \sup\{h_1, h_2\} \in \mathcal{N}$ and, for $A \in \mathcal{M}$, let $B = \{x \in A \mid h_1(x) \geq h_2(x)\}$. Then

$$\int_A h \, d\mu = \int_B h_1 \, d\mu + \int_{A-B} h_2 \, d\mu \leq \nu(B) + \nu(A - B) = \nu(A).$$

This proves the claim for $\sup\{h_1, h_2\}$, which can be extended to all finite suprema.

Existence of f: Let $M = \sup\{\int h \, d\mu \mid h \in \mathcal{N}\}$. Since ν is finite, by definition of $\mathcal{N}$, we have $M \leq \nu(X) < \infty$. Choose $h_1, h_2, \ldots$ in $\mathcal{N}$ with $\lim_k \int h_k \, d\mu = M$. For each k, define $f_k = \sup\{h_1, h_2, \ldots, h_k\}$. Note that f_k is monotonically increasing and, by the above claim, $f_k \in \mathcal{N}$ for all k. Let f be the pointwise limit of the sequence f_k. By Lebesgue's monotone convergence theorem (Theorem 4.8), we have for every $A \in \mathcal{M}$,

$$\int_A f \, d\mu = \lim_k \int_A f_k \, d\mu \leq \nu(A). \tag{4.10}$$

In particular, $f \in \mathcal{N}$ and $\int f \, d\mu = M$.

The opposite inequality of (4.10) will complete the proof of (4.9). Define a measure μ_0 on $\mathcal{M}$ by $\mu_0(A) = \nu(A) - \int_A f \, d\mu$. That μ_0 is nonnegative follows from (4.10) and it is clear that $\mu_0 \ll \mu$.

It remains to be shown that $\mu_0 = 0$. Assume that there exists a set $B \in \mathcal{M}$ with $\mu_0(B) > 0$. By Theorem 4.24, there exists $\epsilon > 0$ such that $\mu_0(B) > \epsilon\mu(B)$. That is, $(\mu_0 - \epsilon\mu)(B) > 0$. The set B then contains a positive set C for the signed measure $\mu_0 - \epsilon\mu$ with $(\mu_0 - \epsilon\mu)(C) > 0$. Since $\mu_0 - \epsilon\mu \ll \mu$ we conclude that $\mu(C) > 0$. Let f_C be the indicator function of C. Then $f + \epsilon f_C \in \mathcal{N}$ because for each $A \in \mathcal{M}$, we have

$$\int_A (f + \epsilon f_C)\, d\mu$$
$$= \int_{A-C} f\, d\mu + \int_{A\cap C} (f + \epsilon f_C)\, d\mu \le \nu(A - C) + \nu(A \cap C) = \nu(A).$$

This contradicts the definition of M because

$$\int (f + \epsilon f_C)\, d\mu = M + \epsilon\mu(C) > M.$$

This means that μ_0 is identically zero on $\mathcal{M}$. This contradiction proves (4.9) for finite μ.

Uniqueness: Suppose that g is another function that satisfies (4.9). Let $A_k = \{x \in X \mid f(x) - g(x) > 1/k\}$. Then

$$\frac{\mu(A_k)}{k} \le \int_{A_k} \big(f(x) - g(x)\big)\, d\mu = \nu(A_k) - \nu(A_k) = 0.$$

From this it follows that $\mu(A_k) = 0$. Consider the union

$$A = \{x \in X \mid f(x) > g(x)\} = \bigcup_k A_k.$$

It follows that $\mu(A) = 0$. Similarly, the set for which $f(x) < g(x)$ also has measure zero. Thus $f = g$, μ-almost everywhere. ∎

Definition 4.19 The function f in the Radon–Nikodym Equation (4.9) is called the **Radon–Nikodym derivative** or the **density function** of the measure ν with respect to μ. We write

$$f = \frac{d\nu}{d\mu}.$$

It has similar properties to a derivative of a function. However, note that the function f is unique only μ-a.e.

Theorem 4.26 Chain rule: *Suppose ν, μ, λ are finite positive measures on X with $\nu \ll \mu \ll \lambda$. The Radon–Nikodym derivative has the following property λ-a.e.:*

$$\frac{d\nu}{d\lambda} = \frac{d\nu}{d\mu} \cdot \frac{d\mu}{d\lambda}.$$

Proof Write $f = \frac{d\nu}{d\mu}$ and $g = \frac{d\mu}{d\lambda}$. Given $\nu(A) = \int_A f\,d\mu$ and $\mu(A) = \int_A g\,d\lambda$ for every measurable $A \subset X$, we want to show $\nu(A) = \int_A fg\,d\lambda$.

Let $0 \le s_1 \le s_2 \le \cdots$ be increasing simple functions converging to f on A. By the definition of integration, we have for every measurable $A \subset X$,

$$\lim_k \int_A s_k\,d\mu = \int_A f\,d\mu = \nu(A). \tag{4.11}$$

Since $0 \le s_1 g \le s_2 g \le \cdots$ is monotonically increasing and converges to fg on A, by Lebesgue's monotone convergence theorem (Theorem 4.8), we have

$$\lim_k \int_A s_k g\,d\lambda = \int_A fg\,d\lambda. \tag{4.12}$$

Now consider each measurable simple function s_k on A, say, $s_k = \sum_{j=1}^N \alpha_j f_{A_{kj}}$ with indicator functions $f_{A_{kj}}$ and $A_k = \bigcup_{j=1}^N A_{kj}$. We have

$$\int_A s_k g\,d\lambda = \sum_{j=1}^N \alpha_j \int_{A_{kj}} g\,d\lambda = \sum_{j=1}^N \alpha_j \mu(A_{kj}) = \int_A s_k\,d\mu. \tag{4.13}$$

By (4.11), the right side of (4.13) converges to $\nu(A)$. By (4.12), the left side converges to $\int_A fg\,d\lambda$. This completes the proof. ∎

Finally we show the relationship between absolute continuity of functions (Definition 3.9) and absolute continuity of measures (Definition 4.18).

Theorem 4.27 *Let F be monotonically increasing on $\mathbb{R}$ and let μ_F be the Lebesgue–Stieltjes measure of F. Then μ_F is absolutely continuous with respect Lebesgue measure m if and only if F is an absolutely continuous function.*

Proof Suppose $\mu_F \ll m$. Given $\epsilon > 0$, there exists $\delta > 0$ such that $m(A) < \delta$ implies $\mu_F(A) < \epsilon$. Consider a finite set of pairwise disjoint open intervals $I_k = (a_k, b_k)$ with $-\infty < a_1 < b_1 \le a_2 < b_2 \le \cdots \le a_N < b_N < \infty$ for which $\sum_{k=1}^N (b_k - a_k) < \delta$. For $A = \bigcup A_k$, we have $m(A) = \sum |I_k| < \delta$. Then by absolute continuity of μ_F, we have

$$\sum_{k=1}^N \Big|F(b_k) - F(a_k)\Big| \le \sum_{k=1}^N \Big|F(b_k-) - F(a_k+)\Big| = \mu_F(A) < \epsilon.$$

This means that F is an absolutely continuous function.

Conversely, suppose F is an absolutely continuous monotonically increasing function. Given $\epsilon > 0$, there exists $\delta > 0$ such that

$$\sum_{k=1}^{N} |F(b_k) - F(a_k)| < \epsilon \quad \text{whenever} \quad \sum_{k=1}^{N} (b_k - a_k) < \delta.$$

Let $A \subset \mathbb{R}$ with $m(A) = 0$. There exist intervals $I'_k = (a'_k, b'_k)$ covering A with $\sum_{k=1}^{\infty} |I'_k| = \sum_{k=1}^{\infty} (b'_k - a'_k) < \delta$.

By continuity of F, we have $\mu_F(I'_k) = F(b'_k) - F(a'_k)$, for all k. Then

$$\mu_F(A) \le \sum_{k=1}^{\infty} \mu_F(I'_k) = \sup_N \left\{ \sum_{k=1}^{N} \left(F(b'_k) - F(a'_k) \right) \right\} \le \epsilon.$$

Taking the infimum with respect to all $\epsilon > 0$, we see $\mu_F(A) = 0$. By Theorem 4.23, we have $\mu_F \ll m$. ∎

Exercises for Section 4.6

4.6.1 Show that the Radon–Nikodym derivative is linear. That is,

$$\frac{d(\alpha_1 \nu_1 + \alpha_2 \nu_2)}{d\mu} = \alpha_1 \frac{d\nu_1}{d\mu} + \alpha_2 \frac{d\nu_2}{d\mu}.$$

4.6.2 Show that if $\nu \ll \mu$ and $\mu \ll \nu$, then $\frac{d\mu}{d\nu} \neq 0$ and

$$\frac{d\nu}{d\mu} = \left(\frac{d\mu}{d\nu} \right)^{-1} \quad \mu\text{-a.e.}$$

4.6.3 Show that if $\nu \ll \mu$ and f is ν-integrable, then

$$\int f \, d\nu = \int f \cdot \frac{d\nu}{d\mu} \, d\mu.$$

4.6.4 Let the Lebesgue–Stieltjes measure μ_F be given as in Theorem 4.27. Show that if μ_F is absolutely continuous with respect to Lebesgue measure, then $\frac{d\mu_F}{dm} = F'$ almost everywhere.

4.6.5 Continuing Exercise 4.6.4, show that $\int f \, d\mu_F = \int f F' \, dm$ for all integrable functions f.

PART II

Elements of Classical Functional Analysis

5

Metric and Normed Spaces

Analysis is primarily concerned with infinite processes, including limits. So far, we have primarily done this in the setting of Euclidean spaces $\mathbb{R}^n$. We are now ready to develop a theory of limits in sets M more general than $\mathbb{R}^n$. All we need is the concept of distance as enumerated for Euclidean spaces in Section 0.5. Most of the properties of Euclidean space carry over, but not all. For example, Cauchy's inequality (Property (**E3**) of Theorem 0.6) and the Heine–Borel theorem (Theorem 0.23) do not hold in all general settings.

5.1 Metric Spaces

A **metric space** is a nonempty set M with a **distance function** or **metric** d having the following properties for all $x, y, z \in M$:

(M0) $d(x, y) \geq 0$;
(M1) $d(x, y) = 0$ if and only if $x = y$;
(M2) $d(x, y) = d(y, x)$ (symmetry);
(M3) $d(x, y) \leq d(x, z) + d(z, y)$ (triangle inequality).

Property **(M0)** can actually be proven from the others as follows:

$$0 = d(x, x) \leq d(x, y) + d(y, x) = d(x, y) + d(x, y) = 2d(x, y).$$

Here are some examples of metric spaces.

Example 5.1 The familiar Euclidean Space: Let $M = \mathbb{R}^n$, with the usual Euclidean metric as given in Section 0.5. For $x = (x_1, \ldots, x_n)$ and $y = (y_1, \ldots, y_n)$ the metric is given by

$$d(x, y) = \sqrt{(x_1 - y_1)^2 + \cdots + (x_n - y_n)^2}.$$

Example 5.2 Discrete or Trivial Metric: Let M be any nonempty set and define

$$d(x,y) = \begin{cases} 1 & \text{if } x \neq y \\ 0 & \text{if } x = y. \end{cases}$$

The metric conditions are easily seen to be true. If you consider the discrete metric on the set $\mathbb{R}$, the distance between any two distinct points is 1, which is very different from the usual metric on $\mathbb{R}$. A major use of the discrete metric is to obtain examples to illustrate specific concepts and to create counterexamples for conjectures. When considering a conjecture about metric spaces, it is a good idea to first test it on the discrete metric.

Example 5.3 Hamming Metric: Let $n \in \mathbb{N}$ be fixed, say $n = 7$, and let M be the set of all n-tuples of 0s and 1s. For $x, y \in M$, let $d(x,y)$ be the number of places in which x and y differ. For example, $d(0010110, 1010011) = 3$. This metric is used in the decoding of messages from error correcting codes.

Example 5.4 Let M consist of the natural numbers $\mathbb{N}$ along with the object ∞. Define the metric $d(x,y)$ as follows:

$$\begin{aligned} d(x,y) &= \left|\frac{1}{x} - \frac{1}{y}\right| \quad \text{if } x, y \in \mathbb{N}, \\ d(x,\infty) &= d(\infty,x) = \frac{1}{x} \quad \text{if } x \in \mathbb{N}, \\ d(\infty,\infty) &= 0. \end{aligned}$$

This metric d is bounded on $\mathbb{N}$ by $d(x,y) < 1$. Adjoining the point ∞ to the space lets the metric attain its upper bound $d(1,\infty) = 1$.

Example 5.5 The function space $C[a,b]$ of continuous functions on $[a,b]$, with $a < b$, is a metric space under the distance function

$$d(f,g) = \int_a^b |f(x) - g(x)|\, dx.$$

The properties **(M2)** and **(M3)** are easy to prove. The proof of **(M1)** is left as Exercise 5.1.14. It follows from an important property of continuous functions: If a function $h \geq 0$ is continuous and $h(x_0) > 0$ for some $x_0 \in [a,b]$, then $\int_a^b h(x)\, dx > 0$.

Example 5.6 Fréchet metric: Recall that ω is the space of all real- or complex-valued sequences. The Fréchet[1] metric on ω is defined as follows. For $x = (x_1, x_2, \ldots)$ and $y = (y_1, y_2, \ldots)$, let

$$d(x,y) = \sum_{k=1}^{\infty} \frac{|x_k - y_k|}{2^k(1 + |x_k - y_k|)}.$$

The properties **(M0)**, **(M1)**, and **(M2)** are clearly satisfied. The triangle inequality **(M3)** follows from the inequality of Lemma 5.1. The details are left to Exercise 5.1.17. A generalization of this metric is given in Exercise 5.3.6.

Lemma 5.1 *For any $a, b \in \mathbb{C}$, we have* $\frac{|a+b|}{1+|a+b|} \le \frac{|a|}{1+|a|} + \frac{|b|}{1+|b|}$.

Proof Consider the function $f(t) = t/(1+t)$ on $\mathbb{R}$. For $t > -1$, we have $f'(t) = 1/(1+t)^2 > 0$. Thus means that f is increasing on $t > -1$. Applying this function to the triangle inequality for $\mathbb{R}$, we have

$$\begin{aligned} \frac{|a+b|}{1+|a+b|} &= f(|a+b|) \le f(|a|+|b|) \\ &= \frac{|a|+|b|}{1+|a|+|b|} \le \frac{|a|}{1+|a|} + \frac{|b|}{1+|b|}. \end{aligned}$$

∎

Theorem 5.2 Subspace metric: *Suppose M is a metric space and N is any nonempty subset. Let the distance between elements of N be defined just as in M. Then N is also a metric space.*

The proof is left as Exercise 5.1.11.

Exercises for Section 5.1

5.1.1 Show that Euclidean space $\mathbb{R}^n$ is a metric space with the Euclidean metric.

5.1.2 **Taxicab metric:** Let $M = \mathbb{R}^2$. Show that $d_1(x,y) = |x_1 - y_1| + |x_2 - y_2|$ is a metric on M when $x = (x_1, x_2)$ and $y = (y_1, y_2)$. Example of use: Whereas the Euclidean metric is the distance "as the crow flies," the fare of a taxicab in an urban grid of streets is based on this metric d_1.

[1] Maurice René Fréchet (1878–1973) was a French mathematician who introduced the concept of metric spaces. He made many contributions to functional analysis, topology, probability, and statistics.

5.1.3 Generalize the taxicab metric of the previous exercise to the space ℓ_1 (defined in Section 0.4). For $x = (x_1, x_2, \ldots), y = (y_1, y_2, \ldots) \in \ell_1$, show that $d(x, y) = \sum_{k=1}^{\infty} |x_k - y_k|$ is a metric on the space ℓ_1.

5.1.4 Show that in Example 5.3, the Hamming function d is a metric.

5.1.5 For Example 5.4, show that d is a metric on $M = \{\mathbb{N}, \infty\}$.

5.1.6 Continuing with Exercise 5.1.5, show that the subspace metric on $\mathbb{N}$ is bounded above by 1 but never attains its upper bound 1.

5.1.7 Consider a nonempty set M satisfying the conditions **(M0)**, **(M1)**, and

$$\textbf{(M3}'\textbf{)}: d(x,z) \leq d(x,y) + d(z,y) \text{ for all } x, y, z \in M.$$

Note the order of terms on the right-hand side. Prove that these conditions **(M0)**, **(M1)**, and **(M3′)** imply conditions **(M0)**, **(M1)**, **(M2)**, and **(M3)**.

5.1.8 Show that $d(x, y) = \sqrt{|x - y|}$ is a metric on $\mathbb{R}$.

5.1.9 Show that $d(x, y) = (x - y)^2$ is not a metric on $\mathbb{R}$.

5.1.10 If M is a metric space with metric d, show that $d'(x, y) = \frac{d(x,y)}{1+d(x,y)}$ is also a metric on M. Show that M is bounded under the metric d'.

5.1.11 Prove Theorem 5.2.

5.1.12 Recall (from Section 0.4) that ℓ_∞ is the set of bounded real (or complex) sequences $x = (x_1, x_2, \ldots)$. Show that ℓ_∞ is a metric space under the metric

$$d(x, y) = \sup_{1 \leq k < \infty} |x_k - y_k|.$$

5.1.13 Let N consist of all sequences of zeros and ones. Note that N is a subset of ℓ_∞. Show that the subspace metric of N is the discrete metric as defined in Example 5.2.

5.1.14 Prove the assertion made in Example 5.5 that if a continuous function $h \geq 0$ has the property $h(x_0) > 0$ for some $a \leq x_0 \leq b$, then $\int_a^b h(x)\,dx > 0$. Hint: Let $\epsilon = h(x_0)/2$. Then there exists $\delta > 0$ such that $h(x) > \epsilon/2$ whenever $x_0 - \delta < x < x_0 + \delta$. Show that $\int_a^b h(x)\,dx \geq \epsilon\delta > 0$.

5.1.15 Show that if d is a metric on a nonempty set M, then so is $2d$.

5.1.16 Show that if d_1 and d_2 are metrics on a nonempty set M, then so is $d_1 + d_2$.

5.1.17 Use Lemma 5.1 to prove Property **(M3)** for the Fréchet metric. First show that

$$\frac{|a-b|}{1+|a-b|} \leq \frac{|a|}{1+|a|} + \frac{|b|}{1+|b|}.$$

Then let $a = x - z$ and $b = z - y$ to derive **(M3)**.

5.1.18 The notion of the Fréchet metric as described in Example 5.6 can be used to combine a countable number of metrics. Let M have a countable number of metrics $\{d_1, d_2, \ldots\}$. Show that the following is a metric on M:

$$d(x, y) = \sum_{k=1}^{\infty} \frac{d_k(x, y)}{2^k\big(1 + d_k(x, y)\big)} \text{ for all } x, y \in M.$$

5.1.19 Show that for any metric d, we have $|d(x, z) - d(z, y)| \leq d(x, y)$.

5.1.20 Recall Definition (0.15) given in Section 0.4,

$$\ell_p = \left\{ x \in \omega \,\middle|\, \sum_{k=1}^{\infty} |x_k|^p < \infty \right\}, \quad \text{where} \quad 1 \leq p < \infty.$$

Use the inequalities in Section 0.13 to show that ℓ_p is a metric space under the distance function

$$d(x, y) = \left(\sum_{k=1}^{\infty} |x_k - y_k|^p \right)^{1/p}.$$

5.2 Normed Spaces

In Section 0.5 we considered the Euclidean norm on $\mathbb{R}^n$, and in Section 0.13 we introduced the p-norm on $\mathbb{R}^n$ and $\mathbb{C}^n$. Here we consider norms in general linear spaces.

A **normed space** (for emphasis, **normed linear space**) consists of a linear space V and a real valued "norm function" $\|\cdot\|$ that satisfies the following properties for all $x, y \in V$ and all scalars α:

(N0) $\|x\| \geq 0$.
(N1) $\|x\| = 0$ if and only if $x = 0$.
(N2) $\|\alpha x\| = |\alpha| \|x\|$ (absolute homogeneity).
(N3) $\|x + y\| \leq \|x\| + \|y\|$ (triangle inequality; illustrated in Figure 5.1).

Actually, **(N0)** can be proven from the other conditions as follows:

$$0 = \|0\| = \|x - x\| \leq \|x\| + |-1|\|x\| = 2\|x\|.$$

Every normed space is a metric space with the metric

$$d(x, y) = \|x - y\|. \tag{5.1}$$

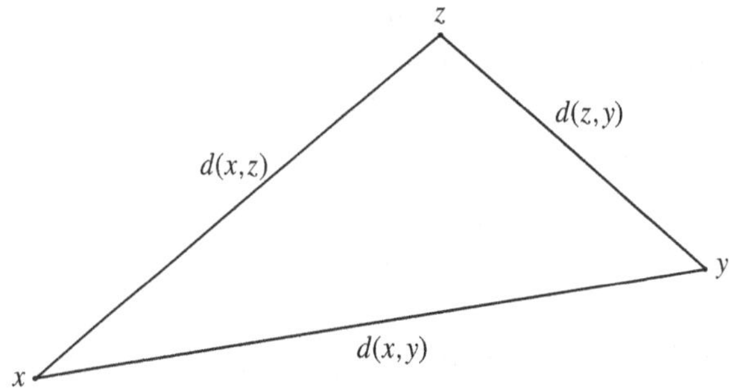

Figure 5.1 The triangle inequality Property **(N3)** in $\mathbb{R}^2$.

It is easy to verify that this satisfies metric properties **(M0)–(M3)**.[2] In fact most metrics found in analysis are derived from norms. However, not *all* metrics arise from norms. Indeed, every normed space must be a linear space but not all metric spaces are linear spaces. Moreover, a metric on a linear space need not arise from a norm (Exercise 5.2.8).

Below are some examples of normed spaces.

Example 5.7 Usual norm on $\mathbb{R}$: For each $x \in \mathbb{R}$, let $\|x\| = |x|$.

Example 5.8 Usual norm on $\mathbb{C}$: For each $z = (x, y) \in \mathbb{C}$, let $\|z\| = |z| = \sqrt{x^2 + y^2}$.

Example 5.9 Euclidean norm: For each $x = (x_1, x_2, \ldots, x_n) \in \mathbb{R}^n$, let

$$\|x\|_2 = \sqrt{x_1^2 + \cdots + x_n^2}.$$

Example 5.10 Unitary norm: For each $x = (z_1, z_2, \ldots, z_n) \in \mathbb{C}^n$, let

$$\|x\|_2 = \sqrt{|z_1|^2 + \cdots + |z_n|^2}.$$

Example 5.11 The space $\ell_p(n)$: For each $x = (x_1, x_2, \ldots, x_n) \in \mathbb{R}^n$ (or $\mathbb{C}^n$), define

$$\|x\|_p = \left(\sum_{k=1}^{n} |x_k|^p\right)^{1/p} \quad \text{for } 1 \le p < \infty \quad \text{and} \quad \|x\|_\infty = \sup_{1 \le k \le n} |x_k|.$$

We denote this n-dimensional normed space by $\ell_p(n)$ to distinguish it from the infinite-dimensional normed space ℓ_p. The triangle inequality follows from Minkowski's inequality (Theorem 0.56).

[2] In addition, all normed spaces satisfy the translation property $d(x + z, y + z) = d(x, y)$ for all $x, y, z \in V$.

Example 5.12 The sequence space ℓ_p: Let $V = \ell_p \subset \omega$ for $1 \le p \le \infty$ with the norm

$$\|x\|_p = \left(\sum_{k=1}^{\infty} |x_k|^p\right)^{1/p} \quad \text{if } 1 \le p < \infty \quad \text{and} \quad \|x\|_\infty = \sup_k |x_k|,$$

as defined in Section 0.13. The triangle inequality follows from Minkowski's inequality (Theorem 0.56).

Example 5.13 $C[a,b]$: This is the space of all continuous real value functions on the closed interval $[a,b]$. The usual norm on $C[a,b]$ is the **uniform norm**

$$\|f\|_\infty = \sup_{a \le x \le b} |f(x)|.$$

Although this norm has the same notation as the supremum norm $\|\cdot\|_\infty$ in the previous two examples, there should be no confusion because we use the notation f for functions and x for sequences.

Example 5.14 The same function space $C[a,b]$ is also normed under the L^1 norm

$$\|f\|_1 = \int_a^b |f(x)|\, dx.$$

As in Example 5.5, the positivity condition **(N1)** follows from the fact that, if a function $h \ge 0$ is continuous and $h(x_0) > 0$ for some $a \le x_0 \le b$, then $\int_a^b h(x)\, dx > 0$. The other norm conditions are easy to prove.

Example 5.15 Subspace norm: Any subspace W of a normed linear space V is also a normed linear space under the same norm as V. For example, the space of bounded sequences ℓ_∞ is a normed space under the supremum norm $\|x\|_\infty = \sup_k |x_k|$; so the subspace of convergent sequences c is also a normed space under the same norm $\|x\|_\infty = \sup_k |x_k|$.

Exercises for Section 5.2

5.2.1 Show that discrete metric on a vector space V cannot be obtained from a norm except for the zero-dimensional case $V = \{0\}$.

5.2.2 Show that the real function $f(x,y) = \left(\sqrt{x} + \sqrt{y}\right)^2$ does not define a norm on $\mathbb{R}^2$.

5.2.3 Show that $C[a,b]$ with the uniform norm $\|f\|_\infty = \sup_{a \le x \le b} |f(x)|$ satisfies the conditions for a normed space.

5.2.4 Show that ℓ_∞ is a normed space with the supremum norm as given in Example 5.12.

5.2.5 For $1 \le p < \infty$ show that ℓ_p is a normed space with the norm $\|\cdot\|_p$ as given in Example 5.12.

5.2.6 Show that the space bs (defined in Section 0.4, Equation (0.12)), is a linear space and that $\|x\|_{bs} = \sup_n \left|\sum_{k=1}^n x_k\right|$ is a norm on this space.

5.2.7 Continuing, show that the space cs (defined in Equation (0.11), Section 0.4), is a subspace of bs and hence a normed space under the same norm $\|\cdot\|_{bs}$.

5.2.8 Show that the Fréchet metric (Example 5.6) on ω does not arise from a norm because the absolute homogeneity property **(N2)** fails.

5.2.9 Show that the space $bv = \{x \in \omega \mid \sum_{k=1}^{\infty} |x_k - x_{k+1}| < \infty\}$ is a linear space and that $\|x\|_{bv} = \sum_{k=1}^{\infty} |x_k - x_{k+1}| + \sup_j |x_j|$ is a norm on this space.

5.2.10 Referring to Exercise 5.2.9, show that $\ell_1 \subset bv \subset c$.

5.2.11 Suppose that V_1 and V_2 are normed spaces under norms $\|\cdot\|_1$ and $\|\cdot\|_2$, respectively. Show that the product space $V_1 \times V_2$ of ordered pairs $x = (x_1, x_2)$ is a normed space under $\|x\| = \max\{\|x_1\|_1, \|x_2\|_2\}$.

5.3 Seminorms

There are important linear spaces that satisfy all conditions for a normed space except for **(N1)**. Such spaces are called **seminorm spaces** defined as follows.

Definition 5.1 A **seminorm space**[3] consists of a vector space V and a function p, called a **seminorm**, satisfying the following, for all $x, y \in V$ and scalars α:

(SN0) $p(x) \ge 0$.
(SN1) $p(\alpha x) = |\alpha| p(x)$.
(SN2) $p(x + y) \le p(x) + p(y)$.

The fact $p(0) = 0$ follows from condition **(SN1)**.

Property **(SN0)** can actually be proven from **(SN1)** and **(SN2)** as follows: $0 = p(0) = p(x - x) \le p(x) + p(-x) = 2p(x)$.

The following example is typical of the concept of seminorm.

[3] One can also define a **semimetric space** as a nonempty set M with a distance function, called a **semimetric** satisfying

$$d(x,x) = 0, \quad d(x,y) = d(y,x), \quad d(x,y) \le d(x,z) + d(z,y),$$

for all $x, y, z \in M$. A semimetric space permits zero distance between distinct elements of M. We will not use semimetric spaces in this book.

Example 5.16 Let V consist of all integrable[4] functions on a fixed interval $[a,b]$ and let

$$p(f) = \int_a^b |f(x)|\,dx.$$

The function $p\colon V \longrightarrow \mathbb{R}$ clearly satisfies conditions for a norm on the space V except for condition **(N1)**. That **(N1)** fails can be seen by looking at the example

$$f(x) = \begin{cases} 0 & \text{if } a < x \le b \\ 1 & \text{if } \; x = a. \end{cases}$$

Clearly $f \neq 0$ but $p(f) = 0$. (Note that f is not continuous. If V were the smaller space $C[a,b]$ of all continuous function on $[a,b]$, then p would be a norm.)

Such examples prompt the following result, which shows how to make a normed space from a seminormed space.

Theorem 5.3 *Let V be a seminorm space and define the relation*

$$x \sim y \quad \textit{to mean} \quad p(x-y) = 0.$$

Then $\sim$ is an equivalence relation on V and the equivalence classes are

$$\langle x \rangle = \{y \in V \mid p(x-y) = 0\}.$$

The equivalence classes form a normed linear space if we define $\|\langle x \rangle\| = p(x)$.

The proof is left as Exercise 5.3.2.

Example 5.17 Lebesgue $L^1[a,b]$: The space $L^1[a,b]$ consists of all functions f on the closed interval $[a,b]$ for which the Lebesgue integral of $|f|$ is finite:

$$\int_a^b |f(x)|\,dx < \infty.$$

An overview of how Lebesgue integration differs from Riemann integration is given in Section 0.11. If we consider equivalence classes of such functions that differ only on a set of measure zero, then by Theorem 2.27, we have

$$f \sim g \text{ if and only if } \int_a^b |f(x) - g(x)|\,dx = 0.$$

[4] This may be with respect to either the Riemann integral or to the more general notion of the Lebesgue integral as covered in Chapter 2.

Thus $L^1[0,1]$ is a normed linear space under

$$\|f\|_1 = \int_a^b |f(x)|\,dx,$$

where it is understood that f is really an equivalence class of functions.

Example 5.18 The sequence space c: In Example 5.12, we considered the sequence space ℓ_∞ with the supremum norm $\|x\|_\infty = \sup_k |x_k|$. The subspace of convergent sequences c is thus also a normed space under the subspace norm. On the other hand, the function

$$p(x) = \lim_{k\to\infty} |x_k|$$

is not a norm. It is only a seminorm. For example, the sequence $x = \left(1, \frac{1}{2}, \frac{1}{3}, \ldots, \frac{1}{k}, \ldots\right) \neq 0$, yet $p(x) = 0$.

Exercises for Section 5.3

5.3.1 Show that any seminorm p satisfies $|p(x) - p(y)| \leq p(x - y)$.

5.3.2 Prove Theorem 5.3. Do not forget to show that $\|\langle x\rangle\|$ is well defined; that is, $x \sim y \implies \|\langle x\rangle\| = \|\langle y\rangle\|$.

5.3.3 Show that the function p in Example 5.18 is a seminorm.

5.3.4 Suppose V is a normed space under norm $\|\cdot\|$. If p is a seminorm on V, show that $\|x\|' = \|x\| + p(x)$ defines a norm on V.

5.3.5 Consider a seminormed space V with seminorm p. Show that the function $d(x,y) = p(x-y)$ defines a semimetric as described in footnote 3 (Section 5.3).

5.3.6 Generalize the Fréchet metric as described in Example 5.6. Let M be a nonempty set with a countable number of seminorms $\{p_1, p_2, \ldots\}$ having the property that

$$p_k(x) = 0\ \forall\, k = 1, 2, \ldots \implies x = 0.$$

Show that the following real- valued-function is a metric on M,

$$d(x,y) = \sum_{k=1}^{\infty} \frac{p_k(x,y)}{2^k\left(1 + p_k(x,y)\right)} \quad \text{for all } x, y \in M.$$

5.3.7 Let X be a semimetric space as described in footnote 3 (Section 5.3). Show how to take equivalence classes to form a metric space. Follow the method used for seminorm spaces in Theorem 5.3.

5.4 Convergence in Metric Spaces

Convergent sequences and Cauchy sequences are defined in metric spaces just as they are in $\mathbb{R}$ and $\mathbb{R}^n$.

Definition 5.2 A sequence $(x^k) = (x^1, x^2, \ldots)$ in a metric space M is said to be **convergent** if there exists $x \in M$ such that $d(x^k, x) \to 0$ as $k \to \infty$. We say that the sequence has **limit** x. A **Cauchy sequence** is one for which $d(x^k, x^j) \to 0$ as $k, j \to \infty$.

Theorem 5.4 *A convergent sequence has a unique limit.*

Proof Suppose that a sequence (x^k) has limits x and y in a metric space M; that is, $d(x^k, x) \to 0$ and $d(x^k, y) \to 0$ as $k \to \infty$. Then

$$0 \le d(x, y) \le d(x, x^k) + d(x^k, y) \to 0 + 0 = 0 \quad \text{as} \quad k \to \infty.$$

Thus $d(x, y) = 0$, which means $x = y$. ∎

Theorem 5.5 *Every convergent sequence is a Cauchy sequence.*

Proof Suppose that a sequence (x^k) has limit x. Then

$$0 \le d(x^k, x^j) \le d(x^k, x) + d(x, x^j) \to 0 + 0 = 0 \quad \text{as} \quad k, j \to \infty.$$

Thus $d(x^k, x^j) \to 0$ as $k, j \to \infty$. ∎

However, unlike the metric spaces $\mathbb{R}$ and $\mathbb{R}^n$, in some metric spaces, not every Cauchy sequence is convergent. This can be seen from the following two examples.

Example 5.19 The space $\mathbb{Q}$ as a metric subspace of $\mathbb{R}$: Define the Cauchy sequence of rationals $(r_0, r_1, r_2, \ldots)$ by $r_0 = 1, r_1 = 1.4, r_2 = 1.41, r_3 = 1.414$, $r_4 = 1.4142, \ldots$. That is, the rule is to let r_k be $\sqrt{2}$ expanded to the kth decimal. Assume that this sequence converges to some $r \in \mathbb{Q}$. Then this sequence also converges to r in the metric space $\mathbb{R}$. But clearly the sequence converges to the irrational $\sqrt{2}$ in $\mathbb{R}$, since $|r_k - \sqrt{2}| < 10^{-k}$. By Theorem 5.4, $r = \sqrt{2}$. Since $\sqrt{2} \notin \mathbb{Q}$, this is a contradiction.

Example 5.20 The open interval $M = (0, 1)$ as a metric subspace of $\mathbb{R}$: The sequence $\left(1, \frac{1}{2}, \frac{1}{3}, \ldots\right)$ is a Cauchy sequence that converges to 0 in $\mathbb{R}$. Since $0 \notin M$, the sequence cannot converge to any other x of M by Theorem 5.4.

Theorem 5.6 *If a Cauchy sequence in a metric space M has a subsequence converging to x, then the entire sequence converges to x.*

Proof Suppose sequence (x^k) has subsequence (x^{k_n}) that converges to x. Let $\epsilon > 0$ be given. Then for some $N_1 > 0$ we have $d(x, x^{k_n}) < \epsilon/2$ whenever $n > N_1$. Since (x^k) is Cauchy, for some $N > 0$ we have $d(x^k, x^j) < \epsilon/2$ whenever $k, j > N$. Let $k > N$. Choose any $k_n > N$ with $n > N_1$. Then

$$d(x, x^k) \leq d(x, x^{k_n}) + d(x^{k_n}, x^k) < \epsilon/2 + \epsilon/2 = \epsilon.$$

This shows that x^k converges to x in M. ∎

Definition 5.3 The **diameter** of a nonempty set A in a metric space M is

$$\operatorname{diam}(A) = \sup_{x,y \in A} d(x, y).$$

A nonempty set A is said to be **bounded** if $\operatorname{diam}(A) < \infty$.

Theorem 5.7 *Every Cauchy sequence (x^k) in a metric space M forms a bounded set in M.*

Proof Let $A = \{x^1, x^2, \ldots\}$ be the set consisting of the elements of the Cauchy sequence (x^k). For $\epsilon = 1$, there exists $N > 0$ such that $d(x^{N+1}, x^j) < 1$ for $j > N$. Define $r = \max\{d(x^1, x^{N+1}), d(x^2, x^{N+1},), \ldots, d(x^N, x^{N+1}), 1\}$. Then $\operatorname{diam}(A) \leq 2r$ because

$$d(x^k, x^j) \leq d(x^k, x^{N+1}) + d(x^{N+1}, x^j) \leq r + r = 2r \quad \text{for all} \quad k, j. \qquad ∎$$

Remark Recall that in $\mathbb{R}^n$, a sequence x^k converges to $x \in \mathbb{R}^n$ if and only if the sequence converges coordinatewise (Theorem 0.7 in Section 0.5). This property does not hold for sequence spaces in general. Consider the sequence space ℓ_∞ of bounded sequences with usual norm $\|x\|_\infty = \sup_j |x_j|$. For each k, let $P^k = (1, 1, \ldots, 1, 0, 0, \ldots)$ be the sequence with values 1 in the first k coordinates and 0 elsewhere. Each P^k is an element of ℓ_∞ with $\|P^k\|_\infty = 1$. This sequence of elements of ℓ_∞ converges coordinatewise to the sequence $e = (1, 1, 1, \ldots)$ because for each fixed coordinate m we have

$$\lim_{k \to \infty} (P^k)_m = 1.$$

But $P^1, P^2, P^3, \ldots$ is not Cauchy because, for all $k \neq j$, $\|P^k - P^j\|_\infty = 1$.

Exercises for Section 5.4

5.4.1 Every Cauchy sequence in a metric space is bounded. Give an example of a metric space M and a bounded sequence in M that is not Cauchy.

5.4.2 Show that in $\mathbb{Q}$, the following sequence is Cauchy but not convergent

$$x_1 = 0.1,\ x_2 = 0.101,\ x_3 = 0.101001,\ x_4 = 0.1010010001, \ldots.$$

5.4.3 Suppose that $x^k \to x$ and $y^k \to y$ in a metric space M. Show that $d(x^k, y^k) \to d(x, y)$ in $\mathbb{R}$.

5.4.4 Suppose that x^k and y^k are Cauchy sequences in a metric space M. Show that the sequence $d(x^k, y^k)$ converges in $\mathbb{R}$.

5.4.5 Suppose d and d' are metrics on M and there exist positive numbers a and b such that

$$ad(x, y) \leq d'(x, y) \leq bd(x, y) \quad \text{for all} \quad x, y \in M.$$

Show that the Cauchy sequences are the same for both metrics.

5.4.6 Show that for the Fréchet metric (defined in Example 5.6) on ω, a sequence $x^k = (x_1^k, x_2^k, \ldots)$ converges to $x = (x_1, x_2, \ldots) \in \omega$ if and only if the sequence converges coordinatewise; that is, for each fixed coordinate $m = 1, 2, 3, \ldots$, we have $x_m^k \to x_m$ as $k \to \infty$. Note that $x^1, x^2, \ldots$ is a sequence of sequences.

5.5 Topological Concepts

Our first task here is to define open sets in metric spaces. Of course, since every normed linear space is a metric space, this applies to normed linear spaces as well.

Definition 5.4 For every x in a metric space M and $0 < r < \infty$, we define with the **open sphere** with **center** x and **radius** r to be the set

$$S(x, r) = \{y \in M \mid d(x, y) < r\}.$$

Theorem 5.8 *In a normed linear space V, the translation of any open sphere is an open spheres. Namely, for any $S(x, r) = \{y \in M \mid d(x, y) = \|x - y\| < r\}$, we have*

$$y + S(x, r) = S(x + y, r).$$

Proof See Figure 5.2. $z \in y + S(x, r) \iff z - y \in S(x, r)$. In a normed space, this is equivalent to $\|x - (z - y)\| = \|(x + y) - z\| < r$, or $z \in S(x + y, r)$. ∎

Example 5.21 Euclidean space $\mathbb{R}^n$: For dimension $n = 1$, the open sphere $S(x, r)$ is the open interval $(x - r, x + r)$. For $n = 2$, it is the open disk with center $x = (x_1, x_2)$ and radius r. For $n = 3$, it is the open ball with center at $x = (x_1, x_2, x_3)$ and radius r.

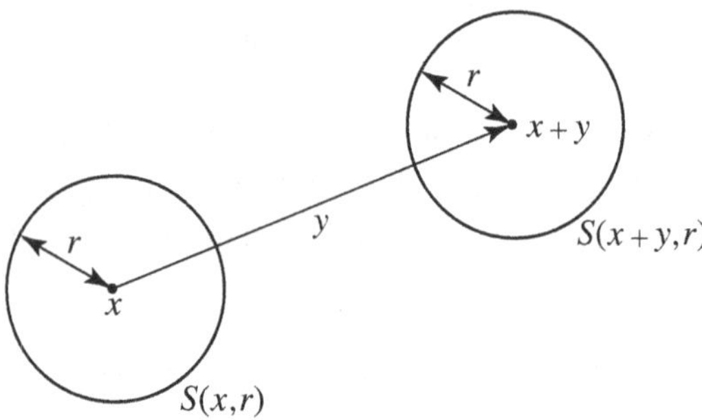

Figure 5.2 Translation of a sphere.

Example 5.22 The space $\ell_p(n)$: For the real case, $n = 2$, and $p = 1, 4/3$, 2, 4 and ∞, the boundaries of the unit spheres $S(0,1)$ are shown in Figure 5.3. The square frame on the outside is the boundary of the sphere for $\ell_p(2)$ with $p = \infty$. The diamond shape on the inside is the boundary of the sphere with $p = 1$. The unit circle is the boundary of the sphere with $p = 2$.

Example 5.23 Function space $C[a,b]$: In the normed space $C[a,b]$ with

$$d(f,g) = \|f - g\|_\infty = \sup_{a \le x \le b} |f(x) - g(x)|,$$

the open sphere $S(f,r)$ consists of all continuous functions g that differ from the function f by less than r on $[a,b]$. That is,

$$S(f,r) = \{g \in C[a,b] \mid f(x) - r < g(x) < f(x) + r \text{ for all } a \le x \le b\}.$$

Example 5.24 An open sphere $S(x,r)$ is always nonempty because it always contains the center x. In a *discrete* metric space M, if $0 < r \le 1$, then the open sphere $S(x,r)$ contains only one point $S(x,r) = \{x\}$ but, if $r > 1$, then $S(x,r)$ consists of the entire space $S(x,r) = M$.

Definition 5.5 A point x in a metric space M is an **interior point** of a set $A \subset M$ if there exists an open sphere $S(x,r)$, with $r > 0$, such that $S(x,r) \subset A$.

A point x is **exterior** to a set A if there exists an open sphere $S(x,r)$, with $r > 0$, such that $S(x,r) \subset A^c = M - A$.

A point x is a **boundary point** of A if it is neither an interior point of A nor exterior to A. That is, for a boundary point x of A, *every* open sphere $S(x,r)$ about x intersects both A and A^c.

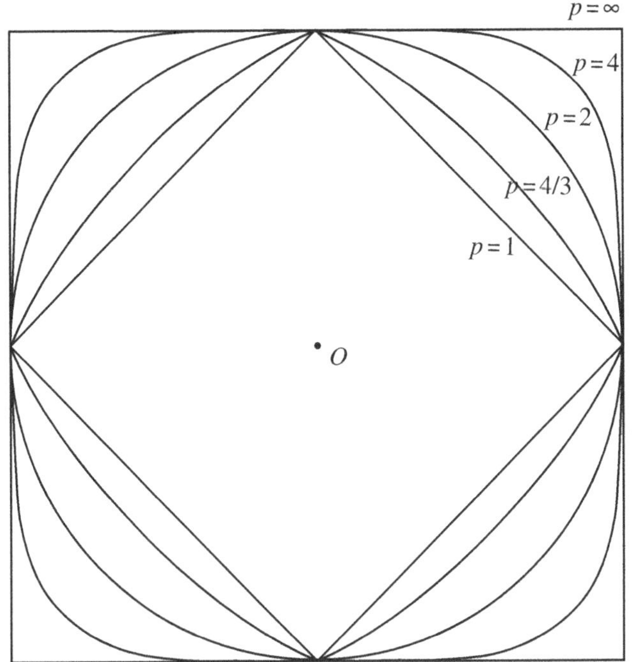

Figure 5.3 Boundaries of spheres $S(0,1) = \{x \in \mathbb{R}^2 \mid \|x\|_p < 1\}$ for $p = 1$, $4/3$, 2, 4, and ∞.

Each set $A \subset M$ is thus associated with three sets, A°, $\mathrm{Ext}(A)$, and ∂A, defined as follows.

- A° is the set of all interior points of A, called the **interior** of A.
- $\mathrm{Ext}(A)$ is the set of points exterior to A, called the **exterior** of A.
- ∂A is the set of boundary points of A, called the **boundary** of A.

Theorem 5.9 *For any set A in a metric space M, we have the following properties:*

(1) *The sets* A°, $\mathrm{Ext}(A)$, *and* ∂A *are pairwise disjoint.*
(2) $A^\circ \cup \mathrm{Ext}(A) \cup \partial A = M$.
(3) $\partial A = \partial(A^c)$.
(4) $\mathrm{Ext}(A) = (A^c)^\circ$.
(5) $A^\circ = A - \partial A$.

The proofs of these properties follow easily from the definitions.

Definition 5.6 An **open set** G in a metric space M is a set for which all of the points of G are interior points; namely $G = G^\circ$.

A **neighborhood** of a point $x \in M$ is any set $A \subset M$ such that there is an open set G with $x \in G \subset A$.

We follow tradition by denoting an arbitrary open set with the letter G.

Theorem 5.10 *A set G is open if and only if it contains none of its boundary points; that is,*

$$G \cap \partial G = \emptyset.$$

Proof This follows from Property **(1)** of Theorem 5.9. ∎

Theorem 5.11 *Every open sphere $S(x,r)$ is an open set; that is, $\big(S(x,r)\big)^\circ = S(x,r)$.*

Proof The proof is essentially the same as that of Theorem 0.12. ∎

The basic properties of open sets are given below. The proofs are omitted since they are essentially the proofs of Theorem 0.13.

Theorem 5.12 *For any metric space M, let $\mathcal{T}$ be the collection of all open subsets of M. The collection $\mathcal{T}$ has the following properties:*

(T1) $\emptyset \in \mathcal{T}$ *and* $M \in \mathcal{T}$.
(T2) *The union of any collection (countable or not) of elements of $\mathcal{T}$ is also an element of $\mathcal{T}$.*
(T3) *The intersection of a* finite *collection of elements of $\mathcal{T}$ is also an element of $\mathcal{T}$.*

Definition 5.7 So far, we have used the term topology to refer to the study of open and closed sets, boundaries and limits, continuity, convergence, and compactness in Euclidean spaces, metric spaces, and normed spaces. Now we give a formal definition. Any collection $\mathcal{T}$ of subsets of a set X satisfying the properties **(T1)–(T3)** is called a **topology** of X. We then refer to X (along with the collection $\mathcal{T}$) as a **topological space**. The sets in $\mathcal{T}$ are called the **open sets** of the topology. By this definition and by Theorem 5.12, every metric space is a topological space. Although, for *metric spaces*, the word 'open' is used in a specific sense, for a *general topological space* X, the 'open' sets are simply the sets in $\mathcal{T}$. The subject of topology is interesting and very abstract. However, in this book we will use the term topology only to refer to *the collection of open sets* in a metric (or normed) space.

Example 5.25 For any set X with the discrete metric, every subset is open. Thus the topology consists of all subsets of X: $\mathcal{T} = 2^X$.

Definition 5.8 If a set X has two topologies $\mathcal{T}$ and $\mathcal{T}'$ (both satisfying properties **(T1)–(T3)**) and if $\mathcal{T}' \subset \mathcal{T}$, we say that $\mathcal{T}'$ is a **weaker topology than $\mathcal{T}$.**

Theorem 5.13 *Suppose that a linear space V has two norms $\|\cdot\|$ and $\|\cdot\|'$. The topology $\mathcal{T}'$ generated by the norm $\|\cdot\|'$ is weaker than the topology $\mathcal{T}$ generated by $\|\cdot\|$ if and only if there exists a number $M > 0$ such that*

$$\|x\|' \leq M\|x\| \quad \textit{for all} \quad x \in V. \tag{5.2}$$

The norm $\|\cdot\|$ is said to **dominate** *the norm $\|\cdot\|'$.*

Proof Let open spheres for the two norms $\|\cdot\|$ and $\|\cdot\|'$ be denoted by $S(x,r)$ and $S'(x,r)$, respectively.

($\Rightarrow$): If $x = 0$, then (5.2) holds for all $M > 0$. So assume $x \neq 0$. Suppose $\mathcal{T}' \subset \mathcal{T}$. The open sphere $S'(0,1)$ is then open in $\mathcal{T}$. This means $S(0,r) \subset S'(0,1)$ for some $r > 0$. Since $\frac{r}{2}\frac{x}{\|x\|} \in S(0,r) \subset S'(0,1)$, we have $\left\|\frac{r}{2}\frac{x}{\|x\|}\right\|' < 1$. Thus $\|x\|' < \frac{2}{r}\|x\|$. This is (5.2) for $M = 2/r$.

($\Leftarrow$): From $\|x-y\|' \leq M\|x-y\|$, we get $S\big(x, \frac{r}{M}\big) \subset S'(x,r)$. So if x is an interior point of a set G under the norm $\|\cdot\|'$, it will be an interior point of G under the norm $\|\cdot\|$. Thus every open set under $\|\cdot\|'$ will be an open set under $\|\cdot\|$; that is, $\mathcal{T}' \subset \mathcal{T}$. ∎

Example 5.26 Clearly, we have $\|x\|_\infty \leq \|x\|_1$ for any $x \in \ell_1$. Thus, the topology on ℓ_1 generated by the ℓ_∞ norm is weaker than the usual topology generated by the ℓ_1 norm.

Theorem 5.14 *Two norms $\|\cdot\|$ and $\|\cdot\|'$ on the same linear space V result in the same topology if and only if there exist numbers $0 < m \leq M$ such that*

$$m\|x\| \leq \|x\|' \leq M\|x\| \ \textit{for all} \ x \in V. \tag{5.3}$$

We say that $\|\cdot\|$ and $\|\cdot\|'$ are ***equivalent norms.***

The proof uses two applications of Theorem 5.13.

Corollary 5.15 *Suppose that $\|\cdot\|$ and $\|\cdot\|'$ are equivalent norms on V. Then the two norms result in the same convergent sequences.*

Theorem 5.16 *The spaces $\ell_p(n)$ under the norms $\|\cdot\|_p$ all result in the same topology on $\mathbb{R}^n$ (or $\mathbb{C}^n$), regardless of the value of $1 \leq p \leq \infty$.*

Proof Let $x \in \mathbb{R}^n$ (or $\mathbb{C}^n$), $1 \le p < \infty$, and $a = \|x\|_\infty = \sup_{1\le k\le n} |x_k|$. Then

$$\|x\|_p = \left(\sum_{k=1}^{n} |x_k|^p\right)^{1/p} \le \left(\sum_{k=1}^{n} |a|^p\right)^{1/p} = n^{1/p} a = n^{1/p}\|x\|_\infty.$$

By Theorem 5.13, $\ell_p(n)$ has a weaker topology than $\ell_\infty(n)$.

Conversely, $\|x\|_\infty \le \|x\|_p$ follows from Jensen's inequality (Theorem 0.58) and Theorem 0.60. So

$$\|x\|_\infty \le \|x\|_p \le n^{1/p}\|x\|_\infty.$$

By Theorem 5.14, all $\ell_p(n)$ norms are equivalent to the $\ell_\infty(n)$ norm. ∎

Looking at Figure 5.3, it is clear that for $\mathbb{R}^2$ we have

$$\|x\|_\infty \le \|x\|_p \le \|x\|_1 \le \sqrt{2}\|x\|_\infty \quad \text{for all} \quad 1 \le p \le \infty.$$

As an extension of Theorem 5.16, we will show in Theorem 5.46 (Section 5.10) that *any* two norms on a finite-dimensional linear space V are equivalent.

Definition 5.9 A set F is said to be **closed** if its complement $G = F^c$ is open.

We follow tradition by denoting an arbitrary closed set with the letter F.

Theorem 5.17 *A set F is closed if and only if it contains all of its boundary; that is, $\partial F \subset F$.*

Proof It follows from Theorem 5.9 (Property **(3)**) and Theorem 5.10. ∎

Corollary 5.18 *In any metric space M, a singleton (that is, a set consisting of a single point) is always closed.*[5]

Proof Consider the set $A = \{x_0\}$ in M. If $y \ne x_0$, let $r = d(x_0, y) > 0$. The open spheres $S(x_0, r/2)$ and $S(y, r/2)$ do not intersect. Thus y is exterior to A. We conclude that $\partial A \subset A$ (actually, $\partial A = \{x_0\} = A$). ∎

Theorem 5.19 *In any metric space M, the closed sets have the following properties:*

(TC1) *The sets $\emptyset$ and M are closed.*
(TC2) *The intersection of any collection (countable or not) of closed sets is closed.*
(TC3) *The union of a* finite *collection of closed sets is closed.*

[5] This theorem is not true for all general topological spaces. A topological space for which every singleton is closed is called a T_1 space. Every metric space is a T_1 space.

The proof, which involves Theorem 5.12 and De Morgan's laws (Theorem 0.1), is left as Exercise 5.5.10.

Theorem 5.20 *Let A be any subset of a metric space M. Then*

$$A^\circ = \bigcup\{G \mid G \text{ is open and } G \subset A\}.$$

Thus the interior A° is open and it is the largest open set contained in A.

The proof is essentially that of Theorem 0.16.

Definition 5.10 The **closure** of a set A is $\overline{A} = A \cup \partial A$.

Theorem 5.21 *Let A be any subset of a metric space M. Then*

$$\overline{A} = \bigcap\{F \mid F \text{ is closed and } A \subset F\}.$$

Thus the closure $\overline{A}$ is closed and the smallest closed set containing A.

The proof is left as Exercise 5.5.21.

Definition 5.11 A set A is said to be **dense** in a metric space M if $\overline{A} = M$. A metric space M is said to be **separable** if it has a countable dense subset.

Example 5.27 Every countable metric space is separable.

Example 5.28 The normed spaces $\mathbb{R}^n$ and $\mathbb{C}^n$ are separable because the subset $\mathbb{Q}^n$ of all points with all rational coordinates is countable and dense in the space.

Example 5.29 ℓ_∞ is not separable: Consider any countable subset $A = \{x^1, x^2, x^3, \ldots\}$ of ℓ_∞. If we can find a point $y \in \ell_\infty$ such that $d(y, x^k) \geq 1$ for all $x^k \in A$, then the set A cannot be dense in ℓ_∞. To construct y, we use a Cantor diagonalization argument. Define the kth coordinate of y as follows.

$$y_k = \begin{cases} x_k^k + 1 & \text{if } |x_k^k| \leq 1 \\ 0 & \text{if } |x_k^k| > 1. \end{cases}$$

Clearly y is a bounded sequence; that is, $y \in \ell_\infty$. Then, for each k,

$$d(y, x^k) = \sup_j |y_j - x_j^k| \geq |y_k - x_k^k| = \left\{ \begin{array}{ll} 1 & \text{if } |x_k^k| \leq 1 \\ |x_k^k| & \text{if } |x_k^k| > 1 \end{array} \right\} \geq 1.$$

Example 5.30 $C[a,b]$ is separable: A real polynomial is a function of the form

$$p(x) = a_0 + a_1 x + a_2 x^2 + \cdots + a_n x^n,$$

where n is a nonnegative integer and $a_0, a_1, \ldots, a_n$ are real numbers. The set of polynomials restricted to closed and bounded interval $[a,b]$ is denoted by $P[a,b]$. The Weierstrass approximation theorem says that $P[a,b]$ is dense in $C[a,b]$, under the uniform norm $\|f\|_\infty = \sup_{a \le x \le b} |f(x)|$ (see Corollary 5.59 in Section 5.12 or Theorem 7.59 in Section 7.8). If we restrict $P[a,b]$ to polynomials with rational coefficients, $a_0, a_1, \ldots, a_n$, then this set is countable and dense in $C[a,b]$.

Definition 5.12 On the opposite extreme, a set A is **nowhere dense** in a metric space M if $(\overline{A})^c$ is dense in M.

Theorem 5.22 *The following statements are equivalent for a set A in a metric space M:*

(a) *A is nowhere dense in M.*
(b) *$\overline{A}$ is nowhere dense in M.*
(c) *$\overline{A}$ has empty interior.*
(d) *$\overline{A}$ contains no open spheres.*

Proof ***(a)*** $\Leftrightarrow$ ***(b)***: This is clear from the definition.
(b) $\Leftrightarrow$ ***(c)***: The set $(\overline{A})^c$ is dense in M if and only if $(\overline{A})^c$ has no exterior points, which is equivalent to $(\overline{A})$ having no interior points.
(c) $\Leftrightarrow$ ***(d)***: This is obvious from the definition of the interior. ∎

Example 5.31 The empty set is nowhere dense in any metric space M.

Example 5.32 In a metric space M with the discrete metric, the empty set is the only nowhere dense set.

Example 5.33 The set of integers $\mathbb{N}$ is nowhere dense in $\mathbb{R}$.

Definition 5.13 Let A be a set in a metric space M. A point $x \in M$ (not necessarily a point of A) is said to be a **limit point** of A if for every $r > 0$, there exists a point $y \neq x$ such that $y \in A \cap S(x,r)$.

Theorem 5.23 *A point $x \in M$ is a limit point of a set A if and only if for every $r > 0$, the open sphere $S(x,r)$ contains infinitely many points of A.*

The proof is left as Exercise 5.5.28.

Theorem 5.24 *A set F is closed if and only if it contains all of its limit points.*

Proof The proof is essentially the same as that of Theorem 0.19 (Section 0.6). ∎

Exercises for Section 5.5

5.5.1 Give an example of a set in $\mathbb{R}^2$ that is neither open nor closed.

5.5.2 Give an example to show that the word 'finite' is needed in Property **(T3)** of Theorem 5.12.

5.5.3 Give an example to show that the word 'finite' is needed in Property **(TC3)** of Theorem 5.19.

5.5.4 Let $f(x) = \sin x$ in $C[-\pi, \pi]$. Describe the open sphere $S(f, 1)$.

5.5.5 Show that for any set A in a metric space, we have $(A^\circ)^\circ = A^\circ$.

5.5.6 A closed sphere is a set of the form $S[x,r] = \{x \in M \mid d(x, y) \leq r\}$. Show that a closed sphere is a closed set.

5.5.7 Referring to Exercise 5.5.6, show that $\overline{S(x,r)} \subset S[x,r]$. Also show that for the discrete metric, the inclusion may be strict.

5.5.8 Prove Corollary 5.15.

5.5.9 For any singleton $A = \{x_0\}$, show that $\partial A = A$.

5.5.10 Prove Theorem 5.19.

5.5.11 Show that if $A \subset B$, then $A^\circ \subset B^\circ$.

5.5.12 Show that $(A \cap B)^\circ = A^\circ \cap B^\circ$.

5.5.13 Show that $A^\circ \cup B^\circ \subset (A \cup B)^\circ$ but that equality is not true in general.

5.5.14 Show that for any set A, we have $A \subset \overline{A}$ and $\overline{A} = \overline{\overline{A}}$.

5.5.15 Show that if $A \subset B$, then $\overline{A} \subset \overline{B}$.

5.5.16 Show that $\overline{A \cup B} = \overline{A} \cup \overline{B}$ and $\overline{A \cap B} \subset \overline{A} \cap \overline{B}$. Give an example where this last inclusion is strict.

5.5.17 Give an example of a set in $\mathbb{R}^2$ that has a boundary point that is not a limit point.

5.5.18 Give an example of a set in $\mathbb{R}^2$ that has a limit point that is not a boundary point.

5.5.19 Prove that for any set A, we have $\partial A = \overline{A} - A^\circ$.

5.5.20 Show that equivalent norms on a normed space V result in the same Cauchy sequences.

5.5.21 Prove Theorem 5.21.

5.5.22 Show that for any set A in a metric space M, we have $x \in \overline{A}$ if and only if there exists a sequence $x^1, x^2, \ldots$ in A that converges to x.

5.5.23 Show that Φ, the sequence space of all sequences with only a finite number of nonzero terms, is not a closed subspace of ℓ_1.

5.5.24 Show that ℓ_1 is not a closed subspace of ℓ_∞ under the usual norm $\|\cdot\|_\infty$.

5.5.25 Show that Φ, the sequence space of all sequences with only a finite number of nonzero terms, is a dense subspace of ℓ_1. What about the density of Φ in ℓ_p, for $1 < p \leq \infty$?

5.5.26 Let V be a normed linear space. Show that for any linear subspace W of V, the closure $\overline{W}$ of W is still a linear subspace of V.

5.5.27 Show that in $\mathbb{R}$, the only sets that are both open and closed are $\emptyset$ and $\mathbb{R}$.

5.5.28 Prove Theorem 5.23

5.5.29 Prove the assertion made in Example 5.30 that the polynomials with rational coefficients form a countable set that is dense in $C[a,b]$. You may assume the Weierstrass approximation theorem.

5.5.30 Show that x is a limit point of a set A if and only if there exists a sequence of distinct $x^k \in A$ such that $x^k \to x$ as $k \to \infty$.

5.5.31 Show that a nonempty set M with the discrete metric is separable if and only if M is countable.

5.5.32 Show that a metric space M is separable if and only if there exists a countable subset A with the following property. For every $\epsilon > 0$ and every $x \in M$, there exists $y \in A$ with $d(x,y) < \epsilon$.

5.5.33 Suppose a metric space M is separable under a metric d. Show that any subset $S \subset M$ is separable under the subspace metric. Be careful: A countable dense subset D of M may be disjoint from S; that is, $D \cap S = \emptyset$.

5.5.34 Show that for each $1 \leq p < \infty$, the space ℓ_p is separable.

5.5.35 Suppose that a sequence $x^1, x^2, x^3, \ldots$ converges to x in a normed space V with norm $\|\cdot\|$. Show that the sequence still converges under any other norm $\|\cdot\|'$ with a weaker topology.

5.5.36 Recall the definition of the sequence space bv as given in Exercise 5.2.9. Show that

$$\|x\|'_{bv} = \sum_{k=1}^{\infty} |x_k - x_{k-1}|$$

is also a norm on bv provided we use the convention in the $k = 1$ term $|x_k - x_{k-1}|$, that $x_0 = 0$. Show that this norm is equivalent to the usual norm

$$\|x\|_{bv} = \sum_{k=1}^{\infty} |x_k - x_{k+1}| + \sup_k |x_k|.$$

5.6 Continuity

Definition 5.14 Let M and M' be a metric spaces with metrics d and d', respectively. A function $f\colon M \longrightarrow M'$ is **continuous at** $x_0 \in M$ if for every $\epsilon > 0$, there exists $\delta = \delta(\epsilon, x_0) > 0$ such that $d'(f(x_0), f(x)) < \epsilon$ (or $f(x) \in S(f(x_0), \epsilon)$) whenever $d(x_0, x) < \delta$ (or $x \in S(x_0, \delta)$). If f is continuous at x_0 for every $x_0 \in M$, then we say f is **continuous on** M. If $\delta = \delta(\epsilon)$, does not depend on x_0, we say that f is **uniformly continuous on** M.

We will now characterize continuity in terms of open sets. This completely avoids ϵ and δ used in the definition above.

Definition 5.15 Let $f\colon M \longrightarrow M'$. For $A \subset M$, the **image** of A is this subset of M'

$$f(A) = \{f(x) \mid x \in A\} \subset M'.$$

For $B \subset M'$, the **inverse image** of B is the subset of M

$$f^{-1}(B) = \{x \in M \mid f(x) \in B\}.$$

Theorem 5.25 *Let $f\colon M \longrightarrow M'$, where M and M' are metric spaces. Then f is continuous on M if and only if, for every open set G in M', the inverse image $f^{-1}(G)$ is open in M.*

Proof (⇒): Suppose $f\colon M \longrightarrow M'$ is continuous and let G be an open subset of M'. We want to show that any $x_0 \in f^{-1}(G)$ is an interior point. Since G is open and $f(x_0) \in G$, there exists an open sphere $S(f(x_0), r) \subset G$. Letting $\epsilon = r > 0$, there exists $\delta > 0$ such that, whenever $x \in S(x_0, \delta)$, then $f(x) \in S(f(x_0), \epsilon) \subset G$. Thus the image of $S(x_0, \delta)$ is a subset of G, or $S(x_0, \delta) \subset f^{-1}(G)$. So every $x_0 \in f^{-1}(G)$ is an interior point.

(⇐): Consider $x_0 \in M$ and $\epsilon > 0$. Then $S(f(x_0), \epsilon)$ is an open subset of M'. Since the inverse image of every open set is open, $f^{-1}\big(S(f(x_0), \epsilon)\big)$ must be open. Then $x_0 \in f^{-1}\big(S(f(x_0), \epsilon)\big)$ is an interior point. This means there exists an open sphere $S(x_0, r) \subset f^{-1}\big(S(f(x_0), \epsilon)\big)$. Letting $\delta = r > 0$ completes the proof. ∎

Remark Now that we have the concept of continuity defined in terms of open sets, its concept can be extended to general topological spaces (Definition 5.7). That is, a function $f\colon X \longrightarrow Y$ from a topological space X to a topological space Y is called continuous if for every open set G in Y, the inverse image $f^{-1}(G)$ is open in X.

Definition 5.16 Let M and M' be metric spaces. A mapping $f: M \longrightarrow M'$ is said to be **sequentially continuous** at a point $x \in M$ if for every convergent sequence $x^k \to x$ in M, we have $f(x^k) \to f(x)$ in M'.

Theorem 5.26 *Let M and M' be metric spaces with $f: M \longrightarrow M'$. Then f is continuous if and only if f is sequentially continuous.*[6]

Proof ($\Rightarrow$): Suppose f is continuous and $x^k \to x$. Given $\epsilon > 0$, the inverse image of $S(f(x), \epsilon)$ is open in M. The inverse image contains x as an interior point. Thus there exists a $\delta > 0$ such that

$$f\big(S(x,\delta)\big) \subset S(f(x),\epsilon). \tag{5.4}$$

Since $x^k \to x$, there exists N such that $x^k \in S(x,\delta)$ whenever $k > N$. By (5.4), we have $f(x^k) \in S(f(x),\epsilon)$ whenever $k > N$. Thus $f(x^k) \to f(x)$.

($\Leftarrow$): Assume f is sequentially continuous on M but f is not continuous at some point $x \in M$. Then, for some $\epsilon > 0$ and every $\delta > 0$, there exists $y_\delta \in S(x,\delta)$ such that $d(f(x), f(y_\delta)) \geq \epsilon$. For each $k = 1, 2, \ldots$, let $\delta = \frac{1}{k}$ and denote y_δ by x^k. That is, for each $k = 1, 2, \ldots$, we have $x^k \in S\big(x, \frac{1}{k}\big)$ but $d(f(x), f(x^k)) \geq \epsilon$. Then x^k converges to x but $f(x^k)$ does not converge to $f(x)$. This contradicts the sequential convergence of f. ∎

Definition 5.17 In a metric space M, the **distance between a point x and a set A** is defined to be

$$d(x, A) = \inf_{y \in A} d(x, y).$$

The **distance between two sets A and B** is

$$d(A, B) = \inf_{x \in A,\ y \in B} d(x, y).$$

Theorem 5.27 *For each nonempty subset A of a metric space M, $f(x) = d(x, A)$ is a continuous function of x.*

The proof is essentially that of Theorem 0.30.

Theorem 5.28 Urysohn lemma: *Let M be a metric space. For every closed set F and open set G with $F \subset G$, there exists a continuous function $f: M \longrightarrow \mathbb{R}$ such that*

$0 \leq f(x) \leq 1$ *for all* $x \in M$,
$f(x) = 1$ *for all* $x \in F$, *and*
$f(x) = 0$ *for all* $x \in G^c$.

[6] This is a result that holds in metric space but *not* in all general topological spaces.

Proof If either F or G^c are empty, the result is trivial. Otherwise, it is easy to verify that the following continuous function satisfies all of the conditions,

$$f(x) = \frac{d(x, G^c)}{d(x, G^c) + d(x, F)}.$$

■

Exercises for Section 5.6

5.6.1 Show that for a fixed x_0 in a metric space M, the real-valued function $f(x) = d(x_0, x)$ is continuous on M.

5.6.2 Give an example of a continuous function and an open set G for which the image $f(G)$ is not open.

5.6.3 Let $f: M \longrightarrow M'$, where M and M' are metric spaces with corresponding metrics d and d'. Show that f is continuous on M if and only if, for every closed set F in M', the inverse image $f^{-1}(F)$ is closed in M.

5.6.4 Show that if $A \cap B \neq \emptyset$, then $d(A, B) = 0$. Yet find examples of two disjoint closed sets A and B in $\mathbb{R}^2$ for which $d(A, B) = 0$.

5.6.5 Let A be a nonempty subset of a metric space M. Show that $x \in \overline{A}$ if and only if $d(x, A) = 0$.

5.6.6 Let A be a nonempty subset of a metric space M. Show that for any $x, y \in M$, we have $|d(x, A) - d(y, A)| \leq d(x, y)$.

5.6.7 Even though the distance function $d(A, B)$ between subsets of a metric space M has some of the properties of a metric, show that d is *not* a metric on the family of subsets of M.

5.6.8 Let M be a metric space and $f: M \longrightarrow M$. We say that f is a **contraction map** if there exists a constant $0 \leq c < 1$ such that

$$d(f(x), f(y)) \leq c d(x, y) \quad \text{for all} \quad x, y \in M.$$

Show that a contraction map is always uniformly continuous.

5.7 Complete Metric Spaces

An important property of $\mathbb{R}$ is the completeness property (Section 0.2), which is equivalent to the Cauchy criterion for convergence (Theorem 0.5). We extend the concept of completeness to metric spaces.

Definition 5.18 A metric space M is called **complete** if every Cauchy sequence converges. That is, if $d(x^k, x^j) \to 0$ as $k, j \to \infty$, then there exists $x \in M$ such that $d(x^k, x) \to 0$ as $k \to \infty$.

Example 5.34 The metric space $\mathbb{R}$ is complete by the Cauchy criterion for convergence. On the other hand, Example 5.19 shows that the subset metric space $\mathbb{Q}$ is not complete.

Example 5.35 Discrete Metric: The discrete metric on any nonempty set M is always complete, because the only Cauchy sequences for the metric

$$d(x,y) = \begin{cases} 1 & : \quad \text{if } x \neq y \\ 0 & : \quad \text{if } x = y \end{cases}$$

are those that are eventually constant. All such sequences converge.

You are likely familiar with the nested interval theorem for $\mathbb{R}$ and $\mathbb{R}^n$ (Theorem 0.3). Below we extend the notion to metric spaces and show that it is equivalent to completeness.

Definition 5.19 Let M be a metric space. A **nest** of closed spheres consists of closed spheres

$$S_k = S[x^k, r_k] = \{x \in M \mid d(x, x^k) \leq r_k\}$$

with $S_1 \supset S_2 \supset \cdots \supset S_k \supset \cdots$ and $r_k \to 0$ as $k \to \infty$.

Theorem 5.29 Nested spheres: *A metric space M is complete if and only if every nest of closed spheres has a nonempty intersection.*

Proof ($\Rightarrow$): Let $S_1 \supset S_2 \supset \cdots$ be a nest of closed spheres as defined above. The centers x^k of the closed spheres clearly form a Cauchy sequence. By completeness, $\lim_k x^k = x$ for some $x \in M$. Since the spheres are closed and nested, we have $x \in S_k$ for all k.

($\Leftarrow$): Consider a Cauchy sequence (x^k). For $\epsilon = 2^{-2}$, find $k_1 > 0$ such that $d(x^k, x^{k_1}) < 2^{-2}$ whenever $k >, k_1$. Similarly, for $n = 1, 2, \ldots$ and $\epsilon = 2^{-(n+1)}$, we can find $0 < k_1 < k_2 < \cdots < k_n < \cdots$, such that $d(x^k, x^{k_n}) < 2^{-(n+1)}$ whenever $k > k_n$. Define the closed spheres $S_n = S[x^{k_n}, 2^{-n}]$. If $x \in S_2$, then $d(x, x^{k_1}) \leq d(x, x^{k_2}) + d(x^{k_2}, x^{k_1}) < 2^{-2} + 2^{-2} = 1/2$. Thus $S_2 \subset S_1$. Repeating the argument, we obtain a nest $S_1 \supset S_2 \supset S_3 \supset \cdots$. By hypothesis, $\bigcap S_n \neq \emptyset$. Let $x \in \bigcap S_n$ (actually $\bigcap S_n$ contains only a single point). The subsequence x^{k_n} converges to x. That is, the sequence (x^k) has a convergent subsequence. By Theorem 5.6, the Cauchy sequence (x^k) converges. ∎

Exercises for Section 5.7

5.7.1 Let a, b be real numbers $-\infty < a < b < \infty$. Show that the closed interval $[a, b]$ is a complete subspace of $\mathbb{R}$, whereas the open interval (a, b) is not.

5.7.2 Show that the space ω with the Fréchet metric (Example 5.6) is complete.

5.7.3 Let c_0 be the sequence space of all null sequences with the usual metric $d(x, y) = \sup_k |x_k - y_k|$. Let Φ be the subspace consisting of all sequences with only a finite number of nonzero terms. Show that Φ is not a complete subspace of c_0.

5.7.4 Suppose M is a complete metric space and A is a nonempty subset of M. Show that A is complete under the subspace metric if and only if A is closed in M.

5.7.5 Suppose M is a complete metric space and A is a nonempty subset of M. Show that the completion of A is its closure in M.

5.7.6 Consider the natural numbers $\mathbb{N}$ with the metric $d(x, y) = \left|\frac{1}{x} - \frac{1}{y}\right|$. Find a Cauchy sequence that does not converge.

5.7.7 Is the metric given in Example 5.4 complete?

5.7.8 In the proof of the nested spheres theorem (Theorem 5.29), prove the assertion that, if the intersection of the nest is nonempty, then the intersection consists of exactly one point.

5.8 Banach Spaces

Definition 5.20 A complete normed linear space V is called a **Banach space.**[7]

Example 5.36 The Euclidean spaces $\mathbb{R}^n$ and the unitary space $\mathbb{C}^n$ are finite-dimensional Banach spaces.

[7] Named for the Polish mathematician Stefan Banach (1892–1945), whose 1920 dissertation contains the axioms for a complete normed linear space. Banach was one of the founders of the general theory of functional analysis and is one of the most important mathematicians of the twentieth century. His 1932 book *Théorie des opérations linéaires* was the first book on functional analysis. He frequented the Scottish Café in Lwów, Poland, where mathematicians met to discuss unsolved problems. They would write in pencil directly onto the marble table tops. Later, Banach's wife presented the group with a notebook that was kept at the café. The notebook with problems and solutions survived the war.

Theorem 5.30 *The space $C[a,b]$ is a Banach space under the supremum norm*

$$\|f\|_\infty = \sup_{a \le x \le b} |f(x)|.$$

Proof Suppose that f_n is a Cauchy sequence under uniform convergence in $C[a,b]$ as defined by the supremum norm. For each $x \in [a,b]$, we have

$$|f_n(x) - f_m(x)| \le \|f_n - f_m\|_\infty \to 0 \quad \text{as} \quad n,m \to \infty.$$

This means that for each $x \in [a,b]$, $f_n(x)$ is a Cauchy sequence in $\mathbb{R}$. By the completeness of $\mathbb{R}$, the limit exists, which we denote by $f(x) = \lim_{n\to\infty} f_n(x)$. Let $\epsilon > 0$ and suppose $\|f_n - f_m\|_\infty < \epsilon$ for all n, $m > N$. Let $m > N$. Then

$$\begin{aligned} |f(x) - f_m(x)| &= \left| \lim_{n\to\infty} f_n(x) - f_m(x) \right| = \lim_{n\to\infty} |f_n(x) - f_m(x)| \\ &\le \lim_{n\to\infty} \|f_n - f_m\|_\infty \le \epsilon. \end{aligned}$$

So $\|f - f_m\|_\infty \le \epsilon$ for $m > N$; or $\|f - f_m\|_\infty \to 0$ as $m \to \infty$. It remains to be shown that f is continuous. It is a standard theorem in real analysis that the uniform limit of continuous functions on $[a,b]$ is continuous. We repeat the argument here. Let $\epsilon > 0$. There exists $N_1 > 0$ such that $\|f - f_n\|_\infty < \epsilon/3$ whenever $n > N_1$. Fix $n > N_1$ and $x \in [a,b]$. Since f_n is continuous there exists $\delta > 0$ such that $|f_n(x) - f_n(y)| < \epsilon/3$ whenever $|x - y| < \delta$. Let $|x - y| < \delta$. Then

$$\begin{aligned} |f(x) - f(y)| &\le |f(x) - f_n(x)| + |f_n(x) - f_n(y)| + |f_n(y) - f(y)| \\ &\le \|f - f_n\| + |f_n(x) - f_n(y)| + \|f_n - f\|_\infty \\ &< \epsilon/3 + \epsilon/3 + \epsilon/3 = \epsilon. \end{aligned}$$

∎

Example 5.37 The space $C[a,b]$ is *not* complete under the L^2 norm

$$\|f\|_2 = \sqrt{\int_a^b |f(x)|^2\, dx}.$$

We give an example of a nonconvergent Cauchy sequence for the case $[a,b] = [-1,1]$ but it is easy to adapt it to the general case $a < b$. Let

$$f_n(x) = \begin{cases} -1 & \text{if } -1 \le x \le -1/n \\ nx & \text{if } -1/n \le x \le 1/n \\ 1 & \text{if } 1/n \le x \le 1. \end{cases}$$

See Figure 5.4 The sequence f_n is a Cauchy sequence of continuous functions under the L^2 norm. It converges in the larger space of square integrable functions $L^2[-1,1]$ to the function f with $f(x) = \frac{x}{|x|}$ for $x \ne 0$ and $f(0) = 0$,

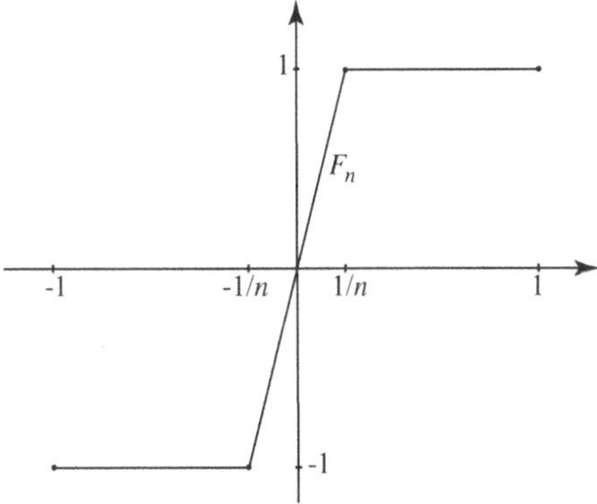

Figure 5.4 The space $C[-1,1]$ is not complete under the $L^2[-1,1]$ norm.

which is not continuous. Furthermore, the discontinuity at $x = 0$ cannot be removed by changing its value there.

Example 5.38 The sequence space c_0 is a Banach space: Here c_0 is assumed to have its usual supremum norm $\|x\|_\infty = \sup_k |x_k|$. Suppose $(x^k) = (x^1, x^2, \ldots)$ is a Cauchy sequence of sequences in c_0. The strategy is to

(1) find a target sequence $a = (a_1, a_2, \ldots)$,
(2) show that $a \in c_0$, and then
(3) show that $x^k \to a$ as $k \to \infty$.

(1): Let $\epsilon > 0$ be given. Then for some $N > 0$ we have $\|x^k - x^j\|_\infty < \epsilon$ whenever $k, j > N$. For each coordinate m the sequence $(x_m^1, x_m^2, \ldots)$ is a Cauchy sequence in $\mathbb{C}$ because

$$|x_m^k - x_m^j| \leq \|x^k - x^j\|_\infty < \epsilon \quad \text{whenever} \quad k, j > N. \tag{5.5}$$

Since $\mathbb{C}$ is complete, for each coordinate m, there exists $\lim_{k\to\infty} x_m^k = a_m$. This forms the target sequence $a = (a_1, a_2, \ldots, a_m, \ldots)$. Taking the limit as $j \to \infty$ in (5.5), we have for each m,

$$|x_m^k - a_m| \leq \epsilon \quad \text{whenever} \quad k > N.$$

Note that in taking the limit above, $<$ is changed to $\leq$. Since this inequality is true for all m, we can take the supremum to obtain

$$\sup_m |x_m^k - a_m| \leq \epsilon \quad \text{whenever} \quad k > N. \tag{5.6}$$

(2): Next, we show that $a \in c_0$. Fix $k > N$. Since $x^k \in c_0$, there exists an $M > 0$ such that $|x_m^k| < \epsilon$ whenever $m > M$. From (5.6) we have

$$|a_m| = |a_m - x_m^k + x_m^k| \le |a_m - x_m^k| + |x_m^k| \le \epsilon + \epsilon \quad \text{whenever} \quad m > M.$$

(3): To show $x^k \to a$, observe from (5.6), $\|x^k - a\|_\infty \le \epsilon$ whenever $k > N$.

Theorem 5.31 *A normed linear space V with a norm $\|\cdot\|$ is complete if and only if $\sum_{k=1}^\infty x^k$ converges (in norm) whenever $\sum_{k=1}^\infty \|x^k\| < \infty$.*

Proof **($\Rightarrow$):** Suppose V is complete and $\sum_{k=1}^\infty \|x^k\| < \infty$. Let $s^n = \sum_{k=1}^n x^k$. Then for $m > n$ we have

$$\begin{aligned} \|s^m - s^n\| &= \|x^{n+1} + \cdots + x^m\| \\ &\le \|x^{n+1}\| + \cdots + \|x^m\| \to 0 \quad \text{as} \quad m, n \to \infty. \end{aligned}$$

This shows that (s^n) is a Cauchy sequence, which means

$$\lim_{n\to\infty} s^n = \sum_{k=1}^\infty x^k$$

converges, since V is complete.

($\Leftarrow$): Consider a Cauchy sequence (x^k) in V. It is sufficient to show that a subsequence converges (Theorem 5.6). Since (x^k) is Cauchy, we can find $k_1 < k_2 < \ldots$ such that

$$\|x^{k_{n+1}} - x^{k_n}\| \le 2^{-n} \quad (n = 1, 2, \ldots).$$

To show that the subsequence (x^{k_n}) converges, note that

$$\sum_{n=1}^\infty \|x^{k_{n+1}} - x^{k_n}\| \le \sum_{n=1}^\infty 2^{-n} = 1 < \infty.$$

By hypothesis $\sum_{n=1}^\infty \left(x^{k_{n+1}} - x^{k_n}\right)$ converges. But the sum telescopes; that is, $\sum_{n=1}^m \left(x^{k_{n+1}} - x^{k_n}\right) = x^{k_{m+1}} - x^{k_1}$. So the subsequence converges. ∎

Definition 5.21 Normed spaces V and V_1 are **isometric** if there exists a linear mapping T from V to V_1 that is one-to-one, onto, and norm preserving $\|Tx\|_{V_1} = \|x\|_V$. The mapping T is called an **isometry** between V and V_1. Isometric spaces can be regarded as identical from the viewpoint of norms.

Definition 5.22 A **completion** of a normed space V is a Banach space V_1 in which V is a dense subspace such that $\|x\|_{V_1} = \|x\|_V$ for every $x \in V$.

Theorem 5.32 *Every normed space V has a completion V_1.*

Proof If V is already complete, let $V_1 = V$. If V is not complete, let V_1 be the set of all Cauchy sequences $x = (x^k)$ in V. For each Cauchy sequence $x = (x^k)$, define

$$\|x\|_{V_1} = \lim_{k\to\infty} \|x^k\|_V.$$

It is routine to show that this is a seminorm on V_1. Define the relation $\sim$ for Cauchy sequences $x = (x^k)$ and $y = (y^k)$ in V by

$$(x^k) \sim (y^k) \iff \|x - y\|_{V_1} = \lim_{k\to\infty} \|x^k - y^k\|_V = 0.$$

We can make V_1 into a normed space with Theorem 5.3. Let V_0 be the set of all constant sequences $(a, a, a, \ldots)$ for $a \in V$. Clearly $V_0 \subset V_1$ and $T(a) = (a, a, a \ldots)$ is an isometry between V and V_0. Showing that V_1 is complete and that V_0 is dense in V_1 is left as Exercise 5.8.13. ∎

Theorem 5.33 *Any two completions of a normed space V are isometric.*

The proof is left as Exercise 5.8.14.

Remark An **isometry** between metric spaces M, M_1 is a mapping T that is one-to-one, onto, and distance preserving $d_{M_1}(Tx, Ty) = d_M(x, y)$. As with normed spaces, we can show that every metric space M has a **completion**.

Example 5.39 It was noted in Example 5.37 that $C[a, b]$ is not complete under the L^2 norm. Since $L^2[a, b]$ is complete and $C[a, b]$ is a dense subspace, we see that $L^2[a, b]$ is the completion of $C[a, b]$.

Exercises for Section 5.8

5.8.1 Show that the sequence f_n defined in Example 5.37 is Cauchy in $L^2[-1, 1]$.

5.8.2 Consider the space $V = P[0, 1]$ of polynomials on the interval $[0, 1]$ under the uniform norm

$$\|p\|_\infty = \sup_{0\le x\le 1} |p(x)|.$$

Show that this space is not complete. Hint: Recall Taylor's theorem from your previous course on real analysis.

5.8.3 Show that every finite-dimensional subspace of a normed space is a Banach space.

5.8.4 Suppose V is a Banach space with norm $\|\cdot\|$. Show that a subspace W of V is a Banach space under the same norm if and only if W is closed in V.

5.8.5 Show that c_0 is a closed subspace of ℓ_∞.

5.8.6 Show that Φ, the sequence space of all sequences with only a finite number of nonzero terms, is not a Banach space under the norm $\|\cdot\|_\infty$.

5.8.7 Let V consist of all functions $f \in C[a,b]$ with $f(a) = f(b)$. Show that V is a complete subspace of $C[a,b]$.

5.8.8 Show that the sequence space c is a Banach space under the usual norm $\|x\|_\infty = \sup_m |x_m|$. Given a Cauchy sequence x^k and $\epsilon > 0$, follow the strategy of Example 5.38 to

(1) find a target sequence $a \in \omega$ with $\sup_m |x^k_m - a_m| \le \epsilon/3$ for large k,

(2) show that $a \in c$ by showing $|a_n - a_m| \le \epsilon$ for sufficiently large n, m,

(3) and then show that $x^k \to a$ as $k \to \infty$.

5.8.9 Show that the sequence space ℓ_∞ is a Banach space under the usual norm $\|x\|_\infty = \sup_k |x_k|$.

5.8.10 Show that the sequence space ℓ_1 is a Banach space under the usual norm $\|x\|_1 = \sum_k |x_k|$.

5.8.11 Show that, for any $1 < p < \infty$, the sequence space ℓ_p is a Banach space under the usual norm $\|x\|_p = \left(\sum_k |x_k|^p\right)^{\frac{1}{p}}$.

5.8.12 Show that $C[0,1]$ and $C[a,b]$ are isometric for every $a < b$.

5.8.13 Finish Theorem 5.32 by showing that V_1 is complete and that V_0 is a dense subset of V_1.

5.8.14 Prove Theorem 5.33. That is, if V_1 and V_2 are both completions of V, find a norm-preserving one-to-one correspondence $f \colon V_1 \longrightarrow V_2$ that is the identity function when restricted to V.

5.8.15 Define the notion of a completion of a metric space M. Let M_1 be the collection of Cauchy sequences in M. For two Cauchy sequences $x = (x^k)$ and $y = (y^k)$ define

$$d_{M_1}(x,y) = \lim_{k\to\infty} d_M(x^k, y^k).$$

Show that M_1 is a semimetric space. It can be made into a metric space by way of equivalence classes (Theorem 5.3). Show that this metric space is a completion of M.

5.8.16 Suppose metric spaces M and M_1 are isometric. Show that if one of them is complete, so is the other.

5.8.17 Suppose f is uniformly continuous on a metric space M and let M_1 be its completion. Show that f can be uniquely extended to M_1 and that this extension is uniformly continuous.

5.9 Compact Sets in Metric Spaces

Compactness in Euclidean spaces $\mathbb{R}^n$ is discussed in Section 0.7. In this section we extend the concept to general metric spaces.

Definition 5.23 Let K be a subset of a metric space M. A collection of sets $\mathcal{A}$ is called a **cover** of K if K is contained in the union of the sets in $\mathcal{A}$; that is, $K \subset \bigcup_{A \in \mathcal{A}} A$. If the sets in $\mathcal{A}$ can be indexed by some set $\mathcal{I}$ (that is, for each $i \in \mathcal{I}$, there corresponds a set $A_i \in \mathcal{A}$ and conversely, for each $A \in \mathcal{A}$, there corresponds $i \in \mathcal{I}$ such that $A = A_i$), we write $K \subset \bigcup_{i \in \mathcal{I}} A_i$. If every set $A \in \mathcal{A}$ is open, we say that $\mathcal{A}$ is an **open cover** of K. If a subcollection of $\mathcal{A}$ still covers K, we say the subcollection is a **subcover** of K.

Definition 5.24 A set $K \subset M$ is said to be **compact** in M if every open cover of K, say $K \subset \bigcup_{i \in \mathcal{I}} G_i$, contains a finite subcover, say $K \subset G_{i_1} \cup G_{i_2} \cup \cdots \cup G_{i_N}$.

Clearly every finite set is compact. Also a finite union of compact sets is compact.

For Euclidean spaces $\mathbb{R}^n$, The Heine–Borel theorem (Theorem 0.23) shows that a set $K \subset \mathbb{R}^n$ is compact if and only if it is closed and bounded. Theorem 5.34 shows that for general metric spaces, compact sets are closed and bounded. But Example 5.40 shows that compactness is a strictly stronger property than being closed and bounded. That is, some closed and bounded sets in metric spaces can fail to be compact.

Theorem 5.34 *If K is a compact subset of a metric space M, then K is closed and bounded.*

Proof Suppose K is compact in metric space M. We first show that K is bounded. Fix $x \in M$ and consider the concentric open spheres

$$S(x,1) \subset S(x,2) \subset \cdots \subset S(x,k) \subset \cdots .$$

Since $\bigcup_{k=1}^{\infty} S(x,k) = M$, the open spheres cover K. Since there exists a finite subcover, K is contained in the largest of them, say $K \subset S(x,N)$. This means that K is bounded.

The proof that K is closed is the same as for Theorem 0.20 (Section 0.7). ∎

Example 5.40 The converse of the above theorem is not generally true. The metric space of natural numbers $\mathbb{N}$ under the discrete metric is closed and bounded but not compact. The proof is left as Exercise 5.9.2.

Later we will show that the converse of Theorem 5.34 actually fails for all infinite dimensional normed spaces (Theorem 5.50 in Section 5.10).

Theorem 5.35 *Let K be a compact set in a metric space M. Every closed subset F of K is compact.*

The proof is the same as for Theorem 0.21 in Section 0.7.

Theorem 5.36 *Let M and M_1 be metric spaces and let $f: M \longrightarrow M_1$ be continuous. If K is a compact subset of M, then the image $f(K)$ is compact in M_1.*

The proof is essentially the same as that of Theorem 0.26 in Section 0.8.

Theorem 5.37 *Suppose f is a continuous real valued function on a compact subset K of metric space M. Then f is bounded on K and attains its bounds.*

The proof is essentially the same as that of Theorem 0.27 in Section 0.8.

Theorem 5.38 *Suppose f is a continuous function from a compact metric space K to a metric space M. Then f is uniformly continuous.*

The proof is essentially the same as that of Theorem 0.28 in Section 0.8.

Definition 5.25 A set K is **sequentially compact** if every sequence in K has a subsequence that converges to a point in K.

Theorem 5.39 *Every sequentially compact metric space K is separable.*

Proof It is sufficient to construct a countable dense subset A of K. For each $n = 1, 2, 3, \ldots$, form sets $A_n = \{x^1, x^2, x^3, \ldots\}$ as follows. Begin with any point $x^1 \in K$. Find, if possible, a point $x^2 \in K$ such that $d(x^1, x^2) > 1/n$. Then, if possible, find a point $x^3 \in K$ such that $d(x^1, x^3) > 1/n$ and $d(x^2, x^3) > 1/n$. Continue this process so that,

$$\text{for all } \ x^i, x^j \in A_n \ \text{ with } \ i \neq j, \ \text{ we have } \ d(x^i, x^j) > 1/n. \tag{5.7}$$

For each n, the process must end after finite steps. For, if not, an infinite sequence would have this property (5.7), so it could not have a convergent subsequence. That is, each A_n, which is maximal with respect to Property (5.7), is finite. Let $A = A_1 \cup A_2 \cup \cdots$, which is countable.

Claim: The set A is dense in K. Let $x \in K$ and consider an open sphere $S(x, r)$. We show that $S(x, r)$ contains a point of A. Let $0 < 1/n < r$. If $S(x, r) \supset S(x, 1/n)$ contained no point of A_n, then $d(x, x^k) > 1/n$ for each $x^k \in A_n$. This contradicts the maximality of the set A_n with respect to Property (5.7). ∎

Theorem 5.40 *A subset K of a metric space M is compact if and only if it is sequentially compact.*[8]

Proof **($\Rightarrow$):** Suppose K is a compact subset of M. Assume that there exists a sequence $x^1, x^2, \ldots$ in K with no convergent subsequence. The sequence must have infinitely many distinct values, for otherwise it would have a constant subsequence. This means that the sequence has a subsequence $y^1 = x^{n_1}, y^2 = x^{n_2}, \ldots$ of distinct values.

Claim: The set $\{y^k \mid k = 1, 2, \ldots\}$ has no limit points. Assume there *is* a limit point y. Then each open sphere $S(y, 1/n)$ contains infinitely many y^k. From this we can construct a subsequence that converges to y, which contradicts the assumption that no subsequence of $x^1, x^2, \ldots$ converges.

The subset $Y = \{y^k \mid k = 1, 2, \ldots\}$ is thus closed in K, and hence compact (Theorem 5.35). For each y^k, there exists an open sphere S_k containing no other element of Y (for otherwise, y^k would be a limit point). Yet the open spheres $S_1, S_2, \ldots$ cover the compact set Y but clearly no subcover does. This contradiction completes the proof of ($\Rightarrow$).

($\Leftarrow$): Suppose K is a sequentially compact subset of M. First we show **(1):** Every open cover $\mathcal{A}$ of $K \subset \bigcup_{G \in \mathcal{A}} G$ has a *countable* subcover $\mathcal{B} \subset \mathcal{A}$,

$$\mathcal{B} = \{G_1, G_2, \ldots\}, \quad K \subset \bigcup_{k=1}^{\infty} G_k. \tag{5.8}$$

Then we show **(2):** Every countable cover $\mathcal{B} = \{G_1, G_2, \ldots\}$ as in (5.8), has a finite subcover.

Proof of (1): Since K is separable (Theorem 5.39), K has a countable dense subset $A = \{x^1, x^2 \ldots\}$. Consider the family $\mathcal{C}$ consisting of all open spheres $S_{nk} = S(x^k, 1/n)$ with centers at the points $x^k \in A$, radii $1/n$, and $S_{nk} \subset G$ for some $G \in \mathcal{A}$. Clearly $\mathcal{C}$ is a countable family of open spheres. For each S_{nk} in $\mathcal{C}$, choose some $G \in \mathcal{A}$ such that $S_{nk} \subset G$. This collection of G we denote by $\mathcal{B}$, which is a countable subset of $\mathcal{A}$.

To show that $\mathcal{B}$ covers K, let $x \in K$. Since $\mathcal{A}$ covers K, for each $x \in K$, there exists an open sphere of radius, say $2r$, such that $S(x, 2r)$ is entirely contained in one of the open sets G in $\mathcal{A}$. Let $1/n < r$. Then the open sphere $S(x, 1/n)$ must contain at least one of the points, say x^k, of the dense set A.

[8] This result is not true in all general topological spaces, in either direction.

Since $d(x,x^n) < 1/n$, we have $x \in S(x^k,1/n) \subset S(x,2/n) \subset S(x,2r) \subset G$; that is $x \in S_{nk} \in \boldsymbol{C}$. Since $\boldsymbol{C}$ covers K, so does $\boldsymbol{B}$.

Proof of (2): Suppose K has a countable cover

$$\boldsymbol{B} = \{G_1, G_2, \ldots\}, \quad K \subset \bigcup_{k=1}^{\infty} G_k.$$

Assume that there is no finite subcover. Then, for each k, the open sets

$$H_k = G_1 \cup \cdots \cup G_k$$

do not cover K. For each k, there exists $x^k \in K - H_k$ (the axiom of choice, as described in Section 0.14, may be needed). Since K is sequentially compact, the sequence $x^1, x^2, \ldots$ has a subsequence converging to some $x \in K$. Since $\boldsymbol{B}$ covers K, there exists N such that $x \in G_N \subset H_N$. By construction, the open set H_N contains $x^1, \ldots, x^N$ but does not contain x^k for $k > N$. This contradicts the convergence of a subsequence to x. ∎

Corollary 5.41 *Every compact metric space is separable.*

Compactness is a stronger property than completeness, as the following theorem shows.

Theorem 5.42 *If M is a compact metric space, then it is complete. The converse is not true, in general.*

Proof Let M be a compact metric space and let (x^k) be a Cauchy sequence. Since every compact set is sequentially compact (Theorem 5.40), the Cauchy sequence has a convergent subsequence. But then the entire sequence converges by Theorem 5.6.

That the converse is false can be seen from the example of $\mathbb{R}$, which is complete but not bounded. Hence, it cannot be compact. ∎

Compactness is related to a property called **total boundedness**. This notion is discussed later in Section 5.11.

Exercises for Section 5.9

5.9.1 Show that a finite union of compact sets is compact.

5.9.2 Complete Example 5.40 by showing that the metric space $\mathbb{N}$ under the discrete metric is closed and bounded yet not compact.

5.9.3 Show that a set that is compact under a given topology will still be compact under any weaker topology.

5.9.4 Prove Corollary 5.41.

$*$5.10 Finite-Dimensional Normed Spaces

In this section we show that the essential properties of $\ell_p(n)$ hold in any finite-dimensional normed space. In particular, the Heine–Borel property of $\ell_p(n)$ that every closed bounded set is compact holds in *all* finite-dimension normed spaces. Yet this property is never true in *any* infinite-dimensional normed space (Theorem 5.51).

Although the following result is a consequence of the Heine–Borel theorem for $\mathbb{R}^{\mathbb{K}} = \mathbb{R}$ (Theorem 0.23), we give an interesting independent proof.

Lemma 5.43 *Every closed and bounded interval $I = [a,b]$ of $\mathbb{R}$ is sequentially compact.*

Proof Consider a sequence (x_k) in $I_0 = I$. If the sequence has only a finite number of distinct values, then there is a convergent subsequence. Otherwise, bisect I_0 into $\left[a, \frac{a+b}{2}\right]$ and $\left[\frac{a+b}{2}, b\right]$. One of these two subintervals, call it I_1, contains a subsequence of (x_k). We next bisect I_1 and continue this process to obtain a nest of closed intervals

$$I = I_0 \supset I_1 \supset I_2 \supset \dots,$$

each containing a subsequence of (x_k). The nest intersects in a point x (Theorem 5.29). The point x is clearly a limit point of the sequence. Thus there exists a subsequence of (x_k) converging to x. Since I is closed, we have $x \in I$. Thus I is sequentially compact. ∎

Corollary 5.44 Bolzano[9]–Weierstrass[10] theorem: *Every bounded sequence in $\mathbb{R}$ has a convergent subsequence.*

Theorem 5.45 Heine–Borel theorem for $\ell_p(n)$: *Let $1 \le p \le \infty$. A subset of $\ell_p(n)$ is compact if and only if it is both closed and bounded.*

Proof **($\Rightarrow$):** This is Theorem 5.34.

($\Leftarrow$): First start with the real case. By Theorem 5.16, all of the $\ell_p(n)$ norms are equivalent on $\mathbb{R}^n$. Suppose K is a closed and bounded subset of $\ell_\infty(n)$.

[9] Bernard Bolzano (1781–1848) was a Bohemian mathematician. He was the first to give the formal ϵ-δ definition of limit and to promote rigor in the study of differential calculus. He proved formally the intermediate value theorem and understood the importance of the least upper bound property of $\mathbb{R}$.

[10] Karl Weierstrass (1815–1897) was a German mathematician who is considered to be the father of modern analysis. For example, he gave the modern ϵ-δ definitions of continuity and uniform continuity. He said, "It is true that a mathematician who is not also something of a poet will never be a perfect mathematician." Georg Cantor was one of his students.

Let (x^k) be a sequence in K. Since K is bounded, K is contained in some n-dimensional *closed* interval of the form

$$I = \{(x_1,\ldots,x_n) \in \mathbb{R}^n \mid a_k \le x_k \le b_k, \text{ for } k = 1,2,\ldots,n\}.$$

The first coordinates $x_1^1, x_1^2, x_1^3, \ldots$ are contained in the interval $[a_1,b_1]$. By the Bolzano–Weierstrass theorem, there is a subsequence of x^k whose first coordinates converge. Similarly, this subsequence has a subsequence whose second coordinates converge (as well as the first coordinates). Continue this process a finite number of times, obtaining each time a subsequence of a previous subsequence, until we obtain a subsequence y^k for which all n coordinates converge. The proof can now easily be extended to $\mathbb{C}^n$. ∎

Theorem 5.46 *Any two norms on a finite-dimensional linear space V determine the same topology.*

Proof Let V be a finite-dimensional normed space with norm $\|\cdot\|$ and an algebraic basis $\{b^1, b^2, \ldots, b^n\}$. For each $x = (x_1,\ldots,x_n) \in \ell_\infty(n)$ define $Tx = \sum_{k=1}^n x_k b^k$. The map T is an isomorphism between $\ell_\infty(n)$ and V.

For each $z = Tx = \sum_{k=1}^n x_k b^k$, we define a new norm $\|\cdot\|'$ by $\|z\|' = \|Tx\|' = \sup_k |x_k| = \|x\|_\infty$ and show it is equivalent to the norm $\|\cdot\|$. By the triangle inequality, for any $x = (x_1,\ldots,x_n) \in \ell_\infty(n)$, we have

$$\|z\| = \left\| \sum_{k=1}^n x_k b^k \right\| \le \sup_k |x_k| \sum_{k=1}^n \|b^k\| = \left(\sum_{k=1}^n \|b^k\| \right) \|z\|'. \tag{5.9}$$

Thus $\|z\| \le M\|z\|'$ for $M = \sum_{k=1}^n \|b^k\|$. This inequality means that the topology under the norm $\|\cdot\|$ is weaker than under $\|\cdot\|'$ (Theorem 5.13).

Let $B = \{x \in \ell_\infty(n) \mid \|x\|_\infty = 1\}$. By the Heine–Borel theorem (Theorem 5.45), B is compact in $\ell_\infty(n)$. If we define $p(x) = \|Tx\|$, then (5.9) shows that for a given $\epsilon > 0$, we have $|p(x-y)| \le M\epsilon$ whenever $\|x-y\|_\infty < \epsilon$. Thus p is continuous on $\ell_\infty(n)$ and attains its lower bound on B (Theorem 5.37). Since $0 \notin B$, $p(x) > 0$ for all $x \in B$, and so the lower bound m of p on B is greater than zero. Therefore,

$$0 < m \le p(x) \le M \quad \text{for all} \quad x \in B.$$

For any $z = Tx \in V$ with $x \in \ell_\infty(n)$ and $x \neq 0$, we have $\frac{x}{\|x\|_\infty} \in B$. Hence

$$0 < m \le p\left(\frac{x}{\|x\|_\infty}\right) = \frac{\|Tx\|}{\|x\|_\infty} = \frac{\|z\|}{\|z\|'} \le M \quad \text{or}$$

$$m\|z\|' \le \|z\| \le M\|z\|'.$$

∎

Corollary 5.47 *Every finite-dimensional normed space is complete.*

The proof is left as Exercise 5.10.4.

Corollary 5.48 *Every finite-dimensional subspace of a normed space is closed.*

Proof A complete subspace must necessarily be closed. ∎

Lemma 5.49 F. Riesz's lemma: *Let V be a normed space and let $B = \{x \in V \mid \|x\| = 1\}$ be the surface of the unit sphere. If F is a proper closed subspace of V, then for all $\epsilon > 0$, there exists $x \in B$ such that*

$$1 - \epsilon < d(x, F) \leq 1.$$

Proof Let $\epsilon > 0$ be given. Since F is a proper subspace of V, there exists $z \in V - F$. Let $d = d(z, F) = \min_{y \in F} \|z - y\|$. Since F is closed, $d > 0$. For any $k > 1$, we can find $y_0 \in F$ such that

$$d \leq \|z - y_0\| < kd. \tag{5.10}$$

Let $x = \dfrac{z - y_0}{\|z - y_0\|}$. For every $y \in F$, we have $\|z - y_0\|y + y_0 \in F$. So

$$\left\|z - \big(\|z - y_0\|y + y_0\big)\right\| \geq d(z, F) = d. \tag{5.11}$$

Hence, for all $y \in F$, applying (5.11) and (5.10), we obtain

$$\begin{aligned}\|x - y\| = \left\|\frac{z - y_0}{\|z - y_0\|} - y\right\| &= \frac{1}{\|z - y_0\|}\left\|z - \big(\|z - y_0\|y + y_0\big)\right\| \\ &\geq \frac{d}{\|z - y_0\|} > \frac{d}{kd} = \frac{1}{k}.\end{aligned}$$

Letting $k = \dfrac{1}{1 - \epsilon}$, gives us the first inequality $1 - \epsilon < d(x, F)$.

The other inequality follows from (5.10) and the linearity of F with $y_0 \in F$,

$$d(x, F) = d\left(\frac{z - y_0}{\|z - y_0\|}, F\right) = \frac{d(z - y_0, F)}{\|z - y_0\|} = \frac{d(z, F)}{\|z - y_0\|} = \frac{d}{\|z - y_0\|} \leq 1. \ ∎$$

Theorem 5.50 *Let V be an infinite-dimensional normed space. The closed unit sphere $S[0, 1] = \{x \in V \mid \|x\| \leq 1\}$ is never compact.*

Proof Let $x^1 \in B = \{x \in V \mid \|x\| = 1\}$ be a point on the surface of the closed unit sphere $S[0, 1]$ and let $F_1 = \text{span}\{x^1\}$. If $F_1 \neq V$, then by Riesz's lemma, there exists $x^2 \in B$ such that $d(x^2, F_1) > 1/2$. Let $F_2 = \text{span}\{x^1, x^2\}$. Continuing this way, if $F_n = \text{span}\{x^1, x^2, \ldots, x^n\} \neq V$, find $x^{n+1} \in B$ such that $d(x^{n+1}, F_n) > 1/2$. The infinite sequence $x^1, x^2, \ldots$ has the property $i \neq j \Rightarrow d(x^k, x^j) > 1/2$, so it has no convergent subsequence. The set B is not sequentially compact. So neither is $S[0,1]$. ∎

Combining the above result with the Heine–Borel theorem (Theorem 5.45) and Theorem 5.46, we obtain the following important result.

Theorem 5.51 *Let V be a normed space. All of its closed bounded sets are compact if and only if it is finite-dimensional.*

Exercises for Section 5.10

5.10.1 Complete the proof of the Heine–Borel theorem for the complex case.

5.10.2 Extend the Bolzano–Weierstrass theorem (Corollary 5.44) to $\mathbb{R}^n$ to show that every bounded sequence in $\mathbb{R}^n$ has a convergent subsequence.

5.10.3 In the proof of Theorem 5.46, show that $\|\cdot\|'$ is indeed a norm.

5.10.4 Prove Corollary 5.47.

*5.11 Total Boundedness

This is a supplement to Section 5.9 that gives further properties of compact sets.

Definition 5.26 A set A in a metric space M is called **totally bounded** if for every given $\epsilon > 0$, there exists a finite collection of ϵ-spheres $S(x^k, \epsilon)$, for $k = 1, 2, \ldots, N$, which covers A:

$$A \subset \bigcup_{k=1}^{N} S(x^k, \epsilon).$$

The set $\{x^1, x^2, \ldots, x^N\}$ is called a **finite ϵ-net** of A. The value of N may depend on ϵ.

Every totally bounded set is bounded (see Exercise 5.11.1). The following example shows that some bounded sets are not totally bounded.

Example 5.41 In the metric space ℓ_∞, consider the subset $A = \{e^1, e^2, \ldots\}$, where, for each k, $e^k = (0, \ldots, 0, 1, 0, \ldots)$ is the sequence with 1 in the kth position and zero elsewhere. We have $\|e^k\| = 1$, for each k, so the set A

is bounded. However, A is not totally bounded, since, for $\epsilon = 1/2$, any open sphere $S(x, 1/2)$ contains at most one of the sequences in A. Thus a finite union of $\frac{1}{2}$-spheres contains only a finite subset of A, and so cannot cover A.

Theorem 5.52 *Every compact set is totally bounded.*

The proof follows directly from the definition of compactness (Definition 5.24).

Theorem 5.53 *Every totally bounded metric space M is separable.*

Proof For each $n = 1, 2, 3, \ldots$, the family $\mathcal{A}_n$ consisting of all open spheres $S(x, \frac{1}{n})$ of all $x \in M$ is an open cover of M. Since there exists a finite subcover,

$$M \subset \bigcup_{k=1}^{N} S(x^k, \epsilon),$$

there is a finite collection of their centers $A_n = \{x^1, \ldots x^N\}$. The union

$$A = A_1 \cup A_2 \cup \cdots$$

is countable. For any $x \in M$ and any $\epsilon > 0$, choose $\frac{1}{n} < \epsilon$. Since the set A_n is a $\frac{1}{n}$-net, there exists an $x^k \in A_n$ such that $x \in S\big(x^k, \frac{1}{n}\big)$. We have

$$x^k \in S\left(x^k, \frac{1}{n}\right) \subset S(x, \epsilon).$$

This shows that the countable set A is dense in M. ∎

Corollary 5.54 *Every compact metric space K is separable.*

Theorem 5.55 *A metric space M is compact if and only if it is totally bounded and complete.*

Proof **($\Rightarrow$):** This follows from Theorems 5.52 and 5.42.

($\Leftarrow$): Let M be totally bounded and complete. We show that M is sequentially compact. Consider a sequence $x^1, x^2, \ldots$ in M. Since M is totally bounded, there exists a finite ϵ-net for $\epsilon = 1$. This means that M is covered by a finite set of open spheres of radii 1. At least one of these spheres, say S_1, must contain a subsequence, say $x^{1_1}, x^{1_2}, \ldots$ of the original sequence. Next, M is also covered by a finite set of open spheres of radii $\epsilon = 1/2$. At least one of these spheres, say S_2, must contain a subsequence $x^{2_1}, x^{2_2}, x^{2_3} \ldots$ of the subsequence $x^{1_1}, x^{1_2}, \ldots$ in S_1. For each k there exists a sphere S_k of radius $1/2^k$ containing a subsequences $x^{k_1}, x^{k_2}, \ldots$ of the subsequence in S_{k-1}.

Let $y^k = x^{k_k}$ for each k. The sequence $y^1, y^2, \ldots$ is a Cauchy subsequence of the original sequence. Since M is complete, the subsequence converges. ∎

Theorem 5.56 *A metric space M is totally bounded if and only if every sequence has a Cauchy subsequence.*

The proof is left to Exercise 5.11.2.

Exercises for Section 5.11

5.11.1 Show that every totally bounded set is bounded.

5.11.2 Prove Theorem 5.56.

*5.12 Stone–Weierstrass Approximation

The Stone–Weierstrass approximation theorem, proved in this section, is a generalization of the Weierstrass approximation theorem, which states that the algebraic polynomials are dense in the space of continuous functions $C[a,b]$. The Weierstrass approximation theorem is proved later as Theorem 7.59 in the chapter on Fourier series. For that reason, the present section may be skipped.

Definition 5.27 A vector space A of real or complex functions on a set S is called an **algebra of functions** whenever the product of two functions in A is also in A. Thus, a set A of scalar valued functions on S is an algebra if and only if for every f and g in A and any scalars a and b, we have $af + bg \in A$ and $fg \in A$.

Theorem 5.57 **Stone[11]–Weierstrass theorem:** *Let K be a compact metric space and consider the algebra $C(K)$ of all continuous* real-valued *functions on K. Suppose that A is a subalgebra of $C(K)$ such that*

(a) *A contains all constant functions, and*
(b) *if $x, y \in K$, $x \neq y$, then there exists $f \in A$ for which $f(x) \neq f(y)$.*

Then A is dense in $C(K)$ with respect to the uniform norm $\|f\| = \sup_{x \in K} |f(x)|$.

[11] Marshall Harvey Stone (1903–1989) was an American mathematician. He contributed to Hilbert spaces, Fourier series, Boolean algebra, topology, and group theoretical methods in quantum theory.

We first prove the Weierstrass approximation theorem for the function $f(x) = |x|$. The idea of basing the proof on the special case of $f(x) = |x|$ is due to Lebesgue.

Lemma 5.58 *There exists a sequence of polynomials that converge uniformly to* $f(x) = |x|$.

Proof The function f is not differentiable at $x = 0$. Since $|x| = \sqrt{x^2}$, the strategy is to approximate $\sqrt{x^2}$ by a function of the form $\sqrt{x^2+c^2}$ for some small $c > 0$ and then use an approximating Taylor polynomial of $\sqrt{x^2+c^2}$.

Let $\epsilon > 0$ be given. Let $P_n(x)$ be the nth Taylor polynomial of $g(x) = \sqrt{x+c^2}$. We leave it as Exercise 5.12.1 to show that the Taylor polynomials converge to g on $[-1, 1]$. Since $[-1, 1]$ is compact, the convergence is uniform. So there exists $N > 0$ such that

$$|\sqrt{x+c^2} - P_n(x)| < \epsilon/2 \quad \text{whenever} \quad n > N \quad \text{and} \quad x \in [-1, 1].$$

Then $\left|\sqrt{x^2+c^2} - P_n(x^2)\right| < \epsilon/2$ whenever $n > N$ and $x \in [-1, 1]$. Now let $c = \epsilon/2$. Then for all $n > N$ and $x \in [-1, 1]$, we have

$$\left||x| - P_n(x^2)\right| \le \left||x| - \sqrt{x^2+c^2}\right| + \left|\sqrt{x^2+c^2} - P_n(x^2)\right| < \epsilon/2 + \epsilon/2 = \epsilon.$$

∎

Proof (Theorem 5.57) Let $\overline{A}$ be the closure of A. It is left for Exercise 5.12.2 to show that $\overline{A}$ is still a subalgebra of $C(K)$.

Claim 1: $f \in \overline{A}$ implies $|f| \in \overline{A}$. Suppose $\|f\|_\infty \le 1$ and let $\epsilon > 0$ be given. Let $p(x) = a_0 + a_1 x + \cdots + a_n x^n$ be a polynomial such that

$$\left||x| - p(x)\right| < \epsilon \quad \text{for all} \quad -1 \le x \le 1$$

as given by Lemma 5.58. Since $\overline{A}$ is an algebra that contains the constant functions, the function $p(f) = a_0 + a_1 f + \cdots + a_n f^n$ is in $\overline{A}$. Since $-1 \le f(x) \le 1$ for all $x \in K$, it follows that

$$\left||f(x)| - p\big(f(x)\big)\right| < \epsilon \quad \text{for all} \quad x \in K.$$

This shows that $|f|$ is in the closure of $\overline{A}$, and hence $|f| \in \overline{A}$. If $\|f\|_\infty > 1$, divide f by $\|f\|_\infty$ and use this argument to show that $|f|/\|f\|_\infty \in \overline{A}$. Then we can conclude that $|f| \in \overline{A}$.

Claim 2: If $f, g \in \overline{A}$, then $\min\{f,g\}$ and $\max\{f,g\}$ are in $\overline{A}$. This follows since

$$\min\{f,g\} = \frac{1}{2}(f+g) - \frac{1}{2}|f-g| \quad \text{and}$$
$$\max\{f,g\} = \frac{1}{2}(f+g) + \frac{1}{2}|f-g|.$$

Let $f \in C(K)$ and let $\epsilon > 0$ be given. We now construct a function $g \in \overline{A}$ such that $\|f-g\|_\infty < \epsilon$.

Claim 3: For each $x, y \in K$ with $x \neq y$, there exists a function $f_{xy} \in \widehat{A}$ such that $f_{xy}(x) = f(x)$ and $f_{xy}(y) = f(y)$. Consider the constant function g with the value $f(x)$. By hypothesis, $g \in A$. Also by hypothesis, there exists $h \in A$ such that $h(x) \neq h(y)$. Then the following function has the desired property:

$$f_{xy}(t) = g(t) + \Big(f(y) - f(x)\Big)\frac{h(t)-h(x)}{h(y)-h(x)}.$$

Claim 4: For each $x \in K$, there exists $f_x \in \overline{A}$ such that $f_x(x) = f(x)$ and $f_x(t) < f(t) + \epsilon$ for all $t \in K$. Since f and $f_{x,y}$ are continuous and $f_{xy}(y) = f(y)$, there exists an open interval sphere S_y containing y such that $f_{xy}(t) < f(t) + \epsilon$, for all $t \in S_y$. The collection of all such S_y covers K. Since K is compact, a finite number of them cover K, say $\{S_{y_1}, \ldots, S_{y_n}\}$. Let

$$f_x = \min\{f_{x,y_1}, \ldots, f_{x,y_n}\}.$$

Then $f_x \in \overline{A}$ and $f_x(x) = f(x)$.

Finally, for every $x \in K$, there exists an open sphere S_x containing x such that $f_x(t) > f(t) - \epsilon$ for all $t \in S_x$. Since K is compact, there exists a finite number of them that cover K, say $\{S_{x_1}, \ldots, S_{x_m}\}$. Let

$$g = \max\{f_{x_1}, \ldots, f_{x_m}\}.$$

Then $g \in \overline{A}$ and $|f(t) - g(t)| < \epsilon$ for every $t \in K$. ∎

Corollary 5.59 Weierstrass approximation: *Let $-\infty < a < b < \infty$ and consider the space $C[a,b]$ of continuous real (or complex) valued functions on $[a,b] \subset \mathbb{R}$ under the uniform norm. Every $f \in C[a,b]$ can be uniformly approximated by algebraic polynomials.*

Proof The space $C[a,b]$ on the compact interval $[a,b] \subset \mathbb{R}$ is an algebra and the polynomials $P[a,b]$ form a subalgebra of $C[a,b]$ that satisfies the conditions of the Stone–Weierstrass theorem. ∎

Exercises for Section 5.12

5.12.1 Show for Lemma 5.58 that the Taylor polynomials $P_n(x)$ of $g(x) = \sqrt{x+c^2}$ converge to the function on $[-1,1]$. Hint: Use Lagrange's form of the remainder

$$R_n(x) = \frac{f^{(n+1)}(c)}{(n+1)!}x^{n+1} \quad \text{for some } c \text{ between } 0 \text{ and } x.$$

You may need Stirling's formula: $n! \sim \left(\frac{n}{e}\right)^n \sqrt{2\pi n}$. The sign $\sim$ means that the two quantities are asymptotic, that is, their ratio tends to 1 as n tends to ∞.

5.12.2 Show that the closure $\overline{A}$ of a subalgebra A of $C(K)$ is still a subalgebra of $C(K)$. This is needed in the proof of the Stone–Weierstrass theorem.

6
Linear Operators

6.1 Linear Operators

Definition 6.1 Let V and W be linear spaces. A function $T: V \longrightarrow W$ is a **linear operator** or **linear map** if for all $x, y \in V$ and all scalars α, β, we have

$$T(\alpha x + \beta y) = \alpha T(x) + \beta T(y).$$

If the codomain W is the same as the domain V, we say that T is a **linear operator on** V. If the codomain W is the field of scalars $\mathbb{C}$ or $\mathbb{R}$, then we use the term **linear functional on** V**.** By common convention we sometimes write Tx instead of $T(x)$.

Definition 6.2 A linear operator $T: V \longrightarrow W$ on *normed spaces* V and W is said to be **bounded** if there exists a constant $M > 0$ such that

$$\|Tx\| \leq M\|x\| \quad \text{for all} \quad x \in V.$$

Here $\|Tx\|$ refers to the norm on W and $\|x\|$ refers to the norm on V. If there is the potential for confusion, we will write $\|Tx\|_W \leq M\|x\|_V$.

Note that this definition is not consistent with the definition of a bounded real valued function as used in calculus. A function $f: \mathbb{R} \longrightarrow \mathbb{R}$ is bounded if $|f(x)| \leq M$ for all $x \in \mathbb{R}$, whereas a linear functional $f: \mathbb{R} \longrightarrow \mathbb{R}$ is bounded if $|f(x)| \leq M|x|$ for all $x \in \mathbb{R}$.

Example 6.1 Shift operators: Let V be any one of the sequences spaces c, c_0, or ℓ_p, with $1 \leq p \leq \infty$. The following are bounded linear operators from V to V with bound $M = 1$:

$$T_L(x_1, x_2, \ldots) = (x_2, x_3, \ldots) \quad \text{and} \quad T_R(x_1, x_2, \ldots) = (0, x_1, x_2, x_3, \ldots).$$

Example 6.2 Section operators: Let V be any of the sequence spaces c, c_0, or ℓ_p, with $1 \le p \le \infty$. For any $n = 1, 2, \ldots$, the following is a bounded linear operator from V to V with bound $M = 1$:

$$P^n(x_1, x_2, \ldots, x_n, x_{n+1}, \ldots) = (x_1, x_2, \ldots, x_n, 0, 0, \ldots).$$

Example 6.3 Integral operator: Let $V = C[a,b]$ with the norm $\|f\|_\infty = \sup_{a \le x \le b} |f(x)|$. Define $T\colon C[a,b] \longrightarrow C[a,b]$ by the indefinite integral

$$Tf(x) = \int_a^x f(t)\,dt.$$

This is a bounded linear operator with bound $M = b - a$. It is a special case of a Fredholm operator as given in Example 6.9.

Example 6.4 Integral functional: Let $V = C[a,b]$ with the norm $\|f\|_\infty = \sup_{a \le x \le b} |f(x)|$. The linear functional T on $C[a,b]$ defined by the integral

$$T(f) = \int_a^b f(x)\,dx$$

satisfies the inequality $|T(f)| \le (b-a)\max_{a \le x \le b} |f(x)| = (b-a)\|f\|_\infty$. Thus T is a bounded linear functional on $C[a,b]$ with a bound $M = b - a$.

Example 6.5 Limit functional: Let V be the sequence space c with the usual norm $\|x\|_\infty = \sup_k |x_k|$. The limit functional on c is

$$T(x) = \lim_{k \to \infty} x_k.$$

It is a bounded linear functional on c with bound $M = 1$ because $|T(x)| = \left|\lim_{k\to\infty} x_k\right| \le \sup_k |x_k| = \|x\|_\infty$.

Example 6.6 An example of a bounded linear functional on ℓ_∞ is

$$T(x) = \sum_{k=1}^{\infty} \frac{x_k}{k^2}.$$

It has a bound $M = \dfrac{\pi^2}{6}$ because $|T(x)| \le \sum_{k=1}^{\infty} \frac{1}{k^2} \sup |x_k| = \frac{\pi^2}{6}\|x\|_\infty$.

Example 6.7 Coordinate functionals: For the sequence space ℓ_∞ and any $k = 1, 2, \ldots$, the kth coordinate is a bounded linear functional (with bound $M = 1$)

$$T_k(x) = x_k.$$

Example 6.8 Matrix operators: You are familiar from linear algebra that any $m \times n$ matrix $A = (a_{ij})$ determines a linear operator $A\colon \mathbb{R}^n \longrightarrow \mathbb{R}^m$ as follows. For each $x = (x_1, \ldots, x_n) \in \mathbb{R}^n$, define $y = (y_1, \ldots, y_m) = Ax \in \mathbb{R}^m$ by

$$y_i = \sum_{j=1}^{n} a_{ij} x_j \quad (i = 1, \ldots, m).$$

This can be extended to infinite matrices as operators between sequence spaces. Let $A = (a_{ij})$, for $i, j = 1, 2, 3, \ldots$, be an infinite matrix and let $E \subset \omega$ and $F \subset \omega$ be sequence spaces. In order for such a matrix operator $A\colon E \longrightarrow F$ to be defined, we must have convergence for each row

$$y_i = \sum_{j=1}^{\infty} a_{ij} x_j \quad \text{for each} \quad x \in E \quad \text{and each} \quad i = 1, 2, 3, \ldots.$$

And then for each $x \in E$, we must also have $Ax = y = (y_1, y_2, \ldots) \in F$.

For example, the infinite matrix

$$A = \begin{pmatrix} 1 & 0 & 0 & 0 & 0 & \cdots \\ 1 & 1 & 0 & 0 & 0 & \cdots \\ 1 & 1 & 1 & 0 & 0 & \cdots \\ 1 & 1 & 1 & 1 & 0 & \cdots \\ \vdots & \vdots & \vdots & \vdots & \vdots & \ddots \end{pmatrix}$$

defines a linear map from cs to c where, for $x = (x_1, x_2, \ldots) \in cs$, we have $Ax = y = (x_1, x_1 + x_2, x_1 + x_2 + x_3, \ldots) \in c$. This map is actually a **linear isometry** between cs and c. That is, the map is one-to-one, onto, and norm preserving $\|x\|_{cs} = \|Ax\|_c$.

Example 6.9 Fredholm operator: Here we consider continuous versions of matrix operators of the previous example. Let $V = C[a,b]$ and let $K(x,t)$ be an integrable function for $a \le x \le b$, $a \le t \le b$. For each $f \in C[a,b]$, the function

$$g(x) = \int_a^b K(x,t) f(t)\, dt$$

is continuous on $[a,b]$. Thus $Tf = g$ defines a linear operator on $C[a,b]$. This integral operator is called a **Fredholm operator of the first kind.**[1]

[1] Named after the Swedish mathematician Erik Ivar Fredholm (1866–1927). If we define $g(x) = f(x) - \int_a^b K(x,t) f(t)\, dt$, then the operator $Tf = g$ is called a **Fredholm operator of the second kind.**

The function $K(x,t)$ is called the **kernel** of the operator. Fredholm operators are bounded with bound

$$M = \int_a^b \int_a^b |K(x,t)|\, dx\, dt.$$

In the chapter on Fourier analysis (Chapter 7) we will study several kinds of kernels: convolution, Dirichlet, Fejér, and Poisson kernels. Fredholm operators are useful in the theory of signal processing.

Example 6.10 Differential operator is not bounded: Let $V = C^1[0,1]$ be the space of all functions which have a continuous derivative on the interval $[0,1]$. The differential operator

$$D(f) = f'$$

is a linear map $D\colon C^1[0,1] \longrightarrow C[0,1]$. But it is *not* a bounded map under the uniform norms. This can be shown using the power functions $f_n(x) = x^n$ and $Df_n(x) = nx^{n-1}$. We have $\|f_n\|_\infty = 1$ for all n but $\|Df_n\|_\infty = n$. So there is no $M > 0$ with the property that $\|Df\|_\infty \le M\|f\|_\infty$ for all f. However, it is possible to find a different norm on $C^1[0,1]$, for which the differential operator is bounded (see Exercise 6.1.9).

Theorem 6.1 *Let V and W be normed spaces with a linear operator $T\colon V \longrightarrow W$.*

The following statements are equivalent:

(a) *T is continuous at the origin.*
(b) *T is continuous on V.*
(c) *T is uniformly continuous on V.*
(d) *T is a bounded operator $T\colon V \longrightarrow W$.*
(e) *T is continuous at some point x_0.*

Proof ***(a)*** $\Rightarrow$ **(c):** Suppose T is continuous at the origin 0. Let $\epsilon > 0$ be given. Then there exists $\delta > 0$ such that

$$\|T(x) - T(0)\| < \epsilon \quad \text{whenever} \quad \|x - 0\| < \delta.$$

Suppose $\|y - z\| < \delta$. Then $\|T(y - z)\| < \epsilon$. By linearity of T, we have **(c)**:

$$\|T(y) - T(z)\| = \|T(y - z)\| < \epsilon \quad \text{whenever} \quad \|x - y\| < \delta.$$

(c) $\Rightarrow$ ***(b)*** $\Rightarrow$ ***(a)*:** is clear. So we have the equivalence of **(a)** through **(c)**.

(a) $\Rightarrow$ ***(d)*:** Let $\epsilon = 1$. There exists $\delta > 0$ such that

$$\|T(x)\| < 1 \quad \text{whenever} \quad \|x\| < \delta. \tag{6.1}$$

If $x = 0$, then $T(x) = 0$. For any $x \neq 0$, let $y = \dfrac{\delta x}{2\|x\|}$. Then

$$\|y\| = \frac{\delta\|x\|}{2\|x\|} = \frac{\delta}{2} < \delta.$$

Thus, $\|T(y)\| < 1$ and hence $\|T(x)\| < \frac{2}{\delta}\|x\|$.

This makes T a bounded linear operator with bound $M = 2/\delta$.

(d) $\Rightarrow$ ***(a)***: Suppose T is bounded, say $\|T(x)\| \leq M\|x\|$ for all $x \in V$.
Given $\epsilon > 0$, let $\|x - 0\| = \|x\| < \delta = \frac{\epsilon}{M}$. Then

$$\|T(x) - T(0)\| = \|T(x)\| \leq M\|x\| < M\left(\frac{\epsilon}{M}\right) = \epsilon.$$

Finally, the equivalence of **(a)** and **(e)** is left as Exercise 6.1.3. ∎

Corollary 6.2 *A linear functional f on a normed space V is continuous on V if and only if for some $M > 0$, we have*

$$|f(x)| \leq M\|x\| \quad \textit{for all} \quad x \in V.$$

The proof is immediate from Theorem 6.1 ***(b)*** $\Leftrightarrow$ ***(d)***.

Theorem 6.3 *Let V be a normed space. Then the same equivalences of Theorem 6.1 that hold for linear operators hold also for seminorms p. In particular, a seminorm p is continuous if and only if for some $M > 0$, we have*

$$p(x) \leq M\|x\| \quad \textit{for all} \quad x \in V.$$

The proof is left as Exercise 6.1.6.

Corollary 6.4 *In any normed space V, the norm is a continuous function from V to $\mathbb{R}$.*

Theorem 6.5 *In any normed space, the operations of vector addition and scalar multiplication are continuous. That is,*

$$\textit{if } x^k \to x,\ y^k \to y,\ \alpha_k \to \alpha, \textit{ then } \ x^k + y^k \to x + y \ \textit{ and } \ \alpha_k x^k \to \alpha x.$$

Proof The results follow from the triangle inequality

$$\|(x^k + y^k) - (x + y)\| \leq \|x^k - x\| + \|y^k - y\|$$

and absolute homogeneity

$$\|\alpha_k x^k - \alpha x\| \leq |\alpha_k|\|x^k - x\| + |\alpha_k - \alpha|\|x\|.$$ ∎

Example 6.11 Since a linear operator is continuous if and only if it is bounded, all of the Examples 6.1 through 6.9 of bounded linear operators are continuous.

Theorem 6.6 *A linear functional f on a normed space V is continuous if and only if* $\ker(f) = f^{-1}(0)$ *is a closed subspace of* V.

Proof **(⇒):** Suppose f is continuous. Then $\ker(f) = f^{-1}(0)$ is closed since it is the inverse image of the closed set $\{0\}$ (by Corollary 5.18 in Section 5.5).

(⇐): Suppose $\ker(f)$ is closed. If $f^{-1}(0) = \ker(f) = V$, then f is the zero functional, which is continuous. Otherwise, there exists $y \notin \ker(f)$. We may assume $f(y) = 1$, for if not, we could choose $\frac{y}{f(y)}$ instead of y. Since $\ker(f)$ is closed, y must be exterior to $\ker(f)$. Thus there exists an open sphere $S(y,r)$ such that

$$S(y,r) \cap \ker(f) = \emptyset. \tag{6.2}$$

Our goal is to show that

$$|f(x)| < 1 \quad \text{whenever} \quad \|x\| < r, \tag{6.3}$$

which is similar to statement (6.1) in the proof of Theorem 6.1. As in that proof, (6.3) shows that f is continuous because it has a bound of $M = \frac{2}{r}$,

$$|f(x)| < \frac{2}{r}\|x\|.$$

Now to prove (6.3), we assume that it is not true. Then there exists $x \in S(0,r)$ with $|f(x)| \geq 1$. But then $w = \frac{-1}{f(x)}x \in S(0,r)$ as well. Translating by y, we have $y + w \in S(y,r)$. But $y + w \in \ker(f)$. Thus

$$y + w \in S(y,r) \cap \ker(f),$$

which contradicts (6.2). This contradiction proves (6.3). ∎

Exercises for Section 6.1

6.1.1 Let $y \in c_0$ be fixed. Show that for any $x \in \ell_\infty$, the expression $T_y x = (y_1x_1, y_2x_2, y_3x_3, \ldots)$ defines a linear operator $T_y : \ell_\infty \longrightarrow c_0$.

6.1.2 Show that Fredholm operators of the first kind as described in Example 6.9 are linear and bounded operators on $C[a,b]$.

6.1.3 Prove the equivalence of **(a)** and **(e)** in Theorem 6.1.

6.1.4 Show that every linear operator on a finite-dimensional normed space is bounded.

6.1.5 Show that all matrix operators on $\mathbb{R}^n$ are continuous.

6.1.6 Prove Theorem 6.3.

6.1.7 Recall the sequence space bv as given in Exercises 5.2.9 and 5.5.36 (in the Exercises for Sections 5.2 and 5.5, respectively). Find a linear isometry from bv onto ℓ_1; that is, find a one-to-one onto linear operator $T\colon bv \longrightarrow \ell_1$ satisfying $\|Tx\|_1 = \|x\|'_{bv}$.

6.1.8 Show that the sum of two norms on a given space V is a norm.

6.1.9 Let $D\colon C^1[0,1] \longrightarrow C[0,1]$ be defined by $D(f) = f'$ as given in Example 6.10. Find a norm which makes D a bounded operator.

6.2 Operator Spaces

Definition 6.3 The space of *all* linear operators from normed spaces V to W is denoted by

$$L(V,W).$$

The space of all *continuous* linear operators is denoted by

$$B(V,W).$$

For $W \neq \{0\}$ we have $L(V,W) = B(V,W)$ if and only if V is finite-dimensional (see Theorem 6.7 and Exercise 6.2.3). For infinite-dimensional V we have $B(V,W) \subsetneq L(V,W)$ but $L(V,W)$ is generally too large to be useful, so $B(V,W)$ is used to obtain information about V.

Definition 6.4 The space of all linear functions on a normed space V is denoted V^*, called the **algebraic dual** of V. The space of all continuous linear functions is denoted by V', called the **topological dual** (or simply the **dual**) of V.

As noted above, if V is infinite-dimensional, not all linear operators are continuous. That is, V' will always be a proper subset of V^* (Exercise 6.2.2). The example below exhibits a linear functional that is not continuous.

Example 6.12 Let $V = \ell_1$ with the supremum norm $\|x\|_\infty = \sup_k |x_k|$ and let $f(x) = \sum_{k=1}^{\infty} x_k$. Clearly $f \in V^*$. We show that $f \notin V'$. Assume $f \in V'$. Then there exists $M > 0$ such that

$$|f(x)| \leq M\|x\|_\infty. \tag{6.4}$$

Choose any integer N larger than M. Define the sequence $x \in \ell_1$ as follows:

$$x_k = \begin{cases} 1 & \text{if } 1 \leq k \leq N \\ 0 & \text{if } N < k. \end{cases}$$

Then $\|x\|_\infty = 1$ and $f(x) = \sum_{k=1}^{\infty} x_k = N$. Inequality (6.4) implies $N \leq M$, which contradicts our choice of N.

Theorem 6.7 *Let T be a linear operator from a* finite-dimensional *normed space V to any other normed space W. Then T is continuous. That is, for finite-dimensional normed space V, we have $L(V, W) = B(V, W)$.*

Proof Let V be a normed space with an algebraic basis $\{b^1, \ldots, b^n\}$. For each $x = \sum_{k=1}^{n} a_k b^k \in V$, define the norm $\|x\|' = \sup_k |a_k|$. This is equivalent to the original norm of V (Theorem 5.46 in Section 5.10). For $T\colon V \longrightarrow W$, we have

$$\|Tx\| = \left\| \sum_{k=1}^{n} a_k T b^k \right\| \leq \sum_{k=1}^{n} |a_k| \|Tb^k\| \leq \left(\sum_{k=1}^{n} \|Tb^k\| \right) \|x\|'.$$

This shows that T is bounded by $M = \sum_{k=1}^{n} \|Tb^k\|$, and is thus continuous. ∎

Corollary 6.8 *If V is finite-dimensional, then $V' = V^*$.*

Definition 6.5 Let $T \in B(V, W)$. There exists $M > 0$ such that $\|Tx\| \leq M\|x\|$ for all $x \in V$. Then M is an upper bound of $\frac{\|Tx\|}{\|x\|}$ for all $x \neq 0$. Hence, there exists a least upper bound, which we denote the **operator** norm $\|T\|$. That is, the operator norm of T is defined to be

$$\|T\| = \sup_{x \neq 0} \frac{\|Tx\|}{\|x\|}. \tag{6.5}$$

A consequence of this definition is the inequality

$$\|Tx\| \leq \|T\| \cdot \|x\| \quad \text{for all } x \in V. \tag{6.6}$$

Theorem 6.9 *Two equivalent formulations of the operator norm* (6.5) *are*

$$\|T\| = \sup_{\|x\| \leq 1} \|Tx\| \tag{6.7}$$

and

$$\|T\| = \sup_{\|x\| = 1} \|Tx\|. \tag{6.8}$$

Proofs that (6.5), (6.7), and (6.8) are equivalent are left as Exercise 6.2.1.

Theorem 6.10 *The norm as described by* (6.5) *makes $B(V, W)$ a normed linear space.*

The proof is left as Exercise 6.2.4.

Theorem 6.11 *Let V and W be normed linear spaces. Then $B(V,W)$ is a Banach space whenever W is a Banach space (even if V is not a Banach space).*

Proof Let T_n be a Cauchy sequence in $B(V,W)$. For each $x \in V$, $T_n x$ is a Cauchy sequence in W since

$$\|T_n x - T_m x\| \le \|T_n - T_m\| \cdot \|x\| \to 0 \quad \text{as} \quad n, m \to \infty.$$

Since W is complete, the Cauchy sequence $T_n x$ converges. We define T by the equation $Tx = \lim_{n\to\infty} T_n x$. We show that:

(1) $T \in L(V,W)$,
(2) $T \in B(V,W)$, and
(3) $T_n \to T$ as $n \to \infty$ in $B(V,W)$.

(1): By properties of limits, T is clearly a linear operator from V to W.

(2): Next, we show that T is continuous. Since every Cauchy sequence is a bounded sequence, there exists an M such $\|T_n\| \le M$ for all n. We have

$$\|T_n x\| \le \|T_n\| \cdot \|x\| \le M\|x\| \quad \text{for all} \quad x \in V.$$

Letting $n \to \infty$, results in $\|Tx\| \le M\|x\|$; that is, $T \in B(V,W)$.

(3): It remains to be shown that the Cauchy sequence T_n actually converges to T. Let $\epsilon > 0$ be given. There exists $N > 0$ such that

$$\|T_n - T_m\| < \epsilon \quad \text{whenever} \quad n, m > N.$$

Then for each $x \in V$ and $n, m > N$, $\|T_n x - T_m x\| \le \|T_n - T_m\| \cdot \|x\| < \epsilon\|x\|$.

Letting $m \to \infty$, we have $\|T_n x - Tx\| \le \epsilon\|x\|$.

By definition of the operator norm, $\|T_n - T\| \le \epsilon$ for all $n > N$; that is,

$$\lim_{n\to\infty} T_n = T \quad \text{in} \quad B(V,W).$$ ∎

Corollary 6.12 *If V is a normed space, then V' is a Banach space under the norm*

$$\|f\| = \sup_{x\neq 0} \frac{|f(x)|}{\|x\|},$$

even if V is not a Banach space.

Exercises for Section 6.2

6.2.1 Prove that (6.7) and (6.8) are equivalent formulations of the definition of the operator norm (6.5).

6.2.2 Show that for any infinite-dimensional normed space V, the topological dual V' is a proper subset of V^*. Hint: Let B be an algebraic basis of V (see Theorem 0.63 in Section 0.14) and let $C = \{b^1, b^2, \ldots\}$ be a countably infinite subset of B. Define f to be the unique linear functional determined by $f(b^k) = k\|b^k\|$ for $b^k \in C$ $(k = 1, 2, 3, \ldots)$ and $f(b) = 0$ for $b \in B - C$.

6.2.3 Show that for $W \neq \{0\}$ and any infinite-dimensional normed space V, the space of continuous linear operators $B(V, W)$ is a proper subset of $L(V, W)$. Hint: Modify Exercise 6.2.2.

6.2.4 Prove Theorem 6.10.

6.3 Linear Functionals

It is often important in functional analysis to identify the spaces of linear operators $B(V, W)$. In many cases, this is difficult. However, the topological duals for most standard sequence spaces have been characterized. First, we formally define the meaning of equivalent normed spaces.

Definition 6.6 Normed spaces V and W are called **isometric** (or **equivalent**) if there is a **linear isometry** between V and W; that is, there exists a linear operator $T : V \longrightarrow W$ which is one-to-one and onto W, and for which $\|Tx\| = \|x\|$, for all $x \in V$. In a formal sense we should use a symbol such as $V \sim W$ but often we simply write $V = W$.

Definition 6.7 The sequences with 1 in the kth position and 0 elsewhere are denoted by

$$e^k = (0, \ldots, 0, 1, 0, \ldots) \text{ for } k = 1, 2, \ldots.$$

The sequence with 1 in every position is denoted by

$$e = (1, 1, 1, \ldots).$$

The sequence with 1 in the first n positions and 0 elsewhere is denoted by

$$P^n = \sum_{k=1}^{n} e^k = (1, \ldots, 1, 0, 0, \ldots).$$

Now we show that the dual of the sequence space c_0 is isometric to the sequence space ℓ_1.

The usual norm of c_0 is $\|x\|_\infty = \sup_k |x_k|$. For each sequence x, let

$$P^n x = \sum_{k=1}^{n} x_k e^k = (x_1, x_2, \ldots, x_n, 0, 0, \ldots).$$

The space c_0 has the property that $P^n x \to x$ as $n \to \infty$ for all $x \in c_0$ because

$$\|x - P^n x\|_\infty = \|(0,\ldots,0,x_{n+1},x_{n+2},\ldots)\|_\infty = \sup_{k>n} |x_k| \to 0 \text{ as } n \to \infty.$$

Theorem 6.13 *Under the usual norms, we have the isometry* $c_0' = \ell_1$.

Proof The meaning of this isometry $c_0' = \ell_1$ is as follows:

(1): For every $f \in c_0'$, there corresponds a unique $a = (a_1,a_2,\ldots) \in \ell_1$ such that $f(x) = \sum_{k=1}^\infty a_k x_k$ for all $x \in c_0$.

(2): For every $a = (a_1,a_2,\ldots) \in \ell_1$, there corresponds a unique $f_a \in c_0'$ such that $f_a(x) = \sum_{k=1}^\infty a_k x_k$ for all $x \in c_0$.

(3): The operator norm on c_0' satisfies $\|f_a\| = \|a\|_1 = \sum_k |a_k|$.

Proof of (1): Let f be a continuous linear functional on c_0 and let $a_k = f(e^k)$ for each $k = 1,2,\ldots$. Then

$$f(x) = f(\lim_n P^n x) = \lim_n f(P^n x) = \lim_n \sum_{k=1}^n a_k x_k = \sum_{k=1}^\infty a_k x_k \text{ for all } x \in c_0.$$

It remains to be shown that $a = (a_1,a_2,\ldots) \in \ell_1$. For each n define the finite sequence $t^n = \sum_{k=1}^n (\text{sgn } a_k) e^k \in c_0$. Clearly $\|t^n\|_\infty \le 1$. Thus $|f(t^n)| = \sum_{k=1}^n |a_k| \le \|f\| \cdot \|t^n\|_\infty \le \|f\|$. Letting $n \to \infty$ gives us $\|a\|_1 \le \|f\|$.

Proof of (2): If $a \in \ell_1$ and f_a is defined by $f_a(x) = \sum_{k=1}^\infty a_k x_k$, then

$$|f_a(x)| = \left|\sum_{k=1}^\infty a_k x_k\right| \le \sup_k |x_k| \sum_k |a_k| = \|a\|_1 \cdot \|x\|_\infty.$$

Letting $M = \|a\|_1$, it follows that f_a is a bounded linear operator on c_0. This also shows that $\|f_a\| \le \|a\|_1$.

Proof of (3): $\|a\|_1 \le \|f_a\|$ is shown in part **(1)** above and $\|f_a\| \le \|a\|_1$ is shown in part **(2)**. ∎

We can follow the pattern of this theorem to prove the following.

Theorem 6.14 *Under the usual norms, we have the isometry* $\ell_1' = \ell_\infty$.

The proof is left as Exercise 6.3.3.

Theorem 6.15 *If* $1 < p < \infty$ *and* $1 < q < \infty$ *are Hölder conjugates* $\left(\frac{1}{p} + \frac{1}{q} = 1\right)$, *then the dual of* ℓ_p *is isometric to* ℓ_q.

Proof We show that

(1): For every $f \in (\ell_p)'$, there corresponds a unique $a = (a_1,a_2,\ldots) \in \ell_q$ such that $f(x) = \sum_{k=1}^\infty a_k x_k$ for all $x \in \ell_p$.

(2): For every $a = (a_1, a_2, \ldots) \in \ell_q$, there corresponds a unique $f_a \in (\ell_p)'$ such that $f_a(x) = \sum_{k=1}^{\infty} a_k x_k$ for all $x \in \ell_p$.

(3): The operator norm on ℓ_p' satisfies $\|f_a\| = \|a\|_q = \left(\sum_{k=1}^{\infty} |a_k|^q\right)^{1/q}$.

Proof of (1): Let $P^n x = \sum_{k=1}^{n} x_k e^k = (x_1, x_2, \ldots, x_n, 0, 0, \ldots)$. The Banach space ℓ_p has the property that $P^n x \to x$ as $n \to \infty$ for each $x \in \ell_p$ (you should check this fact). Let f be a continuous linear functional on ℓ_p and define the sequence $a = (a_1, a_2, \ldots)$ by $a_k = f(e^k)$ for $k = 1, 2, \ldots$. Then

$$f(x) = f(\lim_n P^n x) = \lim_n f(P^n x) = \lim_n \sum_{k=1}^{n} a_k x_k = \sum_{k=1}^{\infty} a_k x_k \ \forall\, x \in \ell_p.$$

It remains to be shown that $a \in \ell_q$. Fix n and define sequence t as follows:

$$t_k = \begin{cases} \frac{|a_k|^q}{a_k} & : \quad \text{if } a_k \neq 0 \text{ and } k \leq n \\ 0 & : \quad \text{otherwise}\,. \end{cases}$$

You can easily check that $|t_k|^p = |a_k|^q$ for $k = 1, \ldots, n$. Then

$$|f(t)| = \left|\sum_{k=1}^{n} t_k f(e^k)\right| = \left|\sum_{k=1}^{n} t_k a_k\right| = \sum_{k=1}^{n} |a_k|^q \leq \|f\| \cdot \|t\|_p.$$

But $\|t\|_p = \left(\sum_{k=1}^{n} |a_k|^{(q-1)p}\right)^{1/p} = \left(\sum_{k=1}^{n} |a_k|^q\right)^{1/p}$. Division by $\|t\|_p$ results in

$$\left(\sum_{k=1}^{n} |a_k|^q\right)^{1/q} \leq \|f\| \ \text{ for each } \ n = 1, 2, \ldots.$$

Letting $n \to \infty$, we have $\|a\|_q \leq \|f\|$.

Proof of (2): Let $a \in \ell_q$ and $f_a(x) = \sum_{k=1}^{\infty} a_k x_k$. Then $|f_a(x)| = \left|\sum_{k=1}^{\infty} a_k x_k\right| \leq \|a\|_q \|x\|_p$ by Holder's inequality. Letting $B = \|a\|_q$, we have that f_a is a bounded linear operator on ℓ_p. This also shows that $\|f_a\| \leq \|a\|_q$.

Proof of (3): $\|a\|_q \leq \|f_a\|$ is shown in part **(1)** above and $\|f_a\| \leq \|a\|_q$ is shown in part **(2)**. ∎

Definition 6.8 Let E be a normed sequence space. For $x \in E$ and each $= 1, 2, 3, \ldots$, define the n**th section of** x to be

$$P^n x = (x_1, x_2, \ldots, x_n, 0, 0, \ldots) = \sum_{k=1}^{n} x_k e^k.$$

A normed sequence space E is said to have the **property AK** (or E is called an **AK-space**) if

$$P^n x \to x \quad (\text{as} \quad n \to \infty) \quad \text{for all} \quad x \in E.$$

The property AK is discussed in further detail in Section 9.6.

Theorem 6.16 *If E is an AK-space, then every $f \in E'$ is of the form*

$$f(x) = \sum_{k=1}^{\infty} x_k a_k \quad \textit{where} \quad a_k = f(e^k) \quad \textit{for} \quad k = 1, 2, \dots.$$

The spaces c_0 and ℓ_p for $1 \le p < \infty$ are AK-spaces. This fact was used in the proofs above to characterize the duals of these spaces. The space ℓ_∞ does not have the property AK. For example, for the sequence $e = (1, 1, 1, \dots)$, we have

$$\|e - P^n e\|_\infty = \|(0, 0, \dots, 0, 0, 1, 1, \dots)\|_\infty = 1 \quad \text{for all} \quad n.$$

So $P^n e$ does not converge to e in ℓ_∞.

Similarly, the sequence space c of convergent sequences, under the usual sup norm $\|x\|_\infty = \sup_k |x_k|$, does not have the property AK. However, it is still possible to characterize the dual space.

Theorem 6.17 *If $f \in c'$, then:*

(1) *There exists a number δ and a sequence $a \in \ell_1$ such that,*

$$f(x) = \delta \lim_k x_k + \sum_k a_k x_k \quad \textit{for all} \quad x \in c. \tag{6.9}$$

(2) *Every f of this form is continuous on the space c.*
(3) *The dual (operator) norm of f is $\|f\| = |\delta| + \sum_k |a_k|$.*

Proof The usual sup norm of c is the same as for c_0. So if f is a continuous linear functional on c, then it is also a continuous linear functional on c_0. We showed in Theorem 6.13 that for every continuous linear functional on c_0, the sequence $a = (a_k) = \big(f(e^k)\big)$ is in ℓ_1.

For every element $x \in c$ with $L = \lim_k x_k$, the sequence $y = x - Le$ is a null sequence (recall, $e = (1, 1, 1, \dots)$). By Theorem 6.13,

$$f(y) = \sum_k a_k y_k = \sum_k a_k (x_k - L) = \sum_k a_k x_k - L \sum_k a_k.$$

Thus

$$\begin{aligned} f(x) &= f(Le+y) = Lf(e) + f(y) \\ &= Lf(e) + \sum_k a_k x_k - L \sum_k a_k \\ &= L\big(f(e) - \sum_k a_k\big) + \sum_k a_k x_k \\ &= \delta L + \sum_k a_k x_k \end{aligned}$$

with $\delta = f(e) - \sum_k a_k$. This proves **(1)**.

Taking the absolute value of a linear functional of the form (6.9),

$$|f(x)| \le |\delta| \big|\lim_k x_k\big| + \sum_k |a_k| \sup_k |x_k| \le \big(|\delta| + \sum_k |a_k|\big) \|x_k\|_\infty,$$

shows that f is bounded on c, and hence continuous. This proves **(2)**. It also gives us the inequality $\|f\| \le |\delta| + \sum_k |a_k|$ for part of **(3)**. To obtain the opposite inequality for **(3)**, fix n, and define the sequence t as follows:

$$t_k = \begin{cases} \text{sgn } a_k & : \quad \text{if } 1 \le k \le n \\ \text{sgn } \delta & : \quad \text{otherwise .} \end{cases}$$

Clearly $t \in c$ and $\|t\|_\infty \le 1$ and

$$|f(t)| = |\delta| + \sum_{k=1}^{n} |a_k| + \left|(\text{sgn } \delta) \sum_{k=n+1}^{\infty} a_k\right| \le \|f\| \cdot \|t\|_\infty = \|f\|.$$

The term $\left|(\text{sgn } \delta) \sum_{k=n+1}^{\infty} a_k\right|$ tends to zero as $n \to \infty$ since it is bounded by $\sum_{k=n+1}^{\infty} |a_k|$, which tends to 0. Thus $|\delta| + \sum_{k=1}^{\infty} |a_k| \le \|f\|$. ∎

Theorem 6.18 *The Banach spaces c' and ℓ_1 are isometric.*

Proof Take the linear isometry $T: c' \longrightarrow \ell_1$ defined by $T(f) = (\delta, a_1, a_2, \ldots)$, for f given by Equation (6.9). The proof that T is one-to-one, onto ℓ_1, and that $\|T(f)\|_1 = \|f\|$, is left for Exercise 6.3.4. ∎

Exercises for Section 6.3

6.3.1 Show that for any fixed constants α, β, and any $f \in C[a,b]$, the function $T(f) = \alpha f(a) + \beta f(b)$ is a bounded linear functional on $C[a,b]$.

6.3.2 Let g be a fixed function in $C[a,b]$. Show that $T(f) = \int_a^b f(t)g(t)\,dt$ is a bounded linear functional on $C[a,b]$.

6.3.3 Follow the pattern of the proof of Theorem 6.13 to prove Theorem 6.14 that the dual of ℓ_1 is equivalent to ℓ_∞.

6.3.4 Finish the details of the proof of Theorem 6.18.

6.4 The Hahn–Banach Extension Principle

This extension principle shows that every continuous linear functional defined on a subspace U of a normed space V can be extended as a continuous linear functional to the entire space V. It was proved by Hans Hahn,[2] and independently by Stefan Banach, both in 1928. However, it had been proved earlier by Eduard Helly[3] in 1912. The proof depends on Zorn's lemma (Theorem 0.62 in Section 0.14), which is equivalent to the axiom of choice.

Definition 6.9 Let V be a linear space with a subspace U. Suppose that f is a linear functional defined on U. A linear functional g on all of V is called a **linear extension** of f if

$$g(x) = f(x) \quad \text{for all} \quad x \in U.$$

Suppose that p is a seminorm defined on V. We say that f is **dominated by** p on U if

$$|f(x)| \le p(x) \quad \text{for all} \quad x \in U.$$

We first consider the real case of the Hahn–Banach theorem.

Theorem 6.19 Hahn–Banach (real version): *Suppose V is a linear space over the* reals *and p is a seminorm on V. If U is a subspace of V and $f \in U^*$ is a linear functional dominated by p on U, then f has a linear extension g to all of V which is dominated by p on all of V.*

[2] Hans Hahn (1879–1934) was an Austrian mathematician. He was one of the founders of the Vienna Circle of mathematicians, physicists, economists, and philosophical thinkers who met in Viennese coffee houses from 1907 onwards.

[3] Eduard Helly (1884–1943) was an Austrian mathematician who had a difficult life. He was shot in the lungs in 1915 during World War I and suffered from this injury for the rest of his life. He was a prisoner of war in Siberia until 1920. He later returned to Austria but fled from the Nazis to America in 1938. He died of a heart attack shortly after he was offered a Chair of mathematics at Illinois Institute of Technology. His work in 1912 contained a number of important results but these contributions were not generally recognized until much later.

Proof We use Zorn's lemma (Theorem 0.62 in Section 0.14). Suppose that V is a linear space with a seminorm p. Let $\mathcal{A}$ be the family of all linear subspaces W for which

$$U \subset W \subset V$$

and such that f has a linear extension to W dominated by p. Now consider $\mathcal{C}$, a chain of subspaces in $\mathcal{A}$. It is left as Exercise 6.4.2, to show that $\bigcup_{W \in \mathcal{C}} W$ is a linear space and that f has a linear extension to all of $\bigcup_{W \in \mathcal{C}} W$ dominated by p. That is, for any chain $\mathcal{C}$ in $\mathcal{A}$, the union of the chain $\bigcup_{W \in \mathcal{C}} W$ is in $\mathcal{A}$. By Zorn's lemma, $\mathcal{A}$ has a maximal subspace M.

It remains to be shown that this maximal subspace is V. Assume not. Then there exists $z \in V - M$. Consider the strictly larger subspace

$$M_z = \text{span}\{M, z\} = \{x + \alpha z \mid x \in M, \alpha \in \mathbb{R}\}.$$

We will obtain a linear extension g of f to M_z dominated by p, which then contradicts the maximality of M.

Let x, $y \in M$. Since f is dominated by p on M,

$$\begin{aligned} f(x) - f(y) = f(x - y) \le p(x - y) &= p\big((x + z) - (y + z)\big) \\ &\le p(x + z) + p(y + z), \end{aligned}$$

and thus

$$-p(y + z) - f(y) \le p(x + z) - f(x).$$

For each $y \in M$, we take the infimum of the right side to obtain

$$-p(y + z) - f(y) \le d = \inf_{x \in M} \{p(x + z) - f(x)\}.$$

Then we take the supremum of the left side to obtain

$$\sup_{y \in M} \{-p(y + z) - f(y)\} \le d.$$

Thus, for any $y \in M$, we have $-p(y+z) - f(y) \le d \le p(y+z) - f(y)$ or

$$|f(y) + d| \le p(y + z). \tag{6.10}$$

For each $w = x + \alpha z \in M_z$, define $g(w) = f(x) + \alpha d$.

Clearly g is a linear functional on M_z and g is a linear extension of f to M_z. To show that g is dominated by p on M_z, let $y = \frac{x}{\alpha}$ for $\alpha \neq 0$. From (6.10) we obtain

$$\left| \frac{f(x) + \alpha d}{\alpha} \right| \le \frac{1}{|\alpha|} p(x + \alpha z).$$

It follows that, for $w = x + \alpha z$, we have

$$|g(w)| = |f(x) + \alpha d| \leq p(x + \alpha z) = p(w),$$

which shows that the extension g is dominated by p on the larger M_z, contradicting the maximality of M. ∎

Now consider the complex form of the Hahn–Banach theorem. It was proved by H. F. Bohnenblust & A. Sobczyk and G.A. Soukhomlinoff in 1938. It uses the real form. The proof is interesting that over twenty years passed before the complex form was proven.

Theorem 6.20 Hahn–Banach: *This is the same statement as given in the real version except that V is a complex linear space.*

Proof Split f into its real and imaginary parts $f = f_1 + if_2$. Then f_1, f_2 are real linear functionals on U, where U is viewed as a linear space over $\mathbb{R}$. The real version of the Hahn–Banach theorem applies. Since $f(ix) = if(x)$, we have

$$f(ix) = f_1(ix) + if_2(ix) = if(x) = if_1(x) - f_2(x).$$

Thus $f_1(ix) = -f_2(x)$ on U and $|f_1(x)| \leq |f(x)| \leq p(x)$ on U. By the real version of the theorem, there exists an extension g_1 of f_1 to V. Next, let

$$g(x) = g_1(x) - ig_1(ix) \quad \text{for all} \quad x \in V,$$

which is an extension of f to V, since for all $x \in U$, we have

$$g(x) = g_1(x) - ig_1(ix) = f_1(x) - if_1(ix) = f_1(x) + if_2(x) = f(x).$$

Also

$$g(ix) = g_1(ix) - ig_1(-x) = ig_1(x) + g_1(ix) = ig(x).$$

This shows that g is complex and linear. To show that g is dominated by p on V, write $g(x)$ in polar form: $g(x) = re^{i\theta}$ for $x \in V$. Then $e^{-i\theta}g(x) = r$ is real and nonnegative, so that

$$|g(x)| = e^{-i\theta}g(x) = g(e^{-i\theta}x) = g_1(e^{-i\theta}x) \leq p(e^{-i\theta}x) = |e^{-i\theta}|p(x) = p(x).$$

The equality $g(e^{-i\theta}x) = g_1(e^{-i\theta}x)$ holds since $g(e^{-i\theta}x)$ is real and hence equal to its real part $g_1(e^{-i\theta}x)$. ∎

The next form of the Hahn–Banach theorem applies to normed spaces.

Theorem 6.21 *Suppose V is a normed space with a subspace U. Let $f \in U'$ be a continuous linear functional on U (that is, there exists $M > 0$ such that*

$|f(x)| \leq M\|x\|$, for all $x \in U$). Then there exists a linear extension g of f that is continuous on all *of V with $\|g\| = \|f\|$.*

Proof Let $p(x) = B\|x\|$, with $B = \|f\|$ (the norm of f on U). By the Hahn–Banach theorem, $|g(x)| \leq \|f\| \cdot \|x\|$, which proves $\|g\| \leq \|f\|$. The opposite inequality follows from the definition of the operator norms $\|f\|$ and $\|g\|$. ∎

Example 6.13 Banach limit: Consider the continuous linear functional $f(x) = \lim_k x_k$ defined on the space of convergent sequences c. We have

$$|f(x)| \leq \sup_k |x_k| = \|x\|_\infty \quad \text{for all} \quad x \in c.$$

By the Hahn–Banach theorem, $\lim_k x_k$ may be extended to a continuous linear functional L on all of ℓ_∞. This extension is called a **Banach limit.**

Corollary 6.22 *Suppose z is a nonzero vector in a normed space V. Then there exists $g \in V'$ such that $g(z) = \|z\|$ and $\|g\| = 1$. So if V is not the trivial space $\{0\}$, then neither is V'.*

Proof Let $U = \{\alpha z \mid \alpha \in \mathbb{C} \text{ (or } \mathbb{R})\}$ be the span of z. Define the linear functional f on U as follows

$$f(\alpha z) = \alpha\|z\| \quad \text{for} \quad \alpha z \in U.$$

Since $z \neq 0$, clearly $\|f\| = 1$ on U and $f(z) = \|z\|$. By the Hahn–Banach theorem, there exists $g \in V'$ such that $g(z) = f(z) = \|z\|$ and $\|g\| = \|f\| = 1$. ∎

Corollary 6.23 *Suppose V is a normed space with elements $x \neq y$. Then there exists $f \in V'$ such that $f(x) \neq f(y)$. We say that V'* **separates points** *on V.*

The proof of this Corollary is left as Exercise 6.4.3.

Corollary 6.24 *Let V be a normed space with $x \in V$. If $f(x) = 0$ for all $f \in V'$, then $x = 0$.*

The proof of this Corollary is left as Exercise 6.4.4.

Recall that the distance between a point z and a set A is defined to be

$$d(z, A) = \inf_{x \in A} d(x, z).$$

Corollary 6.25 *Let U be a subspace of a normed space V. If $z \in V - U$ and $d(z, U) > 0$, then there exists $g \in V'$ such that $g = 0$ on U and $g(z) = 1$.*

Proof Consider the subspace $U_z = \text{span}\{U, z\} = \{w = x + \alpha z \mid x \in U, \alpha \in \mathbb{C} \text{ (or } \mathbb{R})\}$ and define the linear functional

$$g(w) = g(x + \alpha z) = \alpha.$$

Clearly g is linear on U_z, and $g(w) = 0$ for $w \in U$, and $g(z) = 1$. It is easy to see that, since U is a linear space, we have

$$d = \inf_{y \in U} \|y - z\| = \inf_{y \in U} \| - y - z\| = \inf_{y \in U} \|y + z\|$$

(which is the same as the real numbers d used in the proof of Theorem 6.19). Thus, for each $y \in U$, we have $0 < d \le \|y + z\|$. To show that g is continuous on U_z, let $y = \frac{x}{\alpha}$ for $\alpha > 0$. Then

$$0 < d = \inf_{y \in U} \|y + z\| \le \frac{1}{|\alpha|} \|x + \alpha z\|.$$

It follows that, for $w = x + \alpha z$, we have $|g(w)| = |\alpha| \le \frac{1}{d} \|x + \alpha z\| = \frac{1}{d} \|w\|$. By the Hahn–Banach theorem, g has a continuous extension to all of V. ∎

Corollary 6.26 *Suppose U is a* closed *subspace of a normed space V. If $z \in V - U$, then there exists $g \in V'$ such that $g(z) = 1$ but $g(x) = 0$ for all $x \in U$.*

The proof follows from Exercise 6.4.5.

Exercises for Section 6.4

6.4.1 Let V be a linear space, U be a proper subspace $\{0\} \subsetneqq U \subsetneqq V$, and let f be a nonzero linear functional f on U. Show that the following extension g of f to all of V is *not* a linear extension:

$$g(x) = \begin{cases} f(x) & \text{if } x \in U \\ 0 & \text{if } x \notin U. \end{cases}$$

6.4.2 In the proof of the real version of the Hahn–Banach theorem, consider any chain $\mathcal{C}$ in $\mathcal{A}$. Suppose f has a linear extension to W dominated by p, for each $W \in \mathcal{C}$.

(1) Show that $\bigcup_{W \in \mathcal{C}} W$ is a linear space.
(2) Show that f has a linear extension to all of $\bigcup_{W \in \mathcal{C}} W$ dominated by p.
(3) Then use Zorn's lemma to show that $\mathcal{A}$ has a maximal element.

6.4.3 Prove Corollary 6.23.

6.4.4 Prove Corollary 6.24.

6.4.5 Show that if A is a closed subset of a metric space M and $z \notin A$, then $d(z, A) > 0$.

6.5 Second Dual Space

Definition 6.10 Let V be a normed space and let V' be its dual. As noted V' is always a Banach space. The dual of the Banach space V' is called the **second dual** of V and written V''.

Clearly each $x \in V$ defines a linear functional F_x on V' given by $F_x(f) = f(x)$. This linear functional F_x on V' is continuous because it is bounded

$$|F_x(f)| = |f(x)| \leq \|f\| \cdot \|x\|.$$

In this sense we often write $V \subset V''$.

We say that V is **reflexive** if V is equivalent to its second dual V''. Since a dual space is always a Banach space, a normed space V which is *not* a Banach space cannot be reflexive.

Theorem 6.27 *If V' is separable, then V is separable as well.*

Proof Suppose V' is separable; that is, V' has a countable dense subset. Then the surface of the unit sphere $B = \{f \in V' \mid \|f\| = 1\}$ must have a countable dense subset as well (see Exercise 5.5.33). Suppose that $\{f_1, f_2, \ldots\}$ is a countable dense subset of B. Since we have

$$\|f_n\| = \sup_{\|x\|=1} |f_n(x)| = 1, \text{ for each } n = 1, 2, \ldots,$$

there exist $x^n \in V$ with $\|x^n\| = 1$, such that $|f_n(x^n)| > \frac{1}{2}$. We show that $Y = \text{span}\{x^1, x^2, \ldots\}$ is dense in V, which shows that V is separable (take all linear combinations of $x^1, x^2, \ldots$ with rational coefficients). Assume that Y is not dense in V. By Corollary 6.26, there exists a nonzero $f \in V'$ such that $f(x^n) = 0$, for all n. Normalize f so that $\|f\| = 1$; then $f \in B$. Since $\{f_1, f_2, \ldots\}$ is dense in V', for some n, we have $\|f_n - f\| < 1/2$. But this contradicts

$$\frac{1}{2} < |f_n(x^n)| = |(f_n - f)(x^n)| \leq \|f_n - f\| \|x^n\| = \|f_n - f\|. \qquad \blacksquare$$

The converse is false as can be seen by the example of ℓ_1 which is separable but its dual ℓ_∞ is not. See Exercise 6.5.3.

Exercises for Section 6.5

6.5.1 Show that V is reflexive if and only if the second dual V'' is reflexive.

6.5.2 Show that a closed subspace of a reflexive normed space is reflexive.

6.5.3 Show that the spaces ℓ_p for $1 < p < \infty$ are reflexive.

6.5.4 Show that the space ℓ_1 is not reflexive.

6.5.5 Show that the space ℓ_∞ is not reflexive.

6.5.6 Show that the space c_0 is not reflexive.

6.6 Category of a Set

Recall that a subset A of a metric space M is nowhere dense if its closure $\overline{A}$ contains no interior points (Theorem 5.22 in Section 5.5). The union of a finite collection of nowhere dense sets is nowhere dense. However, the union of a countable collection of nowhere dense sets need not be nowhere dense.

Example 6.14 The set of rationals $\mathbb{Q}$ is the union of its elements $r_1, r_2, \ldots$, each of which forms a nowhere dense set in $\mathbb{R}$. Since $\mathbb{Q}$ is dense in $\mathbb{R}$, it is an example of a countable union of nowhere dense sets which is not nowhere dense.

Definition 6.11 A set A in a metric space M is said to be of the **first category** in M if it is the union of countable many nowhere dense sets. Otherwise, A is said to be of the **second category** in M.

The notion of category and its consequence on uniform boundedness of collections of functions is important in analysis. The set $\mathbb{Q}$ is of the first category in $\mathbb{R}$. However, Theorem 6.29 will show that $\mathbb{R}$ is of the second category.

Lemma 6.28 *Let M be a complete metric space and let $G_1, G_2, \ldots$ be a sequence of dense open sets in M. Then $\bigcap_k G_k \neq \emptyset$.*

Proof Take $x^1 \in G_1$. Since G_1 is open, there exists an open sphere $S(x^1, r_1) \subset G_1$. Since G_2 is dense in M, there exists $x^2 \in G_2 \cap S(x^1, r_1/2)$. Since G_2 is open, there exists an open sphere $S(x^2, r_2) \subset G_2$ with $r_2 < \frac{1}{2} r_1$. Then

$$S(x^2, r_2) \subset S(x^1, r_1) \cap G_2 \subset G_1 \cap G_2 \quad \text{with} \quad r_2 < r_1/2.$$

Continuing this way we obtain a nest of closed spheres $S_k = S[x^k, r_k/2] \subset S(x^k, r_k)$ with $r_k < \frac{r_1}{2^k}$ and

$$\bigcap S_k \subset \bigcap S(x^k, r_k) \subset \bigcap G_k.$$

Since M is complete, the nest of closed spheres is nonempty (Theorem 5.29). ∎

Remark Actually the intersection $\bigcap_k G_k$ in the above lemma is dense in M.

Theorem 6.29 Baire category: *A complete metric space M is of the second category in M.*

Proof Assume that M is of the first category. Then $M = \bigcup_{k=1}^{\infty} A_k$, where each A_k is nowhere dense. By Theorem 5.22(**b**), each $F_k = \overline{A_k}$ is nowhere dense as well. Taking the complement, we have

$$\emptyset = \bigcap_{k=1}^{\infty} F_k^c = \bigcap_{k=1}^{\infty} G_k, \text{ where } G_k = F_k^c \text{ is open for all } k.$$

By the definition of nowhere dense, $G_k = F_k^c$ is dense for all k. This contradicts the preceding Lemma. Thus M cannot be of first category. ∎

Example 6.15 Continuous Nowhere Differentiable Functions: We illustrate a use of the Baire category theorem to show that there exists a continuous function on $[0, 1]$ which is not differentiable at any point of the interval. We do this by showing that the set A of functions that have a derivative at some point of the interval is of the first category. Hence, the set of functions with no derivative at any point A^c must be of the second category in $C[0, 1]$, which is thus nonempty.

Let F_n be the set of all functions $f \in C[0, 1]$ for which there is some x in $\left[0, 1 - \frac{1}{n}\right]$ such that

$$|f(x+h) - f(x)| \leq nh \quad \text{for all} \quad 0 < h < \frac{1}{n}. \tag{6.11}$$

It is clear that any function in A has a finite derivative from the right at some point. Thus $A \subset \bigcup_n F_n$.

Claim 1: Each F_n is closed. Suppose that g is in the closure of F_n. Then there exists a sequence f_k of functions in F_n that converges to g. Each f_k satisfies (6.11) at some point $x_k \in \left[0, 1 - \frac{1}{n}\right]$. Since $\left[0, 1 - \frac{1}{n}\right]$ is compact, there exists a subsequence x_{k_j} converging, say to $x \in \left[0, 1 - \frac{1}{n}\right]$. By continuity we have

$$|g(x+h) - g(x)| = \lim_{j\to\infty} |f_{k_j}(x_{k_j} + h) - f_{k_j}(x_{k_j})| \leq \lim_{j\to\infty} nh = nh$$

for all $0 < h < \frac{1}{n}$. Thus $g \in F_n$. This shows that F_n is closed.

Claim 2: Each F_n has no interior points. Let $\epsilon > 0$ be given. We show that for each $f \in C[0, 1]$, the open sphere $S(f, \epsilon)$ contains a function not in F_n. Since f is continuous on $[0, 1]$, and hence uniformly continuous on $[0, 1]$, there exists $\delta > 0$ such that $|f(x) - f(y)| < \epsilon/2$ whenever $d(x, y) < \delta$. For a positive integer $M > \max\left\{\frac{2n}{\epsilon}, \frac{2}{\delta}\right\}$ we create a sawtooth-like function g as follows. For each $x_k = k/M$, let $g(x_k) = f(x_k) + \epsilon/2$ for even k, $g(x_k) = f(x_k) - \epsilon/2$ for odd k, and let g be linear in between x_k and x_{k+1}.

To show $g \in S(f, \epsilon)$, note that each $x_k \leq x \leq x_{k+1}$ can be written as $x = ax_k + (1-a)x_{k+1}$, for $0 < a < 1$, and $g(x) = xg(x_k) + (1-a)g(x_{k+1})$. Then

$$|f(x) - g(x)| \leq a|f(x) - g(x_k)| + (1-a)|f(x) - g(x_{k+1})$$
$$< a\epsilon + (1-a)\epsilon = \epsilon.$$

Now we show that $g \notin F_n$. Since g is linear between x_k and x_{k+1}, for $x_k \leq x < x + h \leq x_{k+1}$, we have

$$g(x+h) - g(x) = \frac{g(x_{k+1}) - g(x_k)}{x_{k+1} - x_k} h = \big(g(x_{k+1}) - g(x_k)\big) Mh.$$

Furthermore,

$$g(x_{k+1}) - g(x_k) = (f(x_{k+1}) \pm \epsilon/2) - (f(x_k) \mp \epsilon/2)$$
$$= \pm\epsilon + f(x_{k+1}) - f(x_k).$$

Combining the two and taking the absolute value, we have

$$|g(x+h) - g(x)| \geq \big(\epsilon - |f(x_{k+1}) - f(x_k)|\big) Mh > \Big(\epsilon - \frac{\epsilon}{2}\Big) Mh = \frac{M\epsilon h}{2}.$$

This is larger than nh by our choice of M. Thus $g \notin F_n$.

Exercises for Section 6.6

6.6.1 Show that the set of irrational numbers in $\mathbb{R}$ is not a countable union of closed sets.

6.6.2 Show that if a subset A of a complete metric space is of the first category, then A^c is of the second category.

6.6.3 Let M be a complete metric space and $M = \bigcup_{k=1}^{\infty} A_k$. Prove that for at least one k we have $(\overline{A_k})^\circ \neq \emptyset$.

6.6.4 Cantor's diagonalization was used in Theorem 0.2 to prove that the closed interval $[0,1]$ is uncountable. Use Baire category to prove the same.

6.6.5 Prove the remark after Lemma 6.28 claiming that the intersection of dense open sets $\bigcap_k G_k$ is dense in M.

6.7 Uniform Boundedness

In this section we give two important results that follow from the Baire category theorem.

Theorem 6.30 Uniform boundedness principle: *Suppose that $T_1, T_2, \ldots$ is a sequence of continuous linear operators from a Banach space V into* *a normed space W. If, for every $x \in V$, we have* $\sup_k \|T_k x\| < \infty$, *then* $\sup_k \|T_k\| < \infty$.

Proof For $n = 1, 2, \ldots$ let $B_n = \{x \in V \mid \|T_k(x)\| \leq n \text{ for all } k = 1, 2, \ldots\}$. Since, for every $x \in V$, $\sup_k \|T_k x\| < \infty$, the union $\bigcup_{n=1}^{\infty} B_n$ is the whole space V.

Claim: Each B_n is closed. Since each operator T_k is continuous and the norm is continuous (by Corollary 6.4 in Section 6.1), the set

$$B_{nk} = \{x \in V \mid \|T_k(x)\| \leq n\},$$

being the inverse image of the closed interval $[0, n]$, is closed. Hence, the intersection $B_n = \bigcap_{k=1}^{\infty} B_{nk}$ is also closed. This proves the claim.

Thus by the Baire category theorem, at least one of the sets, say B_N, has an interior point. If $z \in B_N$ is such an interior point, then there exists $r > 0$ such that $S(z, r) \subset B_N$, which implies $\|T_k(x)\| \leq N$ for all $k = 1, 2, \ldots$ and all $x \in S(z, r)$.

Finally, let $x \in V$ with $\|x\| < 1$. Then $z + rx \in S(z, r)$, so that $\|T_k(z + rx)\| \leq N$ for all $k = 1, 2, \ldots$. Thus

$$\|T_k(x)\| = \left\| T_k\left(\frac{z + rx}{r} - \frac{z}{r}\right) \right\| \leq \left\| T_k\left(\frac{z + rx}{r}\right) \right\| + \left\| T_k\left(\frac{z}{r}\right) \right\| \leq \frac{2N}{r},$$

and hence $\sup_k \|T_k\| \leq \frac{2N}{r} < \infty$. ∎

Definition 6.12 A subset B of a normed space is said to be **weakly bounded** if for each $f \in V'$, the set $\{f(x) \mid x \in B\}$ is bounded.

Corollary 6.31 Banach–Mackey: *A subset B of a normed space V is bounded if and only if it is weakly bounded.*

Proof **(⇒):** Clearly every bounded set is weakly bounded.

(⇐): We use the uniform boundedness principle on the Banach space V'. As noted in Section 6.5, each $x \in V$ defines a continuous linear functional F_x on V' given by $F_x(f) = f(x)$. For each $f \in V'$, the set

$$\{F_x \mid x \in B\}$$

of continuous linear functions on V' is bounded. Applying the uniform boundedness principle on V', we see that there exists $M > 0$ such that $\|F_x\| \leq M$ for all $x \in B$. Since $\|F_x\| = \|x\|$, the result follows. ∎

The following theorem is an often-used application of the uniform boundedness principle.

Theorem 6.32 Banach–Steinhaus: *Let V be a Banach space, W a normed space, and $T_1, T_2, \ldots$, a sequence of continuous linear operators from V into W. If*

$$\lim_{k\to\infty} T_k(x) = T(x)$$

for each $x \in V$, then T is a continuous linear operator from V to W.

Proof That T is linear is left as Exercise 6.7.3. By the uniform boundedness principle, there exists $M > 0$ such that $\sup_k \|T_k\| \leq M$; that is, $\|T_k x\| \leq M\|x\|$ for all $x \in V$. Thus

$$\|Tx\| = \|\lim_{k\to\infty} T_k(x)\| \leq \sup_k \|T_k(x)\| \leq M\|x\|$$

for all $x \in V$, so that T is continuous. ∎

Here is an alternative version of this theorem.

Theorem 6.33 Banach–Steinhaus: *Let V and W be Banach spaces and let $T_1, T_2, \ldots$, be a uniformly bounded sequence of continuous linear operators from V into W. Then the set*

$$B = \{x \in V \mid \lim_{k\to\infty} T_k(x) \text{ exists}\}$$

is a closed subspace of V.

Proof By the uniform boundedness principle, we have $\sup_n \|T_n\| = M < \infty$. Let x be in the closure of B and let $\epsilon > 0$ be given. Then there exists $y \in B$ such that $\|x - y\| < \epsilon/3M$. Since $T_n y \to y$, there exists $N > 0$ such that $\|T_n y - T_m y\| < \epsilon/3$ for $n, m > N$. Then

$$\begin{aligned}
\|T_n x - T_m x\| &\leq \|T_n(x-y)\| + \|T_n y - T_m y\| + \|T_m(x-y)\| \\
&\leq M\|x-y\| + \|T_n y - T_m y\| + M\|x-y\| \\
&< \epsilon/3 + \epsilon/3 + \epsilon/3 = \epsilon.
\end{aligned}$$

That is, $T_n x$ is Cauchy. Since W is a Banach space, $T_n x$ converges. By the definition of B, $x \in B$. This shows that B is closed. ∎

Corollary 6.34 *Let V and W be Banach spaces and let $T_1, T_2, \ldots$ be a uniformly bounded sequence of continuous linear operators from V into W. If $\lim_{k\to\infty} T_k x$ exists in a set whose span is dense in V, then the limit exists* *for all $x \in V$. Furthermore, Tx defined by the equation $Tx = \lim_{k\to\infty} T_k x$ is a continuous linear operator from V to W.*

Proof According to Theorem 6.33, $\lim_n T_n x$ exists for every $x \in V$. Denote $\lim_n T_n x$ by Tx. It is routine to show that T is linear.

Also $\|Tx\| \le \sup_n \|T_n x\| \le M\|x\|$ for every $x \in V$, so T is continuous. ∎

Exercises for Section 6.7

6.7.1 Suppose that $p_1, p_2, \ldots$ is a sequence of continuous seminorms on a Banach space V. Prove that if, for every $x \in V$, we have $p(x) = \sup_k p(x) < \infty$, then p is a continuous seminorm of V. Follow the proof of the uniform boundedness principle.

6.7.2 Consider the norm $\|x\| = \sum_k |x_k|$ on the sequence space with finitely nonzero coordinates $\Phi = \text{span}\{e^k \mid k = 1, 2, 3, \ldots\}$. For each n, define $T_n : \Phi \longrightarrow \Phi$ as follows $T_n x = x + (n+1)x_n e^n$. Show that each T_n is continuous from Φ into Φ with this norm. Also show that $\sup \|T_n x\| < \infty$ for each $x \in \Phi$, but $\sup \|T_n\| = \infty$. This illustrates the necessity of completeness in the uniform boundedness principle.

6.7.3 Show that T, as defined in Theorem 6.32, is a linear operator from V to W.

6.8 Open Mapping Theorem

The open mapping theorem is an application of the Baire category theorem. It is important to the subject of sequence spaces.

Theorem 6.35 Open mapping: *Suppose that T is a continuous linear operator from a Banach space V* onto *a Banach space W. Then, for every open subset $G \subset V$, $T(G)$ is open in W. So, if T is also one-to-one, then T^{-1} is continuous.*

Proof First, consider the open sphere $G_\epsilon = S(0, \epsilon)$ in V where $\epsilon > 0$. Since T maps V onto W, we know $W = \bigcup_{n=1}^{\infty} nTG_\epsilon = \bigcup_{n=1}^{\infty} \{ny \mid y \in T(G_\epsilon)\}$. By the Baire category theorem, at least one of the sets $n\overline{TG_\epsilon}$ must contain an open sphere. Hence $\overline{TG_\epsilon}$ must contain an open sphere, say $S(y, r) \subset \overline{TG_\epsilon}$.

Claim 1: $S(0, r/2) \subset \overline{TG_\epsilon}$. Since $S(y, r) = S(0, r) + y = \{x + y \mid x \in S(0, r)\}$, we have

$$S(0, r) = S(y, r) - y \subset \overline{TG_\epsilon} - y \subset \overline{TG_\epsilon} - \overline{TG_\epsilon}. \tag{6.12}$$

Since $x \in G_\epsilon \iff -x \in G_\epsilon$ and T is linear, $z \in TG_\epsilon \iff -z \in TG_\epsilon$. Using limits, the same applies to $\overline{TG_\epsilon}$.

Similarly, we can show that $z, w \in \overline{TG_\epsilon} \Longrightarrow \frac{z}{2} + \frac{w}{2} \in \overline{TG_\epsilon}$. Thus

$$\overline{TG_\epsilon} - \overline{TG_\epsilon} \subset 2\overline{TG_\epsilon}.$$

Combined with (6.12), we have $S(0, r/2) \subset \overline{TG_\epsilon}$. Thus $S(0, r/2) \subset \overline{TG_\epsilon}$ whenever $S(y, r) \subset \overline{TG_\epsilon}$. This proves the claim.

Claim 2: $S(0, r/2^{n+1}) \subset \overline{TG_{\epsilon/2^n}}$ for all $n = 1, 2, \ldots$. This follows from Claim 1 and the linearity of T.

Claim 3: $S(0, r/4) \subset TG_\epsilon$. Let $y \in S(0, r/2^2) \subset \overline{TG_{\epsilon/2}}$. Then there exists $u_1 \in G_{\epsilon/2}$ such that $\|y - Tu_1\| < r/2^3$. So $y_1 = y - Tu_1 \in S(0, r/2^3)$. Having chosen $u_1, u_2, \ldots, u_n$, and $y_n = y - Tu_1 - \cdots - Tu_n$, choose $u_{n+1} \in G_{\epsilon/2^{n+1}}$ such that $\|y_n - Tu_{n+1}\| < r/2^{n+2}$. Let $y_{n+1} = y_n - Tu_{n+1}$. We have

$$\sum_{n=1}^{\infty} \|u_n\| < \sum_{n=1}^{\infty} \frac{\epsilon}{2^n} = \epsilon.$$

Thus the series $\sum_{n=1}^{\infty} u_n$ converges to some $x \in S(0, \epsilon)$ (Theorem 5.31). Since $\|y - \sum_{n=1}^{\infty} u_n\| = \|y - Tu_1 - Tu_2 - \cdots - Tu_n\| < \frac{r}{2^{n+2}}$, by continuity of T, we have $Tx = y$. Therefore, $y \in TG_\epsilon$. So $S(0, r/4) \subset TG_\epsilon$.

Finally, if G is open and $x \in G$, then 0 is an interior point of $G - x$. By Claim 2, there exists an open sphere $S(0, r/4) \subset TG - Tx$ or $S(Tx, r/4) = Tx + S(0, r/4) \subset TG$. So every Tx in TG is an interior point of TG. ∎

6.9 Closed Graph Theorem

The closed graph theorem is another application of the Baire category theorem. It is also important to the subject of sequence spaces.

Let V and W be normed spaces. The set of ordered pairs

$$V \times W = \{(x, y) \mid x \in V, y \in W\}$$

is a linear space and a normed space under the norm

$$\|(x, y)\| = \|x\| + \|y\|.$$

If V and W are Banach spaces, then so is $V \times W$.

Definition 6.13 Let V and W be Banach spaces and consider a linear operator $T: V \longrightarrow W$. The **graph** of T is the set of ordered pairs

$$\Gamma(T) = \{(x, Tx) \mid x \in V\}.$$

The space $\Gamma(T)$ is clearly a linear space subspace of $V \times W$.

Theorem 6.36 Closed graph: *Let V and W be Banach spaces and consider a linear operator $T: V \longrightarrow W$. Then T is continuous if and only if $\Gamma(T)$ is closed in the Banach space $V \times W$.*

Proof **($\Leftarrow$):** Suppose $\Gamma(T)$ is closed in $V \times W$. The mapping

$$T_1: (x, Tx) \longrightarrow x$$

is a linear operator from $\Gamma(T)$ onto V. It is continuous because the following is bounded:

$$\|T_1(x, Tx)\| = \|x\| \leq \|x\| + \|Tx\| = \|(x, Tx)\|.$$

By the open mapping theorem (Theorem 6.35) the inverse mapping

$$T_1^{-1}: x \longrightarrow (x, Tx)$$

is continuous. Thus there exists $M > 0$ such that $\|T_1^{-1}x\| \leq M\|x\|$. But $\|T_1^{-1}x\| = \|(x, Tx)\| = \|x\| + \|Tx\|$. Thus

$$\|Tx\| \leq \|x\| + \|Tx\| = \|(x, Tx)\| = \|T_1^{-1}x\| \leq M\|x\|$$

for all $x \in V$. Since T is a bounded linear operator, it is continuous.

($\Rightarrow$): This part of the proof is left as Exercise 6.9.2. ∎

Exercises for Section 6.9

6.9.1 Prove the claim at the beginning of this section that $V \times W$ is a Banach space whenever V and W are Banach spaces.

6.9.2 Prove the ($\Rightarrow$) part of Theorem 6.36. Hint: Consider a sequence $(x^n, y^n) \in \Gamma(T)$ which converges to $(x, y) \in V \times W$. Show that $(x, y) \in \Gamma(T)$.

6.10 The Lebesgue Spaces L^p

6.10.1 The Case $1 \leq p < \infty$

Definition 6.14 Let $1 \leq p < \infty$ and $-\infty < a < b < \infty$. The function space $L^p[a, b]$ is defined to be the set of functions for which

$$\int_a^b |f(x)|^p \, dx < \infty.$$

If the context is clear, we sometimes simply write L^p instead of $L^p[a,b]$. We must be careful and remember that we are thinking of f as the equivalence class of all functions equal to f almost everywhere on $[a,b]$. That is, $L^p[a,b]$ consists of equivalence classes of functions, not functions.

Definition 6.15 Let $1 \le p < \infty$. The **p-norm** of f is the real number

$$\|f\|_p = \left(\int_a^b |f(x)|^p \, dx\right)^{1/p}.$$

We first obtain Hölder's and Minkowski's inequalities for integrals.

Theorem 6.37 Hölder's Inequality for integrals: *Suppose* $1 < p < \infty$, $1 < q < \infty$ *with* $\frac{1}{p} + \frac{1}{q} = 1$. *Then we have*

$$\int_a^b |f(x)g(x)| \, dx \le \left(\int_a^b |f(x)|^p \, dx\right)^{1/p} \left(\int_a^b |g(x)|^q \, dx\right)^{1/q}$$

for all $f \in L^p[a,b]$, $g \in L^q[a,b]$. *The inequality is equivalent to*

$$\|fg\|_1 \le \|f\|_p \|g\|_q.$$

Proof If $\|f\|_p = 0$ or $\|g\|_q = 0$, then $f = 0$ or $g = 0$ almost everywhere, and the result is clear. Otherwise, the proof follows the one for Hölder's inequality for sums (Theorem 0.54 in Section 0.13). Instead of inequality (0.24) we use

$$\frac{|f(x)g(x)|}{\|f\|_p \|g\|_q} \le \frac{|f(x)|^p}{p(\|f\|_p)^p} + \frac{|g(x)|^q}{q(\|g\|_q)^q} \tag{6.13}$$

obtained from Young's inequality. Integrating both sides we get

$$\frac{\int_a^b |f(x)g(x)|}{\|f\|_p \|g\|_q} \le \frac{\int_a^b |f(x)|^p}{p \cdot (\|f\|_p)^p} + \frac{\int_a^b |g(x)|^q}{q \cdot (\|g\|_q)^q} = \frac{1}{p} + \frac{1}{q} = 1,$$

which leads to Hölder's inequality for integrals. ∎

Corollary 6.38 *For each* $1 \le p < q < \infty$ *we have* $L^q[a,b] \subset L^p[a,b] \subset L^1[a,b]$. *In particular, all functions in* $L^p[a,b]$ *are Lebesgue integrable. Furthermore,*

$$\|f\|_p \le (b-a)^{\left(\frac{1}{p}-\frac{1}{q}\right)} \|f\|_q.$$

Proof The proof follows by applying Hölder's inequality to $|f|^p$ and $g = 1$. Let $r = \frac{q}{p} > 1$ and $\frac{1}{r} + \frac{1}{s} = 1$. Then $s = \frac{q}{q-p}$ and

$$\int_a^b |f(x)|^p\,dx \le \left(\int_a^b |f(x)|^{pr}\,dx\right)^{1/r}\left(\int_a^b 1^s\,dx\right)^{1/s}$$
$$= \left(\int_a^b |f(x)|^q\,dx\right)^{p/q}(b-a)^{1-(p/q)}. \qquad \blacksquare$$

Warning The above inclusions do *not* hold for the infinite interval $\mathbb{R} = (-\infty,\infty)$. Exercise 6.10.1 shows that $L^1(\mathbb{R}) \not\subset L^2(\mathbb{R})$ and $L^2(\mathbb{R}) \not\subset L^1(\mathbb{R})$.

Theorem 6.39 Minkowski's inequality for integrals: *Suppose $p \ge 1$. Then we have*

$$\left(\int_a^b |f(x)+g(x)|^p\,dx\right)^{1/p} \le \left(\int_a^b |f(x)|^p\,dx\right)^{1/p} + \left(\int_a^b |g(x)|^p ag\,dx\right)^{1/p},$$

for all $f,g \in L^p[a,b]$. In terms of norms, this is the p-norm triangle inequality

$$\|f+g\|_p \le \|f\|_p + \|g\|_p.$$

The proof follows the pattern of Minkowski's inequality (Theorem 0.56).

Lemma 6.40 *Suppose $1 \le p < \infty$. For any sequence of nonnegative functions $f_1, f_2, \ldots$ in L^p, we have*

$$\left\|\sum_{k=1}^{\infty} f_k\right\|_p \le \sum_{k=1}^{\infty} \|f_k\|_p.$$

Proof Define $g_n(x) = \sum_{k=1}^{n} |f_k(x)|$ for each $n = 1,2\ldots$. By Minkowski's inequality,

$$\|g_n\|_p \le \sum_{k=1}^{n} \|f_k\|_p \le \sum_{k=1}^{\infty} \|f_k\|_p.$$

Since g_n is increasing, Lebesgue's monotone convergence theorem (Theorem 2.38 in Section 2.4) gives us

$$\left\|\sum_{k=1}^{\infty} f_k\right\|_p = \left(\int_a^b \lim_{n\to\infty} g_n^p\,dx\right)^{1/p} = \left(\lim_{n\to\infty}\int_a^b g_n^p\,dx\right)^{1/p} \le \sum_{k=1}^{\infty} \|f_k\|_p. \qquad \blacksquare$$

Theorem 6.41 *For each $1 \le p < \infty$, the space $L^p[a,b]$ is a Banach space.*

Proof Using Minkowski's inequality, the verifications that $\|\cdot\|_p$ is a norm and that $L^p[a,b]$ is a normed space are straightforward.

It remains to be shown that each space $L^p[a,b]$ is complete. We use Theorem 5.31 (Section 5.8). Suppose $\sum_{k=1}^{\infty}\|f_k\|_p = M < \infty$ for $f_1, f_2, \ldots \in L^p[a,b]$. We now show that $\sum_{k=1}^{\infty} f_k$ converges in norm to a function f in L^p.

Claim: $\sum_{k=1}^{\infty} f_k$ converges absolutely almost everywhere on $[a,b]$. Let

$$g(x) = \sum_{k=1}^{\infty} |f_k(x)|.$$

By the previous lemma, $g \in L^p[a,b] \subset L^1[a,b]$. Since g is integrable, $g < \infty$ almost everywhere. Thus $\sum_{k=1}^{\infty} f_k$ converges absolutely a.e., say to f.

Claim 1: $f \in L^p$. Define $s_n(x) = \sum_{k=1}^{n} f_k(x)$. Since $|s_n(x)| \le g(x)$ for all n, we have $|f(x)| \le g(x)$. Since $g \in L^p$, it follows that $f \in L^p$.

Claim 2: $\sum_{k=1}^{\infty} f_k$ converges in norm to f in L^p. We use Lebesgue's dominated convergence theorem (Theorem 2.41).

$$\begin{aligned}\lim_{n\to\infty}\|s_n - f\|_p &= \lim_{n\to\infty}\left(\int_a^b |s_n(x) - f(x)|^p\,dx\right)^{1/p}\\ &= \left(\int_a^b \lim_{n\to\infty}|s_n(x) - f(x)|^p\,dx\right)^{1/p} = 0.\end{aligned}$$

∎

Theorem 6.42 *The space $C[a,b]$ is dense in $L^p[a,b]$ for each $1 \le p < \infty$.*

Proof First, we show that the nonnegative measurable simple functions are dense in the set of nonnegative functions of $L^p[a,b]$. By Theorem 2.16 (Section 2.1) every measurable function $0 \le f \in L^p[a,b]$ is a limit of measurable simple functions s_n, such that

$$0 \le s_1 \le s_2 \le \cdots \le f.$$

We have

$$|f - s_1|^p \ge |f - s_2|^p \ge \cdots \ge 0$$

and

$$\lim_{n\to\infty}|f(x) - s_n(x)|^p = 0, a.e.$$

By Lebesgue's dominated convergence theorem (Theorem 2.41), we have

$$\lim_{n\to\infty}\|f - s_n\|_p = \lim_{n\to\infty}\left(\int_a^b |f(x) - s_n(x)|^p\,dx\right)^{1/p} = 0.$$

Each such simple function is of the form

$$s = \sum \alpha_k f_{A_k}, \quad \text{with each } A_k \text{ measurable and each } \alpha_k > 0. \tag{6.14}$$

Next we show that for a simple function of the form (6.14) and for each given $\epsilon > 0$, there exists a continuous function g such that $\|g - s\|_p < \epsilon$. By Theorem 1.32 (Section 1.6), there exist closed F_k and open G_k such that

$$F_k \subset A_k \subset G_k \quad \text{and} \quad m(G_k - F_k) < \left(\frac{\epsilon}{\alpha_k 2^k}\right)^p.$$

For each k, let g_k be the Urysohn function (Theorem 5.28). Then g_k is continuous, $0 \le g_k \le 1$, $g_k = 1$ on F_k, and $g_k = 0$ on $[a,b] - G_k$. Also, the function $|g_k - f_{A_k}|$ takes values between 0 and 1. It has the value 0 on both F_k and $[a,b] - G_k$. Thus we have

$$\|g_k - f_{A_k}\|_p \le \left(\int_a^b (f_{G_k} - f_{F_k})\right)^{1/p} = \left(m(G_k - F_k)\right)^{1/p} < \frac{\epsilon}{\alpha_k 2^k}.$$

Let $g = \sum \alpha_k g_k$. Then

$$\|g - s\|_p \le \sum_k \alpha_k \|g_k - f_{A_k}\|_p < \sum_k \alpha_k \frac{\epsilon}{\alpha_k 2^k} < \epsilon.$$

Finally, since every $f \in L^p$ is the difference of two nonnegative functions in L^p, the result follows. ∎

Theorem 6.43 *The spaces $L^p[a,b]$ are separable for $1 \le p < \infty$.*

Proof In the proof of Theorem 6.42, we showed that the set of nonnegative measurable simple functions is dense in the set of nonnegative functions of $L^p[a,b]$. We may clearly require the α_k in (6.14) to be rational as well as positive.

Suppose $s = \sum \alpha_k f_{A_k}$ is such a simple function with each A_k measurable and each α_k positive and rational. For each $\epsilon > 0$, there exists a closed F_k and an open G_k with

$$F_k \subset A_k \subset G_k \quad \text{and} \quad m(G_k - F_k) < \left(\epsilon/\alpha_k 2^{k+1}\right)^p.$$

By the Heine–Borel theorem (Theorem 5.45), F_k is compact. Each G_k is a union of open intervals with rational centers and rational radii. This countable union covers F_k from which we can extract a finite subcover. Let G_k^* be the union of this finite subcover and let

$$s^* = \sum \alpha_k f_{G_k^*}. \tag{6.15}$$

There are only countably many such simple functions. If we let $t = \sum \alpha_k f_{G_k}$, then

$$\begin{aligned}
\|s^* - s\|_p &\leq \|s^* - t\|_p + \|t - s\|_p \\
&\leq \sum \alpha_k \|f_{G_k^*} - f_{G_k}\|_p + \sum \alpha_k \|f_{G_k} - f_{A_k}\|_p \\
&\leq \sum \alpha_k \left(\int_a^b (f_{G_k} - f_{G_k^*}) \right)^{1/p} + \sum \alpha_k \left(\int_a^b (f_{G_k} - f_{A_k}) \right)^{1/p} \\
&\leq \sum \alpha_k \left(\int_a^b (f_{G_k} - f_{F_k}) \right)^{1/p} + \sum \alpha_k \left(\int_a^b (f_{G_k} - f_{F_k}) \right)^{1/p} \\
&\leq 2 \sum \frac{\epsilon}{2^{k+1}} = \epsilon.
\end{aligned}$$

So every $0 \leq f \in L^p$ is within ϵ of one of the countable collection of simple functions of the form s^*. Since every $f \in L^p[a,b]$ is the difference of two nonnegative functions in $L^p[a,b]$, the result follows. ∎

6.10.2 The Case $p = \infty$

Definition 6.16 A function f is said to be **essentially bounded** on $I = [a,b]$, for $a < b$, if there exists a positive $M > 0$ and a set $A \subset I$ of measure zero such that

$$|f(x)| \leq M \quad \text{for all} \quad x \in I - A.$$

We define the norm $\|f\|_\infty = \min\{M\}$.

Let $L^\infty[a,b]$ be the set of functions that are essentially bounded on $[a,b]$.

It is easy to show that Hölder's and Minkowski's inequalities hold for the case $p = \infty$. The Hölder conjugate of $p = \infty$ is $q = 1$.

Theorem 6.44 *The space $L^\infty[a,b]$ is a Banach space.*

Proof Suppose that $\{f_k\}$ is a Cauchy sequence in $L^\infty[a,b]$. By Definition 6.16 of the essential bound $\|\cdot\|_\infty$, for each k, m, n, the sets

$$A_k = \{x \in [a,b] \mid |f_k(x)| > \|f_k\|_\infty\} \quad \text{and}$$

$$A_{mn} = \{x \in [a,b] \mid |f_m(x) - f_n(x)| > \|f_m - f_n\|_\infty\}$$

are of measure zero (see Exercise 6.10.3). Let B be the union of all of these sets A_k and A_{mn}. Being a countable union of sets of measure zero, B is of measure zero. For all $x \in [a,b] - B$, we have

$$|f_m(x) - f_n(x)| \le \|f_m - f_n\|_\infty \to 0 \quad \text{as } m,n \to \infty, \tag{6.16}$$

which thus converges. Define the function f as follows.

$$f(x) = \lim_{n\to\infty} f_k(x) \;\text{ for } x \notin B \;\text{ and }\; f(x) = 0 \;\text{ for } x \in B.$$

Claim: $f \in L^\infty$. For each $x \in [a,b] - B$ and $\epsilon = 1$, there exists $N > 0$ such that $|f(x) - f_k(x)| < 1$ whenever $k \ge N$. In particular, we have

$$|f(x)| < 1 + |f_N(x)| \le 1 + \|f_N\|_\infty \le 1 + \sup_k \|f_k\|_\infty \;\text{ for all }\; x \in [a,b] - B.$$

Since Cauchy sequences are bounded, f is bounded by $1 + \sup_k \|f_k\|_\infty$ almost everywhere. Thus $\|f\|_\infty < \infty$.

Claim: $\|f - f_n\|_\infty \to 0$. Let $\epsilon > 0$ be given. By (6.16) we see that, for $x \in [a,b] - B$ and sufficiently large m and n, we have $|f_m(x) - f_n(x)| \le \|f_m - f_n\|_\infty < \epsilon$. Let $m \to \infty$. For sufficiently large n, it follows that $|f(x) - f_n(x)| \le \epsilon$. So $\|f - f_n\|_\infty \le \epsilon$. ∎

Exercises for Section 6.10

6.10.1 Show that the inclusions of Corollary 6.38 do not hold for infinite intervals, such as $\mathbb{R} = (a,b)$. In particular $L^1(\mathbb{R}) \not\subset L^2(\mathbb{R})$ and $L^2(\mathbb{R}) \not\subset L^1(\mathbb{R})$. Let

$$f(x) = \begin{cases} x^{-2/3} & \text{if } 0 < x < 1 \\ 0 & \text{otherwise,} \end{cases}$$

$$g(x) = \begin{cases} x^{-2/3} & \text{if } x > 1 \\ 0 & \text{otherwise.} \end{cases}$$

Show that $f \in L^1(\mathbb{R})$ but $f \notin L^2(\mathbb{R})$, and $g \in L^2(\mathbb{R})$ but $g \notin L^1(\mathbb{R})$.

6.10.2 Although $L^1(\mathbb{R}) \not\subset L^2(\mathbb{R})$, show that $L^1(\mathbb{R}) \cap L^\infty(\mathbb{R}) \subset L^2(\mathbb{R})$.

6.10.3 Show that for each $f \in L^\infty[a,b]$, the following set is of measure zero.

$$A = \{x \in [a,b] \;\big|\; |f(x)| > \|f\|_\infty\}.$$

6.10.4 Show that for $a < b$, the space $L^\infty[a,b]$ is not separable.

*6.11 The Dual of $C[a,b]$.

Theorem 6.45 below shows that the dual of $C[a,b]$ is the Banach space $BV[a,b]$ of functions of bounded variation on $[a,b]$. This section assumes that you have read Section 0.12 on functions of bounded variation and Sections 4.2 and 4.4 which cover the topic of Riemann–Stieltjes integrals.

Theorem 6.45 Riesz Theorem:[4] *Every continuous linear functional g on $C[a,b]$ can be represented by a Riemann–Stieltjes integral*

$$g(x) = \int_a^b x(t)\,df(t), \quad (\textit{for } x \in C[a,b]), \tag{6.17}$$

where f is of function of bounded variation on $[a,b]$ and has total variation

$$V_f[a,b] = \|g\|.$$

Proof The space $C[a,b]$ is a subspace of $L^\infty[a,b]$. Recall that both are Banach spaces under the sup norm $\|g\| = \sup_{a\le x\le b} |g(x)|$. By the Hahn–Banach theorem (Theorem 6.20), every continuous linear functional g on $C[a,b]$ can be extended to a continuous linear functional G on all of $L^\infty[a,b]$ such that $g(x) = G(x)$ for every $x \in C[a,b]$ and $\|g\| = \|G\|$. For any point $a < s \le b$, let $K_s = \chi_{[a,s]}$ be the characteristic (indicator) function of the interval $[a,s]$ (see Definition 2.4). And let $K_a \equiv 0$ in the case $s = a$.

Note that $K_s \in L^\infty[a,b]$ for all $a \le s \le b$. So we can define $f(s) = G(K_s)$ for $a \le s \le b$. We first show that f is a function of bounded variation on $[a,b]$ with total variation $V_f[a,b] \le \|g\|$.

Consider any partition $\mathcal{P}$: $\{a = t_0 < t_1 < \cdots < t_n = b\}$. Then

$$\begin{aligned}
\sum_{k=1}^n |f(t_k) - f(t_{k-1})| &= \sum_{k=1}^n \{f(t_k) - f(t_{k-1})\}\operatorname{sgn}\big(f(t_k) - f(t_{k-1})\big)\\
&= \sum_{k=1}^n \{G(K_{t_k}) - G(K_{t_{k-1}})\}\operatorname{sgn}\big(f(t_k) - f(t_{k-1})\big)\\
&= G\left(\sum_{k=1}^n \{K_{t_k} - K_{t_{k-1}}\}\operatorname{sgn}\big(f(t_k) - f(t_{k-1})\big)\right)\\
&\le \|G\| \left\|\sum_{k=1}^n \{K_{t_k} - K_{t_{k-1}}\}\operatorname{sgn}\big(f(t_k) - f(t_{k-1})\big)\right\|\\
&= \|G\| = \|g\|.
\end{aligned}$$

[4] This was proved by F. Riesz in 1909. It has led to various generalizations of duals of spaces $C(K)$ of continuous function on compact sets K of various topological spaces. Here $K = [a,b] \subset \mathbb{R}$. The general case of compact $K \subset \mathbb{R}$ is left as Exercise 6.11.2.

The inequality in the above display follows from $|G(x)| \leq \|G\| \cdot \|x\|$ (Equation (6.6)). Taking the supremum with respect to all partitions $\mathcal{P}$ shows that f is a function of bounded variation on $[a,b]$ with $V_f[a,b] \leq \|g\|$.

Since f is a function of bounded variation, the Riemann–Stieltjes integral $\int_a^b x(t)\,df(t)$ exists, for every $x \in C[a,b]$. So for any $\epsilon > 0$, we have

$$\left| \int_a^b x(t)\,df(t) - \sum_{k=1}^{n} x(t_{k-1})\{f(t_k) - f(t_{k-1})\} \right| < \epsilon, \tag{6.18}$$

provided the mesh $\|\mathcal{P}\| = \max_{1 \leq k \leq n} (t_k - t_{k-1})$ is sufficiently small.

Next, before proving the other inequality $\|g\| \leq V_f[a,b]$, we show the validity of Equation (6.17). For each $x \in C[a,b]$ and partition $\mathcal{P}$, define the bounded function $z \in L^{\infty}[a,b]$ as follows:

$$z = \sum_{k=1}^{n} x(t_{k-1})\{K_{t_k} - K_{t_{k-1}}\}.$$

Using the functions $f(s) = G(K_s)$ as defined above, we have

$$G(z) = \sum_{k=1}^{n} x(t_{k-1})\{G(K_{t_k}) - G(K_{t_{k-1}})\} = \sum_{k=1}^{n} x(t_{k-1})\{f(t_k) - f(t_{k-1})\}.$$

By (6.18) we see that $G(z) \to \int_a^b x(t)\,df(t)$ as the mesh $\|\mathcal{P}\| \to 0$.

Since each continuous function on $[a,b]$ is uniformly continuous, we have $\|z - x\| < \epsilon$ provided the mesh $\|\mathcal{P}\|$ is sufficiently small. Since G is continuous on $L^{\infty}[a,b]$, we have $G(z) \to G(x)$ as $\|\mathcal{P}\| \to 0$. Thus

$$G(x) = \int_a^b x(t)\,df(t).$$

This is (6.17) because G is an extension of g and $x \in C[a,b]$.

Finally, we have the inequality $\|g\| \leq V_f[a,b]$ because, for Riemann–Stieltjes integrals,

$$|g(x)| = \left| \int_a^b x(t)\,df(t) \right| \leq \|x\| V_f[a,b].$$

∎

The function of bounded variation f is not unique. For example, in the proof of Theorem 6.45, if $F(t) = f(t) + c$ where c is any constant, then

$$g(x) = \int_a^b x(t)\,df(t) = \int_a^b x(t)\,dF(t) \quad \text{and} \quad \|g\| = V_f[a,b] = V_F[a,b].$$

So the dual of $C[a,b]$ cannot be identified with the space of functions of bounded variation $BV[a,b]$. However, if we restrict to functions f that are continuous from the right (that is, $f(t+) = f(t)$ for $a \le t < b$) with $f(0) = 0$, then this subspace of $BV[a,b]$ is isometric with the dual of $C[a,b]$.

Exercises for Section 6.11

6.11.1 Prove the last step in the proof of the Riesz theorem:

$$|g(x)| = \left| \int_a^b x(t)\, df(t) \right| \le \|x\| V_f[a,b].$$

Consider the Riemann–Stieltjes sums $\sum_{k=1}^{n} x(t_{k-1})\{f(t_k) - f(t_{k-1})\}$ appearing in (6.18).

6.11.2 Prove the Riesz theorem for $C(K)$, where K is a compact subset of $\mathbb{R}$.

*6.12 Contraction Mappings

Definition 6.17 Let M be a metric space. A mapping $T\colon M \longrightarrow M$ is called a **contraction** in M if there exists a number $0 \le c < 1$ such that, for all x, $y \in M$,

$$d(Tx, Ty) \le cd(x, y).$$

Theorem 6.46 *A contraction T is uniformly continuous on M.*

The proof is left as Exercise 6.12.1.

Example 6.16 Let $f : \mathbb{R} \longrightarrow \mathbb{R}$ be differentiable with $|f'(x))| \le c < 1$. Then f is a contraction on $\mathbb{R}$ because the mean value theorem shows that, for some $x < a < y$, we have

$$|f(x) - f(y)| = |f'(a)(x - y)| \le c|x - y|.$$

Definition 6.18 Let M be a metric space and let $T\colon M \longrightarrow M$. A **fixed point** for the map T is a point $x \in M$ such that $T(x) = x$.

Theorem 6.47 Banach fixed point principle: *Let M be a complete metric space and let T be a contraction on M. Then T has a unique fixed point.*

Proof

Uniqueness: Assume we have fixed points $x \ne y$ with $Tx = x$ and $Ty = y$. This leads to the contradiction $d(x, y) = d(Tx, Ty) \le cd(x, y) < d(x, y)$.

Existence of a fixed point: Start with any $x^0 \in M$ and let

$$x^1 = Tx^0, x^2 = Tx^1 = T^2x^0, \ldots, x^n = T^n x^0.$$

Note that $d(x^{n+1}, x^n) \leq cd(x^n, x^{n-1}) \leq \cdots \leq c^n d(x^1, x^0)$. This implies that the $\{x^n\}$ is a Cauchy sequence in M because, for $m > n$, we have

$$\begin{aligned} d(x^m, x^n) &\leq d(x^m, x^{m-1}) + d(x^{m-1}, x^{m-2}) + \cdots + d(x^{n+1}, x^n) \\ &\leq c^{m-1} d(x^1, x^0) + c^{m-2} d(x^1, d_0) + \cdots c^n d(x^1, x^0) \\ &\leq \left(c^{m-1} + c^{m-2} + \cdots c^n\right) d(x^1, x^0) \\ &\leq \left(\frac{c^n}{1-c}\right) d(x^1, x^0) \to 0 \text{ as } n \to \infty \text{ (since } c < 1). \end{aligned}$$

Since M is complete, the Cauchy sequence converges, say to $x = \lim_n x^n$. Finally, since T is continuous, we have

$$T(x) = T\left(\lim_n x^n\right) = \lim_n T(x^n) = \lim_n x^{n+1} = x. \qquad \blacksquare$$

The result does not depend on the starting point x^0. However, to keep numerical procedures short, one must be judicious in choosing a good starting point.

Exercises for Section 6.12

6.12.1 Prove Theorem 6.46.

6.12.2 Use the Banach fixed point principle to prove that there is a unique x such that $\cos x = x$.

6.12.3 Show that there is a unique function y with continuous derivative y' such that $y(0) = 0$ and $y'(t) = y^2(t)$ for $0 \leq t \leq 1$. Hint: Consider the integral equation $y(t) = \int_0^t y^2(s)\,ds$ on the space $C[0,r]$ for each $0 < r < 1$.

PART III

Discrete Functional Analysis

7
Fourier Series

This is the beginning of the study of discrete functional analysis. It is the part of functional analysis dealing with linear spaces that can be represented as discrete structures, such as sequences, series, and matrices. This is an area of functional analysis with many applications, such as digital recording of music and video (CD and DVD), magnetic resonance imaging (MRI), computerized tomography (CT) scans, atomic theory of matter, quantum theory, and options trading.

We start in this chapter with inner product spaces (Section 7.1). These spaces permit representations of their elements by series called **Fourier series**. Function spaces of this type are isometric to sequence spaces. Initially, we deal with the inner product function space $L^2[a,b]$ and the corresponding sequence space ℓ_2 of Fourier coefficients. Afterwards, we expand Fourier analysis to more general space without an inner product, such as that of integrable functions $L^1[a,b]$.

7.1 Inner Product Spaces

An **inner product** on a linear space V is a function $\langle \cdot, \cdot \rangle$ from $V \times V$ to the field of scalars that satisfies the following four axioms for all $x, y, z \in V$ and all scalars α:

(IP1) $\langle x + y, z \rangle = \langle x, z \rangle + \langle y, z \rangle$;
(IP2) $\langle \alpha x, y \rangle = \alpha \langle x, y \rangle$;
(IP3) $\langle x, y \rangle = \overline{\langle y, x \rangle}$ (if V is a *real* vector space, then $\langle x, y \rangle = \langle y, x \rangle$);
(IP4) $\langle x, x \rangle$ is real and positive except for $x = 0$.

We can conclude the following property to be true from the above axioms:

(IP5) $\langle x, y+z\rangle = \langle x, y\rangle + \langle x, z\rangle$ and $\langle x, \alpha y\rangle = \overline{\alpha}\langle x, y\rangle$.

An **inner product space** is a linear space V with an inner product defined on it. Here are some examples of inner product spaces.

Example 7.1 The space $\mathbb{R}^n$ along with the inner product $\langle x, y\rangle = x_1y_1 + \cdots + x_ny_n$ is called the ***n*-dimensional Euclidean space.** This inner product is also called the **dot product** and often written $x \cdot y = x_1y_1 + \cdots + x_ny_n$.

Example 7.2 The space $\mathbb{C}^n$ over the field of scalars $\mathbb{C}$, along with the inner product

$$\langle x, y\rangle = x_1\overline{y_1} + \cdots + x_n\overline{y_n},$$

is called the ***n*-dimensional complex Euclidean space** or the ***n*-dimensional unitary space.** This is also called the **dot product**, written $x \cdot y$.

Example 7.3 The space ℓ_2 is an inner product space with $\langle x, y\rangle = \sum_{k=1}^{\infty} x_k\overline{y_k}$.

This series converges by Hölder's inequality (Theorem 0.55) for the case $p = q = 2$. Here it also makes sense to use the term **dot product** and write $x \cdot y$.

Example 7.4 The space $L^2[a,b]$, for $-\infty < a < b < \infty$, is an inner product space with

$$\langle f, g\rangle = \frac{1}{b-a}\int_a^b f(t)\overline{g(t)}\,dt.$$

Division by $b-a$ in the definition is not essential but it simplifies some results in the theory of this function space.

Definition 7.1 Every inner product on a space V gives rise to a **Euclidean norm** for V defined by

$$\|x\| = \sqrt{\langle x, x\rangle}.$$

The Euclidean norm on $\mathbb{R}^n$ (Section 0.5), defined by $\|x\|_2 = \sqrt{x_1^2 + \cdots + x_n^2}$, is a special case.

Every Euclidean norm satisfies the norm properties **(N0)** through **(N3)** as given in Section 5.2. However, not all norms arise from an inner product as will be shown in Example 7.5 below.

If a normed space is *known* to be an inner product space, the inner product can be recovered from the norm using the following polarization identities.

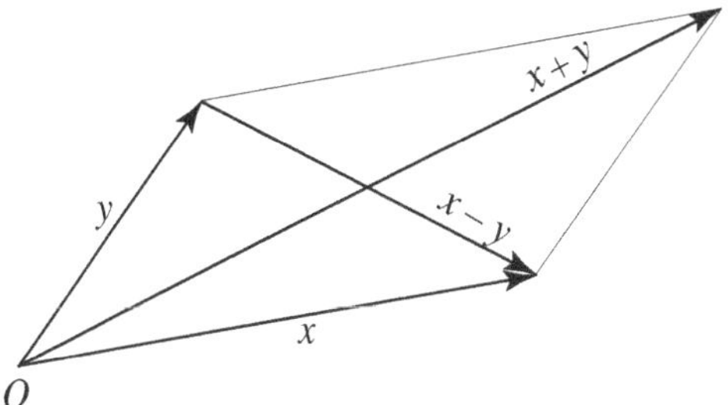

Figure 7.1 A parallelogram in the plane.

Theorem 7.1 Real Polarization Identity: *Suppose V is a real inner product space with Euclidean norm $\|\cdot\|$. Then for all $x, y \in V$, we have*

$$\langle x, y\rangle = \frac{1}{4}\|x + y\|^2 - \frac{1}{4}\|x - y\|^2. \tag{7.1}$$

The verification follows by direct simplification of the expression on the right-hand side (the proof is left as Exercise 7.1.5).

Theorem 7.2 Complex Polarization Identity: *Suppose V is a complex inner product space with Euclidean norm $\|\cdot\|$. Then for all $x, y \in V$, we have*

$$\langle x, y\rangle = \frac{1}{4}\|x + y\|^2 - \frac{1}{4}\|x - y\|^2 + \frac{i}{4}\|x + iy\|^2 - \frac{i}{4}\|x - iy\|^2. \tag{7.2}$$

Again, the verification follows by direct simplification of the expression on the right-hand side (the proof is left as Exercise 7.1.6).

Theorem 7.3 Parallelogram identity: *Every Euclidean norm satisfies the identity*

$$2\|x\|^2 + 2\|y\|^2 = \|x + y\|^2 + \|x - y\|^2. \tag{7.3}$$

The name is derived from the parallelogram law of plane geometry as illustrated in Figure 7.1. The proof is straightforward geometry and left as Exercise 7.1.7.

Example 7.5 Not every norm is a Euclidean norm because not every norm satisfies the parallelogram identity. For example, the usual norm on ℓ_1

$$\|x\|_1 = \sum_{k=1}^{\infty} |x_k|$$

satisfies the norm properties **(N0)** through **(N3)** but it does not arise from an inner product because it does not satisfy the parallelogram identity. Just consider the vectors $x = e^1 = (1,0,0,\ldots)$ and $y = e^2 = (0,1,0,0,\ldots)$.

Theorem 7.4 *The only space $L^p[a,b]$ whose norm arises from an inner product is $L^2[a,b]$.*

The proof is left as Exercise 7.1.12.

Actually, inner product spaces are exactly those normed spaces for which the parallelogram identity holds. This is Theorem 7.9 below. In order to prove it we use the following Cauchy–Schwarz inequality.

Theorem 7.5 Cauchy–Schwarz inequality: *In an inner product space V with Euclidean norm $\|\cdot\|$, we have*

$$|\langle x, y\rangle| \leq \|x\| \cdot \|y\| \quad \textit{for all} \quad x, y \in V. \tag{7.4}$$

Proof Let $x, y \in V$ and let α be any scalar. Then we have

$$0 \leq \langle x - \alpha y, x - \alpha y\rangle = \langle x, x\rangle - \overline{\alpha}\langle x, y\rangle - \alpha\overline{\langle x, y\rangle} + |\alpha|^2\langle y, y\rangle. \tag{7.5}$$

Inequality (7.4) is obvious for the case $y = 0$. If $y \neq 0$, let $\alpha = \langle x, y\rangle/\langle y, y\rangle$. Then (7.5) becomes

$$0 \leq \langle x, x\rangle - \frac{|\langle x, y\rangle|^2}{\langle y, y\rangle},$$

which is equivalent to (7.4). ∎

Corollary 7.6 Inner product functionals are continuous: *In any inner product space V, and for every $y \in V$, the linear functional defined by*

$$f_y(x) = \langle x, y\rangle$$

is continuous. Furthermore, we have the norm identity

$$\|y\| = \|f_y\| = \sup_{\|x\| \leq 1} |\langle x, y\rangle|. \tag{7.6}$$

Proof We know from Corollary 6.2 that a linear functional f on a normed space V is continuous if and only if there exists some $M > 0$ such that

$$|f(x)| \leq M\|x\| \quad \text{for all } x \in V.$$

The Cauchy–Schwarz inequality clearly shows that for functionals of the form $f_y(x) = \langle x, y\rangle$, we have continuity with $M = \|y\|$. Furthermore, this gives us the norm inequality

$$\|f_y\| \leq \|y\|. \tag{7.7}$$

To prove (7.6), the opposite inequality of (7.7) needs to be proven. The case $y = 0$ is clear. But $\|y\|^2 = \langle y, y\rangle = f_y(y) \le \|f_y\| \cdot \|y\|$. So, for the case $y \neq 0$, division by $\|y\| > 0$ gives us the opposite inequality $\|y\| \le \|f_y\|$. ∎

Corollary 7.7 *In any inner product space, if* $x^n \to x$, $y^n \to y$, $\alpha_n \to \alpha$ *(as* $n \to \infty$*), then*

$$\langle x^n, y^n\rangle \to \langle x, y\rangle \text{ and } \langle \alpha_n x, y\rangle \to \alpha\langle x, y\rangle \ (\text{as } n \to \infty).$$

Corollary 7.8 *In any inner product space, if* $x^n \to x$, *then* $\|x^n\| \to \|x\|$ *(as* $n \to \infty$*).*

We can now characterize the inner product spaces as exactly those for which the parallelogram identity holds.

Theorem 7.9 *A norm arises from an inner product if and only if the parallelogram identity holds for all x and y.*

Proof **(⇒):** A norm arising from an inner product satisfies the parallelogram identity by Theorem 7.3.

(⇐): For the implication in the other direction, we consider the real and complex cases separately.

Proof of the real case: We show that the real polarization identity

$$\langle x, y\rangle = \frac{1}{4}\|x+y\|^2 - \frac{1}{4}\|x-y\|^2,$$

as given in Theorem 7.1, actually defines an inner product. That **(IP3)** and **(IP4)** hold are obvious. We need **(IP1)** and **(IP2)**.

Proof of (IP1): $\langle x, z\rangle + \langle y, z\rangle = \langle x + y, z\rangle$. By the parallelogram identity, we have the following two equalities:

$$2\|x+z\|^2 + 2\|y\|^2 = \|x+y+z\|^2 + \|x-y+z\|^2,$$
$$2\|x-z\|^2 + 2\|y\|^2 = \|x+y-z\|^2 + \|x-y-z\|^2.$$

Subtracting the two, we obtain

$$2\|x+z\|^2 - 2\|x-z\|^2 = \|x+y+z\|^2 + \|x-y+z\|^2 - \|x+y-z\|^2 - \|x-y-z\|^2.$$

By the real polarization identity, the left-hand side is $8\langle x, z\rangle$. So

$$8\langle x, z\rangle = \|x+y+z\|^2 + \|x-y+z\|^2 - \|x+y-z\|^2 - \|x-y-z\|^2. \tag{7.8}$$

Interchanging x and y, we also obtain

$$\begin{aligned}8\langle y,z\rangle = \|x+y+z\|^2 + \|-x+y+z\|^2 \\ - \|x+y-z\|^2 - \|-x+y-z\|^2.\end{aligned}$$

Using the fact $\|w\| = \|-w\|$, this equation is also equal to

$$8\langle y,z\rangle = \|x+y+z\|^2 + \|x-y-z\|^2 - \|x+y-z\|^2 - \|x-y+z\|^2. \tag{7.9}$$

Adding equations (7.8) and (7.9), we obtain

$$8\langle x,z\rangle + 8\langle y,z\rangle = 2\|(x+y)+z\|^2 - 2\|(x+y)-z\|^2,$$

which according to the real polarization identity is

$$8\langle x,z\rangle + 8\langle y,z\rangle = 8\langle x+y,z\rangle.$$

This proves **(IP1)**. It remains to prove **(IP2)** to complete the real case.

Proof of (IP2): $\langle \alpha x, y\rangle = \alpha\langle x,y\rangle$. From the real polarization identity, it is clear that $\langle \alpha x, y\rangle = \alpha\langle x,y\rangle$ for $\alpha = -1, 0, 1$. Also, from **(IP1)** we have $\langle 2x,y\rangle = \langle x+x,y\rangle = \langle x,y\rangle + \langle x,y\rangle = 2\langle x,y\rangle$. By induction we obtain $\langle mx,y\rangle = m\langle x,y\rangle$ for all $m = 0, \pm 1, \pm 2, \ldots$. For any rational number $r = m/n$, we thus have

$$n\langle rx,y\rangle = \langle nrx,y\rangle = \langle mx,y\rangle = m\langle x,y\rangle.$$

Dividing by n, we obtain $\langle rx,y\rangle = r\langle x,y\rangle$ for all rational numbers r. By continuity of the norm (Corollary 7.8), $\langle \alpha x,y\rangle = \alpha\langle x,y\rangle$ for $\alpha \in \mathbb{R}$.

Proof of the complex case: The complex polarization identity as given by Equation (7.2) (Theorem 7.2) can be written

$$Re\langle x,y\rangle = \frac{1}{4}\|x+y\|^2 - \frac{1}{4}\|x-y\|^2,$$

$$Im\langle x,y\rangle = \frac{1}{4}\|x+iy\|^2 - \frac{i}{4}\|x-iy\|^2.$$

To show that $\langle x,y\rangle = \mathrm{Re}\langle x,y\rangle + i\,\mathrm{Im}\langle x,y\rangle$ actually defines an inner product, first observe from this definition that $\langle ix,y\rangle = i\langle x,y\rangle$ and $\overline{\langle x,y\rangle} = \langle y,x\rangle$. Then apply the real case of this theorem to both real and imaginary parts of the complex polarization identity. The details are left as Exercise 7.1.19. ∎

Inner products give us a way of defining orthogonality and the angle between nonzero vectors.

Definition 7.2 In any inner product space V, two nonzero vectors x, y are said to be **orthogonal**, written $x \perp y$, if $\langle x, y\rangle = 0$. A set of nonzero vectors $S \subset V$ is called an **orthogonal set** if $x \perp y$ for all distinct $x, y \in S$. If, in addition, $\|x\| = \sqrt{\langle x, x\rangle} = 1$ for all $x \in S$, the set is called an **orthonormal set**. In a real inner product space V, the **angle** θ between two nonzero vectors x and y can be defined by the formula

$$\cos\theta = \frac{\langle x, y\rangle}{\|x\|\|y\|}.$$

Orthogonality is then equivalent to $\cos\theta = 0$.

Example 7.6 In Example 7.3, if we write $e^k = (0, \ldots, 0, 1, 0, \ldots)$ with 1 in the kth place, then $S = \{e^1, e^2, \ldots\}$ is an orthonormal set in ℓ_2.

Example 7.7 The function space $C[-\pi, \pi]$ is a subspace of the inner product space $L^2[-\pi, \pi]$ of Example 7.4. We can use the same inner product

$$\langle f, g\rangle = \frac{1}{2\pi}\int_{-\pi}^{\pi} f(t)\overline{g(t)}\,dt.$$

The functions $\ldots, e^{-3it}, e^{-2it}, e^{-it}, 1, e^{it}, e^{2it}, e^{3it}, \ldots.$ form an orthonormal set. The proof is left as Exercise 7.1.24.

In the *real* case, the functions 1, $\sin t$, $\cos t$, $\sin 2t$, $\cos 2t$, $\sin 3t$, $\cos 3t, \ldots$ form an orthogonal set. We can make the set orthonormal by dividing each function by its norm. The proof is left as Exercise 7.1.25.

Theorem 7.10 Pythagorean theorem: *If $x \perp y$, then $\|x+y\|^2 = \|x\|^2 + \|y\|^2$. More generally, if $x^1, x^2, \ldots, x^n$ are orthogonal, then*

$$\left\|\sum_{k=1}^{n} x^k\right\|^2 = \sum_{k=1}^{n} \|x^k\|^2.$$

Proof For two vectors x and y, the proof follows immediately from the definition of inner product. The general case follows from the identity

$$\langle x^1 + \cdots + x^n, x^1 + \cdots + x^n\rangle = \langle x^1, x^1\rangle + \cdots + \langle x^n, x^n\rangle. \qquad \blacksquare$$

Theorem 7.11 Linear independence: *Any orthogonal set S is linearly independent.*

Proof Consider a finite subset $x^1, x^2, \ldots, x^n$ of an orthogonal set S. Suppose $\sum \alpha_k x^k = 0$. For any $j = 1, 2, \ldots, n$, we have $\langle x^k, x^j\rangle = 0$ unless $k = j$.

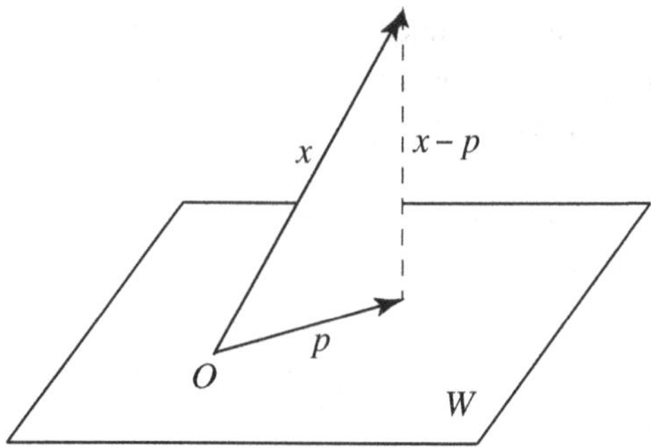

Figure 7.2 An orthogonal projection p of x onto the plane W.

Thus $0 = \left\langle \sum_{k=1}^{n} \alpha_k x^k, x^j \right\rangle = \sum_{k=1}^{n} \alpha_k \langle x^k, x^j \rangle = \alpha_j \langle x^j, x^j \rangle = \alpha_j \|x^j\|^2,$

which implies $\alpha_j = 0$. This is true for all $j = 1, 2, \ldots, n$. ∎

Theorem 7.12 Orthogonal projection: *Let V be an inner product space and let W be a* complete *subspace of V. For every fixed $x \in V$, there exists a unique $p \in W$ which minimizes the distance between x and W. That is,*

$$\|x - p\| = d(x, W) = \min_{w \in W} \|x - w\|.$$

The vector $p \in W$ is called the **orthogonal projection** *of x onto W. This is illustrated in Figure 7.2.*

Proof

Existence: Define $d = \inf_{w \in W} \|x - w\|$. By definition of infimum, there exists a sequence $w^n \in W$ such that

$$\|x - w^n\| \to d \quad \text{as} \quad n \to \infty.$$

We first show that w^n is a Cauchy sequence. Applying the parallelogram identity to two vectors $x - w^n$ and $x - w^m$, we have

$$\begin{aligned} &\|(x - w^m) - (x - w^n)\|^2 + \|(x - w^m) + (x - w^n)\|^2 \\ &\quad = 2\|x - w^m\|^2 + 2\|x - w^n\|^2 \end{aligned}$$

or

$$\begin{aligned} \|w^n - w^m\|^2 &= 2\|x - w^m\|^2 + 2\|x - w^n\|^2 - \|2x - (w^n + w^m)\|^2 \\ &= 2\|x - w^m\|^2 + 2\|x - w^n\|^2 - 4\left\|x - \frac{w^n + w^m}{2}\right\|^2. \end{aligned}$$

Since $\dfrac{w^n + w^m}{2} \in W$, we have $\|x - \frac{w^n+w^m}{2}\|^2 \geq d^2$, and hence

$$\|w^n - w^m\|^2 \leq 2\|x - w^m\|^2 + 2\|x - w^n\|^2 - 4d^2.$$

Taking the limit results in $\|w^n - w^m\|^2 \to 2d^2 + 2d^2 - 4d^2 = 0$ as $n, m \to \infty$. This shows that the sequence w^n is Cauchy in W.

Since W is complete, w^n must converge to some $p \in W$. Since $p \in W$, we have $\|x - p\| \geq d(x, W) = d$. Thus $d \leq \|x - p\| \leq \|x - w^n\| + \|w^n - p\| \to d + 0$, which shows us that $\|x - p\| = d$. Existence of p is proven.

Uniqueness: Suppose there were two vectors p, $p' \in W$ satisfying $\|x - p\| = \|x - p'\| = d$. Using the parallelogram identity and $\left\|x - \frac{p+p'}{2}\right\|^2 \geq d^2$, we have

$$\begin{aligned} 0 \leq \|p - p'\|^2 &= \|(x - p') - (x - p)\|^2 \\ &= 2\|x - p\|^2 + 2\|x - p'\|^2 - \|(x - p') + (x - p)\|^2 \\ &= 2d^2 + 2d^2 - 4\left\|x - \frac{p + p'}{2}\right\|^2 \leq 2d^2 + 2d^2 - 4d^2 = 0. \end{aligned}$$

This shows that $\|p - p'\|^2 = 0$, or $p = p'$. ∎

Theorem 7.13 *Let V be an inner product space and let W be a complete subspace of V. For a fixed $x \in V$, let p be the projection of x onto W. If $x \in W$, then $p = x$.*

Otherwise, $x - p \neq 0$ is orthogonal to every element of W. Write $x - p \perp W$.

Conversely, if for some $w \in W$, $x - w \neq 0$ and $x - w \perp W$, then w is the orthogonal projection of x onto W.

Proof For a given fixed $x \in V$, let p be the orthogonal projection of x onto W with $\|x - p\| = d = d(x, W)$, as given by the preceding orthogonal projection theorem. Let $z = x - p$ as shown in Figure 7.2. If $x \in W$, then $p = x$, since $\|x - p\| = d = 0$. If $x \in V - W$, then $z = x - p \neq 0$ since $x \notin W$.

We show orthogonality $z \perp W$ by contradiction. Assume not. Then there must exist a vector $w \in W$ such that $\langle z, w \rangle = \alpha \neq 0$. For any scalar t, we have $\|z - tw\| = \|(x - p) - tw\| = \|x - (p + tw)\| \geq d = \|x - p\| = \|z\|$. In other words, the following is nonnegative:

$$\begin{aligned} \|z - tw\|^2 - \|z\|^2 &= \langle z - tw, z - tw \rangle - \langle z, z \rangle \\ &= -\bar{t}\langle z, w \rangle - t\langle w, z \rangle + |t|^2\|w\|^2 \\ &= -\bar{t}\alpha - t\bar{\alpha} + |t|^2\|w\|^2. \end{aligned}$$

But if we choose $t = \frac{\alpha}{\|w\|^2}$, the right side has the negative value $-\frac{|\alpha|^2}{\|w\|^2}$. This contradiction proves the orthogonality of z with W.

Conversely, suppose $x \in V - W$ and that there exists a $w \in W$ such that $x - w \perp W$. We show that w must be the orthogonal projection p of x onto W. It is sufficient to show that

$$\|x - p\| = \|x - w\|$$

since the orthogonal projection theorem shows that the projection p is the unique vector in W with the property

$$\|x - p\| = d = d(x, W).$$

That $\|x - p\| \leq \|x - w\|$, is immediate by this definition of p. To obtain the opposite inequality, we note

$$\begin{aligned}\|x - p\|^2 &= \|(x - w) + (w - p)\|^2 \\ &= \|x - w\|^2 + \langle x - w, w - p\rangle + \langle w - p, x - w\rangle + \|w - p\|^2.\end{aligned}$$

Both of these inner products are zero since $w - p \in W$ and $x - w \perp W$. Thus

$$\|x - p\|^2 = \|x - w\|^2 + \|w - p\|^2 \geq \|x - w\|^2,$$

which completes the proof. ∎

Theorem 7.14 *Let V be an inner product space and let W be a finite-dimensional subspace with an orthogonal basis $\{w^1, w^2, \ldots, w^n\}$. For every $x \in V$, the orthogonal projection of x onto W is given by the formula*

$$p = \sum_{k=1}^{n} \frac{\langle x, w^k\rangle}{\langle w^k, w^k\rangle} w^k. \tag{7.10}$$

If a basis $\{e^1, e^2, \ldots, e^n\}$ of W is actually orthonormal $\left(e^k = \frac{w^k}{\|w_k\|}\right)$, then

$$p = \sum_{k=1}^{n} \langle x, e^k\rangle e^k. \tag{7.11}$$

Proof Since W is a finite-dimensional subspace, it is complete (Corollary 5.47). By Theorem 7.12, the projection p of x exists in W, say $p = \sum_{j=1}^{n} c_j w^j$. We show that $c_k = \frac{\langle x, w^k\rangle}{\langle w^k, w^k\rangle}$ for $k = 1, 2, 3, \ldots$. By Theorem 7.13, $x - p \perp w$ for all $w \in W$. In particular, $x - p \perp w^k$ for all k. So by orthogonality of the basis $\{w^1, w^2, \ldots, w^n\}$, we have

$$0 = \langle x - p, w^k\rangle = \left\langle x - \sum_{j=1}^{n} c_j w^j, w^k \right\rangle = \langle x, w^k\rangle - c_k\langle w^k, w^k\rangle,$$

which yields the desired formula for each c_k. In the orthonormal case, Equation (7.11) can easily be verified. ∎

Lemma 7.15 *Let A be an $n \times m$ matrix of real (or complex) numbers with $m \leq n$. Denote by A^T its transpose matrix and by $A^H = \overline{A^T} = (\overline{A})^T$ its conjugate transpose (often called the* **Hermitian conjugate***). In the real case, $A^H = A^T$. Then the $m \times m$ matrix $(A^H A)$ is invertible if and only if the m columns of A are linearly independent in $\mathbb{R}^n$ (or $\mathbb{C}^n$).*

Proof First we show that A and $(A^H A)$ have the same null space. For each $m \times 1$ column vector x, $Ax = 0$ clearly implies $A^H(Ax) = (A^H A)x = 0$. That is, any x which is in the null space of A is also in the null space of $A^H A$. Conversely, suppose $(A^H A)x = 0$. Then, for any $1 \times m$ row vector $x^H = (\overline{x_1}, \overline{x_2}, \ldots, \overline{x_m})$, we have $x^H A^H Ax = x^H 0 = 0$, which implies $(Ax)^H(Ax) = 0$. But this can happen only when $Ax = 0$, since the Ax consists of the column of dot products of x^H with the rows of A (so that $(Ax)^H(Ax)$ is the sum of their Euclidean norms). Any x in the null space of $A^H A$ is thus also in the null space of A.

To have linearly independent columns is equivalent to the statement that $Ax = 0 \implies x = 0$, or that the null space of A (and hence null space of $A^H A$) consists only of $x = 0$. In that case, the $m \times m$ matrix $(A^H A)$ must have rank m, and so it must be invertible. ∎

Theorem 7.16 Projection matrix: *Consider $V = \mathbb{R}^n$ (or $\mathbb{C}^n$) with the usual dot product $x \cdot y = x_1\overline{y}_1 + \cdots + x_n\overline{y}_n$. Suppose $X = \{x^1, x^2, \ldots, x^m\}$ is a linearly independent sequence in V and denote by A the $n \times m$ matrix with columns consisting of these linearly independent vectors. For a given element $x \in \mathbb{R}^n$ (or $x \in \mathbb{C}^n$), let p be its projection onto the span of X. We write the projection p, which is a linear combination of X, as $p = \widehat{p}_1 x^1 + \cdots + \widehat{p}_m x^m$. This may be written in the form $p = A\widehat{p}$ where $\widehat{p}$ is an $m \times 1$ column vector. The projection p of x satisfies the matrix equation*

$$(A^H A)\widehat{p} = A^H x.$$

Since $(A^H A)$ is invertible by Lemma 7.15, we can write $p = Px$, where P is the $n \times n$ **projection matrix** *defined by*

$$P = A(A^H A)^{-1} A^H.$$

Proof The projection p of x onto the span of X has two properties.

First: The projection p is in the subspace spanned by X; hence it is a linear combination of the vectors $x^1, x^2, \ldots, x^m$. In column form, this means that $p = A\widehat{p}$ for some $m \times 1$ column vector $\widehat{p}$.

Second: The vector $x - p$ is orthogonal to every vector in the span of X. In particular $(x - p) \perp x^k$ for $k = 1, 2, \dots m$. For the dot product, we have $x^k \cdot (x - p) = 0$ for $k = 1, 2, \dots, m$. Or, in column form, the matrix product $A^H(x - p) = 0$ is the zero $m \times 1$ column.

Combining the two properties, we have $A^H(x - A\widehat{p}) = 0$ or

$$\begin{aligned}(A^H A)\widehat{p} &= A^H x,\\ \widehat{p} &= (A^H A)^{-1} A^H x,\\ p = A\widehat{p} &= A(A^H A)^{-1} A^H x.\end{aligned}$$

∎

Example 7.8 In $\mathbb{R}^4$, let $x^1 = (1,0,0,1), x^2 = (0,1,0,1)$, and $x = (1,0,1,0)$. Let us find the projection p of x onto the span of $X = \{x^1, x^2\}$. It is clear that x^1 and x^2 are linearly independent and x is not in the span of X. Here

$$A = \begin{pmatrix} 1 & 0 \\ 0 & 1 \\ 0 & 0 \\ 1 & 1 \end{pmatrix} \quad \text{and} \quad A^H = A^T = \begin{pmatrix} 1 & 0 & 0 & 1 \\ 0 & 1 & 0 & 1 \end{pmatrix}.$$

$$\text{So} \quad A^T A = \begin{pmatrix} 2 & 1 \\ 1 & 1 \end{pmatrix} \quad \text{and} \quad A^T x = \begin{pmatrix} 1 \\ 0 \end{pmatrix}.$$

We write the equation $(A^T A)\widehat{p} = A^T x$ in augmented matrix form

$$\left(\begin{array}{cc|c} 2 & 1 & 1 \\ 1 & 2 & 0 \end{array}\right)$$

and apply Gaussian elimination to obtain

$$\left(\begin{array}{cc|c} 1 & 0 & 2/3 \\ 0 & 1 & -1/3 \end{array}\right).$$

So $\widehat{p} = \left(\frac{2}{3}, \frac{-1}{3}\right)$ and $p = A\widehat{p} = \frac{2}{3}x^1 + \frac{-1}{3}x^2 = \frac{1}{3}(2, -1, 0, 1)$.

We have the correct solution because clearly p is in the span of X, and $x - p$ is orthogonal to both x^1 and x^2.

Instead of Gaussian elimination we could have computed the inverse of $(A^T A)$ to compute $\widehat{p}$ (in column form)

$$(A^T A)^{-1} A^T x = \frac{1}{3}\begin{pmatrix} 2 & -1 \\ -1 & 2 \end{pmatrix}\begin{pmatrix} 1 \\ 0 \end{pmatrix} = \begin{pmatrix} 2/3 \\ -1/3 \end{pmatrix}.$$

The 4×4 projection matrix $P = A(A^T A)^{-1} A^T$ is then

$$P = \frac{1}{3}\begin{pmatrix} 1 & 0 \\ 0 & 1 \\ 0 & 0 \\ 1 & 1 \end{pmatrix}\begin{pmatrix} 2 & -1 \\ -1 & 2 \end{pmatrix}\begin{pmatrix} 1 & 0 & 0 & 1 \\ 0 & 1 & 0 & 1 \end{pmatrix} = \frac{1}{3}\begin{pmatrix} 2 & -1 & 0 & 1 \\ -1 & 2 & 0 & 1 \\ 0 & 0 & 0 & 0 \\ 1 & 2 & 0 & 1 \end{pmatrix}.$$

Theorem 7.17 Gram–Schmidt orthogonalization: *Suppose $X = \{x^1, x^2, \ldots\}$ is a linearly independent sequence in an inner product space V. Then there is an orthonormal sequence $E = \{e^1, e^2, \ldots\}$ with the same span as X.*

Proof

Step 1: Since X is linearly independent, every x^k is nonzero. Start with

$$z^1 = x^1$$

and normalize z^1 to obtain e^1:

$$e^1 = \frac{z^1}{\|z^1\|} = \frac{x^1}{\|x^1\|}.$$

Step 2: To obtain e^2, let

$$p^2 = \langle x^2, e^1 \rangle e^1$$

and set

$$z^2 = x^2 - p^2.$$

It follows that $z^2 \perp e^1$ (Exercise 7.1.23). Figure 7.3 shows p^2 as the **orthogonal projection** of x^2 onto the span of e^1. Normalize z^2 to obtain

$$e^2 = \frac{z^2}{\|z^2\|}.$$

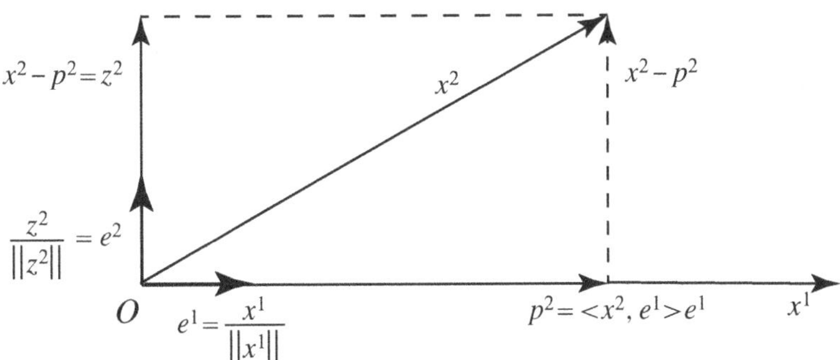

Figure 7.3 Gram–Schmidt orthogonalization. Steps 1 and 2.

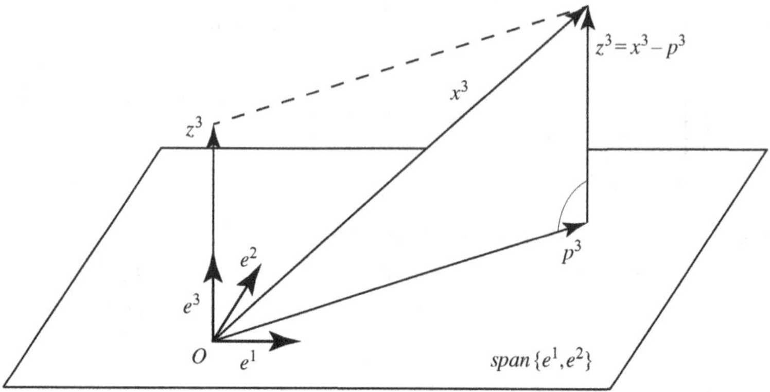

Figure 7.4 Gram–Schmidt orthogonalization. Step 3.

Clearly $\{e^1, e^2\}$ has the same span as $\{x^1, x^2\}$.

Step n: Please see Figure 7.4. Continuing this process, let p^n be the projection of x^n onto the span of $\{e^1, e^2, \ldots, e^{n-1}\}$

$$p^n = \sum_{k=1}^{n-1} \langle x^n, e^k \rangle e^k.$$

If we set

$$z^n = x^n - p^n,$$

then $z^n \perp \{e^1, e^2, \ldots, e^{n-1}\}$. Finally, normalize z^n to obtain

$$e^n = \frac{z^n}{\|z^n\|}.$$

Since each e^n is a linear combination of the vectors $x^1, x^2, \ldots, x^n$, we have $\text{span}\{e^1, e^2, \ldots, e^n\} \subset \text{span}\{x^1, x^2, \ldots, x^n\}$. We obtain equality of these spans since both finite sets $\{e^1, \ldots, e^n\}$ and $\{x^1, \ldots, x^n\}$ are linearly independent and have the same number of elements. Since every linear combination of the e^k is a linear combination of the x^k, and conversely, it follows that the spans of X and E are equal. ∎

Remark The Gram–Schmidt procedure may be summarized as follows. Suppose $X = \{x^1, x^2, \ldots\}$ is a linearly independent sequence in an inner product space V. Define $Z = \{z^1, z^2, \ldots\}$ as follows.

$$
\begin{aligned}
z^1 &= x^1 \\
z^2 &= x^2 - \frac{\langle x^2, z^1\rangle}{\langle z^1, z^1\rangle} z^1 = x^2 - p^2 \\
z^3 &= x^3 - \frac{\langle x^3, z^1\rangle}{\langle z^1, z^1\rangle} z^1 - \frac{\langle x^3, z^2\rangle}{\langle z^2, z^2\rangle} z^2 = x^3 - p^3 \\
&\vdots \\
z^n &= x^n - \frac{\langle x^n, z^1\rangle}{\langle z^1, z^1\rangle} z^1 - \cdots - \frac{\langle x^n, z^{n-1}\rangle}{\langle z^{n-1}, z^{n-1}\rangle} z^{n-1} = x^n - p^n.
\end{aligned}
$$

Then Z is an orthogonal set with the same span as X. Further, if we define $e^k = \frac{z^k}{\|z^k\|}$ for $k = 1, 2, \ldots$, then $E = \{e^1, e^2, \ldots\}$ is an orthonormal set with the same span as X.

We can use this remark along with the projection matrix theorem (Theorem 7.16) to obtain the following. Details of the proof are left as Exercise 7.1.31.

Corollary 7.18 Orthogonalization by matrices: *Consider $V = \mathbb{R}^n$ with the usual dot product $x \cdot y$. Suppose $X = \{x^1, x^2, \ldots, x^m\}$ is a linearly independent set in V. Let A be the $n \times m$ matrix with columns consisting of the vectors $x^1, x^2, \ldots, x^m$. Apply the Gram–Schmidt procedure on X to obtain the orthonormal set $E = \{e^1, e^2, \ldots, e^n\}$ and let Q be the $n \times m$ matrix consisting as the vectors of E as its columns. Then the matrix A may be written in the form*

$$A = QR,$$

where R is an upper triangular matrix.

Example 7.9 Laguerre polynomials:[1] Consider $L^2[0, \infty]$ under the real inner product

$$\langle f, g\rangle = \int_0^\infty f(t)g(t)e^{-t}\,dt.$$

The power functions

$$1,\ t,\ t^2,\ t^3, \ldots$$

[1] Named in honor of the French mathematician Edmond Laguerre (1834–1886).

form a linearly independent set. If one applies the Gram–Schmidt process to these functions, one obtains an orthonormal sequence of polynomials called the Laguerre polynomials. The first few polynomials are as follows:

$$L_0(t) = 1,$$
$$L_1(t) = t - 1,$$
$$L_2(t) = \frac{1}{2}(t^2 - 4t + 2),$$
$$L_3(t) = \frac{1}{6}(t^3 - 9t^2 + 18t - 6).$$

Computation of the polynomials is aided by the **gamma function** formula

$$\Gamma(n+1) = \int_0^\infty t^n e^{-t}\,dt = n! \quad \text{for} \quad n = 0, 1, \ldots.$$

It can be shown that

$$L_n(t) = \frac{(-1)^n e^t}{n!}\frac{d^n}{dt^n}\left(t^n e^{-t}\right), \quad n = 0, 1, 2, \ldots.$$

The polynomials can also be generated by the recursive equation

$$L_{n+1}(t) = \frac{1}{n+1}\Big((t - 2n - 1)L_n(t) + nL_{n-1}(t)\Big).$$

The Laguerre polynomials are used in the analysis of the hydrogen atom in quantum mechanics.

Exercises for Section 7.1

7.1.1 Prove Property **(IP5)** from the axioms **(IP1)–(IP4)** at the beginning of Section 7.1.

7.1.2 Prove the equality $\langle \alpha x + \beta y, z\rangle = \alpha\langle x, z\rangle + \beta\langle y, z\rangle$ in an inner product space.

7.1.3 Prove the equality $\langle x, \alpha y + \beta z\rangle = \overline{\alpha}\langle x, y\rangle + \overline{\beta}\langle x, z\rangle$ in an inner product space.

7.1.4 Prove that any subspace W of an inner product space V is still an inner product space under the same inner product.

7.1.5 Prove the real polarization identity (7.1) as given in Theorem 7.1. Use the definition of a Euclidean norm and the axioms of an inner product.

7.1.6 Prove the complex polarization identity (7.2) as given in Theorem 7.2. Use the definition of a Euclidean norm and the axioms of an inner product.

7.1.7 Prove the parallelogram identity Theorem 7.3.

7.1.8 Show that the space ℓ_2 satisfies the parallelogram identity.

7.1.9 Show that if $1 \le p \le \infty$ but $p \ne 2$, then ℓ_p is not an inner product space.

7.1.10 Which of the axioms for an inner product space **(IP1)**, **(IP2)**, **(IP3)**, **(IP4)** fails for the space ℓ_1 in Example 7.5?

7.1.11 Show that $L^2[a,b]$ is an inner product space as stated in Example 7.4.

7.1.12 Theorem 7.4 states that $L^p[a,b]$ is not an inner product space when $p \ne 2$. Prove this in the case $[a,b] = [-1,1]$.

7.1.13 Show that the function space of continuous functions $C[0,1]$ under the norm $\|f\|_\infty = \sup_{1\le t\le 1} |f(t)|$ is not an inner product space.

7.1.14 Show that in an inner product space, if $\langle x,y\rangle = \langle x,z\rangle$ for all x, then $y = z$.

7.1.15 Suppose that $\langle x^n,x\rangle \to \langle x,x\rangle$ and $\|x^n\| \to \|x\|$. Show that $x^n \to x$.

7.1.16 Suppose $y \perp x^n$ for all $n = 1,2,\ldots$ and $x^n \to x$. Show that $y \perp x$.

7.1.17 Prove Corollary 7.7.

7.1.18 Prove Corollary 7.8.

7.1.19 Prove in detail the complex case of Theorem 7.9.

7.1.20 Show that $x \perp y$ if and only if $\|x+\alpha y\| = \|x-\alpha y\|$ for all scalars α.

7.1.21 Show that $x \perp y$ if and only if $\|x\| \le \|x+\alpha y\|$ for all scalars α.

7.1.22 Prove the Pythagorean theorem (Theorem 7.10) in detail.

7.1.23 In the proof of Theorem 7.17, show that $y^2 \perp e^1$. Then show $e^1 \perp e^2$.

7.1.24 Prove that the set $\{e^{ikt}\}$ for $k = 0, \pm 1, \pm 2, \ldots$ forms an orthonormal set in $C[-\pi,\pi]$ as given in Example 7.7.

7.1.25 Prove that the set of functions $\{1, \sin t, \cos t, \sin 2t, \cos 2t, \sin 3t, \cos 3t, \ldots\}$, as given in Example 7.7, forms an orthogonal set in $C[-\pi,\pi]$. One way of doing this is to use Exercise 7.1.24 along with the equalities

$$\cos nt = \frac{e^{int}+e^{-int}}{2} \quad \text{and} \quad \sin mt = \frac{e^{imt}-e^{-imt}}{2i}.$$

7.1.26 Use Exercise 7.1.25 and the Pythagorean theorem to show that

$$\int_{-\pi}^{\pi} (\sin t + 2\sin 4t + 3\sin 6t)^2\,dt = 14\pi.$$

7.1.27 For the space $C[a,b]$ with real inner product $\langle f,g\rangle = \frac{1}{b-a}\int_a^b f(t)g(t)\,dt$, show that the following functions are orthogonal:

$$1,\ \sin\frac{2\pi}{b-a}t,\ \cos\frac{2\pi}{b-a}t,\ \sin\frac{2\pi}{b-a}2t,\ \cos\frac{2\pi}{b-a}2t,$$

$$\sin\frac{2\pi}{b-a}3t,\ \cos\frac{2\pi}{b-a}3t,\ \ldots.$$

7.1.28 Show that in Lemma 7.15, A^H cannot be replaced by A^T in the complex case. Give an example of a nonzero 2×1 complex matrix A for which $A^TA = 0$.

7.1.29 Consider the two vectors $x^1 = (1,2,2)$ and $x^2 = (-1,0,2)$ in $\mathbb{R}^3$. Use the Gram–Schmidt procedure to find an orthonormal set $E = \{e^1, e^2\}$ with the same span as $X = \{x^1, x^2\}$. Find the orthogonal projection p^3 of the vector $x^3 = (0,0,1)$ onto E.

7.1.30 Given the vectors $x^1 = (1,2,2), x^2 = (-1,0,2)$, and $x^3 = (0,0,1)$ in $\mathbb{R}^3$ of Exercise 7.1.29, use the Gram–Schmidt procedure to find an orthonormal basis $\{e^1, e^2, e^3\}$. Write the matrix equation $A = Q\cdot R$ as explained in Corollary 7.18.

7.1.31 Recall your linear algebra to prove Corollary 7.18. The triangular form of the matrix R is apparent from Theorem 7.1 whereby each vector e^k is derived from the first k linearly independent vectors $x^1,\ldots,x^k$.

7.1.32 Show that $\langle f,g\rangle = \int_0^\infty f(t)g(t)e^{-t}\,dt$ satisfies the conditions for a real inner product on the space $L^2[0,\infty]$ as given in Example 7.9.

7.1.33 Consider the inner product given in Example 7.9 and the linearly independent sequence of power functions $1, t, t^2, t^3, \ldots$. Use the Gram–Schmidt orthogonalization process to find the first four Laguerre polynomials

$$L_0(t),\ L_1(t),\ L_2(t),\ L_3(t).$$

7.1.34 Use Gram–Schmidt orthogonalization to find orthonormal functions with the same span as the functions $\{1, t, t^2\}$ in the space $C[0,1]$ of continuous real valued functions on $[0,1]$. Assume the real inner product

$$\langle f,g\rangle = \int_0^1 f(t)g(t)\,dt.$$

7.1.35 Repeat Exercise 7.1.34 for the same set of functions but in the order $\{t^2, t, 1\}$.

7.1.36 Prove the uniqueness part of Theorem 7.12.

7.1.37 Every normed space V has a completion V_1 which is unique up to isometry (Theorems 5.32 and 5.33 on p. 199). If V satisfies the parallelogram identity, show that the completion V_1 does also.

7.2 Fourier Series in Inner Product Spaces

In this section we consider only spaces with norms that arise from an inner product. Later we will extend Fourier series to spaces of functions with norms that do not satisfy the parallelogram identity.

Recall that the Gram–Schmidt process generates an orthonormal sequence from any linearly independent sequence.

Definition 7.3 Let $E = \{e^1, e^2, \ldots\}$ be an orthonormal sequence in an inner product space V. The sequence may be finite or infinite. Consider any $x \in V$.

(a) For each k, $\widehat{x}(k) = \langle x, e^k \rangle$ is called the kth **Fourier coefficient** of x with respect to E.

(b) $\widehat{x} = \big(\widehat{x}(k)\big) = \big(\widehat{x}(1), \widehat{x}(2), \widehat{x}(3), \ldots\big)$ is called the **sequence of Fourier coefficients** of x with respect to E.

(c) $\widehat{V} = \{\widehat{x} \mid x \in V\}$ is called the **sequence space of Fourier coefficients** of V with respect to E.

(d) And $\sum_k \widehat{x}(k)e^k$ is called the **Fourier series** of x with respect to E. At this point, the Fourier series is merely a formal expression with no assumptions about its convergence. This association is written

$$x \sim \sum_k \widehat{x}(k)e^k. \tag{7.12}$$

Example 7.10 We now show that if V is an inner product space of finite dimension n, then $\widehat{V}$ is isometric to $\ell_2(n)$. We can use Gram–Schmidt orthogonalization to obtain an orthonormal basis $\{e^1, \ldots, e^n\}$ for V. Let

$$x = x_1 e^1 + \cdots + x_n e^n \quad \text{and} \quad y = y_1 e^1 + \cdots + y_n e^n.$$

Direct calculations show that

$$\langle x, y \rangle = \langle x_1 e^1 + \cdots + x_n e^n, y_1 e^1 + \cdots + y_n e^n \rangle = x_1 \overline{y_1} + \cdots + x_n \overline{y_n}.$$

So $\|x\| = \sqrt{\langle x, x \rangle} = \sqrt{|x_1|^2 + \cdots + |x_n|^2} = \|(x_1, \ldots, x_n)\|_2 = \|\widehat{x}\|_2$, which is an isometry between $x \in V$ and $\widehat{x} \in \ell_2(n)$. Further, for each $x \in V$,

$$x = \sum_{k=1}^{n} \langle x, e^k \rangle e^k = \sum_{k=1}^{n} \widehat{x}(k)e^k.$$

In the coordinate system of the basis $\{e^1, \ldots, e^n\}$, we thus have $x = \widehat{x}$.

Example 7.11 $\widehat{\ell_2} = \ell_2$**:** The inner product space ℓ_2 has an orthonormal sequence $E = \{e^1, e^2, e^3, \ldots\}$, where $e^k = (0, \ldots, 0, 1, 0, \ldots)$ with 1 in the kth position. For each $x = (x_1, x_2, \ldots) \in \ell_2$, the kth Fourier coefficients $\widehat{x}(k) = \langle x, e^k \rangle = x_k$ are the kth coordinates of $x \in \ell_2$. Hence, $\widehat{x} = (x_1, x_2, \ldots) = x$.

Example 7.12 The space $L^2[-\pi, \pi]$ is an inner product space with the inner product

$$\langle f, g \rangle = \frac{1}{2\pi} \int_{-\pi}^{\pi} f(t)\overline{g(t)}\, dt.$$

That the functions $e_k(t) = e^{ikt}$, for $k = 0, \pm 1, \pm 2, \pm 3, \ldots$ form an orthonormal sequence was observed in Example 7.7.

For each $f \in L^2[-\pi, \pi]$, the Fourier series is of the form

$$f \sim \sum_{k=-\infty}^{\infty} c_k e^{ikt} \text{ where } c_k = \langle f, e_k \rangle = \frac{1}{2\pi} \int_{-\pi}^{\pi} f(t) e^{-ikt}\, dt.$$

Theorem 7.19 *Suppose that* $E = \{e^1, e^2, e^3, \ldots\}$ *is an orthonormal sequence in an inner product space* V*. If* $x \in V$ *can be written as a convergent series* $x = \sum_{k=1}^{\infty} c_k e^k$*, then it is the Fourier series of* x*; that is,* $c_k = \widehat{x}(k) = \langle x, e^k \rangle$ *for all* k*.*

Proof Given $x \in V$, let $s^n = \sum_{k=1}^{n} c_k e^k$ and $m = 1, 2, \ldots$. By the Cauchy–Schwarz inequality, $|\langle s^n, e^m \rangle - \langle x, e^m \rangle| = |\langle s^n - x, e^m \rangle| \le \|s^n - x\| \cdot \|e^m\| = \|s^n - x\|$. By hypothesis, $\|s^n - x\| \to 0$. Thus $\langle s^n, e^m \rangle \to \langle x, e^m \rangle$ as $n \to \infty$. But $\langle s^n, e^m \rangle = \sum_{k=1}^{n} c_k \langle e^k, e^m \rangle = c_m$ for $n \ge m$. So $c_m = \langle x, e^m \rangle$ for all m. ∎

Remark The above result shows the advantage of expressing a series $x = \sum_k c_k e^k$ in terms of an *orthonormal* sequence $E = \{e^1, e^2, \ldots\}$. The coefficients c_k are then the easily obtained Fourier coefficients $c_k = \langle x, e^k \rangle = \widehat{x}(k)$. If a sequence $G = \{y^1, y^2, \ldots\}$ is merely orthogonal, the Fourier series of x is

$$x \sim \sum_k \frac{\langle x, y^k \rangle}{\langle y^k, y^k \rangle}\, y^k. \tag{7.13}$$

It can be made orthonormal by replacing each y^k with $e^k = \frac{y^k}{\|y^k\|}$.

Example 7.13 Consider the space $C[0, \pi]$ under the inner product $\langle f, g \rangle = \int_0^{\pi} f(t)g(t)\, dt$. The sequence $\sin t$, $\sin 2t$, $\sin 3t, \ldots,$ is orthogonal on the interval $[0, \pi]$. This can be shown using the identity

$$\sin kt \sin jt = \frac{1}{2}\big\{\cos(k-j)t - \cos(k+j)t\big\}$$

to obtain, for $k \neq j$,

$$\int_0^\pi \sin kt \sin jt\, dt = \frac{1}{2}\int_0^\pi \cos(k-j)t\, dt - \frac{1}{2}\int_0^\pi \cos(k+j)t\, dt = 0.$$

Also, for $k = j$, we have $\int_0^\pi \sin^2 kt\, dt = \int_0^\pi \frac{1-\cos 2kt}{2}\, dt = \frac{\pi}{2}$. By Equation (7.13), the Fourier series of $f \in C[0,\pi]$ can thus be written

$$f \sim \sum_{k=1}^{\infty} A_k \sin kt \quad \text{with} \quad A_k = \frac{\int_0^\pi f(t)\sin kt\, dt}{\int_0^\pi \sin^2 kt\, dt} = \frac{2}{\pi}\int_0^\pi f(t)\sin kt\, dt. \tag{7.14}$$

Theorem 7.20 Bessel's inequality: *Suppose that $e^1, e^2, \ldots$ is an orthonormal sequence in an inner product space V. Then for every $x \in V$, we have $\widehat{x} \in \ell_2$ and*

$$\sum_{k=1}^{\infty} |\langle x, e^k\rangle|^2 = \sum_{k=1}^{\infty} |\widehat{x}(k)|^2 \;\leq\; \|x\|^2 = \langle x, x\rangle. \tag{7.15}$$

Proof Let $x \in V$ and $s^n = \sum_{k=1}^n \widehat{x}(k)e^k$. Then, for $1 \leq m \leq n$, we have

$$\langle x - s^n, e^m\rangle = \langle x, e^m\rangle - \sum_{k=1}^{n} \widehat{x}(k)\langle e^k, e^m\rangle = 0. \tag{7.16}$$

Thus $x - s^n \perp e^m$ for all $m = 1, \ldots, n$. Hence, $x - s^n \perp s^n$. By the Pythagorean theorem (Theorem 7.10),

$$\begin{aligned} \|x\|^2 = \|x - s^n\|^2 + \|s^n\|^2 &= \|x - s^n\|^2 + \sum_{k=1}^{n} \|\widehat{x}(k)e^k\|^2 \\ &= \|x - s^n\|^2 + \sum_{k=1}^{n} |\widehat{x}(k)|^2. \end{aligned} \tag{7.17}$$

So $0 \leq \|x - s^n\|^2 = \|x\|^2 - \sum_{k=1}^{n} |\widehat{x}(k)|^2$, or $\|x\|^2 \geq \sum_{k=1}^{n} |\widehat{x}(k)|^2$, for all n. ∎

Corollary 7.21 *Suppose E is an orthonormal set in an inner product space V. Every $x \in V$ has at most countably many nonzero Fourier coefficients with respect to E.*

Proof If the orthonormal set E is countable, then the statement is clearly true. Consider the case where V has an uncountable orthonormal set E. For a fixed $x \in V$ and $m \in \mathbb{N}$, let $E_m(x) = \{e \in E \mid |\langle x, e\rangle| > 1/m\}$. By Inequality (7.15), this set must be finite. The union $E(x) = \bigcup_{m=1}^{\infty} E_m(x)$ is thus countable and consists of the nonzero Fourier coefficients of x with respect to the orthonormal set E. ∎

Theorem 7.22 Best approximation: *Suppose that $E = \{e^1, e^2, e^3, \ldots\}$ is an orthonormal sequence in an inner product space V. For a given $x \in V$, let*

$$s^n = \sum_{k=1}^{n} \widehat{x}(k) e^k$$

be the partial sum of the Fourier series of x, and let

$$t^n = \sum_{k=1}^{n} \beta_k e^k$$

be any other linear combination of $\{e^1, e^2, \ldots, e^n\}$. Then

$$\|x - s^n\| \leq \|x - t^n\|,$$

and equality holds if and only if $\beta_k = \widehat{x}(k)$ for $k = 1, \ldots, n$.

That is, among all the linear combinations t^n, the best approximation of x is given by s^n. In other words, the orthogonal projection of x onto the linear combinations t^n is s^n. The orthogonal projection s^n is often called the (best) ***least squares approximation*** *of x by $\{e^1, e^2, \ldots, e^n\}$.*

Proof Start with $x - t^n = x - s^n + \sum_{k=1}^{n} \gamma_k e^k$, where $\gamma_k = \widehat{x}(k) - \beta_k$ for $k = 1, \ldots, n$. Equation (7.16), shows that $x - s^n \perp e^m$ for $m = 1, \ldots, n$. By the Pythagorean theorem (Theorem 7.10),

$$0 \leq \|x - t^n\|^2 = \|x - s^n\|^2 + \sum_{k=1}^{n} \|\gamma_k e^k\|^2 = \|x - s^n\|^2 + \sum_{k=1}^{n} |\gamma_k|^2, \tag{7.18}$$

which gives us the desired inequality $\|x - s^n\| \leq \|x - t^n\|$. Clearly (7.18) is minimized if and only if $\gamma_k = 0$ $\big($or $\beta_k = \widehat{x}(k)\big)$ for all k. ∎

Theorem 7.23 Parseval: *Suppose that $e^1, e^2, e^3, \ldots$ is an orthonormal sequence in an inner product space V and let $x \in V$. The Fourier series $\sum_{k=1}^{\infty} \widehat{x}(k) e^k$ converges to x if and only if Bessel's inequality (Theorem 7.20) is an equality*

$$\|\widehat{x}\|_2 = \sqrt{\sum_{k=1}^{\infty} |\widehat{x}(k)|^2} = \|x\| = \sqrt{\langle x, x \rangle}.$$

Proof Let $x \in V$. Going back to Equation (7.17) in the proof of Bessel's inequality

$$\|x\|^2 = \|x - s^n\| + \sum_{k=1}^{n} |\widehat{x}(k)|^2, \tag{7.19}$$

we see that $\|x\|^2 = \sum_{k=1}^{\infty} |\widehat{x}(k)|^2$ if and only if $s^n \to x$ as $n \to \infty$. ∎

Corollary 7.24 *Suppose that $e^1, e^2, e^3, \ldots$ is an orthonormal sequence in an inner product space V and let $x \in V$. If the Fourier series $\sum_{k=1}^{\infty} \widehat{x}(k)e^k$ converges to x, we still have convergence for any rearrangement of the orthonormal sequence.*

Proof This follows from absolute convergence of the sum in (7.19). ∎

Definition 7.4 Suppose that $E = \{e^1, e^2, e^3, \ldots\}$ is an orthonormal sequence in an inner product space V. We say that E is an **orthonormal basis** of V if, for *every* $x \in V$, the Fourier series $\sum_{k=1}^{\infty} \widehat{x}(k)e^k$ converges to x. That is,

$$x = \sum_{k=1}^{\infty} \widehat{x}(k)e^k \quad \text{for every} \quad x \in V.$$

We also say that E is a **complete orthonormal sequence** in V.

Example 7.14 The space ℓ_2 has an orthonormal basis: Since the space ℓ_2 has the property AK (Definition 6.8), it is easy to see that the Fourier series of every $x \in \ell_2$ *converges* to x in the ℓ_2 norm. That is,

$$x = \sum_{k=1}^{\infty} \widehat{x}(k)e^k = \sum_{k=1}^{\infty} x_k e^k.$$

Example 7.15 Warning: An inner product space with a complete orthonormal sequence is *not* necessarily complete as a normed space. For example, the subspace Φ of ℓ_2 consisting of all sequences with only a finite number of nonzero coordinates has the same complete orthonormal sequences $\{e^1, e^2, \ldots\}$ as ℓ_2. But not all Cauchy sequences in Φ converge.

Theorem 7.25 *If V is a separable inner product space and E is any orthonormal set in V, then E is countable.*

Proof By the Pythagorean theorem (Theorem 7.10), the distance between any two points $x, y \in E$ must be $\sqrt{2}$ because

$$\|x - y\|^2 = \|x\|^2 + \|y\|^2 = 2.$$

If we take spheres of radii $\frac{\sqrt{2}}{2}$ about the points in E, the spheres are pairwise disjoint. Any dense subset must have points in each of the $(\frac{\sqrt{2}}{2})$-spheres. If V has a countable dense subset, then there can only be countably many $(\frac{\sqrt{2}}{2})$-spheres, and hence countably many elements of E. ∎

Caution There are nonseparable inner product spaces. They have uncountable orthonormal sets E. Such a space is given in Example 7.19. The good news given by Corollary 7.21 is that, even if an orthogonal set E is uncountable, each $x \in V$ has at most countably many nonzero Fourier coefficients $E(x) = \{e \in E \mid \langle x, e\rangle \neq 0\}$ with respect to E. This permits us to always associate a Fourier *series*

$$x \sim \sum_{e^\kappa \in E(x)} \langle x, e^\kappa \rangle e^\kappa \text{ for every } x \in V.$$

In the nonseparable case, the orthonormal sequence $e^\kappa \in E(x)$ used in the series depends on x. In the separable case, a Fourier series of x uses the same orthonormal sequence $E = \{e^1, e^2, e^3, \ldots\}$ for all $x \in V$.

Theorem 7.26 *Let V be an inner product space. Then V is separable if and only if there exists a countable orthonormal set E whose span is dense in V.*

Proof (⇒): If $A = \{x^1, x^2, \ldots\}$ is a countable dense subset of V, we can extract a linearly independent subset $B = \{y^1, y^2, \ldots\}$ with the same span as follows: Let y^1 be the first nonzero element of A. Let y^2 be the first x^k which is not in the span of y^1. Continue to let y^n be the first x^k which is not in the span of $\{y^1, y^2, \ldots, y^{n-1}\}$. By the Gram–Schmidt process (Theorem 7.17), we can find an orthonormal sequence $E = \{e^1, e^2, \ldots\}$ with the same span as B, which is also the same span as A. The span of E is thus dense in V.

(⇐): Suppose $E = \{e^1, e^2, \ldots\}$ is an orthonormal sequence whose span is dense in V. Consider the set of all finite sums of E having rational coefficients (in the complex case, let the coefficients have rational real and complex parts). Routine calculations will show that this set of all finite sums $\sum a_k e^k$, with a_k rational, is a countable set which is dense in V. ∎

Theorem 7.27 *Let V be an inner product space. Then V is separable if and only if V has a complete orthonormal sequence.*

Proof Suppose V is separable. Theorem 7.26 shows that there exists an orthonormal sequence $E = \{e^1, e^2, \ldots\}$ whose span is dense in V. Let $x \in V$. Since the span of E is dense, there exists a sequence $t^n = \sum_{k=1}^{N_n} t_{nk} e^k$ in the span of E such that $\|x - t^n\| \to 0$ as $n \to \infty$. We may assume that the number of terms N_n is nondecreasing. For each n, let s^n be the nth partial sum of the Fourier series of x with respect to E. By the best approximation theorem (Theorem 7.22), $\|x - s^1\| \geq \|x - s^2\| \geq \cdots$ and $\|x - s^{N_n}\| \leq \|x - t^n\|$. Since $\|x - t^n\| \to 0$ as $n \to \infty$, the decreasing sequence $\|x - s^1\| \geq \|x - s^2\| \geq \cdots$ tends to 0. Thus the Fourier series of each $x \in V$ converges to x.

The converse follows from the proof of Theorem 7.26. Since V has a complete orthonormal sequence, every $x \in V$ has a convergent Fourier expansion. Each $x \in V$ can be approximated by the partial sums of the Fourier series, and each partial sum can be approximated by ones having rational coefficients. ∎

Example 7.16 Legendre polynomials:[2] The Legendre polynomials are solutions $y = P_n(t)$ of the differential equation

$$(1 - t^2)y'' - 2ty' + n(n+1)y = 0, \quad \text{for } n = 0, 1, 2, \ldots.$$

The first few Legendre polynomials are

$$P_0(t) = 1,\ P_1(t) = t,\ P_2(t) = \frac{1}{2}(3t^2 - 1),\ P_3(t) = \frac{1}{2}(5t^3 - 3t), \ldots.$$

If one applies the Gram–Schmidt process to the functions $1,\ t,\ t^2,\ t^3, \ldots$ on the interval $[-1, 1]$ under the real inner product $\langle f, g \rangle = \int_{-1}^{1} f(t)g(t)\,dt$, one obtains the polynomials $e_n(t) = \sqrt{\frac{2n+1}{2}}\,P_n(t)$, which form an orthonormal sequence in $L^2[-1, 1]$.

The Legendre polynomials span all polynomials on $[-1, 1]$. By the Weierstrass approximation theorem (see Corollary 5.59 or Theorem 7.59), the polynomials are dense in $C[-1, 1]$ under the supremum norm. The space $C[-1, 1]$ is also dense in $L^2[-1, 1]$ under the L^2 norm. So the Legendre polynomials form a complete orthonormal sequence in the inner product space $L^2[-1, 1]$. Legendre polynomials are used in engineering and physics (for example, electrostatics, and quantum mechanics).

Example 7.17 Chebyshev polynomials of the first kind: The Chebyshev[3] polynomials of the first kind are solutions $y = T_n(t)$ of the differential equation

$$(1 - t^2)y'' - ty' + n^2 y = 0, \quad n = 0, 1, 2, \ldots.$$

The first few polynomials are

$$T_0(t) = 1, T_1(t) = t, T_2(t) = 2t^2 - 1, T_3(t) = 4t^3 - 3t, \ldots.$$

[2] Named in honor of the French mathematician Adrien-Marie Legendre (1752–1833).

[3] Named in honor of Pafnuty Lvovich Chebyshev (1821–1894). He was Russian but did not like to be called a Russian mathematician, preferring to be a world-wide mathematician.

We can also obtain the relationship $T_{n+1}(t) = 2tT_n(t) - T_{n-1}(t)$. If one applies the Gram–Schmidt process to the functions $1,\ t,\ t^2,\ t^3, \ldots.$ on the interval $[-1, 1]$ under the real inner product

$$\langle f, g\rangle = \int_{-1}^{1} f(t)g(t)w(t)\,dt \text{ with function } w(t) = \frac{1}{2\sqrt{1-t^2}},$$

then one obtains the normalized Chebyshev polynomials $e_0(t) = \frac{1}{\sqrt{\pi}} T_0(t)$ and $e_n(t) = \sqrt{2/\pi}\, T_n(t)$ for $n = 1, 2, \ldots.$ The Chebyshev polynomials of the first kind form a complete orthogonal sequence in the inner product space $L^2[-1, 1]$. The function $w(t)$ is called the **weight function**. Note that the weight function for the Legendre polynomials above is $w(t) = 1$.

Example 7.18 Hermite polynomials:[4] The Hermite functions are defined on the real line and are given by the formulas

$$H_0(t) = 1, \qquad H_n(t) = (-1)^n e^{t^2} \frac{d^n}{dt^n} e^{-t^2}, \quad n = 1, 2, 3, \ldots.$$

These functions $H_n(t)$ turn out to be polynomials of degree n with

$$H_0(t) = 1,\ H_1(t) = 2t,\ H_2(t) = 4t^2 - 2,\ H_3(t) = 8t^3 - 12t,\ \ldots.$$

In general, $H_n(t) = n! \sum_{k=0}^{N} (-1)^k \frac{2^{n-2k}}{k!(n-2k)!} t^{n-2k}$, where $N = n/2$ if n is even and $N = (n-1)/2$ if n is odd.

We can also obtain the relationship

$$H_{n+1}(t) = 2tH_n(t) - H_n'(t).$$

By applying the Gram–Schmidt process to the functions $1,\ t,\ t^2,\ t^3, \ldots$ on the real line under the real inner product

$$\langle f, g\rangle = \int_{-\infty}^{\infty} f(t)g(t)w(t)\,dt \quad \text{with weight function} \quad w(t) = e^{-t^2},$$

one obtains the normalized Hermite functions

$$e_n(t) = \frac{1}{(2^n n! \sqrt{\pi})^{1/2}} H_n(t).$$

The Hermite functions form a complete orthogonal sequence in $L^2(\mathbb{R})$. They are used to analyze harmonic oscillators in quantum mechanics.

[4] Named in honor of the French mathematician Charles Hermite (1822–1901).

Exercises for Section 7.2

7.2.1 Give an example of an inner product space V with an orthonormal sequence $E = \{e^1, e^2, e^3, \ldots\}$ which is not complete.

7.2.2 Let $x^1, x^2, x^3, \ldots$ be a sequence in an inner product space V. Define the partial sums $s^n = x^1 + x^2 + \cdots + x^n$.

Show that s^n is Cauchy whenever $\sum_{k=1}^{\infty} \|x^k\| < \infty$.

7.2.3 Consider $C[0,\pi]$ under the inner product $\langle f, g\rangle = \int_0^\pi f(t)g(t)\,dt$. Follow Example 7.13 to show that the sequence 1, $\cos t$, $\cos 2t$, $\cos 3t, \ldots,$ is orthogonal on the interval $[0,\pi]$. Then, for this orthogonal sequence, find the formulation of the Fourier series of a function $f \in C[0,\pi]$.

7.2.4 Consider $C[0,\pi/2]$ with the inner product $\langle f, g\rangle = \int_0^{\pi/2} f(t)g(t)\,dt$. Follow Example 7.13 to show that the sequence $\sin t$, $\sin 3t$, $\sin 5t, \ldots,$ is orthogonal on the interval $[0,\pi/2]$. Then, for this orthogonal sequence, find the formulation of the Fourier series of a function $f \in C[0,\pi/2]$.

7.2.5 Use a Fourier partial sum to find the best least squares approximation of $f(t) = t^3$ on the interval $[-1,1]$ by the Legendre polynomials $P_0(t) = 1, P_1(t) = t, P_2(t) = \frac{1}{2}(3t^2 - 1)$. Use the inner product $\langle f, g\rangle = \int_0^1 f(t)g(t)\,dt$.

7.2.6 Use a Fourier partial sum to find the best least squares approximation of $f(t) = t^3$ on the interval $[-1,1]$ by the Chebyshev polynomials $T_0(t) = 1, T_1(t) = t, T_2(t) = 2t^2 - 1$. Use the inner product

$$\langle f, g\rangle = \frac{1}{2}\int_{-1}^{1} f(t)g(t)\frac{1}{\sqrt{1-t^2}}\,dt.$$

7.2.7 Use a Fourier partial sum to find the best least squares approximation of the function $f(t) = \sqrt{t}$ by a quadratic polynomial $P(t) = a + bt + ct^2$ on the interval $[0,1]$ under the inner product

$$\langle f, g\rangle = \int_0^1 f(t)g(t)\,dt.$$

7.2.8 Use Legendre polynomials to find the least squares approximation of the function $f(t) = t^3 + t^2$ on $[-1,1]$ by a quadratic polynomial.

7.2.9 Use the Legendre polynomials to find the least squares approximation of the function $f(t) = e^t$ on the interval $[-1,1]$ by a linear polynomial $p_1(t) = b_0 + b_1 t$. Compare $p_1(0.5)$ with $f(0.5)$.

7.2.10 Use the Legendre polynomials to find the least squares approximation of the function $f(t) = e^t$ on the interval $[-1, 1]$ by a quadratic polynomial $p_2(t) = A + Bt + Ct^2$. Compare $p_2(0.5)$ with $f(0.5)$.

7.2.11 Use the Chebyshev polynomials of Example 7.17 with weight $w(t) = \frac{1}{\sqrt{1-t^2}}$ to find a least squares approximation of $f(t) = e^t$ on $[-1, 1]$ by a linear polynomial $p_1(t)$. Compare $p_1(0.5)$ with $f(0.5)$.

7.3 Hilbert Space

Definition 7.5 A **Hilbert**[5] **space** H is an inner product space which is complete (that is, it is a Banach space) under the Euclidean norm $\|x\| = \sqrt{\langle x, x\rangle}$.

Theorem 7.28 *The completion of an inner product space V is a Hilbert space H.*

Proof Every normed space has a completion which is unique up to isometry (Theorems 5.32 and 5.33 in Section 5.8). Since an inner product space is normed, it remains to be shown that the completion is an inner product space. This can be done with the parallelogram identity and Exercise 7.1.37. ∎

Theorem 7.29 *If $x^1, x^2, \ldots,$ is an orthogonal sequence in a Hilbert space H, then*

$$\sum x^k \quad \textit{converges if and only if} \quad \sum \|x^k\|^2 < \infty. \tag{7.20}$$

Proof Let $s^n = x^1 + \cdots + x^n$. The series $\sum x^n$ converges if and only if (s^n) is Cauchy. By the Pythagorean theorem (Theorem 7.10), for $m > n$, we have

$$\|s^m - s^n\|^2 = \left\| \sum_{k=n+1}^{m} x^k \right\|^2 = \sum_{k=n+1}^{m} \|x^k\|^2.$$

This tends to 0 (as $n, m \to \infty$) if and only if $\sum \|x^k\|^2 < \infty$. ∎

Bessel's inequality shows that for every inner product space V, we have

$$\widehat{V} \subset \ell_2.$$

The following is a converse in the case of a complete orthonormal sequence.

[5] The German mathematician David Hilbert (1862–1943) formulated the theory of complete inner product spaces. He is considered to be one of the foremost mathematicians of the late nineteenth and early twentieth centuries. At the Congress of Mathematicians held in Paris in 1900, he set out a list of 23 problems that were the focus of much of mathematics for the next century. Many of the problems have been resolved but some remain open today.

Theorem 7.30 Riesz–Fischer:[6] *Let $e^1, e^2, e^3 \dots$ be a complete orthonormal sequence in a Hilbert space H. Then, for each $x \in \ell_2$, there exists $y \in H$ such that $\widehat{y}(k) = x_k$ for all k. That is,*

$$\ell_2 \subset \widehat{H}.$$

Proof Suppose $x \in \ell_2$. Then $\sum |x_k|^2 = \sum \|x_k e^k\|^2 < \infty$. By Theorem 7.29, $\sum x_k e^k$ converges to some $y \in H$. For $m < n$ we have, by the Cauchy–Schwarz inequality (Theorem 7.5),

$$\begin{aligned} \left|\left\langle y - \sum_{k=1}^{n} x_k e^k, e^m \right\rangle\right| &= |\langle y, e^m\rangle - x_m| \le \left\| y - \sum_{k=1}^{n} x_k e^k \right\| \cdot \|e^m\| \\ &= \left\| y - \sum_{k=1}^{n} x_k e^k \right\|, \end{aligned}$$

which tends to 0 as $n \to \infty$. Thus $x_m = \langle y, e^m\rangle = \widehat{y}(m)$ for all m. ∎

Definition 7.6 A set A is **total** in an inner product space V if there is no nonzero $x \in V$ which is orthogonal to all elements of A.

Theorem 7.31 *Suppose that $e^1, e^2, \dots$ is an orthonormal sequence in a Hilbert space H. The orthonormal sequence is complete in H if and only if it is total in H.*

Proof **(⇒):** Given a complete orthonormal sequence $e^1, e^2, \dots$ and $x \in H$ with $x \perp e^k$ for each k. Then $x = \sum_{k=1}^{\infty} \widehat{x}(k) e^k = 0$ since $\widehat{x}(k) = \langle x, e^k\rangle = 0$, for all k.

(⇐): Suppose no nonzero element of H is orthogonal to all e^k. Let $x \in H$. By Bessel's inequality (Theorem 7.20), $\widehat{x} \in \ell_2$. Thus

$$\sum \|\widehat{x}(k) e^k\|^2 = \sum |\widehat{x}(k)|^2 < \infty.$$

By Theorem 7.29 and Theorem 7.19, $\sum \widehat{x}(k) e^k$ converges to some $y \in H$ with $\langle y, e^k\rangle = \widehat{x}(k)$. But $x - y$ is orthogonal to all e^k since

$$\langle x - y, e^k\rangle = \widehat{x}(k) - \widehat{x}(k) = 0, \forall\, k = 1, 2, \dots.$$

Thus $x - y = 0$. ∎

The above theorem does *not* hold for general inner product spaces, as is shown in the example given in Exercise 7.3.8.

[6] This theorem was proved independently by F. Riesz and E. Fischer in 1907.

Theorem 7.32 *Suppose that $H \neq \{0\}$ is a separable Hilbert space.*

(a) *If H is n-dimensional, then H is isometric to $\widehat{H} = \ell_2(n)$.*

(b) *If H is infinite-dimensional, then H is isometric to $\widehat{H} = \ell_2$.*

Proof By Theorem 7.27, a separable inner product space has a complete orthonormal sequence E, which means that H is isometric to $\widehat{H}$. For the finite-dimensional case, $E = \{e^1, e^2, \ldots, e^n\}$ must be finite. It was already shown in Examples 7.10 that we have $\widehat{H} = \ell_2(n)$. For the infinite-dimensional case, we show $\widehat{H} = \ell_2$. That $\widehat{H} \subset \ell_2$, follows from Bessel's inequality (Theorem 7.20). To prove the other inclusion, let $x \in \ell_2$. For $n > m$,

$$\left\| \sum_{k=m}^{n} x_k e^k \right\|^2 = \sum_{k=m}^{n} \|x_k\|^2 \to 0,$$

which means the Cauchy sequence $\sum_{k=1}^{n} x_k e^k$ converges in H. So $x \in \widehat{H}$. ∎

Theorem 7.33 *$L^2[a,b]$ is a separable Hilbert space.*

Proof The proof follows from these facts.

(1) $L^2[a,b]$ is an inner product space under $\langle f,g \rangle = \frac{1}{b-a} \int_a^b f(x)\overline{g(x)}\, dx$.

(2) $L^2[a,b]$ is a Banach space by Theorem 6.41 (Section 6.10).

(3) [$L^2[a,b]$ is separable by Theorem 6.43 (Section 6.10). ∎

Corollary 7.34 $\widehat{L^2[a,b]} = \ell_2$.

Definition 7.7 The spaces $\ell_2(X)$: Let X be any nonempty set. Define $\ell_2(X)$ to be the set of all functions $f \colon X \longrightarrow \mathbb{C}$ such that $f(x) = 0$ except for a countable many x and for which $\sum_{x \in X} |f(x)|^2 < \infty$. Define the inner product

$$\langle f,g \rangle = \sum_{x \in X} f(x)\overline{g(x)}.$$

The set $\ell_2(X)$ is clearly a linear space of functions with an inner product.

Theorem 7.35 *For any nonempty set X, the space $\ell_2(X)$ is a Hilbert space.*

Proof We must show that $\ell_2(X)$ is complete. Let $f_1, f_2, \ldots$ be a Cauchy sequence and let $M = \{x \in X \mid f_k(x) \neq 0 \text{ for some } k = 1,2,3,\ldots\}$. M is countable. The subspace of functions in $\ell_2(X)$ restricted to the countable set $M \subset X$ contains the Cauchy sequence and is clearly isometric to ℓ_2 (or $\ell_2(n)$, in case M is finite with n elements). Since ℓ_2 and $\ell_2(n)$ are complete, there exists $f \in \ell_2(X)$ to which the Cauchy sequence $f_1, f_2, \ldots$ converges. Thus $\ell_2(X)$ is complete, which shows that it is a Hilbert space. ∎

Example 7.19 The Hilbert space $\ell_2(\mathbb{R})$ is nonseparable: For each $x \in \mathbb{R}$, define the function

$$f_x(t) = \begin{cases} 1 & : \quad \text{if } t = x \\ 0 & : \quad \text{if } t \neq x. \end{cases} \tag{7.21}$$

The set of all these f_x form an uncountable orthonormal set in the Hilbert space $\ell_2(\mathbb{R})$ as described in Theorem 7.35. By Theorem 7.25, an orthonormal set in a separable inner product space *must* be countable. Hence, $\ell_2(\mathbb{R})$ is nonseparable.

The same notion works for any uncountable set X. The cardinality of the orthonormal set $\{f_x\}$ as given in (7.21) is the same as the cardinality of X. Actually, any Hilbert space H with a complete orthonormal set of the same cardinality as X will be isometric to $\ell_2(X)$.

Caution A Banach space with an inner product does not make a Hilbert space: The space $C[a,b]$ is a Banach space under the uniform norm $\|f\|_\infty = \sup|f(x)|$, as shown in Theorem 5.30 (Section 5.8). It is also separable (Example 5.30, Section 5.5). Also as a subspace of $L^2[a,b]$, it has the inner product

$$\langle f,g\rangle = \int_a^b f(t)\overline{g(t)}\,dt.$$

However, the space $C[a,b]$ is not a Hilbert space. As noted in Example 5.37 (Section 5.8) the space $C[a,b]$ is not complete under the Euclidean L^2 norm. Indeed, there are continuous functions whose Fourier series do not converge in the supremum norm.

Theorem 7.36 Riesz representation: *Every continuous linear functional f on a Hilbert space H is an inner product functional of the form*

$$f(x) = f_y(x) = \langle x, y\rangle \quad \textit{for a unique} \quad y \in H.$$

Furthermore, we have the isometric identity $\|f_y\| = \|y\|$.

Proof We have already shown in Corollary 7.6 that in an inner product space, every linear functional f_y of the form

$$f_y(x) = \langle x, y\rangle$$

is continuous and that $\|f_y\| = \|y\|$. We now show that in Hilbert spaces, these are the *only* continuous linear functionals.

Suppose f is a continuous linear functional on H. If $f = 0$, we simply take $y = 0$ and we have $f(x) = 0 = \langle x, 0\rangle$ for all $x \in H$. Now suppose $f \neq 0$. Let

$H_0 = \{x \in H | f(x) = 0\}$ be the null space of f. Since f is continuous, H_0 is a closed subspace of H, which means that H_0 is also complete. Since $f \neq 0$, there exists $x \neq 0$ such that $f(x) \neq 0$. By the orthogonal projection theorem (Theorem 7.12), there exists a nonzero z which is orthogonal to H_0. We may assume $\|z\| = 1$. Set

$$v = f(x)z - f(z)x.$$

Using the linearity of f, it is clear that v is also in the null space H_0, which means that z is also orthogonal to v. Hence,

$$0 = \langle v, z \rangle = f(x)\langle z, z \rangle - f(z)\langle x, z \rangle = f(x) - \langle x, \overline{f(z)}z \rangle,$$

which shows that $f(x) = \langle x, \overline{f(z)}z \rangle$ for every $x \in H$. If we take $y = \overline{f(z)}z$, we have now proven that every continuous linear functional f on a Hilbert space is an inner product functional f_y.

It remains to show that y is unique. Assume that

$$f(x) = \langle x, y \rangle = \langle x, y' \rangle \quad \text{for all} \quad x \in H.$$

Then $\langle x, y - y' \rangle = 0$ for all $x \in H$. In particular, if we take $x = y - y'$, we obtain $\|y - y'\|^2 = 0$, which implies $y = y'$. ∎

Corollary 7.37 *Every continuous linear functional T on $L^2[a,b]$ is of the form*

$$T(f) = \frac{1}{b-a} \int_a^b f(x)\overline{g(x)}\, dx \quad \textit{for some fixed} \quad g \in L^2[a,b].$$

Exercises for Section 7.3

7.3.1 Show that any linear functional f on $\mathbb{R}^n$ is a dot product

$$f(x) = x_1 y_1 + \cdots + x_n y_n \quad \text{for some} \quad y \in \mathbb{R}^n.$$

7.3.2 Show that any continuous linear functional f on ℓ_2 is a dot product

$$f(x) = \sum_{k=1}^{\infty} x_k \overline{y_k} \quad \text{for some} \quad y \in \ell_2.$$

7.3.3 Let $x^1, x^2, x^3, \ldots$ be a sequence in a Hilbert space H. Show that convergence of $\sum \|x^k\|$ implies convergence of $\sum x^k$.

7.3.4 Compare (7.20) of Theorem 7.29 in Section 7.3 with Theorem 5.31 in Section 5.8. If an inner product space V satisfies condition (7.20), is it necessarily a Hilbert space?

7.3.5 Let $e^1, e^2, e^3, \ldots$ be a complete orthonormal sequence in a Hilbert space H, and let $d^1, d^2, d^3, \ldots$ be another orthonormal sequence in H such that

$$\sum_{k=1}^{\infty} \|e^k - d^k\| < \infty.$$

Show that $d^1, d^2, d^3, \ldots$ is also a complete orthonormal sequence in H.

7.3.6 Let $e^1, e^2, e^3, \ldots$ be an orthonormal sequence in a Hilbert space H. For every $x \in H$, show that

$$y = \sum_{k=1}^{\infty} \widehat{x}(k) e^k \quad \text{converges in} \quad H.$$

Furthermore, show that $x - y \perp e^k$ for all $k = 1, 2, 3, \ldots$.

7.3.7 Prove Theorem 7.29 directly as follows. Let H be the completion of an inner product space V. For x and y in H, there exist Cauchy sequences x^n and y^n in V converging to x and y respectively. Define an inner product by setting

$$\langle x, y \rangle = \lim_{n \to \infty} \langle x^n, y^n \rangle.$$

Show that this definition actually satisfies the conditions **(IP1)** through **(IP4)** for an inner product.

7.3.8 Let Φ consist of all sequences $x = (x_1, x_2, \ldots) \in \omega$ with a finite number of nonzero coordinates; that is $\Phi = \operatorname{span}\{e^1, e^2, \ldots\}$. Let $F = \{f^1, f^2, \ldots\}$, where $f^k = e^k - e^{k+1}$ for $k = 1, 2, \ldots$. Show that F is linearly independent but $M = \operatorname{span}(F)$ is a proper subspace of Φ. The Gram–Schmidt process may be applied to F to obtain an orthonormal basis of M. Show that this orthonormal basis of M is total in Φ but it is not a complete orthonormal basis of Φ. Does this contradict Theorem 7.31?

7.3.9 Show that in every separable Hilbert space, every orthonormal sequence can be extended to a total orthonormal sequence.

7.3.10 Suppose that the inner product space V has a finite total orthonormal set $E = \{e^1, \ldots, e^n\}$. Show that the linear space V has dimension n.

7.3.11 Let A be a total set in an inner product space V. Show that if $\langle x, u \rangle = \langle x, v \rangle$ for all $x \in A$, then $u = v$.

7.3.12 Let A be a subset of an inner product space V with the property that $\langle x, u \rangle = \langle x, v \rangle$ for all $x \in A$, implies that $u = v$. Show that A is a total set.

7.3.13 Use Zorn's lemma (Theorem 0.62) to show that every Hilbert space, whether separable or not, has a total orthonormal set.

7.3.14 Prove the assertion made after Example 7.19 that any Hilbert space H with a complete orthonormal set of the same cardinality as X will be isometric to $\ell_2(X)$.

∗7.4 Adjoint Operators

As a result of the Riesz representation theorem, we can now show that every bounded (continuous) operator $T\colon H \longrightarrow K$ between Hilbert spaces has a corresponding **adjoint operator** $T^*\colon K \longrightarrow H$. In quantum theory, **self-adjoint operators**, wherein $H = K$ and $T = T^*$, represent real-valued **observable magnitudes**; that is, physical quantities that can be measured.

Definition 7.8 Let $T\colon H \longrightarrow K$ be a bounded linear operator between Hilbert spaces H and K. Its **adjoint operator** $T^*\colon K \longrightarrow H$ is the operator T^* defined by

$$\langle Tx, y\rangle = \langle x, T^*\rangle \quad \text{for all } x \in H \text{ and } y \in K.$$

The first inner product is the one for the space K and second is for H.

We first show that the adjoint T^* of T exists.

Theorem 7.38 *Suppose $T\colon H \longrightarrow K$ is a bounded linear operator between Hilbert spaces. Then the adjoint $T^*\colon K \longrightarrow H$ exists, is unique, and is a bounded linear operator with operator norm*

$$\|T^*\| = \|T\|.$$

Proof

T^* **exists:** For each $y \in K$, the composite map $f\colon x \longrightarrow \langle Tx, y\rangle$ is a continuous linear functional on H. By the Riesz representation theorem (Theorem 7.36), for each $y \in K$, there exists a unique $z \in H$ such that

$$f(x) = \langle x, z\rangle = \langle Tx, y\rangle,\ \forall\, x \in H.$$

We write the unique z as $z = T^*y$. This defines T^* uniquely.

T^* **is linear:** For $x \in H$, and $y_1, y_2 \in K$, and scalars α, β, we have

$$\begin{aligned}\langle x, T^*(\alpha y_1 + \beta y_2)\rangle &= \langle Tx, \alpha y_1 + \beta y_2\rangle \\ &= \overline{\alpha}\langle x, T^* y_1\rangle + \overline{\beta}\langle x, T^* y_2\rangle \\ &= \langle x, \alpha T^* y_1 + \beta T^* y_2\rangle.\end{aligned}$$

Since this is true for all $x \in H$, we conclude $T^*(\alpha y_1 + \beta y_2) = \alpha T^* y_1 + \beta T^* y_2$.

$\|T\| = \|T^*\|$: By the definition of an operator norm and by (7.6) in Corollary 7.6,

$$\begin{aligned}\|T\| &= \sup_{\|x\|\le 1} \|Tx\| = \sup_{\|x\|\le 1} \sup_{\|y\|\le 1} \{|\langle Tx, y\rangle|\} = \sup_{\|x\|\le 1} \sup_{\|y\|\le 1} \{|\langle x, T^* y\rangle|\} \\ &= \sup_{\|y\|\le 1} \|T^* y\| = \|T^*\|. \end{aligned}$$

∎

Theorem 7.39 *Let S and T be bounded linear operators between Hilbert spaces H and K.*

(a) $(T^*)^* = T$,
(b) $(S+T)^* = S^* + T^*$,
(c) $(ST)^* = T^*S^*$, *provided* $H = K$,
(d) $\|T^*T\| = \|T^*T\| = \|T\|^2$,
(e) $T^*T = 0 \iff T = 0$,
(f) *If T is invertible, so is T^*, with* $(T^*)^{-1} = (T^{-1})^*$.

The proofs are straightforward and are left as Exercise 7.4.4.

Example 7.20 Adjoint Fredholm operator: Let operator $T : L^2[a,b] \longrightarrow L^2[a,b]$ be defined by $(Tf)(x) = \int_a^b K(x,t) f(t)\,dt$, with a Fredholm kernel $K(x,t)$ which is integrable on $a \le x \le b, a \le t \le b$ (see Example 6.9 in Section 6.1). We show that the adjoint T^* is the Fredholm operator S defined by

$$(Sf)(x) = \int_a^b \overline{K(t,x)} f(t)\,dt.$$

Let $f, g \in L^2[a,b]$. By the definition of T^* and Fubini's theorem (Theorem 3.2 in Section 3.1),

$$\begin{aligned}\langle T^* f, g\rangle = \langle f, Tg\rangle &= \int_a^b f(x) \overline{\left\{\int_a^b K(x,t) g(t)\,dt\right\}}\,dx \\ &= \int_a^b f(x) \left\{\int_a^b \overline{K(x,t) g(t)}\,dt\right\}\,dx \\ &= \int_a^b \left\{\int_a^b \overline{K(x,t)} f(x)\,dx\right\} \overline{g(t)}\,dt = \langle Sf, g\rangle.\end{aligned}$$

Since $\langle (T^* - S)f, g\rangle = \langle T^* f - Sf, g\rangle = 0$ for all $f, g \in L^2[a,b]$, we have $T^* f - Sf = (T^* - S)f = 0$ for all $f \in L^2[a,b]$. Hence $T^* - S = 0$.

Note that if the kernel satisfies $\overline{K(x,t)} = K(t,x)$, then $T = T^*$. This happens when K is real and symmetric about the diagonal $x = t$. This is a special case of a self-adjoint operator as defined in Definition 7.9.

Definition 7.9 Suppose $T: H \longrightarrow H$ is a bounded linear operator on a Hilbert space H. Then T is said to be **self-adjoint** (or **Hermitian**) if $T^* = T$.

Theorem 7.40 *Let T be a bounded linear operator on the Hilbert space H over the complex field $\mathbb{C}$.*

(a) *$T = 0$ if and only if $\langle Tz, z\rangle = 0$ for all $z \in H$.*

(b) *T is self-adjoint if and only if $\langle Tz, z\rangle$ is real for all $z \in H$. It is essential in this theorem that H be complex.*

Proof **(a):** The proof of ($\Rightarrow$) is clear. Conversely, suppose $\langle Tz, z\rangle = 0$ for all $z \in H$. Let x and y be arbitrary elements of H. If $z = x + y$, we have $0 = \langle Tx, y\rangle + \langle Ty, x\rangle$. And if $z = ix + y$, we have $0 = \langle Tx, y\rangle - \langle Ty, x\rangle$. Adding these equations results in $\langle Tx, y\rangle = 0$ for all $x, y \in H$. But $\langle Tx, y\rangle = 0$ for all $y \in H$ proves $Tx = 0$. Then $Tx = 0$ for all $x \in H$, proves $T = 0$.

(b): The proof is left as Exercise 7.4.7. Here H must be complex, for if H is a Hilbert space over the reals, then $\langle Tz, z\rangle$ is *always* real. ∎

Exercises for Section 7.4

7.4.1 Let $T: H \longrightarrow K$ be a bounded linear operator between Hilbert spaces. Show that $T = 0$ if and only if $\langle Tx, y\rangle = 0$ for all $x \in H$, $y \in K$.

7.4.2 Let $T: H \longrightarrow K$ be a bounded linear operator between Hilbert spaces. Prove the uniqueness of the adjoint as follows. Suppose S_1 and S_2 satisfy $\langle Tx, y\rangle = \langle x, S_1 y\rangle = \langle x, S_2 y\rangle$ for all $x \in H, y \in K$. Show that $S_1 - S_2 = 0$.

7.4.3 Prove $0^* = 0$ and $I^* = I$.

7.4.4 Prove all six parts of Theorem 7.39.

7.4.5 Suppose H and K are the finite dimensional spaces $\mathbb{C}^n$ and $\mathbb{C}^m$, respectively. Then a linear operator T from H to K has a matrix representation A (with respect to the standard bases). What is the matrix representation of the adjoint operator T^*? What about the real case $H = \mathbb{R}^n$ and $K = \mathbb{R}^m$? Under what conditions on the matrix is such a transformation self-adjoint?

7.4.6 Suppose $T_1, T_2, \ldots$ is a sequence of bounded linear operators on H. Show that if $T_k \to T$ as $k \to \infty$, then $T_k^* \to T^*$.

7.4.7 Prove Theorem 7.40(**b**). Use part (**a**) and $\langle (T - T^*)z, z\rangle = 0$ for all $z \in H$.

7.4.8 Give an example where Theorem 7.40(**a**) fails in the real case.

7.4.9 Let $H = \mathbb{C}^2$. Use the definition to show that the operator defined by the matrix $\begin{pmatrix} 1 & 2+i \\ 2-i & 3 \end{pmatrix}$ is self-adjoint. Verify that Theorem 7.40(**b**) holds.

7.4.10 Let $V = \ell_2$. Consider the right shift operator $T_R(x_1, x_2, \ldots) = (0, x_1, x_2, \ldots)$ as defined in Example 6.1. Find the adjoint operator $(T_R)^*$.

7.4.11 Suppose S, T are two self-adjoint operators on Hilbert space H. Show that ST is self-adjoint if and only if they commute ($ST = TS$).

7.5 Trigonometric Fourier Series

As seen in the previous section, $L^2[a,b]$ has many nice properties, such as:

(H1) $L^2[a,b]$ is a Hilbert space.
(H2) $f \in L^2[a,b] \iff \widehat{f} \in \ell_2$.
(H3) The Fourier series of every $f \in L^2[a,b]$ converges in norm to f.

However, for many applications, limiting ourselves to functions that are square integrable is too restrictive. For this reason, we now deal with the larger space of integrable functions $L[a,b]$, which will be our universal space of functions.[7] Unfortunately, none of the properties **(H1)**, **(H2)**, **(H3)** hold for $L[a,b]$. In this section we will derive other properties, some of which are almost as good.

The values of a and b in the interval $[a,b]$ are not important. However, there is some technical simplicity in using an interval of width 2π.

Generally, it is convenient to extend these functions f to all of $\mathbb{R}$ by making them 2π-periodic. Namely,

$$f(x) = f(x \pm 2\pi) = f(x \pm 4\pi) = f(x \pm 6\pi) = \cdots.$$

[7] Other even larger spaces are important, such as those composed of generalized functions, Radon measures, and distributions. We only deal with integrable functions $L[a,b] = L^1[a,b]$ here. A good source for reading about these more general spaces is the second volume of R.E. Edwards, *Fourier series: A Modern Introduction*, Springer-Verlag, New York (1979). Most of the properties carry over except for Riemann–Lebesgue lemma where, instead of $\widehat{f}(n) \to 0$ we may have $\widehat{f}(n)/n^k \to 0$ (as $n \to \infty$) for some k.

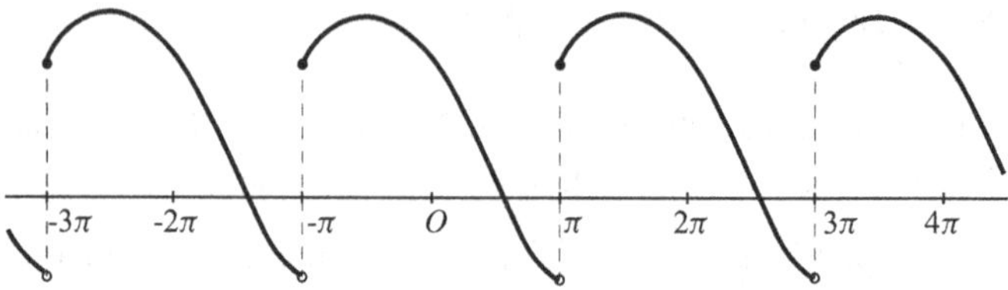

Figure 7.5 Periodic extension of a function.

In particular, the integral of f over any interval of length 2π has the same value as the integral over any other integral of the same length.[8] So

$$\int_a^{a+2\pi} f(x)\,dx = \int_b^{b+2\pi} f(x)\,dx \quad \text{for any} \quad -\infty < a,b < \infty.$$

For example, $\int_{-\pi}^{\pi} f(x)\,dx = \int_0^{2\pi} f(x)\,dx$.

Definition 7.10 We use the notation $L_{2\pi}$, $L^p_{2\pi}$, $C_{2\pi}$, etc. to indicate the spaces of 2π-periodic functions, as distinguished from $L^1[-\pi,\pi]$, $L^p[-\pi,\pi]$, $C[-\pi,\pi]$, etc. of spaces of functions defined only on the interval $[-\pi,\pi]$ (as illustrated in Figure 7.5).

A major advantage of using periodic extensions is that it permits translation operators.

Definition 7.11 The **translation operators** T_a (for $a \in \mathbb{R}$) are defined by the formula

$$T_a f(x) = f(x-a).$$

Clearly

$$\int_{-\pi}^{\pi} |T_a f(x)|\,dx = \int_{-\pi}^{\pi} |f(x-a)|\,dx = \int_{-\pi-a}^{\pi-a} |f(x)|\,dx = \int_{-\pi}^{\pi} |f(x)|\,dx.$$

So $\|T_a f\|_1 = \|f\|_1$.

For spaces of periodic functions, it is convenient to divide the norm by the length of the periodic interval. For example, the norm on $L_{2\pi}$ is

$$\|f\|_1 = \frac{1}{2\pi}\int_{-\pi}^{\pi} |f(x)|\,dx.$$

[8] Here is another convenient way to regard these functions. Map $\mathbb{R}$ onto the unit circle $\{z \in \mathbb{C} \mid |z| = 1\}$ by the function $x \longrightarrow e^{ix}$. Then $x + 2\pi \longrightarrow e^{i(x+2\pi)} = e^{ix}$. And f is 2π-periodic if and only if there exists a function f_0 such that $f(x) = f_0(e^{ix})$.

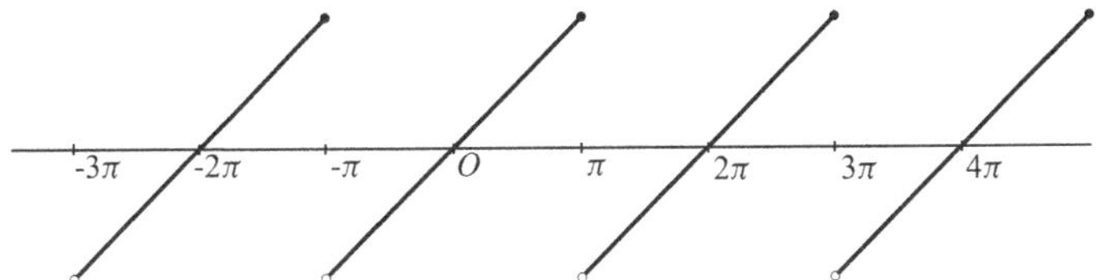

Figure 7.6 A periodic extension of $f(x) = x$ is not everywhere continuous.

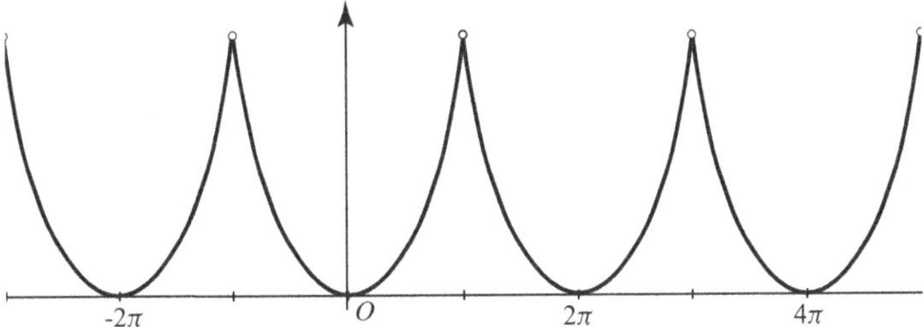

Figure 7.7 The graph of $f(x) = x^2$ $(-\pi < x < \pi)$ with its periodic extension.

There are some differences between the function spaces defined on $[-\pi, \pi]$ and the spaces of their periodic extensions.

(1) To define a periodic extension, the values at the endpoints $-\pi$ and π have to be equal. The values at these points may have to be resolved, for example, by restricting the original interval to $(-\pi, \pi]$.

(2) The continuous functions $f \in C_{2\pi}$ must be continuous on all of $\mathbb{R}$. Thus the function $f(x) = x$ is continuous on $(\pi, \pi]$ but a periodic extension has a sawtooth shape. So $f \notin C_{2\pi}$. See Figure 7.6.

(3) The same goes for differentiability. The function $f(x) = x^2$ defined on $[-\pi, \pi]$ has a continuous periodic extension but the extension is not differentiable at the points $\pm\pi, \pm 3\pi, \ldots$. See Figure 7.7 (Section 7.6).

(4) All translation operators $T_a f(x) = f(x - a)$ are possible for periodic functions but not for functions defined merely on an interval.

Remark We will deal with representations of functions $f \in L_{2\pi}$ using *trigonometric* polynomials and series. There are good reasons to choose the trigonometric series among other possible expansions in terms of orthonormal sets. One major reason: For each $k = 0, \pm 1, \pm 2, \ldots$, the space $V_k = \text{span}\{e^{ikx}\}$ is one-dimensional and translation invariant. That is, for each $f \in V_k$, and each $a \in \mathbb{R}$, we have $T_a f \in V_k$. This is important because translations play an important role in Fourier series. These V_k

are the *only* one-dimensional translation invariant subspaces of $L_{2\pi}$. This is discussed further in Section 8.11.

Definition 7.12 The **complex form of a trigonometric polynomial** on $\mathbb{R}$ is a finite sum of the form

$$s_n(x) = \sum_{k=-n}^{n} c_k e^{ikx} \quad \text{for} \quad c_k \in \mathbb{C}. \tag{7.22}$$

The **real form of a trigonometric polynomial** on $\mathbb{R}$ is

$$s_n(x) = \frac{a_0}{2} + \sum_{k=1}^{n} (a_k \cos kx + b_k \sin kx) \quad \text{for} \quad a_k, b_k \in \mathbb{C} \text{ (or } \mathbb{R}). \tag{7.23}$$

For most purposes, the complex form is more convenient. We can show that the two forms are equivalent with Euler's identity (Equation (0.6) in Section 0.2):

$$e^{i\theta} = \cos\theta + i\sin\theta.$$

This results in

$$\cos kx = \frac{e^{ikx} + e^{-ikx}}{2} \text{ and } \sin kx = \frac{e^{ikx} - e^{-ikx}}{2i} = i\frac{-e^{ikx} + e^{-ikx}}{2}. \tag{7.24}$$

Substituting these expressions into Equation (7.23), we obtain

$$s_n(x) = \frac{a_0}{2} + \sum_{k=1}^{n} \left(\frac{a_k - ib_k}{2} e^{ikx} + \frac{a_k + ib_k}{2} e^{-ikx} \right).$$

Letting

$$c_0 = \frac{a_0}{2}, \quad c_k = \frac{a_k - ib_k}{2}, \quad \text{and} \quad c_{-k} = \frac{a_k + ib_k}{2}, \tag{7.25}$$

we obtain Equation (7.22) : $s_n(x) = \sum_{k=1}^{n} (c_k e^{ikx} + c_{-k} e^{-ikx}) = \sum_{k=-n}^{n} c_k e^{ikx}$.

Theorem 7.41 *Consider a trigonometric polynomial* $s_n(x) = \sum_{k=-n}^{n} c_k e^{ikx}$.

(a) *The polynomial is a* ***real*** *function if and only if* $c_{-k} = \overline{c_k}$ *for* $k = 0, 1, \ldots, n$.

(b) *It is* ***even*** *if and only if* $c_{-k} = c_k$ *for* $k = 1, 2, \ldots$.
This means that the real form has only cosine terms.

(c) *It is* ***odd*** *if and only if* $c_{-k} = -c_k$ *for* $k = 1, 2, \ldots$.
This means that the real form has only sine terms.

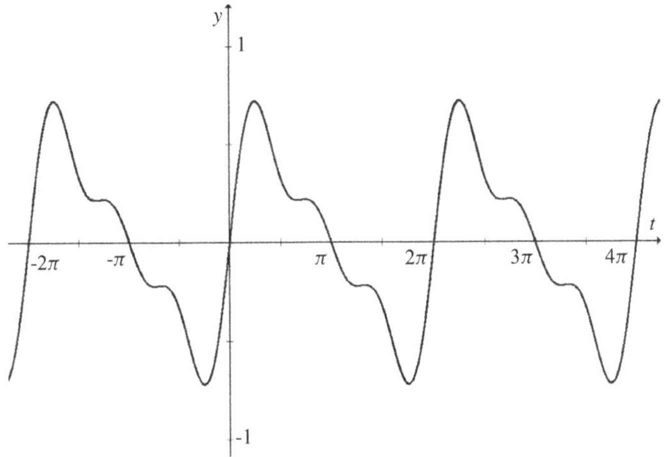

Figure 7.8 The graph of $y = \sin t + \frac{1}{2}\sin 2t + \frac{1}{3}\sin 3t$.

Proof Equation (7.25) shows that a_k and b_k are real if and only if $c_{-k} = \overline{c_k}$. The other statements follow easily from Equations (7.22) and (7.25). ■

If we multiply (7.22) by e^{-imx}, where $m = 0, 1, \ldots, n$, and then integrate, we obtain the mth **Fourier coefficient**

$$c_m = \frac{1}{2\pi}\int_{-\pi}^{\pi} s_n(x)e^{-imx}\,dx \quad \text{for} \quad m = 0, 1, \ldots, n. \tag{7.26}$$

For the real form,

$$a_k = \frac{1}{\pi}\int_{-\pi}^{\pi} s_n(x)\cos kx\,dx \quad \text{and} \quad b_k = \frac{1}{\pi}\int_{-\pi}^{\pi} s_n(x)\sin kx\,dx. \tag{7.27}$$

Figure 7.8 shows an example of a trigonometric polynomial written in real form. It is the sum of sine terms, which makes it an odd function of period 2π.

Definition 7.13 A **trigonometric series** is a series of the form

$$\sum_{k=-\infty}^{\infty} c_k e^{ikx} \quad \text{for} \quad x \in \mathbb{R}. \tag{7.28}$$

Convergence is always understood to be with respect to the symmetric partial sums (trigonometric polynomials)

$$s_n(x) = \sum_{k=-n}^{n} c_k e^{ikx}.$$

Definition 7.14 If $f \in L_{2\pi}$ and $c_k = \widehat{f}(k) = \frac{1}{2\pi}\int_{-\pi}^{\pi} f(x)e^{-ikx}\,dx$, then the trigonometric series (7.28) is called the **Fourier series** of f and we write $f \sim \sum \widehat{f}(k)e^{ikx}$. We use the notation $s_n f(x) = \sum_{k=-n}^{n} \widehat{f}(k)e^{ikx}$ for partial sums of the Fourier series whereas we use the notation $s_n(x) = \sum_{k=-n}^{n} c_k e^{ikx}$ for general trigonometric polynomials.

Definition 7.15 Given two 2π-periodic integrable functions f and g, the **convolution** $f * g$ is defined by

$$(f * g)(x) = \frac{1}{2\pi}\int_{-\pi}^{\pi} f(t)g(x-t)\,dt.$$

Below are properties of the convolution.

Theorem 7.42 *For all $f, g, h \in L_{2\pi}$, we have*

(C1) $f * g = g * f$;
(C2) $(f * g) * h = f * (g * h)$;
(C3) $\widehat{f * g}(k) = \widehat{f}(k) \cdot \widehat{g}(k)$;
(C4) $\|f * g\|_1 \le \|f\|_1 \|g\|_1$.

Proof
(C1): By making the change of variable $u = x - t$, followed by a translation of the interval of integration, we can show that

$$\begin{aligned}(f * g)(x) &= \frac{1}{2\pi}\int_{-\pi}^{\pi} f(t)g(x-t)\,dt \\ &= \frac{1}{2\pi}\int_{-\pi}^{\pi} f(x-u)g(u)\,du = (g * f)(x).\end{aligned}$$

(C2): The proof is left as Exercise 7.5.3.

(C3): With an interchange in the order of integration below, we obtain

$$\begin{aligned}\widehat{f * g}(k) &= \frac{1}{2\pi}\int_{-\pi}^{\pi} (f * g)(x)e^{-ikx}\,dx \\ &= \frac{1}{2\pi}\int_{-\pi}^{\pi} e^{-ikx}\left\{\frac{1}{2\pi}\int_{-\pi}^{\pi} f(t)g(x-t)\,dt\right\}dx \\ &= \frac{1}{2\pi}\int_{-\pi}^{\pi} f(t)e^{-ikt}\left\{\frac{1}{2\pi}\int_{-\pi}^{\pi} g(x-t)e^{ik(x-t)}\,dx\right\}dt \\ &= \frac{1}{2\pi}\int_{-\pi}^{\pi} f(t)e^{-ikt}\left\{\frac{1}{2\pi}\int_{-\pi}^{\pi} g(x)e^{ikx}\,dx\right\}dt \\ &= \frac{1}{2\pi}\int_{-\pi}^{\pi} f(t)e^{-ikt}\{\widehat{g}(k)\}\,dt = \widehat{f}(k)\cdot\widehat{g}(k).\end{aligned}$$

(C4): Let $h(x) = (f * g)(x)$. Then

$$\begin{aligned}\|h\|_1 &= \frac{1}{2\pi}\int_{-\pi}^{\pi} |h(x)|\,dx = \frac{1}{2\pi}\int_{-\pi}^{\pi}\left|\frac{1}{2\pi}\int_{-\pi}^{\pi} f(t)g(x-t)\,dt\right| dx \\ &\le \frac{1}{(2\pi)^2}\int_{-\pi}^{\pi}\left\{\int_{-\pi}^{\pi} |f(t)||g(x-t)|\,dt\right\} dx \\ &= \frac{1}{(2\pi)^2}\int_{-\pi}^{\pi} |f(t)|\,dt \int_{-\pi}^{\pi} |g(x-t)|\,dx \\ &= \frac{1}{(2\pi)^2}\int_{-\pi}^{\pi} |f(t)|\,dt \int_{-\pi}^{\pi} |g(x)|\,dx \\ &= \|f\|_1\|g\|_1.\end{aligned}$$

∎

Property **(C3)** shows that the convolution is a type of multiplication.

Property **(C4)** can be generalized to $\|f * g\|_p \le \|f\|_p\|g\|_1$ for $1 \le p \le \infty$.

Next we show that the convolution is a smoothing operation. For example, for $f \in L_{2\pi}$ and g essentially bounded, the convolution is continuous; that is, $f * g \in C_{2\pi}$. First, we prove a lemma.

Lemma 7.43 *Translations are continuous in $L_{2\pi}^p$ for $1 \le p < \infty$. That is, if $f \in L_{2\pi}^p$, then*

$$\lim_{a \to 0} \|T_a f - f\|_p = 0.$$

Proof Let $f \in L_{2\pi}^p$ and let $\epsilon > 0$ be given. By Theorem 6.42 (Subsection 6.10.1), there exists a continuous $g \in C_{2\pi}$ such that

$$\|f - g\|_p < \epsilon/3.$$

Then also $\|T_a f - T_a g\|_p = \|f - g\|_p < \epsilon/3$.

Since g is continuous on $[-\pi, \pi]$, it is uniformly continuous. So there exists $\delta > 0$ such that, for all $x \in [-\pi, \pi]$ and $|a| < \delta$, we have

$$|T_a g(x) - g(x)| = |g(x + a) - g(x)| < \epsilon/3.$$

That is, $\|T_a g - g\|_p = \frac{1}{2\pi}\left(\int_{-\pi}^{\pi} |g(x + a) - g(x)|^p\,dx\right)^{1/p} < \epsilon/3$.

Then, for $|a| < \delta$, we have

$$\begin{aligned}\|T_a f - f\|_p &\le \|T_a f - T_a g\|_p + \|T_a g - g\|_p + \|g - f\|_p \\ &< \frac{\epsilon}{3} + \frac{\epsilon}{3} + \frac{\epsilon}{3} = \epsilon.\end{aligned}$$

∎

Theorem 7.44 *Let $1 \le p < \infty$, and let q be its Hölder dual; that is, $\frac{1}{p} + \frac{1}{q} = 1$.*

$$\text{If } f \in L^p_{2\pi} \text{ and } g \in L^q_{2\pi}, \text{ then } f * g \in C_{2\pi}.$$

Proof Let $h(x) = (f * g)(x)$. By Hölder's inequality for integrals (Theorem 6.37), we have

$$\begin{aligned} |h(x+a) - h(x)| &= \left| \frac{1}{2\pi} \int_{-\pi}^{\pi} g(t)\big(f(x+a-t) - f(x-t)\big)\, dt \right| \\ &\le \|g\|_q \|T_a f - f\|_p. \end{aligned}$$

By Theorem 7.43, $\|T_a f - f\|_p \to 0$ as $a \to 0$. Thus $\lim_{a \to 0} h(x+a) = h(x)$. ∎

Theorem 7.45 *For $n = 0, \pm 1, \pm 2, \ldots$ and $e^n(x) = e^{inx}$, we have $e^n * e^n(x) = e^n(x)$ and $e^n * e^m(x) = 0$, if $n \ne m$.*

The proof is left as Exercise 7.5.5.

Theorem 7.46 *For every $f \in L_{2\pi}$, we have $|\widehat{f}(n)| \le \|f\|_1$.*

The proof of this is left as Exercise 7.5.7.

Corollary 7.47 $\ell_2 = \widehat{L^2_{2\pi}} \subset \widehat{L_{2\pi}} \subset \ell_\infty$.

We can improve this inclusion to $\ell_2 \subset \widehat{L_{2\pi}} \subset c_0$, by the use of the Riemann–Lebesgue lemma below. First we prove the following.

Lemma 7.48 *For every $f \in L_{2\pi}$, and every $n = 0, \pm 1, \pm 2, \ldots$, we have $\widehat{T_a f}(n) = e^{-ina} \widehat{f}(n)$.*

Proof Let $g(x) = f(x)e^{-inx}$. Then $T_a g(x) = e^{ina} T_a f(x) e^{-inx}$. By integral invariance of translations,

$$\begin{aligned} \widehat{f}(n) &= \frac{1}{2\pi} \int_{-\pi}^{\pi} f(x) e^{-inx}\, dx = \frac{1}{2\pi} \int_{-\pi}^{\pi} g(x)\, dx \\ &= \frac{1}{2\pi} \int_{-\pi}^{\pi} T_a g(x)\, dx = e^{ina} \frac{1}{2\pi} \int_{-\pi}^{\pi} T_a f(x) e^{-inx}\, dx \\ &= e^{ina} \widehat{T_a f}(n). \end{aligned}$$

∎

Theorem 7.49 Riemann–Lebesgue lemma: *For every $f \in L_{2\pi}$, we have $\lim_{|n| \to \infty} \widehat{f}(n) = 0$. Also for the real form, the Fourier coefficients a_n, b_n tend to zero as $n \to \infty$.*

Proof By Lemma 7.48 we have $\widehat{T_{\pi/n} f}(n) = e^{i\pi} \widehat{f}(n) = -\widehat{f}(n)$ or $\widehat{f}(n) = -\widehat{T_{\pi/n} f}(n)$. Then

$$\widehat{f}(n) = \frac{1}{2}\left\{\widehat{f}(n) + \widehat{f}(n)\right\} = \frac{1}{2}\left\{\widehat{f}(n) - \widehat{T_{\pi/n} f}(n)\right\} = \frac{1}{2}\left\{\widehat{f - T_{\pi/n} f}\right\}(n). \tag{7.29}$$

Thus by Theorem 7.46 above, we have $|\widehat{f}(n)| \leq \frac{1}{2}\|f - T_{\pi/n} f\|_1$. By continuity of translations (Lemma 7.43), we have $\lim_{n\to\infty} \widehat{f}(n) = 0$. And since $\|f - T_{-\pi/n} f\|_1 = \|f - T_{\pi/n} f\|_1$, we also have $\lim_{n\to-\infty} \widehat{f}(n) = 0$. The result applies to the real form if we use Equations (7.25) in Definition 7.12. ∎

Property **(C3)** shows that $\widehat{f * g} = \widehat{f} \cdot \widehat{g}$. That is, convolution of functions corresponds to coordinatewise product of Fourier coefficients.

The result goes in the opposite direction as well, as the following theorem shows. We define the convolution $x * y$ of two sequences $x, y \in \ell_2$ as the *sequence* defined by the equation

$$(x * y)_n = \sum_{k=-\infty}^{\infty} x_k y_{n-k}.$$

Note that $x, y \in \ell_2 \Longrightarrow x * y \in \ell_2$.

Theorem 7.50 *Suppose that $f, g \in L^2_{2\pi}$, and let $h = fg$. Then $h \in L^2_{2\pi}$ and $\widehat{h} = \widehat{f} * \widehat{g}$.*

Proof Suppose that f and g are trigonometric polynomials. Then

$$\begin{aligned}
\sum_{k=-\infty}^{\infty} \widehat{f}(k)e^{ikx} \cdot \sum_{j=-\infty}^{\infty} \widehat{g}(j)e^{ijx} &= \sum_{k=-\infty}^{\infty} \widehat{f}(k) \sum_{j=-\infty}^{\infty} \widehat{g}(j)e^{i(k+j)x} \\
&= \sum_{k=-\infty}^{\infty} \widehat{f}(k) \sum_{m=-\infty}^{\infty} \widehat{g}(m-k)e^{imx} \\
&= \sum_{m=-\infty}^{\infty} \sum_{k=-\infty}^{\infty} \widehat{f}(k)\widehat{g}(m-k)e^{imx} \\
&= \sum_{m=-\infty}^{\infty} (\widehat{f} * \widehat{g})_m e^{imx}.
\end{aligned}$$

This proves the result for trigonometric polynomials. The extension to functions in $L^2_{2\pi}$ is left as Exercise 7.5.9. ∎

Exercises for Section 7.5

7.5.1 Prove $\dfrac{1}{2\pi}\displaystyle\int_{-\pi}^{\pi} e^{ikx}\,dx = \begin{cases} 1 & : \quad \text{if } k = 0 \\ 0 & : \quad \text{if } k = \pm 1, \pm 2, \ldots. \end{cases}$

Then use this to prove Equation (7.26).

7.5.2 Prove that the trigonometric system

$$1,\ \cos x,\ \sin x,\ \cos 2x,\ \sin 2x, \ldots, \cos kx,\ \sin kx, \ldots$$

is an orthogonal sequence under $\langle f, g\rangle = \int_{-\pi}^{\pi} f(x)\overline{g(x)}\,dx$. Hint: Compute $e^{ikx}e^{-imx} + e^{ikx}e^{imx}$ and $e^{ikx}e^{-imx} - e^{ikx}e^{imx}$ and use Exercise 7.5.1.

7.5.3 Prove Property **(C2)** of Theorem 7.42 by interchanging integrals and making a change of variable.

7.5.4 If we define the translation operator $T_a f(x) = f(x+a)$, prove that, as in the proof of Property **(C4)** (Theorem 7.42), that $T_a(f * g) = T_a f * g = f * T_a g$.

Also show $T_{a+b}(f * g) = T_a f * T_b g$.

7.5.5 Prove Theorem 7.45.

7.5.6 We say that a function f is an **idempotent** for the convolution if $f = f * f$. Show that finite sums of distinct e^k are idempotent.

7.5.7 Prove Theorem 7.46.

7.5.8 Prove the assertion made above that $x, y \in \ell_2 \Longrightarrow x * y \in \ell_2$.

7.5.9 Complete the proof of Theorem 7.50.

7.6 Examples of Trigonometric Series

Example 7.21 $f(x) = x$ for $-\pi < x \le \pi$. Figure 7.6 (Section 7.5) shows a graph and its periodic extension.

A calculation of the coefficients uses integration by parts.
If $k = 0$, then $\widehat{f}(0) = \frac{1}{2\pi}\int_{-\pi}^{\pi} x\,dx = 0$. If $k \ne 0$, we integrate by parts,

$$\widehat{f}(k) = \frac{1}{2\pi}\int_{-\pi}^{\pi} xe^{-ikx}\,dx = -\frac{x}{2\pi ik}e^{-ikx}\Big]_{-\pi}^{\pi} + \frac{1}{2\pi ik}\int_{-\pi}^{\pi} e^{-ikx}\,dx = \frac{(-1)^{k+1}}{ik}.$$

Thus the Fourier series is

$$f(x) \sim \sum_{k\ne 0}\frac{(-1)^{k+1}}{ik}e^{ikx} = 2\left(\sin x - \frac{\sin 2x}{2} + \frac{\sin 3x}{3} - \frac{\sin 4x}{4} + \cdots\right).$$

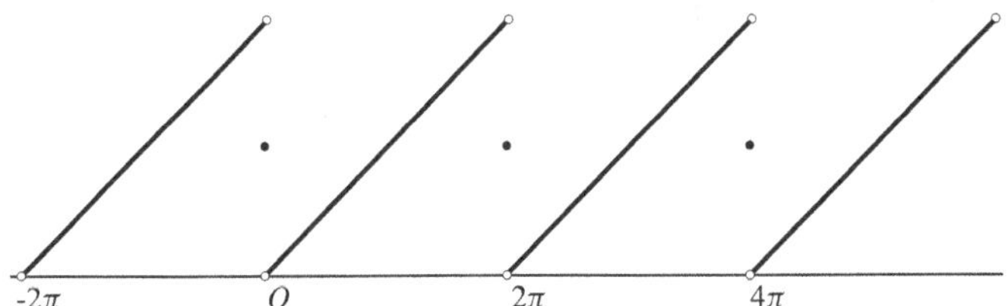

Figure 7.9 The graph of $f(x) = x$ $(0 < x < 2\pi)$ with its periodic extension.

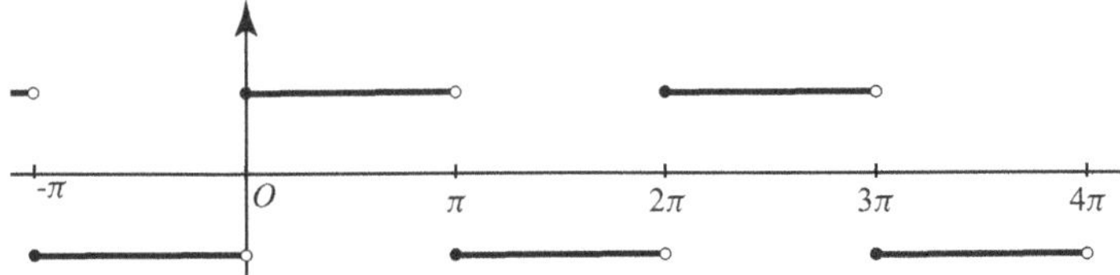

Figure 7.10 The graph of $f(x) = -1$ $(-\pi < x < 0)$ and $f(x) = 1$ $(0 \leq x < \pi)$ with its periodic extension.

The function is odd, so there are no cosine terms in the series.

Example 7.22 $f(x) = x$ for $0 < x < 2\pi$, $f(2\pi) = \pi$. Figure 7.9 shows a graph and its periodic extension.

This function is similar to the one in Example 7.21, but it is neither an even function nor an odd function. Here we integrate between 0 and 2π, performing a similar integration by parts, to obtain the Fourier series

$$f(x) \sim \pi - 2\left(\sin x + \frac{\sin 2x}{2} + \frac{\sin 3x}{3} + \frac{\sin 4x}{4} + \cdots\right).$$

Example 7.23 Figure 7.10 illustrates the function

$$f(x) = \begin{cases} -1 & : \quad \text{for } -\pi < x < 0 \\ 1 & : \quad \text{for } \quad 0 \leq x < \pi. \end{cases}$$

This is an odd function, so the cosine terms are zero. Using Equation (7.26), the sine terms have coefficients,

$$b_k = \frac{1}{\pi}\int_{-\pi}^{\pi} f(x)\,dx = \frac{2}{\pi}\int_0^{\pi} \sin kx\,dx = \frac{2}{k\pi}\big(1 - \cos k\pi\big).$$

The Fourier series of this function is thus

$$f(x) \sim \frac{4}{\pi}\left(\sin x + \frac{\sin 3x}{3} + \frac{\sin 5x}{5} + \cdots\right).$$

Example 7.24 $f(x) = x^2$ for $-\pi < x < \pi$.

Figure 7.7 illustrates this function. This is an even function, so the sine terms are zero. For $k = 0$, the cosine term has coefficient $a_0 = \frac{1}{\pi}\int_{-\pi}^{\pi} x^2\,dx = \frac{x^3}{3\pi}\Big]_{-\pi}^{\pi} = \frac{2\pi^2}{3}$. For $k \neq 0$, we integrate by parts twice

$$\begin{aligned} a_k &= \frac{1}{\pi}\int_{-\pi}^{\pi} x^2 \cos kx\,dx = -\frac{2}{k\pi}\int_{-\pi}^{\pi} x \sin kx\,dx \\ &= \frac{2x\cos kx}{k^2}\Big]_{-\pi}^{\pi} - \frac{2}{k^2\pi}\int_{-\pi}^{\pi} \cos kx\,dx \\ &= \frac{4}{k^2}\cos k\pi = (-1)^n \frac{4}{k^2}. \end{aligned}$$

The Fourier series is thus

$$f(x) \sim \frac{\pi^2}{3} - 4\left(\cos x - \frac{\cos 2x}{2^2} + \frac{\cos 3x}{3^2} - \frac{\cos 4x}{4^2} + \cdots\right).$$

Example 7.25 $f(x) = ax^2 + bx + c$ for $-\pi < x < \pi$. We use the results from previous examples to obtain the Fourier series:

$$f(x) \sim \frac{a\pi^2}{3} + c + 4a\sum_{k=1}^{\infty}(-1)^k \frac{\cos kx}{k^2} - 2b\sum_{k=1}^{\infty}(-1)^k \frac{\sin kx}{k}.$$

Example 7.26 A trigonometric series that is not a Fourier series: The series

$$\sum_{k=2}^{\infty} \frac{\sin kx}{\log k}$$

is a trigonometric series that converges everywhere but it is not the Fourier series of any function $f \in L_{2\pi}$. (For a proof, please see R.E. Edwards, *Fourier Series: A Modern Introduction*, Springer-Verlag, New York (1979).)

Exercises for Section 7.6

7.6.1 Compute the Fourier series of the zigzag function

$$f(x) = \begin{cases} x & : \text{ if } \frac{-\pi}{2} \leq x < \frac{\pi}{2} \\ \pi - x & : \text{ if } \frac{\pi}{2} \leq x < \frac{3\pi}{2}. \end{cases}$$

7.6.2 Compute the Fourier series of the step function

$$f(x) = \begin{cases} 0 & : \text{ if } -\pi \leq x < 0 \\ 1 & : \text{ if } 0 \leq x < \pi. \end{cases}$$

7.6.3 Compute the Fourier series of $f(x) = x + \frac{x^2}{4}$ for $-\pi < x < \pi$. Sketch a graph of f with its periodic extension.

7.6.4 Compute the Fourier series of $f(x) = \cos x$ for $-\pi < x < \pi$.

7.6.5 Compute the Fourier series of $f(x) = x^2$ for $0 < x < 2\pi$. Sketch a graph of f with its periodic extension.

7.6.6 Use Exercise 7.6.5 and examples in this section to find the Fourier series of $f(x) = ax^2 + bx + c$ for $0 < x < 2\pi$.

7.6.7 Find and sketch the function with the Fourier series $\sum_{k=1}^{\infty}(-1)^k \frac{\cos kx}{k^2}$.

7.6.8 Find and sketch the function with the Fourier series $\sum_{k=1}^{\infty}(-1)^k \frac{\sin kx}{k}$.

7.6.9 Compute the Fourier series of $f(x) = |x|$ for $-\pi < x < \pi$.
Sketch a graph of f with its periodic extension.

7.6.10 Compute the Fourier series of $f(x) = |\sin x|$ for $-\pi < x < \pi$.
Sketch a graph of f with its periodic extension.

7.7 Arbitrary Periods

Replacing x by $\left(\frac{2\pi}{A}\right)x$ in $\sum_k c_k e^{ikx}$ and $\frac{a_0}{2} + \sum_k (a_k \cos kx + b_k \sin kx)$ will transform trigonometric functions of period 2π into period A:

$$\sum_k c_k e^{i\left(\frac{2\pi}{A}\right)kx} \quad \text{and} \quad \frac{a_0}{2} + \sum_k \left(a_k \cos\left(\frac{2\pi}{A}\right)kx + b_k \sin\left(\frac{2\pi}{A}\right)kx\right).$$

For example, the following trigonometric functions will have periods 1,

$$\sum_k c_k e^{i2\pi kx} \quad \text{and} \quad \frac{a_0}{2} + \sum_k (a_k \cos 2\pi kx + b_k \sin 2\pi kx).$$

For a function f of period $A = 2a$, the Fourier coefficients of f are

$$a_k = \frac{1}{a}\int_{-a}^{a} f(x)\cos\frac{\pi}{a}kx\,dx \quad \text{and} \quad b_k = \frac{1}{a}\int_{-a}^{a} f(x)\sin\frac{\pi}{a}kx\,dx,$$

and the Fourier series of f is $\frac{a_0}{2} + \sum_k \left(a_k \cos\frac{\pi}{a}kx + b_k \sin\frac{\pi}{a}kx\right)$.

Exercises for Section 7.7

7.7.1 Suppose $f(x) = x^2$ on the interval $[-3, 3]$ and has period 6. Find the Fourier series of f.

7.7.2 Suppose $f(x) = \sin x$ on $\left[-\frac{\pi}{2}, \frac{\pi}{2}\right]$ and has period π. Find the Fourier series of f and sketch the graph of its periodic extension.

7.7.3 Suppose $f(x) = x$ on $[0,1)$. Find the Fourier series of f and sketch the graph of its periodic extension.

7.7.4 Suppose

$$f(x) = \begin{cases} -2 & : \quad \text{for } -3 < x < 0 \\ 2 & : \quad \text{for } \quad 0 < x < 3. \end{cases}$$

Find the Fourier series of f and sketch the graph of its periodic extension.

7.8 Fourier Series and Summability

Definition 7.16 In the function space $L_{2\pi}$, it is useful to consider (Fredholm) **integral operators** of the form

$$Tf(x) = \frac{1}{2\pi}\int_{-\pi}^{\pi} K(x,t) f(t)\, dt.$$

The function $K(x,t)$ is called the **kernel** of the operator. Note that the convolution as an integral operator of f has the kernel $K(x,t) = g(x-t)$,

$$Tf(x) = (f * g)(x) = \frac{1}{2\pi}\int_{-\pi}^{\pi} f(t) g(x-t)\, dt.$$

In this section we examine kernels $K(x,t) = g(x-t)$ for the following specific functions g.

$\mathbf{D}_n$: The Dirichlet kernel with $g(x) = D_n(x) = \sum_{k=-n}^{n} e^{ikx}$.

$\mathbf{F}_n$: The Fejér kernel with $g(x) = F_n(x) = \dfrac{D_0(x) + D_1(x) + \cdots + D_n(x)}{n+1}$.

$\mathbf{P}_r$: The Poisson kernel with $g(x) = P_r(x) = \sum_{k=-\infty}^{\infty} r^{|k|} e^{ikx} \quad (0 \le r < 1)$.

Each of these kernels is associated with some form of summability of Fourier series.

Definition 7.17 The **Dirichlet**[9] **kernel** is the sequence (D_n) of trigonometric polynomials

$$D_n(x) = \sum_{k=-n}^{n} e^{ikx}.$$

[9] Named for the German mathematician Johann Peter Lejeune-Dirichlet (1805–1859). He is credited with giving the modern definition of a function. He is also considered the founder of the rigorous theory of Fourier series.

Theorem 7.51 *For $n = 0, 1, \ldots,$ we have*

$$D_n(x) = \begin{cases} \dfrac{\sin(n+\frac{1}{2})x}{\sin\frac{x}{2}} & : \text{ if } x \neq 0, \pm 2\pi, \pm 4\pi, \ldots \\ 2n+1 & : \text{ if } x = 0, \pm 2\pi, \pm 4\pi, \ldots. \end{cases}$$

Proof The case $x = 0 \pmod{2\pi}$ follows immediately from the definition of D_n. Otherwise, let $z = e^{ix}$ and sum a geometric progression,

$$D_n(x) = \sum_{k=-n}^{n} e^{ikx} = z^{-n} \sum_{k=0}^{2n} z^k = z^{-n}\left(\frac{1 - z^{2n+1}}{1 - z}\right) = \frac{z^{-n} - z^{n+1}}{1 - z}.$$

Then multiply numerator and denominator by $z^{-\frac{1}{2}}$ and use Equation (7.24) to obtain,

$$D_n(x) = \frac{z^{-(n+\frac{1}{2})} - z^{n+\frac{1}{2}}}{z^{-\frac{1}{2}} - z^{\frac{1}{2}}} = \frac{\sin\left(n + \frac{1}{2}\right)x}{\sin\frac{x}{2}}.$$

∎

Theorem 7.52 *For every $f \in L_{2\pi}$ and $n = 0, 1, \ldots,$ we have*

$$s_n f(x) = (f * D_n)(x) = \frac{1}{2\pi}\int_{-\pi}^{\pi} f(t) D_n(x - t)\, dt.$$

Proof

$$\begin{aligned} s_n f(x) &= \sum_{k=-n}^{n} \widehat{f}(k) e^{ikx} = \sum_{k=-n}^{n} \frac{1}{2\pi}\int_{-\pi}^{\pi} f(t) e^{-ikt} e^{ikx}\, dt \\ &= \frac{1}{2\pi}\int_{-\pi}^{\pi} f(t) \sum_{k=-n}^{n} e^{ik(x-t)}\, dt = \frac{1}{2\pi}\int_{-\pi}^{\pi} f(t) D_n(x - t)\, dt \\ &= (f * D_n)(x). \end{aligned}$$

∎

Definition 7.18 An **approximate identity** for $L_{2\pi}$ is a sequence (K_n) in $L_{2\pi}$ satisfying

$$\lim_{n\to\infty} \|K_n * f - f\|_1 = 0 \text{ for all } f \in L_{2\pi}.$$

The Dirichlet kernel does not form an approximate identity as will be shown in Theorem 7.64. However, a general class of kernels, called summability kernels, are approximate identities for $L_{2\pi}$.

Definition 7.19 A **summability kernel** is a sequence of functions (K_n) in $L_{2\pi}$ satisfying the following properties:

(SK1) $\sup_n \|K_n\|_1 = M < \infty$.

(SK2) $\frac{1}{2\pi}\int_{-\pi}^{\pi} K_n(x)\,dx = 1$ for all n.

(SK3) For every $0 < \delta < \pi$, we have $\lim_n \frac{1}{2\pi}\int_{\delta\le|x|\le\pi} |K_n(x)|\,dx = 0$.

Lemma 7.53 *Suppose that (K_n) is a summability kernel as defined in Definition 7.19. Then for each $f \in C_{2\pi}$, the sequence $K_n * f$ converges uniformly to f. That is,*

$$\lim_{n\to\infty} \|K_n * f - f\|_\infty = 0.$$

Proof Suppose $f \in C_{2\pi}$ and let $\epsilon > 0$ be given. Choose $\delta > 0$ such that

$$|f(x-t) - f(x)| < \epsilon/2M \quad \text{whenever} \quad |t| < \delta,$$

where $M > 0$ is given by Property **(SK1)**. By Property **(SK2)**, we have

$$\begin{aligned}(K_n * f)(x) - f(x) &= \frac{1}{2\pi}\int_{-\pi}^{\pi} K_n(t) f(x-t)\,dt \; - \; f(x)\\ &= \frac{1}{2\pi}\int_{-\pi}^{\pi} K_n(t)\big(f(x-t) - f(x)\big)\,dt.\end{aligned}$$

Then, for any x, we have

$$\begin{aligned}|(K_n * f)(x) - f(x)| &\le \frac{1}{2\pi}\int_{-\pi}^{\pi} |K_n(t)|\big|f(x-t) - f(x)\big|\,dt\\ &= \frac{1}{2\pi}\int_{|t|<\delta} |K_n(t)|\big|f(x-t) - f(x)\big|\,dt\\ &\quad + \frac{1}{2\pi}\int_{\delta\le|t|\le\pi} |K_n(t)|\big|f(x-t) - f(x)\big|\,dt\\ &\le \frac{\epsilon}{2M}\frac{1}{2\pi}\int_{-\pi}^{\pi} |K_n(t)|\,dt + 2\|f\|_\infty\\ &\quad \times \frac{1}{2\pi}\int_{\delta\le|t|\le\pi} |K_n(t)|\,dt\\ &\le \epsilon/2 \; + \; \frac{\|f\|_\infty}{\pi}\int_{\delta\le|t|\le\pi} |K_n(t)|\,dt.\end{aligned}$$

By Property **(SK3)**, the last term is less than $\epsilon/2$ for sufficiently large n. Thus $\|K_n * f - f\|_\infty < \epsilon$ for sufficiently large n. ∎

Theorem 7.54 *Every summability kernel (K_n) is an approximate identity for $L_{2\pi}$.*

Proof Let $f \in L_{2\pi}$ and let $\epsilon > 0$ be given. Since $C_{2\pi}$ is dense in $L_{2\pi}$ (Theorem 6.42), there exists $g \in C_{2\pi}$ such that $\|f - g\|_1 < \epsilon/3$ and $\|f - g\|_1 < \epsilon/3M$, for $M > 0$ given by Property **(SK1)**. By Property **(C4)** (Theorem 7.42), we have

$$\|K_n * f - K_n * g\|_1 = \|K_n * (f - g)\|_1 \leq \|K_n\|_1 \|f - g\|_1 < \epsilon/3.$$

By Lemma 7.53, for sufficiently large n, we have

$$\|K_n * g - g\|_\infty < \epsilon/3 \quad \text{and hence} \quad \|K_n * g - g\|_1 < \epsilon/3. \quad \text{Thus}$$

$$\|K_n * f - f\|_1 \leq \|K_n * f - K_n * g\|_1 + \|K_n * g - g\|_1 + \|f - g\|_1 < \epsilon. \quad \blacksquare$$

7.8.1 The Fejér Kernel and Cesáro Summability

Definition 7.20 The **Fejér**[10] **kernel** is the arithmetic mean of the Dirichlet kernel

$$F_n(x) = \frac{D_0(x) + D_1(x) + \cdots + D_n(x)}{n+1}.$$

Theorem 7.55 *For $n = 0, 1, \ldots,$ we have*

$$F_n(x) = \begin{cases} \dfrac{1}{n+1} \dfrac{\sin^2 \frac{1}{2}(n+1)x}{\sin^2 (x/2)} & : \quad \text{if } x \neq 0, 2\pi, 4\pi, \ldots \\ n+1 & : \quad \text{if } x = 0, 2\pi, 4\pi, \ldots. \end{cases}$$

The proof is left as Exercise 7.8.2.

Theorem 7.56 *The Fejér kernel (F_n) forms a summability kernel.*

Proof The Fejér kernel has the three properties of a summability kernel.

$$\textbf{(SK2)}: \frac{1}{2\pi} \int_{-\pi}^{\pi} D_n(x)\, dx = \frac{1}{2\pi} \int_{-\pi}^{\pi} \sum_{k=-n}^{n} e^{ikx}$$

$$= \frac{1}{2\pi} \int_{-\pi}^{\pi} e^0\, dx + \frac{1}{2\pi} \sum_{k=1}^{n} \int_{-\pi}^{\pi} \left(e^{ikx} + e^{-ikx}\right) dx$$

$$= 1 + \frac{1}{2\pi} \sum_{k=1}^{n} 0 = 1 \text{ for all } n.$$

[10] Lipót Fejér (1880–1959) was a Hungarian mathematician who worked in the areas of Fourier series and approximation theory.

Thus $$\frac{1}{2\pi}\int_{-\pi}^{\pi} F_n(x)\,dx = \frac{1}{2\pi}\int_{-\pi}^{\pi} \frac{D_0(x)+D_1(x)+\cdots+D_n(x)}{n+1}\,dx = 1.$$

(SK1): By Theorem 7.55, $F_n \geq 0$. The result then follows from Property **(SK2)** with $M = 1$.

(SK3): Since $\sin^2(x/2) \geq \sin^2(\delta/2) \geq (\delta/\pi)^2$ for $0 < \delta \leq |x| \leq \pi$, we have, by Theorem 7.55,

$$0 \leq F_n(x) \leq \frac{1}{(n+1)\sin^2(\delta/2)} \leq \frac{\pi^2}{(n+1)\delta^2} \quad \text{for} \quad 0 < \delta \leq |x| \leq \pi. \tag{7.30}$$

Hence $$\frac{1}{2\pi}\int_{\delta\leq|x|\leq\pi} |F_n(x)|\,dx \leq \frac{1}{(n+1)\sin^2\delta/2} \to 0 \quad \text{as } n \to \infty.$$ ∎

Definition 7.21 The nth **Cesàro**[11] **mean** of a series $\sum x_k$ is the average of the first n partial sums

$$\sigma_n = \frac{s_1+\cdots+s_n}{n}, \text{ where } s_1 = x_1, s_2 = x_1+x_2, s_3 = x_1+x_2+x_3, \ldots.$$

Recall that trigonometric partial sums are *symmetric*

$$s_k f(x) = \sum_{j=-k}^{k} \widehat{f}(j)e^{ijx} = \widehat{f}(-k)e^{-ikx} + \cdots + \widehat{f}(0) + \cdots + \widehat{f}(k)e^{ikx}.$$

The nth **Cesàro mean of a function** $f \in L_{2\pi}$ is then defined to be

$$\sigma_n f(x) = \frac{s_0 f(x) + s_1 f(x) + \cdots + s_n f(x)}{n+1}, \quad \text{where} \quad n = 0, 1, 2, \ldots.$$

Theorem 7.57 *For every $f \in L_{2\pi}$, $F_n * f$ is the nth Cesàro mean of the function f.*

The proof is immediate from the definition of the Fejér kernel and Theorem 7.52. Then the following corollary follows from Lemma 7.53.

Corollary 7.58 *If $f \in C_{2\pi}$, f can be uniformly approximated by trigonometric polynomials.*

Theorem 7.59 Weierstrass approximation: *Every continuous function on the interval $[-\pi,\pi]$ can be uniformly approximated by algebraic polynomials.*

[11] These means were investigated by Ernesto Cesàro (1859–1906).

Proof By Taylor's theorem, $\cos x$ and $\sin x$ can be uniformly approximated by algebraic polynomials on $[-\pi, \pi]$ as follows:

$$\cos x = 1 - \frac{x^2}{2!} + \frac{x^4}{4!} - \cdots + (-1)^n \frac{x^{2n}}{(2n)!} + R_{2n}(x),$$
$$\text{where } |R_{2n}(x)| \leq \frac{\pi^{2n+1}}{(2n+1)!}.$$

Similarly,

$$\sin x = x - \frac{x^3}{3!} + \frac{x^5}{5!} - \cdots + (-1)^n \frac{x^{2n+1}}{(2n+1)!} + R_{2n+1}(x),$$
$$|R_{2n+1}(x)| \leq \frac{\pi^{2n+2}}{(2n+2)!}.$$

Combining the two, we see that each of the exponential functions

$$e^{ikx} = \cos kx + i \sin kx$$

can be uniformly approximated by algebraic polynomials. Thus every trigonometric polynomial can be uniformly approximated by algebraic polynomials. The result then follows from Corollary 7.58. ∎

Section 5.12, has a proof of the Weierstrass approximation theorem that does not involve Fourier series. The Weierstrass approximation theorem again shows that $L_{2\pi}$ is separable.

Theorem 7.60 Fejér: *If $f \in L_{2\pi}$, then the Cesàro means of f converge to f in norm:*

$$\lim_{n\to\infty} \|\sigma_n f - f\|_1 = 0.$$

This is immediate from Theorems 7.56 and 7.54. The following corollaries are easy consequences of Fejér's theorem.

Corollary 7.61 *If $f \in L_{2\pi}$ with $\widehat{f} = 0$, then $\|f\|_1 = 0$. In that case, $f = 0$ almost everywhere.*

Corollary 7.62 *Each $f \in L_{2\pi}$ is completely determined by its Fourier coefficients $\widehat{f}$.*

Theorem 7.63 *If we define $\|\widehat{f}\|_1 = \|f\|_1$ then $\widehat{L_{2\pi}}$ is isometric to $L_{2\pi}$.*

Finally we show that the Dirichlet kernel is not an approximate identity.

Theorem 7.64 *It is* not *the case that $\lim_{n\to\infty} \|s_n f - f\|_1 = 0$ for all $f \in L_{2\pi}$.*

Proof For each n, define the linear operator $T_n : L_{2\pi} \longrightarrow L_{2\pi}$ by $T_n f = s_n f$. If it were the case that $T_n f \to f$ on $L_{2\pi}$ then by the uniform boundedness principle (Theorem 6.30), there exists $M > 0$ such that for all n,

$$\|T_n\| = \sup_{\|f\|_1 \le 1} \|s_n f\|_1 = \sup_{\|f\|_1 \le 1} \|f * D_n\|_1 \le M < \infty.$$

We show that this is not the case by showing that $\lim_n \|T_n\| = \infty$.

Claim: $\|T_n\| \ge \|D_n\|_1$. By Theorem 7.56 we have $\|F_N\|_1 = 1$ for all N. Thus $\|T_n\| \ge \|F_N * D_n\|_1$ for all N. By Fejér's theorem, we have $\|F_N * D_n - D_n\|_1 \to 0$ as $N \to \infty$. Thus $\|T_n\| \ge \|D_n * F_N\|_1 \ge \|D_n\|_1$.

Claim: $\lim_n \|D_n\|_1 = \infty$. Use the change of variable $u = x/2$. Then

$$\|D_n\|_1 = \frac{1}{2\pi} \int_{-\pi}^{\pi} \left| \frac{\sin\left(n + \frac{1}{2}\right) x}{\sin \frac{x}{2}} \right| dx = \frac{2}{\pi} \int_0^{\pi/2} \frac{|\sin(2n+1)u|}{|\sin u|}\, du.$$

Using $|\sin u| \le |u|$ and another change of variable $y = (2n+1)u$, we have

$$\begin{aligned}
\|D_n\|_1 &\ge \frac{2}{\pi} \int_0^{\pi/2} \frac{|\sin(2n+1)u|}{|u|}\, du = \frac{2}{\pi} \int_0^{(2n+1)\pi/2} \frac{|\sin y|}{y}\, dy \\
&= \frac{2}{\pi} \sum_{k=0}^{2n} \int_{k\pi/2}^{(k+1)\pi/2} \frac{|\sin y|}{y}\, dy \\
&\ge \frac{2}{\pi} \sum_{k=0}^{2n} \left\{ \frac{2}{(k+1)\pi} \int_{k\pi/2}^{(k+1)\pi/2} |\sin y|\, dy \right\} \\
&= \frac{2}{\pi} \sum_{k=0}^{2n} \frac{4}{(k+1)\pi} \sim \frac{8}{\pi^2} \log n, \quad \text{which proves the claim.}
\end{aligned}$$

∎

7.8.2 The Poisson Kernel and Abel Summability

The Fejér kernel is a sequence (F_n), indexed by $n = 0, 1, 2, \ldots$, that is associated with Cesàro summability. The Poisson kernel (P_r) is indexed by a continuous r with $0 \le r < 1$, as defined below. We will show that it is associated with Abel summability. The Poisson kernel arose in the study of the heat equation and has many applications to engineering and physics.

Definition 7.22 The **Poisson**[12] **kernel** consists of the trigonometric series

$$P_r(x) = \sum_{k=-\infty}^{\infty} r^{|k|} e^{ikx} \quad \text{where} \quad 0 \le r < 1. \tag{7.31}$$

Theorem 7.65 *The Poisson kernel satisfies the following identity:*

$$P_r(x) = \frac{1 - r^2}{1 - 2r\cos x + r^2} \quad \textit{for} \quad 0 \le r < 1. \tag{7.32}$$

Proof Recall the Weierstrass M-test which states that a series $\sum_{k=1}^{\infty} u_k(x)$ converges uniformly on I whenever whenever $|u_k(x)| \le M_k$ on I ($k = 1, 2, \ldots$) and $\sum_k M_k < \infty$. For the series (7.31), we have $u_k(x) = r^k e^{-ikx} + r^k e^{ikx}$ and $|u_k(x)| \le 2r^k \equiv M_k$. Since $\sum_k M_k < \infty$ for $0 \le r < 1$, the series (7.31) of the Poisson kernel converges uniformly. The partial sums are

$$s_n(x) = 1 + \sum_{k=1}^{n} \left(re^{ix}\right)^k + \sum_{k=1}^{n} \left(re^{-ix}\right)^k.$$

Summing the two geometric series, we obtain the desired result:

$$P_r(x) = 1 + \frac{re^{ix}}{1 - re^{ix}} + \frac{re^{-ix}}{1 - re^{-ix}} = \frac{1 - r^2}{1 - 2r\cos x + r^2}. \qquad \blacksquare$$

The kernels we have studied up to now have been sequences of functions K_n ($n = 0, 1, 2, \ldots$). For the Poisson kernel, we are dealing with functions P_r indexed by a continuous $r \to 1^-$ instead of by a discrete $n \to \infty$. This calls for an extension of the definition of summability kernel, which can be done by simply replacing n with r and taking the limit $r \to 1^-$ as follows.

A family of functions $\{K_r\}_{0 \le r < 1}$ in $L_{2\pi}$ is a **summability kernel** if it satisfies the following three conditions.

(SK1′) $\sup_{0 \le r < 1} \|K_r\|_1 = M < \infty.$

(SK2′) $\frac{1}{2\pi} \int_{-\pi}^{\pi} K_r(x)\, dx = 1$ for all $0 \le r < 1$.

(SK3′) For every $0 < \delta < \pi$, we have $\lim_{r \to 1^-} \int_{\delta \le |x| \le \pi} |K_r(x)|\, dx = 0.$

[12] Siméon-Denis Poisson (1781–1840) was a French mathematician who made many discoveries in probability theory and analysis. It is reported that he said, "Life is good for only two things, discovering mathematics and teaching mathematics."

Theorem 7.66 *The Poisson kernel* $\{P_r\}_{0 \le r < 1}$ *forms a summability kernel.*

Proof Verification of properties (**SK1′**) and (**SK2′**) are left as Exercise 7.8.4.

(**SK3′**): We showed in Theorem 7.65 that $P_r(x) = \dfrac{1-r^2}{1-2r\cos x + r^2}$. The denominator may be rewritten $1 - 2r\cos x + r^2 = (1-r)^2 + 2r(1-\cos x)$. Let $0 < \delta < \pi$. If $\frac{1}{2} \le r \le 1$ and $\delta \le |x| \le \pi$, then for some $c = c_\delta > 0$, we have

$$1 - 2r\cos x + r^2 = (1-r)^2 + 2r(1-\cos x) \ge c > 0.$$

This means $P_r(x) \le \frac{1-r^2}{c}$ whenever $\delta \le |x| \le \pi$. Thus

$$0 \le \frac{1}{2\pi}\int_{\delta \le |x| \le \pi} P_r(x)\,dx \le \frac{1-r^2}{c} \to 0 \quad \text{as} \quad r \to 1^-. \qquad \blacksquare$$

Definition 7.23 The Abel means of a Fourier series are $A_r f(x) = \sum_{k=-\infty}^{\infty} r^{|k|}\widehat{f}(k)e^{ikx}$.

Since Fourier coefficients $\widehat{f}(k)$ for $f \in L_{2\pi}$ converge to zero as $k \to \infty$ (Riemann–Lebesgue lemma), Abel means converge absolutely and uniformly for each $0 \le r < 1$.

Theorem 7.67 *The Abel means of a function* $f \in L_{2\pi}$ *are convolutions with the Poisson kernel:*

$$A_r f(x) = (f * P_r)(x).$$

Proof By uniform convergence we can interchange integrals and infinite sums:

$$\begin{aligned}
A_r f(x) &= \sum_{k=-\infty}^{\infty} r^{|k|}\widehat{f}(k)e^{ikx} \\
&= \sum_{k=-\infty}^{\infty} r^{|k|}\left(\frac{1}{2\pi}\int_{-\pi}^{\pi} f(t)e^{-ikt}\,dt\right)e^{ikx} \\
&= \frac{1}{2\pi}\int_{-\pi}^{\pi} f(t)\left(\sum_{k=-\infty}^{\infty} r^{|k|}e^{-ik(t-x)}\right)dt = (f * P_r)(x). \qquad \blacksquare
\end{aligned}$$

Theorem 7.68 *For all* $f \in L_{2\pi}$, *we have* $\lim_{r \to 1^-} \|A_r f - f\|_1 = 0.$

The proof follows directly from Theorem 7.54.

*7.8.3 The Gauss Kernel

A related summability kernel is the **Gauss kernel**. It has applications to heat dispersion and the atomic theory of matter, as we will see in the next chapter (Sections 8.3 and 8.6). The Gauss kernel, defined on $\mathbb{R}$, is

$$G_t(x) = \frac{1}{\sqrt{4\pi t}} e^{\frac{-x^2}{4t}} \quad \text{for} \quad t > 0. \tag{7.33}$$

Here we deal with the space of integrable functions L^1 on the real line $\mathbb{R}$ instead of 2π-periodic functions of $L_{2\pi}$. Additionally, the kernel is indexed by a continuous $t \to 0^+$ instead of a discrete $n \to \infty$ (as in the Fejér kernel) or a continuous $r \to 1^-$ (as in the Poisson kernel). This again calls for a modification of the definition of a summability kernel $\{K_t\}_{t>0}$.

A family of real valued functions $\{K_t\}_{t>0}$ is a **summability kernel** on $\mathbb{R}$ as $t \to 0^+$ if the following conditions are satisfied:

(SK1″) $\sup_{t>0} \|K_t(x)\|_1 = \sup_{t>0} \int_{-\infty}^{\infty} |K_t(x)|\, dx = M < \infty.$

(SK2″) $\int_{-\infty}^{\infty} K_t(x)\, dx = 1$ for all $t > 0$.

(SK3″) For every $0 < \delta < \infty$, $\lim_{t\to 0^+} \int_{\delta \le |x|} |K_t(x)|\, dx = 0.$

We have already defined the convolution $f * g$ of two functions in $L^2_{2\pi}$. Now we extend the notion to $L^1(\mathbb{R})$.

Definition 7.24 For integrable functions $f, g \in L^1(\mathbb{R})$, define the **convolution** $f * g$ on $\mathbb{R}$ as $(f * g)(x) = \int_{-\infty}^{\infty} f(u)\overline{g(x-u)}\, du$, provided the integral exists.

For example, if g is bounded and $f \in L^1(\mathbb{R})$, then $f * g$ is defined. In particular, for fixed t, the Gauss function G_t is bounded. So the convolution $f * G_t$ is defined.

Theorem 7.69 *A summability kernel $\{K_t\}_{t>0}$ on $\mathbb{R}$, is an approximate identity on $L^1(\mathbb{R})$. That is, $\lim_{t\to 0^+} \|K_t * f - f\|_1 = 0$ for all $f \in L^1(\mathbb{R})$.*

Proof Let $f \in L^1(\mathbb{R})$ and let $\epsilon > 0$ be given. Since $\int_{-\infty}^{\infty} K_t(u)\, du = 1$, we have

$$\begin{aligned}K_t * f(x) - f(x) &= \int_{-\infty}^{\infty} K_t(u) f(x-u)\,du - f(x)\\ &= \int_{-\infty}^{\infty} K_t(u) f(x-u)\,du - \int_{-\infty}^{\infty} K_t(u) f(x)\,du\\ &= \int_{-\infty}^{\infty} K_t(u)\big[f(x-u) - f(x)\big]\,du.\end{aligned}$$

By Lemma 7.43, the translation operator $T_u f(x) = f(x-u)$ is continuous on $L^1(-\pi,\pi)$. Choose $\delta > 0$ such that $\|T_u f - f\|_1 < \epsilon/2M$ for $|u| \le \delta < \pi$, where M is the bound given by **(SK1″)**. By **(SK2″)**, we have

$$\begin{aligned}\|K_t * f(x) - f(x)\|_1 &\le \int_{-\infty}^{\infty} |K_t(u)| \big\|T_u f - f\big\|_1\,du\\ &\le \int_{|u|\le\delta} |K_t(u)| \big\|T_u f - f\big\|_1\,du\\ &\quad + 2\|f\|_1 \int_{\delta\le|u|} |K_t(u)|\,du\\ &\le (\epsilon/2)\int_{|u|\le\delta} |K_t(u)|\,du + 2\|f\|_1 \int_{\delta\le|u|} |K_t(u)|\,du\\ &\le (\epsilon/2M)\cdot(M) \;+\; 2\|f\|_1 \int_{\delta\le|u|} |K_t(u)|\,du.\end{aligned}$$

By **(SK3″)**, this is less than ϵ for $t > 0$ sufficiently small. ∎

Definition 7.25 The **Gaussian curves** are defined by the equation $f(x) = \frac{1}{\sigma\sqrt{2\pi}} e^{-\frac{1}{2}\left(\frac{x-\mu}{\sigma}\right)^2}$, for various $-\infty < \mu < \infty$ and $\sigma > 0$. The **standard Gaussian curve** is $f(x) = \frac{1}{\sqrt{2\pi}} e^{\frac{-x^2}{2}}$, with $\mu = 0$ and $\sigma = 1$. This is the classic bell-shaped curve of statistics, also called the **standard normal curve.**

The Gaussian curves have (Lebesgue) integral 1, $\int_{-\infty}^{\infty} f(x)\,dx = 1$. This follows from the substitution $u = \frac{x-\mu}{\sigma}$ in the well-known identity

$$\int_{-\infty}^{\infty} e^{-u^2}\,du = \sqrt{\pi}. \tag{7.34}$$

A proof of (7.34) is outlined in Exercise 7.8.5. For each $t > 0$, the **Gauss kernel** is the Gaussian curve with $\mu = 0$ and $\sigma = \sqrt{2t}$.

Theorem 7.70 *The Gauss kernel $\{G_t\}_{t>0}$ on $\mathbb{R}$ forms a summability kernel as $t \to 0^+$.*

Proof
(SK2″): Since, for all $t > 0$, G_t is a Gaussian curve, $\int_{-\infty}^{\infty} G_t(x)\,dx = 1$.

(SK1″): The Gauss kernel is positive, so we also get **(SK1″)** with $M = 1$.

(SK3″): Here we want to show that for each $\delta > 0$, we have

$$\lim_{t\to 0^+} \int_{\delta\le|x|} G_t(x)\,dx = \lim_{t\to 0^+} \frac{2}{\sqrt{4\pi t}} \int_{\delta}^{\infty} e^{\frac{-x^2}{4t}}\,dx = 0.$$

First, given $\epsilon > 0$, find $N > 0$ such that $\int_N^\infty e^{-u^2}\,du < \frac{\sqrt{\pi}}{2}\epsilon$. Then for each fixed $\delta > 0$, let $0 < t < \frac{\delta^2}{4N^2}$. By the substitution $u = \frac{x}{\sqrt{4t}}$, we have

$$\int_{\delta\le|x|} G_t(x)\,dx = \frac{2\sqrt{4t}}{\sqrt{4\pi t}} \int_{\delta/\sqrt{4t}}^{\infty} e^{-u^2}\,du = \frac{2}{\sqrt{\pi}} \int_{\delta/\sqrt{4t}}^{\infty} e^{-u^2}\,du.$$

Since $t < \frac{\delta^2}{4N^2}$, we have $\delta/\sqrt{4t} > \frac{2N\delta}{\sqrt{4\delta^2}} = N$. Thus

$$\int_{\delta\le|x|} G_t(x)\,dx = \frac{2}{\sqrt{\pi}} \int_{\delta/\sqrt{4t}}^{\infty} e^{-u^2}\,du < \frac{2}{\sqrt{\pi}} \int_{N}^{\infty} e^{-u^2}\,du < \epsilon. \qquad \blacksquare$$

A generalization of the Gauss kernel is the **heat kernel** defined on $\mathbb{R}$ for any fixed $D > 0$, as

$$H_t(x) = \frac{1}{\sqrt{4\pi Dt}} e^{\frac{-x^2}{4Dt}} \quad \text{for } t > 0. \tag{7.35}$$

The Gauss kernel and heat kernel are the same except for a change of scale. It is not difficult to show that the heat kernel is a summability kernel (Exercise 7.8.8) and an approximate identity (Exercise 7.8.10) and satisfies the following **heat equation** on $\mathbb{R}$ (Exercise 7.8.9):

$$\frac{\partial u}{\partial t} = D\frac{\partial^2 u}{\partial x^2}.$$

Exercises for Section 7.8

7.8.1 Use the Weierstrass approximation theorem (Theorem 7.59) to show that $C[0,1]$ is dense in $L^2[0,1]$.

7.8.2 Prove Theorem 7.55. Hint: Let $z = e^{ix}$ and use an identity in the proof of Theorem 7.51

$$(n+1)F_n(x) = \sum_{k=0}^{n} D_k(x) = \sum_{k=0}^{n} \left(\frac{z^{-k} - z^{k+1}}{1-z}\right).$$

7.8.3 **Dilation Kernel:** Let $f \in L_{2\pi}$ with $\frac{1}{2\pi}\int_{-\pi}^{\pi} f(x)\,dx = 1$ and define $K_n(x) = nf(nx)$ for $n = 1, 2, \ldots$. Show that $\{K_n\}$ forms a summability kernel.

7.8.4 Finish the proof of Theorem 7.66 by proving that the Poisson kernel satisfies conditions **(SK1′)** and **(SK2′)** of a summability kernel as given at the beginning of Subsection 7.8.2.

7.8.5 Prove the identity $\int_{-\infty}^{\infty} e^{-x^2}\,dx = \sqrt{\pi}$. Let $I = \int_{-\infty}^{\infty} e^{-x^2}\,dx = \int_{-\infty}^{\infty} e^{-y^2}\,dy$. Then $I^2 = \left(\int_{-\infty}^{\infty} e^{-x^2}\,dx\right)\left(\int_{-\infty}^{\infty} e^{-y^2}\,dy\right) = \int_{-\infty}^{\infty}\int_{-\infty}^{\infty} e^{-(x^2+y^2)}\,dx\,dy$.

Change to polar coordinates, $x = r\cos\theta, y = r\sin\theta$, and show $I^2 = \pi$.

7.8.6 Show that convolution on $L^1(\mathbb{R})$ is commutative.

7.8.7 Show that for the Gauss kernel defined by Equation (7.33), $u(x,t) = G_t(x)$ is a solution of the differential equation

$$\frac{\partial u}{\partial t} = \frac{\partial^2 u}{\partial x^2}.$$

7.8.8 Show that the heat kernel on $\mathbb{R}$, as defined by Equation (7.35), forms a summability kernel as $t \to 0^+$.

7.8.9 Continuing with Exercise 7.8.8, show that $u(x,t) = H_t(x)$ satisfies the following differential equation (called the **heat equation**):

$$\frac{\partial u}{\partial t} = D\frac{\partial^2 u}{\partial x^2}.$$

7.8.10 Continuing with Exercise 7.8.8, show that $\{H_t\}_{t>0}$ is an approximate identity as $t \to 0^+$ for $f \in L^1(\mathbb{R})$.

7.8.11 Define the Gauss–Weierstrass kernel on the reals by $W_t(x) = \frac{1}{t}e^{-\frac{\pi x^2}{t^2}}$ for $t > 0$. Show that this kernel forms a summability kernel as $t \to 0^+$.

7.9 Convergence of Fourier Series

Theorem 7.71 *Consider a trigonometric series* $\sum_{k=-\infty}^{\infty} c_k e^{ikx}$ *with coefficients that converge absolutely; namely,* $\sum_{k=-\infty}^{\infty} |c_k| < \infty$. *Then* $\sum_{k=-\infty}^{\infty} c_k e^{ikx}$ *is the Fourier series of a function* $f \in C_{2\pi}$, *and the Fourier series converges uniformly to* f.

Proof Recall the Weierstrass M-test which states that a series $\sum_{k=1}^{\infty} u_k(x)$ converges uniformly to some function f on I whenever whenever $|u_k(x)| \leq M_k$ on I $(k = 1, 2, \ldots)$ and $\sum_k M_k < \infty$. Here $u_k(x) = c_k e^{ikx}$ and $M_k = c_k$, so the trigonometric polynomials $s_n(x) = \sum_{k=-n}^{n} c_k e^{ikx}$ converge uniformly to some function f. Since trigonometric polynomials are continuous, the uniform limit f must be continuous. A uniform convergent series can be integrated term by term. Thus

$$\widehat{f}(m) = \frac{1}{2\pi}\int_{-\pi}^{\pi} f(x)e^{-imx}\,dx = \sum_{k=-\infty}^{\infty} \frac{1}{2\pi}\int_{-\pi}^{\pi} c_k e^{ikx}e^{-imx}\,dx = c_m. \quad ∎$$

Corollary 7.72 $\ell_1 \subset \widehat{C_{2\pi}} \subset \ell_2 = \widehat{L^2_{2\pi}} \subset L_{2\pi} \subset c_0$.

This Corollary follows from Theorem 7.71 and Corollary 7.34.

Theorem 7.73 *Suppose that $f \in L_{2\pi}$ is continuously differentiable. Then for all k, we have*

$$\widehat{f'}(k) = ik\widehat{f}(k).$$

Proof

$$\text{Integrate by parts:}\quad \widehat{f'}(k) = \frac{1}{2\pi}\int_{-\pi}^{\pi} f'(x)e^{-ikx}\,dx$$
$$= \frac{1}{2\pi} f(x)e^{-ikx}\Big]_{-\pi}^{\pi} + \frac{1}{2\pi}\int_{-\pi}^{\pi} f(x)e^{-ikx}ik\,dx$$
$$= ik\widehat{f}(k).$$

The first term on the second line vanishes because of 2π-periodicity. ∎

Theorem 7.74 *Suppose $f \in L_{2\pi}$ is twice continuously differentiable. Then the Fourier series of f converges absolutely and uniformly to f.*

Proof Applying the above theorem we obtain $\widehat{f''}(k) = (ik)^2\widehat{f}(k)$. Hence

$$k^2|\widehat{f}(k)| \le |\widehat{f''}(k)| = \left|\frac{1}{2\pi}\int_{-\pi}^{\pi} f''(x)e^{ikx}\,dx\right| \le \frac{1}{2\pi}\int_{-\pi}^{\pi} |f''(x)|\,dx \le M,$$

where $M = \sup_x |f''(x)|$. Since $\sum_k 1/k^2$ converges, the theorem follows. ∎

Corollary 7.75 $\widehat{C^2_{2\pi}} \subset \ell_1 \subset \widehat{C_{2\pi}} \subset \ell_2$.

The importance of the next theorem shows that the convergence of $s_n f(x)$ depends only on the values of f in some arbitrarily small neighborhood of x.

Theorem 7.76 Localization: *For every $f \in L_{2\pi}$ and $0 < \delta < \pi$ we have*

$$\lim_{n\to\infty}\int_{|t|>\delta} f(x-t)D_n(t)\,dt = 0.$$

Proof By Theorem 7.51, $D_n(t) = \frac{\sin\left(n+\frac{1}{2}\right)t}{\sin\frac{t}{2}}$. Using the trigonometric addition formula $\sin\left(n+\frac{1}{2}\right)t = \sin\frac{t}{2}\cos nt + \cos\frac{t}{2}\sin nt$, we obtain

$$D_n(t) = \cos nt + \cot\frac{t}{2}\sin nt.$$

Fix x and define

$$g_x(t) = \begin{cases} 0 & : \text{ if } |t| < \delta \\ f(x-t) & : \text{ if } \delta \le |t| \le \pi, \end{cases} \tag{7.36}$$

$$h_x(t) = \begin{cases} 0 & : \text{ if } |t| < \delta \\ f(x-t)\cot\frac{t}{2} & : \text{ if } \delta \le |t| \le \pi. \end{cases} \tag{7.37}$$

The function g_x is integrable and, because $\cot\frac{t}{2}$ is continuous when $\delta \le |t| \le \pi$, so is h_x. Also,

$$\int_{|t|>\delta} f(x-t)D_n(t)\,dt = \int_{-\pi}^{\pi} g_x(t)\cos nt\,dt + \int_{-\pi}^{\pi} h_x(t)\sin nt\,dt.$$

Each of these integrals is a Fourier coefficient (of the real form). By the Riemann–Lebesgue lemma (Theorem 7.49), the result follows. ∎

Corollary 7.77 *Suppose $f, g \in L_{2\pi}$ and $f = g$ in some open interval containing x. If the Fourier series of one of them converges at x, then so does the Fourier series of the other, and the two Fourier series are equal at x.*

Proof The function $f - g$ vanishes on some interval (a, b) containing x. Choose $\delta > 0$ such that $(x-\delta, x+\delta) \subset (a, b)$. Then

$$\frac{1}{2\pi}\int_{-\delta}^{\delta} \left[f(x-t) - g(x-t)\right]D_n(t)\,dt = 0.$$

Applying the localization theorem above to the function $f - g$ we have

$$\sum_{k=-\infty}^{\infty} \left[\widehat{f}(k) - \widehat{g}(k)\right]e^{ikx} = 0.$$ ∎

Definition 7.26 A function $f \in L_{2\pi}$ is said to satisfy the **Lipschitz condition of order $\alpha > 0$** if there exists $M > 0$ such that

$$|f(x+h) - f(x)| \le M|h|^{\alpha} \text{ for all } x \text{ and } h.$$

The space of all such functions is denoted by Lip α.

Theorem 7.78 *If $0 < \alpha \le \alpha' \le 1$, then $C^1_{2\pi} \subset \text{Lip}\,1 \subset \text{Lip}\,\alpha' \subset \text{Lip}\,\alpha \subset C_{2\pi}$.*

The proof is left as Exercise 7.9.1.

Definition 7.27 For $f \in L_{2\pi}$, let $\phi_x(t) = \dfrac{\left[f(x+t)-f(x)\right]+\left[f(x-t)-f(x)\right]}{2}$.

Lemma 7.79 *For every $f \in L_{2\pi}$ and $n = 0, 1, \ldots,$ we have*

$$s_n f(x) - f(x) = \frac{1}{2\pi}\int_{-\pi}^{\pi} \phi_x(t) D_n(t)\,dt.$$

Proof Since $\frac{1}{2\pi}\int_{-\pi}^{\pi} D_n(t)\,dt = 1$ for all n, Theorem 7.52 takes the form

$$\begin{aligned} s_n f(x) - f(x) &= (D_n * f)(x) - f(x) \\ &= \frac{1}{2\pi}\int_{-\pi}^{\pi} f(x-t) D_n(t)\,dt - f(x) \\ &= \frac{1}{2\pi}\int_{-\pi}^{\pi} \left[f(x-t) - f(x)\right] D_n(t)\,dt. \end{aligned}$$

Noting that $D_n(t) = D_n(-t)$, we have

$$\begin{aligned} s_n f(x) - f(x) &= (D_n * f)(x) - f(x) \\ &= \frac{1}{2\pi}\int_{-\pi}^{\pi} f(x+t) D_n(-t)\,dt - f(x) \\ &= \frac{1}{2\pi}\int_{-\pi}^{\pi} \left[f(x+t) - f(x)\right] D_n(t)\,dt. \end{aligned}$$

Take the average of these two to obtain the desired result. ∎

Theorem 7.80 *If $f \in \mathrm{Lip}\,\alpha$ for $\alpha > 0$, then the Fourier series of f converges to f for all x.*

Proof Since $f \in \mathrm{Lip}\,\alpha$ there exists $M > 0$ such that $|f(x+t) - f(x)| \leq M|t|^\alpha$ for all x. Clearly $|\phi_x(t)| \leq M|t|^\alpha$ as well. Fix the point x and let

$$F_x(t) = \frac{\phi_x(t)}{t} \quad \text{for} \quad t \neq 0,\ |t| < \pi \quad \text{and} \quad F(0) = 0.$$

Claim: F_x is integrable. For $\alpha > 0$, the function $g(t) = M|t|^{-1+\alpha}$ is integrable. Using Lebesgue's dominated convergence theorem (Theorem 2.41) on the functions $f_n(t) = \min_t\{F_x(t), n\} \leq g(t)$ proves the claim that $F_x \in L_{2\pi}$.

Let $\epsilon > 0$ be given. Choose $0 < \delta < \pi$ such that $\frac{2}{\pi}\int_{-\delta}^{\delta} M|t|^{-1+\alpha}\,dt < \epsilon/2$. By Theorem 7.79 we have

$$\begin{aligned} s_n f(x) - f(x) &= \frac{1}{2\pi} \int_{-\pi}^{\pi} \phi_x(t) D_n(t)\, dt \\ &= \frac{1}{2\pi} \int_{-\delta}^{\delta} \phi_x(t) D_n(t)\, dt + \frac{1}{2\pi} \int_{|t| \geq \delta} \phi_x(t) D_n(t)\, dt. \end{aligned}$$

The second integral tends to zero by the Localization theorem above. So for sufficiently large n, the second integral will be $< \epsilon/2$. For the first integral, we have

$$\begin{aligned} \frac{1}{2\pi} \left| \int_{-\delta}^{\delta} \phi_x(t) D_n(t)\, dt \right| &= \frac{1}{2\pi} \left| \int_{-\delta}^{\delta} \left[F_x(t) t \right] D_n(t)\, dt \right| \\ &\leq \frac{1}{2\pi} \int_{-\delta}^{\delta} \left| F_x(t) \frac{t}{\sin \frac{1}{2} t} \sin \left(n + \frac{1}{2} \right) t \right| dt \\ &\leq \frac{4}{2\pi} \int_{-\delta}^{\delta} \left| F_x(t) \right| dt \leq \frac{2}{\pi} \int_{-\delta}^{\delta} M |t|^{-1+\alpha}\, dt < \epsilon/2, \end{aligned}$$

because the function $\frac{t}{\sin \frac{1}{2} t}$ is continuous (if we define it to have the value 2 at $t = 0$) and it has an upper bound of 4 for $-\pi < t < \pi$. ∎

Remark The proof above can be modified to obtain not only convergence of the Fourier series everywhere, but uniform convergence as well.

Corollary 7.81 *If $f \in L_{2\pi}$ has a continuous derivative, its Fourier series converges to f.*

Although Theorems 7.71, 7.74, and 7.80 deal with functions $f \in L_{2\pi}$ whose Fourier series converge to f, a convergent Fourier series need not actually converge to f. A result dealing with such a situation is the following.

Recall the definition of one-sided limits, when they exist,

$$f(x+) = \lim_{t \to x^+} f(t) \quad \text{and} \quad f(x-) = \lim_{t \to x^-} f(t).$$

Theorem 7.82 *Consider a function $f \in L_{2\pi}$ and $x \in \mathbb{R}$ where $f(x+)$ and $f(x-)$ both exist. Suppose that for some $\alpha > 0$, there exists $M > 0$ such that*

$$\begin{aligned} &|f(x+t) - f(x+)| \leq M|t|^{\alpha} \quad \text{and} \\ &|f(x-t) - f(x-)| \leq M|t|^{\alpha} \quad \text{for} \quad t > 0. \end{aligned}$$

Then $\lim_{n \to \infty} s_n f(x) = \dfrac{f(x+) + f(x-)}{2}$.

The proof is a modification of the proofs of Lemma 7.79 and Theorem 7.80. It is left as Exercise 7.9.2.

Example 7.27 All of the functions considered in Section 7.6 have convergent Fourier series everywhere. In particular the function $f(x) = x$ on $(-\pi, \pi]$ (see Figure 7.6 in Section 7.5) has the Fourier series (see Example 7.21)

$$\sum_{k\neq 0} \frac{(-1)^{k+1}}{ik} e^{ikx} = f(x) = x \quad \text{for} \quad -\pi < x < \pi.$$

We have $f(\pi) = 1$, but the Fourier series converges to the value

$$\sum_{k\neq 0} \frac{(-1)^{k+1}}{ik} e^{ik\pi} = \frac{f(\pi+) + f(\pi-)}{2} = \frac{-\pi + \pi}{2} = 0.$$

Similarly, $f(-\pi) = 1$ but the Fourier series also converges to 0 there.

Although all functions f satisfying a Lipschitz condition of order $\alpha > 0$ are continuous with convergent Fourier series, this property cannot be extended to all continuous functions.

Theorem 7.83 *There exists a continuous function whose Fourier series diverges at $x = 0$.*

Proof Assume that the Fourier series of every continuous function converges at $x = 0$; that is, $\lim_{n\to\infty} s^n f(0)$ exists for all $f \in C_{2\pi}$. Consider the linear functional $T_n f = s^n f(0)$ on $C_{2\pi}$.

Since $|D_n(t)| = |1 + \sin t + \cdots + \sin nt| \leq 2n + 1$, we have

$$\begin{aligned} |T_n f| &= |s^n f(0)| = |(f * D_n)(0)| \\ &\leq \frac{1}{2\pi} \int_{-\pi}^{\pi} |f(t) D_n(t)|\, dt \leq (2n+1)\|f\|_\infty. \end{aligned}$$

Thus T_n is continuous for each n. If $T_n(f)$ converges for all $f \in C_{2\pi}$, then by the uniform boundedness principle (Theorem 6.30), $\sup_n \|T_n\| = M < \infty$.

We now show that $\|T_n\| = \|D_n\|_1$ for each $n = 1, 2, \ldots$. The inequality $\|T_n\| \leq \|D_n\|_1$ is clear since

$$|T_n f| = |s^n f(0)| = |(f * D_n)(0)| \leq \frac{1}{2\pi} \int_{-\pi}^{\pi} |f(t) D_n(t)|\, dt \leq \|D_n\|_1 \|f\|_\infty.$$

To prove the other inequality, let $h_n = \operatorname{sgn} D_n$ and let $\epsilon > 0$ be given. We can use Urysohn functions, as was done in the proof of Theorem 6.42 (Subsection 6.10.1), to find a continuous function $g \in C_{2\pi}$ such that $\|g\|_\infty = 1$ and $\|h_n - g\|_1 < \dfrac{\epsilon}{2n+1}$. Then

$$\begin{aligned}|T_n g| &= \left|\frac{1}{2\pi}\int_{-\pi}^{\pi} h_n(t)D_n(t)\,dt + \frac{1}{2\pi}\int_{-\pi}^{\pi}\big[g(t)-h_n(t)\big]D_n(t)\,dt\right| \\ &\geq \left|\frac{1}{2\pi}\int_{-\pi}^{\pi} h_n(t)D_n(t)\,dt\right| - \left|\frac{1}{2\pi}\int_{-\pi}^{\pi}\big[g(t)-h_n(t)\big]D_n(t)\,dt\right| \\ &\geq \|D_n\|_1 - (2n+1)\|g-h_n\|_1 \geq \|D_n\|_1 - \epsilon.\end{aligned}$$

Hence, $|T_n g| \geq \|D_n\|_1 = \|D_n\|_1\|g\|_\infty$, which proves $\|T_n\| \geq \|D_n\|_1$. Since we have shown in the proof of Theorem 7.64 that $\lim_{n\to\infty}\|D_n\| = \infty$, this completes the proof. ∎

Remark There is nothing special about the point $x = 0$. In fact one can find continuous functions whose Fourier series diverge in an uncountable set. However, Lennart Carleson[13] in 1966 showed that the points of divergence must be of measure zero.

For functions $f \in L_{2\pi}$, the situation is worse. It may happen that the Fourier series diverges everywhere.[14]

Exercises for Section 7.9

7.9.1 Prove Theorem 7.78.

7.9.2 Prove Theorem 7.82.

7.9.3 Use the result from Exercise 7.6.9 to compute the sum of $\frac{1}{k^2}$ for odd k

$$1 + \frac{1}{9} + \frac{1}{25} + \frac{1}{49} + \frac{1}{81} + \cdots.$$

7.10 Square Wave Fourier Series

Here we consider orthogonal systems consisting of step functions that assume only the values ± 1. Unlike the trigonometric system, these functions are not continuous and therefore not differentiable. Because square waves assume only two values, ± 1, they are useful tools in any kind of digital processing, for example, coding theory, pattern recognition, and digital recording of audio and video. In this section we consider the incomplete Rademacher system,

[13] Lennart Carleson (1928–) is a Swedish mathematician. In 1915 Nikolai Lusin asked whether $\widehat{f} \in \ell_2$ (that is, $f \in L^2_{2\pi}$) ensures the convergence almost everywhere of the Fourier series of f. This was considered the most important question in Fourier series for over 50 years. In 1966 Carleson showed that the answer is yes. His proof covers 22 journal pages.

[14] This was shown by Andrey Kolmogorov in 1926.

the complete Walsh system, and the complete Haar system. Each system has advantages depending on the application. For trigonometric functions, 2π-periodicity is notationally convenient, hence the spaces $L^2_{2\pi}$ and $L_{2\pi}$. For square waves, the spaces $L^2[0,1]$ and $L[0,1]$ are more appropriate.

7.10.1 Rademacher System

The Rademacher functions $\{r_n\}$ are discretized versions of sine functions

$$r_n(t) = \text{sign}[\sin(2^n \pi t)], \quad \text{for} \quad n = 0, 1, 2, \ldots, \text{ and } 0 \le t \le 1,$$

where $r_n(t)$ is given the value of the right-hand limit at discontinuities (instead of $\text{sign}\, 0 = 0$). If we partition the interval $[0,1]$ into 2^n equal subintervals, then r_n is the function that alternates between $+1$ and -1 on these subintervals. So, $r_0 \equiv 1$ on $[0,1]$, r_1 has the value 1 on $[0,1/2]$ and -1 on $(1/2,1]$, etc.

It is not difficult to verify (Exercise 7.10.1) that the Rademacher functions are orthonormal over $[0,1]$

$$\int_0^1 r_m(t) r_n(t)\, dt = \delta_{mn}.$$

However, they do not form a *total* orthonormal set (Definition 7.6). For example, $f(t) = \cos 2\pi t$ is a nonzero function in $L^2[0,1]$ which is orthogonal to all of the Rademacher functions. The details are left as Exercise 7.10.2. This means that the Rademacher functions do not form a *complete* orthonormal system. That is, there are functions $g \in L^2[0,1]$ which are *not* expressible in the form

$$g(t) = \sum_{k=0}^{\infty} c_k r_k(t) \quad \text{with} \quad \sum_{k=0}^{\infty} |c_k|^2 < \infty.$$

In spite of this, the Rademacher functions are useful in probability theory. Now we extend the Rademacher system to a complete orthonormal system.

7.10.2 Walsh System

The Walsh functions consist of all finite products of the Rademacher functions. Clearly there are countable many Walsh functions.

A convenient way to enumerate them is as follows. Write $n = 0, 1, 2, \ldots$, in binary form $n = b_k \cdots b_1$, with base 2 digits $b_k, \ldots, b_1$ of either 0 or 1. Then the Walsh function w_n is defined as the product of Rademacher functions $w_n(t) = r_k(t)^{b_k} \cdots r_1(t)^{b_1}$.

$$
\begin{aligned}
w_0 &= r_0 = 1 \\
w_1 &= r_1 \\
w_2 &= r_2^1 r_1^0 = r_2 \\
w_3 &= r_2^1 \cdot r_1^1 = r_2 \cdot r_1 \\
w_4 &= r_3^1 \cdot r_2^0 \cdot r_1^0 = r_3 \\
&\vdots \\
w_n &= r_k^{b_k} \cdots r_1^{b_1}
\end{aligned}
$$

Note that $w_{2^{k-1}} = r_k$ for $k = 1, 2, \ldots$. Also note that a product of two Walsh functions is a Walsh function and $w_n^2 = w_0$ for all $n = 0, 1, 2, \ldots$. It is then not difficult to verify that the Walsh functions are orthonormal (Exercise 7.10.4). For completeness of the orthonormal system, we show that the Walsh system is total (see Theorem 7.31).

Theorem 7.84 *The Walsh functions w_n for $n = 0, 1, 2, \ldots$ form a complete set in $L^2[0,1]$ under the real inner product $\langle f, g\rangle = \int_0^1 f(t)g(t)\,dt$.*

Proof We show that the Walsh functions form a total set in $L^2[0,1]$. Consider any $f \in L^2[0,1]$ and suppose $\langle f, w_n\rangle = \int_0^1 f(t)w_n(t)\,dt = 0$ for all $n = 0, 1, 2, \ldots$. Since f is Lebesgue integrable, we can define $F(x) = \int_0^x f(t)\,dt$, which is continuous on $[0,1]$ (Theorem 3.19). So, if we can show that $F(t) = 0$ on a dense subset of $[0,1]$, then $F \equiv 0$. By the fundamental theorem of calculus, $F'(t) = f(t)$ almost everywhere (Theorem 3.20). This will ensure that $f(t) = 0$ a.e., which makes f the zero function in $L^2[0,1]$. The following claim will complete the proof.

Claim: $F(x) = 0$ for all diadic fractions $x = k/2^m \in [0,1]$ with $m = 0, 1, 2, \ldots$ and $k = 0, \ldots, 2^m$. We use induction on m. Clearly $F(0) = 0$. Since $w_0 \equiv 1$, we also have $\langle f, w_0\rangle = \int_0^1 f(t)w_0(t)\,dt = F(1) = 0$. This completes the proof for the case $m = 0$. Next

$$
\begin{aligned}
\langle f, w_1\rangle = \int_0^1 f(t)w_1(t)\,dt &= \int_0^{1/2} f(t)\,dt - \int_{1/2}^1 f(t)\,dt \\
&= \big[F(1/2) - F(0)\big] - \big[F(1) - F(1/2)\big] \\
&= 2F(1/2) = 0. \quad \text{This completes the case } m = 1.
\end{aligned}
$$

$$\begin{aligned}\langle f, w_2\rangle &= \int_0^1 f(t)w_2(t)\,dt = \big[F(1/4) - F(0)\big] - \big[F(1/2) - F(1/4)\big] \\ &\quad + \big[F(3/4) - F(1/2)\big] - \big[F(1) - F(3/4)\big] \\ &= 2\big[F(1/4) + F(3/4)\big] = 0 \\ \langle f, w_3\rangle &= \int_0^1 f(t)w_3(t)\,dt = 2\big[F(1/4) - F(3/4)\big] = 0\end{aligned}$$

Combining $\langle f, w_2\rangle \pm \langle f, w_3\rangle$, we obtain $F(1/4) = F(3/4) = 0$, which completes case $m = 2$. Continuing this way, we can show that $F(k/2^m) = 0$ for all $k = 0, \ldots, 2^m$ implies $F(k/2^{m+1}) = 0$ for all $k = 0, \ldots, 2^{m+1}$. The details are left as Exercise 7.10.5. ∎

Theorem 7.85 *Every $f \in L^2[0,1]$ has a series representation as a Fourier series of Walsh functions*

$$f(t) = \sum_{k=0}^{\infty} c_k w_k(t) \quad \textit{with} \quad c_k = \int_0^1 f(t)w_k(t)\,dt \quad \textit{and} \quad \sum_{k=0}^{\infty} |c_k|^2 < \infty.$$

7.10.3 Haar System

This is another complete orthonormal sequence of square wave functions on the Hilbert space $L^2[0,1]$. Let H be the constant function $H(t) = 1$ on $[0,1]$ and define the function h on $\mathbb{R}$ to be

$$h(t) = \begin{cases} 1 & : \text{ if } 0 < t \le 1/2 \\ -1 & : \text{ if } 1/2 < t \le 1 \\ 0 & : \text{ otherwise.} \end{cases}$$

Then we define the **Haar functions**[15] $h_{jk}\colon [0,1] \longrightarrow \mathbb{R}$ for integers $j = 0, 1, 2, \ldots$ and $k = 0, 1, 2, \ldots, 2^j - 1$ as follows

$$h_{jk}(t) = 2^{\frac{j}{2}} h(2^j t - k).$$

An alternative definition is $h_{jk}(t) = \begin{cases} 2^{\frac{j}{2}} & : \text{ if } \frac{k}{2^j} < t \le \frac{k+\frac{1}{2}}{2^j} \\ -2^{\frac{j}{2}} & : \text{ if } \frac{k+\frac{1}{2}}{2^j} < t \le \frac{k+1}{2^j} \\ 0 & : \text{ otherwise.} \end{cases}$

[15] Named after the Hungarian mathematician Alfrèd Haar (1885–1933).

For each j and k, the function $h_{jk}(t)$ is a step function with support[16] on the interval $\left[\frac{k}{2^j}, \frac{k+1}{2^j}\right]$, having the value of $2^{j/2}$ on the first half of that interval $\left(\frac{k}{2^j}, \frac{k+1}{2^j}\right]$ and $-2^{j/2}$ on the second half. Note that $h(t) \equiv h_{00}(t)$.

Theorem 7.86 *The family $\{H, h_{jk}\}$ consisting of the constant function H along with the Haar functions h_{jk} for $j = 0, 1, 2, \ldots,\ k = 0, \ldots, 2^j - 1$ is orthonormal under the inner product $\langle f, g\rangle = \int_0^1 f(t)g(t)\,dt$.*

Proof Two different Haar functions h_{jk} and $h_{j'k'}$ have either disjoint supports or one will have support that is contained in an interval on which the other is constant. So $\langle h_{jk}, h_{j'k'}\rangle = 0$ for $j,k \neq j',k'$. Similarly $\langle H, h_{jk}\rangle = 0$ for all j,k. Finally, $\langle h_{jk}, h_{jk}\rangle = \int_{\frac{k}{2^j}}^{\frac{k+1}{2^j}} 2^j\,dt = 1$ for all j,k and $\langle H, H\rangle = 1$. ∎

Definition 7.28 A **diadic fraction** r in $[0,1]$ is a number of the form $r = \frac{k}{2^j}$ where $j = 0, 1, 2, \ldots,\ k = 0, \ldots, 2^j$. A **diadic step function** on $[0,1]$ is a function g for which there is a partition $\mathcal{P} = \{0 = t_0, t_1, \ldots, t_{n-1}, t_n = 1\}$ of diadic fractions such that g is constant on intervals $(0, t_1], (t_1, t_2], \ldots, (t_{n-1}, 1]$.

Theorem 7.87 *The diadic step functions form a vector space, say W. The space $W \subset L^2[0,1]$ is the linear span of the family of Haar functions $\{h_{jk}\}$ for $j = 0, 1, 2, \ldots,\ k = 0, \ldots, 2^j - 1$ along with the constant function H.*

Proof Since W is closed under addition and scalar multiplication, it is clearly a linear space. To show that W is the linear span of the system $\{H, h_{jk}\}$, it is sufficient to show that every $f \in W$ has a finite Fourier representation

$$f = \langle f, H\rangle H + \sum_j \sum_{k=0}^{2^j-1} \langle f, h_{jk}\rangle h_{jk} \text{ for finitely many } j. \tag{7.38}$$

Let f_r be the indicator function of the interval $(0, r]$ for diadic fraction $r = \frac{k}{2^j}$. We first show that every function f_r has such a representation (7.38). Indeed, it is easy to verify that for $j = 0$, we have $f_0 = 0$, $f_1 = H$. Then for $j = 1$, we have $f_{\frac{1}{2}} = \langle f_{\frac{1}{2}}, H\rangle H + \langle f_{\frac{1}{2}}, h_{00}\rangle h_{00} = \frac{1}{2}H + \frac{1}{2}h_{00}$. And for $j = 2$, we have

$$f_{\frac{1}{4}} = \frac{1}{4}H + \frac{1}{4}h_{00} + \frac{\sqrt{2}}{4}h_{10} + \frac{0}{4}h_{11}, \quad f_{\frac{3}{4}} = \frac{3}{4}H + \frac{1}{4}h_{00} + \frac{0}{4}h_{10} + \frac{\sqrt{2}}{4}h_{11}.$$

We then proceed by induction on j using the identity $f_r(t) = f_{2r}(2t)$. The details are left as Exercise 7.10.6. Once we have (7.38) for all functions

[16] The **support** of a function f is the closure of the set $\{t \mid f(t) \neq 0\}$.

$f_r = f_{\left(0,\frac{k}{2^j}\right]}$, we clearly have it for all functions $f_{(r,r']} = f_{r'} - f_r$. And then for all linear combinations, the diadic step functions. ∎

So far we have shown that $\{H, h_{jk}\}$ is a complete orthonormal system in the space of diadic step functions W.

Theorem 7.88 *The Haar functions h_{jk} for $j = 0, 1, 2, \ldots,\ k = 0, \ldots, 2^j - 1$ along with the constant function H, form a complete orthonormal set in the Hilbert space $L^2[0,1]$.*

Proof The diadic fractions are clearly dense in $[0,1]$. In particular, every indicator function $f_{[a,b]}$ of an interval $[a,b] \subset [0,1]$ is an increasing sequence of indicator functions $f_{[r_k, r'_k]}$, where $r_1, r_2, \ldots$ decreases to a and $r'_1, r'_2, \ldots$ increases to b.

Now we show that $\{H, h_{jk}\}$ is total in the Hilbert space $L^2[0,1]$. Suppose $f \in L^2[0,1]$ is orthogonal to H and to every h_{jk} (for $j = 0, 1, 2, \ldots$ and $k = 0, 1, \ldots, 2^j - 1$). Then f is orthogonal to the linear space W, which includes all diadic indicator functions $f_{[r,r']}$. Thus $\langle f, f_{[r,r']}\rangle = \int_r^{r'} f(t)\,dt = 0$ for all diadic fractions r, r'. By Lebesgue's dominated convergence theorem (Theorem 2.41) we have, using $|f|$ as the dominating function,

$$\int_a^b f(t)\,dt = \lim_{k\to\infty} \int_{r_k}^{r'_k} f(t)\,dt = 0 \text{ whenever } [r_k, r'_k] \to [a,b] \subset [0,1].$$

By Corollary 2.43, this proves $f = 0$, almost everywhere. ∎

Corollary 7.89 *Every $f \in L^2[0,1]$ can be written as a series expansion of Haar functions*

$$f(t) = c_0 + \sum_{j=0}^{\infty} \sum_{k=0}^{2^j-1} c_{jk} h_{jk}(t) \quad \textit{with} \quad \sum_{j=0}^{\infty} \sum_{k=0}^{2^j-1} |c_{jk}|^2 < \infty. \tag{7.39}$$

Here $c_0 = \widehat{f}(0) = \int_0^1 f(t)\,dt$ and $c_{jk} = \widehat{f}(jk) = \int_0^1 f(t) h_{jk}(t)\,dt$. Furthermore the convergence of the series is uniform for all continuous functions.

Note that, unlike the Haar series (7.39), not all orthonormal trigonometric series for continuous $f \in L_{2\pi}$ have uniform convergence.

Exercises for Section 7.10

7.10.1 Show that the Rademacher functions $\{r_n\}$ for $n = 0, 1, 2, \ldots$ are orthonormal in $L^2[0,1]$. That is, $\langle r_m, r_n\rangle = \int_0^1 r_m(t) r_n(t)\,dt = \delta_{mn}$.

7.10.2 Show that the sequence of Rademacher functions is not total by showing that the nonzero function $f(t) = \cos 2\pi t$ is in $L^2[0,1]$ and is orthogonal to all of the Rademacher functions.

7.10.3 Show that another nonzero function in $L^2[0,1]$ that is orthogonal to all of the Rademacher functions is the Walsh function $r_1 \cdot r_2$.

7.10.4 Show that the Walsh functions $\{w_n\}$ are orthonormal. You can use the fact that $\int_0^1 w_n(t)\,dt = \delta_{0n}$ for $n = 0, 1, 2, \dots$.

7.10.5 Complete the proof of the claim of Theorem 7.84. Show that $F(k/2^m) = 0$ for all $k = 0, \dots, 2^m$ implies $F(k/2^{m+1}) = 0$ for all $k = 0, \dots, 2^{m+1}$. Note that $F(k/2^m) = 0$ for $k = 0, 1, 2, \dots, 2^m$ implies $F(k/2^{m+1}) = 0$ for even $k = 0, 2, \dots, 2^{m+1}$. A proof of $F((b+1)/2^{m+1}) - F(b/2^{m+1}) = 0$ is then sufficient to complete the implication for odd k.

7.10.6 Prove the identity $f_r(t) = f_{2r}(2t)$ and prove the claim made in Theorem 7.87 that W contains the indicator functions $f_r = f_{(0,r]}$ for every diadic rational r. Hint: Let P_n be the statement that the indicator functions $f_{(0,\frac{k}{2^n}]}$ are finite linear combinations of the family $\{H, h_{jk}\}$ for every $k = 0, 1, \dots, 2^n - 1$. Confirm P_n for $n = 0, 1, 2$ and then show $P_n \Longrightarrow P_{n+1}$.

7.10.7 Show that every continuous function on $[0,1]$ can be uniformly approximated by diadic step functions.

7.10.8 Prove that for continuous functions $f \in C[0,1]$, the convergence of the series (7.39) of Corollary 7.89 is uniform.

*7.11 Fourier Transform

The study of the Fourier transform is an optional section here, not because it is hard, but because it extends beyond the scope of discrete functional analysis. It is included, however, because of its beauty and because it shows where the subject goes after the study of Fourier series.

So far, we have studied how functions can be associated with infinite sequences and series. For example, each $f \in L^2_{2\pi}$ is associated with a unique sequence of Fourier coefficients $\widehat{f} \in \ell_2$. And conversely, from each $c \in \ell_2$, we can recover a function $f \in L^2_{2\pi}$, for which $\widehat{f} = c$.

In this section we study how a function $f \in L^1(\mathbb{R})$, is associated with its Fourier transform, which is not a sequence, but rather another function

$\widehat{f} \in L^1(\mathbb{R})$. Conversely, there is an inversion theorem which permits the recovery of the function f from its Fourier transform $\widehat{f} \in L^1(\mathbb{R})$.

Definition 7.29 The **Fourier transform** of a function $f \in L^1(\mathbb{R})$ is the function $\widehat{f}$ given by the formula

$$\widehat{f}(s) = \int_{-\infty}^{\infty} f(t)e^{-ist}\,dt \text{ for all } s \in \mathbb{R}. \tag{7.40}$$

Theorem 7.90 *For any $f \in L^1(\mathbb{R})$, the Fourier transform $\widehat{f}$ has the following properties.*

(FT1) *$\widehat{f}(s)$ is defined for all $s \in \mathbb{R}$.*
(FT2) *$\widehat{f} \in L^1(\mathbb{R})$ with $\|\widehat{f}\|_1 = \|f\|_1$.*
(FT3) *$\widehat{f}$ is bounded, and $\widehat{f} \in L^2(\mathbb{R})$ with $\|\widehat{f}\|_2 \leq \|f\|_1$.*
(FT4) *$\widehat{f}$ is continuous.*

Proof
(FT1): We have $|f(t)e^{-ist}| = |f(t)|$ for all s, t. If $f \in L^1(\mathbb{R})$, then $f(t)e^{-ist}$ is integrable for all $s \in \mathbb{R}$. So $\widehat{f}(s)$ is defined for all $s \in \mathbb{R}$.

(FT2): $\|\widehat{f}\|_1 = \int_{-\infty}^{\infty} |f(t)e^{-ist}|\,dt = \int_{-\infty}^{\infty} |f(t)|\,dt = \|f\|_1 < \infty.$

(FT3): $|\widehat{f}(s)| = \left|\int_{-\infty}^{\infty} f(t)e^{-ist}dt\right| \leq \int_{-\infty}^{\infty} |f(t)|dt = \|f\|_1$ for all $s \in \mathbb{R}$.

Since $|\widehat{f}| \leq \|f\|_1$, we have $|\widehat{f}|^2 \leq \|f\|_1 \cdot |\widehat{f}|$. So, $\|\widehat{f}\|_2 \leq \|f\|_1 < \infty$.

(FT4): We show that $\widehat{f}$ is sequentially continuous, which is equivalent to continuity (Theorem 5.26). For any $s \in \mathbb{R}$, with $s_k \to s$, define $f_k(t) = f(t)e^{-is_k t}$. Then $\lim_{k\to\infty} f_k(t) = f(t)e^{-ist}$. By Lebesgue's dominated convergence theorem (Theorem 2.41), with $|f|$ as the dominating integrable function, we have

$$\widehat{f}(s) = \int_{-\infty}^{\infty} f(t)e^{-ist}dt = \lim_{k\to\infty} \int_{-\infty}^{\infty} f_k(t)\,dt = \lim_{k\to\infty} \widehat{f}(s_k). \qquad \blacksquare$$

Lemma 7.91 *If $f(t) = e^{-at^2}$ for $a > 0$, then $\widehat{f}(s) = \int_{-\infty}^{\infty} e^{-au^2}e^{-isu}\,du = \sqrt{\frac{\pi}{a}}e^{\frac{-s^2}{4a}}$.*

Proof Start with the integral $\int_{-\infty}^{\infty} e^{-au^2} e^{-isu}\, du$, combine the exponentials, and then complete the square in the exponent of e:

$$\begin{aligned}\int_{-\infty}^{\infty} e^{-au^2} e^{-isu}\, du &= \int_{-\infty}^{\infty} e^{-a\left(u^2 + \frac{iu}{a} s\right)}\, du \\ &= \int_{-\infty}^{\infty} e^{-a\left(u^2 + \frac{iu}{a} s + \left(\frac{i}{2a}\right)^2 s^2\right)} e^{a\left(\frac{i}{2a}\right)^2 s^2}\, du \\ &= e^{\frac{-s^2}{4a}} \int_{-\infty}^{\infty} e^{-a\left(u^2 + \frac{iu}{a} s + \left(\frac{i}{2a}\right)^2 s^2\right)}\, du.\end{aligned}$$

Now make the substitution $v = u + \frac{i}{2a}s$ and $dv = du$ to obtain

$$\int_{-\infty}^{\infty} e^{-au^2} e^{-isu}\, du = e^{\frac{-s^2}{4a}} \int_{-\infty}^{\infty} e^{-av^2}\, dv.$$

This last integral is $\sqrt{\frac{\pi}{a}}$ by identity (7.34) of Definition 7.25. So the theorem follows. ∎

Theorem 7.92 Fourier inversion: *Suppose $f \in L^1(\mathbb{R})$ and $\widehat{f} \in L^1(\mathbb{R})$. If f is continuous, then*

$$f(t) = \frac{1}{2\pi} \int_{-\infty}^{\infty} \widehat{f}(s) e^{ist}\, ds. \tag{7.41}$$

At this point, please notice the beautiful symmetry between the Fourier transform formula (7.40) and the Fourier inversion formula (7.41):

$$\widehat{f}(s) = \int_{-\infty}^{\infty} f(t) e^{-ist} dt, \qquad f(t) = \frac{1}{2\pi} \int_{-\infty}^{\infty} \widehat{f}(s) e^{ist}\, ds.$$

Proof Here we use the Gauss kernel functions $G_a(t) = \frac{1}{\sqrt{4\pi a}} e^{\frac{-t^2}{4a}}$ as defined in Subsection 7.8.3. Let $F(t,a) = \frac{1}{2\pi} \int_{-\infty}^{\infty} \widehat{f}(u) e^{iut} e^{-au^2}\, du$. Then

$$\begin{aligned}F(t,a) &= \frac{1}{2\pi} \int_{-\infty}^{\infty} \left[\int_{-\infty}^{\infty} f(s) e^{-ius}\, ds\right] e^{iut} e^{-au^2}\, du \\ &= \frac{1}{2\pi} \int_{-\infty}^{\infty} \int_{-\infty}^{\infty} f(s) e^{-au^2} e^{-iu(s-t)}\, ds\, du \\ &= \frac{1}{2\pi} \int_{-\infty}^{\infty} f(s) \left[\int_{-\infty}^{\infty} e^{-au^2} e^{-iu(s-t)}\, du\right] ds.\end{aligned}$$

The interchange of integrals is permitted because $|f(s)e^{-au^2}e^{-iu(s-t)}| \le |f(s)|$ for $a > 0$. We now apply Lemma 7.91 to the inner integral,

$$F(t,a) = \int_{-\infty}^{\infty} f(s)\left[\frac{1}{\sqrt{4\pi a}}e^{\frac{-(s-t)^2}{4a}}\right] ds = \int_{-\infty}^{\infty} f(s)G_a(s-t)\, ds.$$

Thus $F(t,a) = f * G_a(t)$. The Gauss kernel is an approximate identity on $\mathbb{R}$ as $a \to 0^+$. That is, $\lim_{a\to 0+} \|f * G_a - f\|_1 = 0$. Since f is continuous, we have $\lim_{a\to 0+} F(t,a) = \lim_{a\to 0+} f * G_a(t) = f(t)$.

Finally, since $|\widehat{f}(u)e^{iut}e^{-au^2}| \le |\widehat{f}(u)|$, it follows that $F(t,a)$ is a continuous function of t and a. Thus

$$f(t) = \lim_{a\to 0+} F(t,a) = F(t,0) = \frac{1}{2\pi}\int_{-\infty}^{\infty} \widehat{f}(u)e^{iut}\, du. \qquad \blacksquare$$

Theorem 7.93 Parseval's theorem: *Suppose $f,g \in L^1(\mathbb{R})$ and $f,g \in L^2(\mathbb{R})$. Then $\langle f,g\rangle = \dfrac{1}{2\pi}\langle \widehat{f},\widehat{g}\rangle$.*

Proof Suppose $f,g \in L^1(\mathbb{R})$ and $f,g \in L^2(\mathbb{R})$. Using the Fourier transform and the inversion theorem, we have

$$\begin{aligned}
\langle f,g\rangle &= \int_{-\infty}^{\infty} f(t)\overline{g(t)}\, dt = \frac{1}{2\pi}\int_{-\infty}^{\infty} f(t)\overline{\int_{-\infty}^{\infty} \widehat{g}(s)e^{ist}\, ds}\, dt \\
&= \frac{1}{2\pi}\int_{-\infty}^{\infty} f(t)\int_{-\infty}^{\infty} \overline{\widehat{g}}(s)e^{-ist}\, ds\, dt \\
&= \frac{1}{2\pi}\int_{-\infty}^{\infty}\left(\int_{-\infty}^{\infty} f(t)e^{-ist}\, dt\right)\overline{\widehat{g}}(s)\, ds \\
&= \frac{1}{2\pi}\int_{-\infty}^{\infty} \widehat{f}(s)\overline{\widehat{g}}(s)\, ds.
\end{aligned} \qquad \blacksquare$$

Theorem 7.94 Plancherel: *Suppose $f \in L^1(\mathbb{R})$ and $f \in L^2(\mathbb{R})$. Then $\|f\|_2 = \frac{1}{\sqrt{2\pi}}\|\widehat{f}\|_2$.*

Proof Using Parseval's theorem (Theorem 7.93) and taking $g = f$, we obtain

$$\|f\|_2 = \sqrt{\langle f,f\rangle} = \sqrt{\frac{1}{2\pi}\langle \widehat{f},\widehat{f}\rangle} = \frac{1}{\sqrt{2\pi}}\|\widehat{f}\|_2. \qquad \blacksquare$$

Exercises for Section 7.11

7.11.1 Show that the Fourier transform is linear. That is, for all α, $\beta \in \mathbb{R}$ and all f, $g \in L^1(\mathbb{R})$, the Fourier transform of $\alpha f + \beta g$ is $\alpha \widehat{f} + \beta \widehat{g}$.

7.11.2 What can be said about the Fourier transform of a function f if it is known that (a) f is even, (b) f is odd, or (c) f is real valued?

7.11.3 Suppose $f \in L^1(\mathbb{R})$ and $\widehat{f}(s) = 0$ for all s. What can be said about f?

7.11.4 Prove $\widehat{f}(0) = \int_{-\infty}^{\infty} f(u)\, du$ for all $f \in L^1(\mathbb{R})$.

7.11.5 Prove $\|\widehat{f}\|_\infty \le \|f\|_1$ for all $f \in L^1(\mathbb{R})$.

7.11.6 Let f be the indicator (characteristic) function of the interval $(-\pi, \pi)$. Find the Fourier transform $\widehat{f}$ of f.

7.11.7 Find the Fourier transform $\widehat{f}(s)$ of

$$f(t) = \begin{cases} 1 - |t| & : \quad \text{if } -\pi \le t \le \pi \\ 0 & : \quad \text{elsewhere} \end{cases}.$$

7.11.8 Find the Fourier transform of $f(t) = \frac{1}{t^2+1}$.

7.11.9 Find the Fourier transform of $f(t) = \frac{t}{t^2+a^2}$ for $a \in \mathbb{R}$.

7.11.10 Find the Fourier transform $\widehat{f}(s)$ of the function $f(t) = e^{-|t|}$.

7.11.11 Find the Fourier transform $\widehat{f}(s)$ of $f(t) = e^{-t^2}$. Without using the inversion theorem, find the Fourier transform of $\widehat{f}(s)$.

7.11.12 Find the Fourier transform $\widehat{f}(s)$ of $f(t) = \frac{\sin t}{t}$. Without using the inversion theorem, find the Fourier transform of $\widehat{f}(s)$.

7.11.13 Show that the Fourier transform of $g(t) = f'(t)$ is $\widehat{g} = is\widehat{f}(s)$.

7.11.14 Show that the Fourier transform of $g(t) = tf(t)$ is $\widehat{g} = i(\widehat{f})'(s)$.

7.11.15 For $f \in L^1(\mathbb{R})$ and $T_a f(t) = f(t-a)$, show that $\widehat{T_a f}(s) = e^{-isa}\widehat{f}(s)$.

7.11.16 Suppose $f \in L^1(\mathbb{R})$ and $g(t) = e^{iat} f(t)$. Show that $\widehat{g}(s) = T_a \widehat{f}(s)$.

7.11.17 Prove $\widehat{f * f} = (\widehat{f})^2$ for all $f \in L^1(\mathbb{R})$.

7.11.18 Suppose $f \in L^1(\mathbb{R})$. Prove that if $\widehat{f * f} = 0$, then $f = 0$.

7.11.19 Prove $\widehat{f * g} = \widehat{f} \cdot \widehat{g}$ for all $f, g \in L^1(\mathbb{R})$.

7.11.20 Give examples of two nonzero functions $f, g \in L^1(\mathbb{R})$ such that $\widehat{f * g} = 0$.

7.11.21 Suppose $f \in L^1(\mathbb{R})$. If $\widehat{f} = 0$, then $f = 0$, almost everywhere. In the case when f is continuous, then $f = 0$ everywhere.

7.11.22 Given $f, g \in L^1(\mathbb{R})$ and $\widehat{f} = \widehat{g}$, prove that $f = g$, almost everywhere.

8
Applications

In this chapter we give some applications of functional analysis. Many of the topics merit entire books by themselves but here we give only an introduction to each.

8.1 Least Squares Methods

8.1.1 Overdetermined Systems

As we know from linear algebra, a system of equations n equations with m unknowns

$$\begin{aligned} a_{11}x_1 + a_{12}x_2 + \cdots + a_{1m}x_m &= y_1 \\ a_{21}x_1 + a_{22}x_2 + \cdots + a_{2m}x_m &= y_2 \\ \vdots \quad &= \vdots \\ a_{n1}x_1 + a_{n2}x_2 + \cdots + a_{nm}x_m &= y_n \end{aligned}$$

can be written in matrix form $AX = Y$ as

$$AX = \begin{pmatrix} a_{11} & a_{12} & \cdots & a_{1m} \\ a_{21} & a_{22} & \cdots & a_{2m} \\ \vdots & \vdots & \vdots & \vdots \\ a_{n1} & a_{n2} & \cdots & a_{nm} \end{pmatrix} \begin{pmatrix} x_1 \\ x_2 \\ \vdots \\ x_m \end{pmatrix} = \begin{pmatrix} y_1 \\ y_2 \\ \vdots \\ y_n \end{pmatrix} = Y.$$

Here, A is an $n \times m$ matrix, X is an $m \times 1$ column vector, and Y is an $n \times 1$ column. Let us use the notation $a^1, a^2, \ldots, a^m$ to denote the m column vectors of A. If the matrix A is invertible (which requires $m = n$), the system has a unique solution $X = A^{-1}Y$.

If there are more equations than unknowns ($n > m$), the system is said to be **overdetermined**. Typically there is no solution to such a system of equations, in which case, $Y \neq AX$ for all possible vectors of unknowns X. When this happens, it is often desirable to find an approximate solution. A standard way is to find $\widehat{Y} = AX$, for some values X, for which the **residual vector** $E = Y - \widehat{Y}$ has the minimum Euclidean (ℓ_2) norm. For this we need a vector X which minimizes the **sum of squares**

$$SOS = \|Y - \widehat{Y}\|_2^2 = \|Y - AX\|_2^2 = \sum_{j=1}^{n} \left(y_j - \sum_{k=1}^{m} a_{jk} x_k \right)^2 .$$

Such an X is called a **least squares** solution of the system of equations.

Theorem 8.1 *Suppose $AX = Y$ represents an overdetermined system of equations, where A is an $n \times m$ matrix with columns $a^1, a^2, \ldots, a^m$ in $\mathbb{R}^n$ (or $\mathbb{C}^n$), X in an $m \times 1$ column vector of unknowns, and Y is an $n \times 1$ column of constants. Using $A^H = \overline{A^T}$ to denote the conjugate (Hermitian) transpose of A, then any least squares solution of the system satisfies the **normal matrix equation***

$$(A^H A)X = A^H Y. \tag{8.1}$$

Further, if the columns of A are linearly independent, then $A^H A$ is invertible and there is a unique least squares solution

$$X = (A^H A)^{-1} A^H Y.$$

*The matrix $(A^H A)^{-1} A^H$ is called the **generalized inverse** of the overdetermined matrix A.*

Proof Two conditions must be satisfied for a least squares solution $\widehat{Y} = AX$.

First: There is a column vector X for which $\widehat{Y} = AX$; that is, $\widehat{Y}$ is a linear combination $\widehat{Y} = x_1 a^1 + \cdots + x_m a^m$ of the column vectors of A.

Second: Of all the linear combinations AX of the columns of A, $\widehat{Y}$ is the one with the minimum Euclidean (ℓ_2) distance to Y.

By the orthogonal projection theorem (Theorem 7.12, Section 7.1), the vector $\widehat{Y}$ is the orthogonal projection of the vector Y onto the space spanned by the column vectors of A. In this case, the residual vector $E = Y - \widehat{Y} = (e_1, \ldots, e_n)$ is orthogonal to the column vectors of A (Theorem 7.13). So the inner (dot) products $\langle E, a^k \rangle = E \cdot a^k = e_1\overline{a_{1k}} + \cdots + e_n\overline{a_{nk}}$ of E with all of the columns a^k of A must be zero. This means $A^H E = 0$.

Combining these two conditions, gives us

$$A^H E = A^H(Y - \widehat{Y}) = A^H(Y - AX) = 0,$$

which results in the normal matrix equation $(A^H A)X = A^H Y$.

If the $m \times m$ matrix $A^H A$ is invertible, as will happen whenever the m columns of A are linearly independent (Lemma 7.15), then the unique least squares solution is

$$X = (A^H A)^{-1} A^H Y. \qquad \blacksquare$$

Example 8.1 Let us find the least squares solution of the system of equations

$$\begin{aligned} x - y &= 0, \\ 2x + y &= 2, \\ x + y &= 1. \end{aligned}$$

The matrix equation is

$$\begin{pmatrix} 1 & -1 \\ 2 & 1 \\ 1 & 1 \end{pmatrix} \begin{pmatrix} x \\ y \end{pmatrix} = \begin{pmatrix} 0 \\ 2 \\ 1 \end{pmatrix}.$$

The normal matrix equation $(A^H A)X = A^H Y$ is

$$\begin{pmatrix} 1 & 2 & 1 \\ -1 & 1 & 1 \end{pmatrix} \begin{pmatrix} 1 & -1 \\ 2 & 1 \\ 1 & 1 \end{pmatrix} \begin{pmatrix} x \\ y \end{pmatrix} = \begin{pmatrix} 1 & 2 & 1 \\ -1 & 1 & 1 \end{pmatrix} \begin{pmatrix} 0 \\ 2 \\ 1 \end{pmatrix},$$

$$\begin{pmatrix} 6 & 2 \\ 2 & 3 \end{pmatrix} \begin{pmatrix} x \\ y \end{pmatrix} = \begin{pmatrix} 5 \\ 3 \end{pmatrix}.$$

Gaussian elimination gives us $\begin{pmatrix} x \\ y \end{pmatrix} = \frac{1}{14} \begin{pmatrix} 9 \\ 8 \end{pmatrix}$.

For this purpose, Gaussian elimination is simpler than actually finding the generalized inverse. However, the generalized inverse matrix can be computed as well:

$$(A^T A)^{-1} A^T = \begin{pmatrix} 6 & 2 \\ 2 & 3 \end{pmatrix}^{-1} \begin{pmatrix} 1 & 2 & 1 \\ -1 & 1 & 1 \end{pmatrix} = \frac{1}{14} \begin{pmatrix} 5 & 4 & 1 \\ -8 & 2 & 4 \end{pmatrix}.$$

Exercises for Subsection 8.1.1

8.1.1 Find the least squares solution of the system of two equations

$$2x = 5 \quad \text{and} \quad 3x = 5.$$

8.1.2 Let $y_1, y_2, \ldots, y_n$ be fixed reals. Find the least squares solution of the system of n equations $x = y_1,\ x = y_2,\ \ldots,\ x = y_n$.

8.1.3 Find the least squares solution of this system with two unknowns:

$$y = 1 - i,\ \ x = 2 + i,\ \ y = 3 - i,\ \ x = 6 - i.$$

8.1.4 Find the least squares solution of this system of three equations:

$$\begin{aligned} x + y &= 1, \\ x - 2y &= 2, \\ 2x + 3y &= 3. \end{aligned}$$

8.1.5 Find the generalized inverse of $A = \begin{pmatrix} 1 & 1 \\ 1 & -2 \\ 2 & 3 \end{pmatrix}$ (from Exercise 8.1.4).

8.1.2 Simple Regression

As another form of least squares approximation, let us now consider n data points $(x_1, y_2), (x_2, y_2), \ldots, (x_n, y_n)$ in the real plane $\mathbb{R}^2$ that we wish to model by a line of the form $y = b_0 + b_1 x$. The variable x is called the **independent variable** (also called the **predictor variable**) and y is called the **response variable**. For each $k = 1, 2, \ldots, n$, the **measured value** of the response is y_k and the **predicted value** (or **fitted value**) is $\widehat{y}_k = b_0 + b_1 x_k$. The accuracy of each of the predictions is measured by the **residual** $e_k = y_k - \widehat{y}_k$. The least squares **fitted line** $\widehat{y} = b_0 + b_1 x$, also called the **regression line of** y **upon** x, is the one which minimizes the sum of squares of the residuals $e_1^2 + e_2^2 + \cdots + e_n^2$.

Solution In simple regression, we are looking for the least squares solution of the overdetermined system of equations

$$\begin{aligned} b_0 + b_1 x_1 &= y_1 \\ b_0 + b_1 x_2 &= y_2 \\ \vdots \quad &= \vdots \\ b_0 + b_1 x_n &= y_n. \end{aligned}$$

Table 8.1. *Age x in months versus height y in centimeters*

Child ID	x	y
1	24	87
2	48	101
3	60	120
4	96	159
5	63	135
6	39	104
7	63	126
8	39	96
Totals	432	928

The matrix form of this system of equations is

$$A\,B = Y$$

$$\begin{pmatrix} 1 & x_1 \\ 1 & x_2 \\ \vdots & \vdots \\ 1 & x_n \end{pmatrix} \begin{pmatrix} b_0 \\ b_1 \end{pmatrix} = \begin{pmatrix} y_1 \\ y_2 \\ \vdots \\ y_n \end{pmatrix}.$$

According to the preceding section, the solution is given by the normal matrix equation (since the data values are real numbers, $A^H = A^T$),

$$(A^T A)B = A^T Y.$$

Alternatively, the generalized inverse is $B = \begin{pmatrix} b_0 \\ b_1 \end{pmatrix} = (A^T A)^{-1} A^T Y.$

Example 8.2 Age and heights of children: Consider Table 8.1 of ages (in months) and heights (in centimeters) of eight children:

Here we consider age to be the independent variable and height the response variable because we want to use age to predict height.[1] Thus we want to find

[1] **Caution:** The regression line of x upon y is a different line than the regression line of y upon x. If we were trying to guess a child's age, we would be doing informal regression by taking into consideration many factors, including height. In that case, height is an independent variable and age is the response variable. The regression line of x upon y is then the fitted line $x = a_0 + a_1 y$ that minimizes the sum of squares of the horizontal residuals $x_k - \widehat{x}_k = x_k - (a_0 + a_1 y_k)$. So, depending on the purpose of the fitted line, it is important to correctly identify one of the variables to be the predictor and the other to be the response.

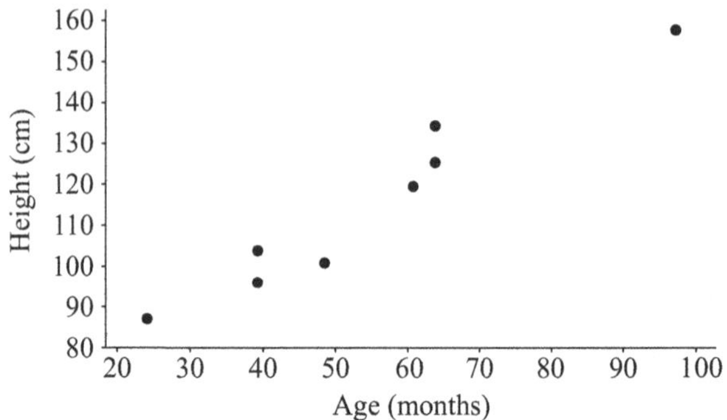

Figure 8.1 Scatter plot of height versus age for eight children

the fitted line $\widehat{y} = b_0 + b_1 x$. We plot the eight pairs of data points (x, y) in Figure 8.1. This is called the **scatter plot** of the data.

The data create an overdetermined system of equations $AB = Y$:

$$\begin{pmatrix} 1 & 24 \\ 1 & 48 \\ 1 & 60 \\ 1 & 96 \\ 1 & 63 \\ 1 & 39 \\ 1 & 63 \\ 1 & 39 \end{pmatrix} \begin{pmatrix} b_0 \\ b_1 \end{pmatrix} = \begin{pmatrix} 87 \\ 101 \\ 120 \\ 159 \\ 135 \\ 104 \\ 126 \\ 96 \end{pmatrix}.$$

The normal equation $(A^T A)B = A^T Y$ can easily be computed to be:

$$\begin{pmatrix} 8 & 432 \\ 432 & 26676 \end{pmatrix} \begin{pmatrix} b_0 \\ b_1 \end{pmatrix} = \begin{pmatrix} 928 \\ 53643 \end{pmatrix}.$$

Triangulation of the augmented matrix by row reduction leads to

$$\left(\begin{array}{cc|c} 1 & 54 & 116 \\ 0 & 3348 & 3531 \end{array}\right),$$

which has solution $b_1 = \frac{3531}{3348} = 1.055$ and $b_0 = 116 - 54b_1 = \frac{3661}{62} = 59.048$.

Table 8.2 shows the data along with the fitted (predicted) values and residual values, and the minimized sum of squares (SOS).

Note that no other values of b_0 and b_1 will give a smaller SOS. Figure 8.2 shows the regression line $\widehat{y} = 59.048 + 1.055x$ of y upon x superimposed on the scatter plot.

Table 8.2. *Ages x in months versus heights y in centimeters*

Child ID	x	y	$\hat{y}$	$e = y - \hat{y}$	e^2
1	24	87	84.36	+2.64	6.97
2	48	101	109.67	−8.67	75.20
3	60	120	122.33	−2.33	5.42
4	96	159	160.30	−1.20	1.68
5	63	135	125.49	+9.51	90.40
6	39	104	100.18	−3.82	14.59
7	63	126	125.49	+0.51	0.26
8	39	96	100.18	−4.18	17.47
Totals	432	928	928	0	$SOS = 212.00$

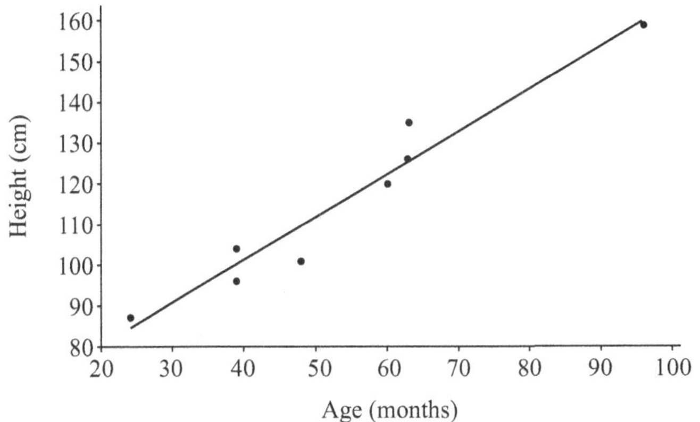

Figure 8.2 Scatter plot of age versus height of children, with regression line superimposed

Exercises For Subsection 8.1.2

8.1.6 Consider Example 8.2. The regression line $\widehat{y} = 59.048 + 1.055x$ is the line $y = b_0 + b_1x$ which gives the smallest possible SOS (as computed in Table 8.2). Compute the SOS for the line $y = 59 + x$ and confirm that its SOS is larger than 212.00.

8.1.7 Find the normal matrix equation $(A^T A)B = A^T Y$ for the three data points

$$(0, 2), (2, 1), (4, 3).$$

8.1.8 Use row reduction to solve the normal matrix equation for the data points given in Exercise 8.1.7. Then give an equation for the regression line (of y upon x). Superimpose a graph of the regression line on a scatter plot of the three data points.

8.1.9 Find the normal matrix equation $(A^T A)B = A^T Y$ for the four data points

$$(-1,0),\ (0,1),\ (1,1),\ (2,2).$$

8.1.10 Use row reduction to solve the normal matrix equation for the four data points given in Exercise 8.1.9. Then give an equation for the regression line (of y upon x). Superimpose a graph of the regression line on a scatter plot of the four data points.

8.1.11 Given the three data points $(0,6)$, $(1,0)$, $(3,0)$, find an equation for the regression line of y upon x. Superimpose a graph of the regression line on the scatter plot of these three data points.

8.1.12 Consider the following scores of six students in a math class, where x denotes the score on the first test and y that of the second test:

	Student ID					
	1	2	3	4	5	6
x:	30	50	60	70	60	90
y:	65	70	60	65	50	80

Set x to be the independent variable and y to be the response variable.

a. Draw a scatter plot. **b.** Find the normal matrix equation. **c.** Compute the regression line of y upon x. **d.** Superimpose the graph of the regression line onto the scatter plot.

8.1.13 Consider the following data points, with independent x and response y:

$x:$	10	20	25	25	30	40
$y:$	7	5	4	3	4	1

a. Draw a scatter plot. **b.** Find the normal matrix equation. **c.** Compute the regression line of y upon x. **d.** Superimpose the graph of the regression line onto the scatter plot.

8.1.3 Multiple Regression

The principles of simple regression as presented in Subsection 8.1.2 can be applied to many other types of data. An important case is when the data consists of one response variable y and several independent (predictor) variables. For example, we may want to model the systolic blood pressures y of male subjects using the independent variables of weight w, age x, and serum cholesterol z. Given n data points $(w_1, x_1, z_1, y_1), \ldots, (w_n, x_n, z_n, y_n)$, we want to fit the data to an equation

$$b_0 + b_1 w + b_2 x + b_3 z = y$$

predicting the blood pressure (response) y for independent variables w, x, and z.

In this example of three independent variables $\{x, y, z\}$, we have an overdetermined system of equations (provided $n > 4$)

$$\begin{aligned} b_0 + b_1 w_1 + b_2 x_1 + b_3 z_1 &= y_1 \\ b_0 + b_1 w_2 + b_2 x_2 + b_3 z_2 &= y_2 \\ \vdots \quad &= \vdots \\ b_0 + b_1 w_n + b_2 x_n + b_3 z_n &= y_n. \end{aligned}$$

This system can be written in matrix form $AB = Y$ as follows

$$AB = \begin{pmatrix} 1 & w_1 & x_1 & z_1 \\ 1 & w_2 & x_2 & z_2 \\ \vdots & \vdots & \vdots & \vdots \\ 1 & w_n & x_n & z_n \end{pmatrix} \begin{pmatrix} b_0 \\ b_1 \\ b_2 \\ b_3 \end{pmatrix} = \begin{pmatrix} y_1 \\ y_2 \\ \vdots \\ y_n \end{pmatrix} = Y.$$

The normal matrix equation is then

$$(A^T A)B = A^T Y \text{ (Note: } A^H = A^T \text{ because data values are real).}$$

Typically for real-world data, the sample size n is large and the columns of A are linearly independent, so row reduction can be used on the normal matrix equation to solve for the unique coefficient matrix B.

8.1.4 Curve Fitting

Another important application of least squares solutions of overdetermined systems is in fitting a curve to data. Given a set of data points (x_k, y_k) for $k = 1, 2, \ldots, n$, and a class of functions F, we want to solve the system of equations $f(x_k) = y_k$ for $k = 1, 2, \ldots, n$ for functions $f \in F$ in the class. In an overdetermined system, the residuals are the values $e_k = y_k - f(x_k)$ for $k = 1, 2, \ldots, n$. The least squares fit is a function $\widehat{f}$ in the class F which minimizes the sum of squares of residuals

$$SOS = \sum_k e_k^2 = \sum_k (y_k - \widehat{y}_k)^2, \quad \text{where} \quad \widehat{y}_k = \widehat{f}(x_k).$$

First, let us consider approximations by the class $F = \mathcal{P}_m$ of polynomials of degree $\leq m$. The case of $\mathcal{P}_1$ has already been covered in Subsection 8.1.2, using the principles of simple regression.

For polynomials of higher degrees, recall that three points determine a quadratic polynomial, four points determine a cubic, and so on. But if we have n data points and we want to fit a polynomial of degree m to the data points, where $n > m + 1$, then we have an overdetermined system for which we can obtain a least squares approximate solution.

The overdetermined system of equations is

$$\begin{aligned} b_0 + b_1 x_1 + b_2 x_1^2 + \cdots + b_m x_1^m &= y_1 \\ b_0 + b_1 x_2 + b_2 x_2^2 + \cdots + b_m x_2^m &= y_2 \\ \vdots \quad &= \vdots \\ b_0 + b_1 x_n + b_2 x_n^2 + \cdots + b_m x_n^m &= y_n, \end{aligned}$$

which can be written in matrix form as

$$AB = \begin{pmatrix} 1 & x_1 & x_1^2 & \cdots & x_1^m \\ 1 & x_2 & x_2^2 & \cdots & x_2^m \\ \vdots & \vdots & \vdots & \cdots & \vdots \\ 1 & x_n & x_n^2 & \cdots & x_n^m \end{pmatrix} \begin{pmatrix} b_0 \\ b_1 \\ b_2 \\ \vdots \\ b_m \end{pmatrix} = \begin{pmatrix} y_1 \\ y_2 \\ \vdots \\ y_n \end{pmatrix} = Y. \tag{8.2}$$

The normal matrix equation is then

$$(A^T A)B = A^T Y.$$

The Examples and Exercises below use made-up data. Real-world problems generally involve large data sets with decimals and are solved with the aid of software packages.

Example 8.3 Consider four data points $(-2, 0)$, $(0, 4)$, $(1, 3)$, $(2, -1)$. Figure 8.3 shows the least squares fit by a straight line. It is $\widehat{y} = \frac{1}{35}(53 - 2x)$ with a sum of squares of residuals $SOS = \sum_k e_k^2 = 19.01878$.

Now let us find a least squares fit with a quadratic curve $y = b_0 + b_1 x + b_2 x^2$. The system of equations written in matrix form (using Equation (8.2)) as

$$AB = \begin{pmatrix} 1 & -2 & 4 \\ 1 & 0 & 0 \\ 1 & 1 & 1 \\ 1 & 2 & 4 \end{pmatrix} \begin{pmatrix} b_0 \\ b_1 \\ b_2 \end{pmatrix} = \begin{pmatrix} 0 \\ 4 \\ 3 \\ -1 \end{pmatrix} = Y.$$

The normal form $(A^T A)B = A^T Y$ is

$$\begin{pmatrix} 4 & 1 & 9 \\ 1 & 9 & 1 \\ 9 & 1 & 33 \end{pmatrix} \begin{pmatrix} b_0 \\ b_1 \\ b_2 \end{pmatrix} = \begin{pmatrix} 6 \\ 1 \\ -1 \end{pmatrix}.$$

Writing as an augmented matrix and row reducing to triangular form:

$$\left(\begin{array}{ccc|c} 1 & 9 & 1 & 1 \\ 0 & 5 & -7 & 7 \\ 0 & 0 & 44 & -51 \end{array}\right),$$

has solution $b_2 = -51/44 = -255/220, b_1 = -49/220, b_0 = 916/220$. The least squares quadratic solution (also shown in Figure 8.3) is thus

$$\widehat{y} = \frac{1}{220}(916 - 49x - 255x^2).$$

We have the following analysis of the residuals for the quadratic fit:

x	y	$\hat{y}$	$e = y - \hat{y}$	e^2
-2	0	$-6/220 = -.02727$	0.02727	0.00074
0	4	$916/220 = 4.16364$	-0.16364	0.02678
1	3	$612/220 = 2.78182$	0.21818	0.04760
2	-1	$-202/220 = -0.91818$	-0.08182	0.00670
Totals	0.00000	0.08182		

The quadratic curve is clearly a much better fit of the four data points.

Example 8.4 The data from Example 8.3 can also be fitted using a cubic curve $y = b_0 + b_1 x + b_2 x^2 + b_3 x^3$. Since there are four data points, and since four points determine a cubic, the fit would be exact. This means that the sum of squares of the residuals would be zero. For this example, the least squares cubic fit comes out to be

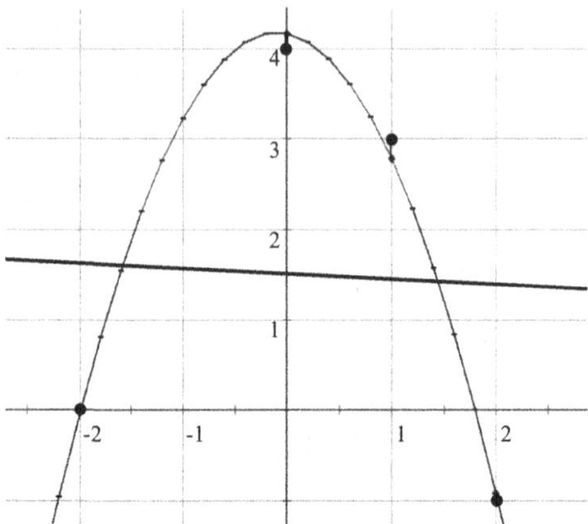

Figure 8.3 Data plot of four points along with the least squares fit by a line $\widehat{y} = \frac{1}{35}(53 - 2x)$ with $SOS = 19.01878$ and by a quadratic curve $\widehat{y} = \frac{1}{220}(916 - 49x - 255x^2)$ with $SOS = 0.08182$.

$$\widehat{y} = 4 + \frac{1}{4}x - \frac{9}{8}x^2 - \frac{1}{8}x^3 = \frac{1}{8}(32 + 2x - 9x^2 - x^3). \tag{8.3}$$

However, the degree n of the fitting polynomial can be too large, in the sense that the polynomial fit may have twists and turns not evident in the data. This cubic equation (8.3) has a local minimum and a change of direction to the left of the body of the data points. Finding the right value of the polynomial degree n of the fit requires the art of eyeing the shape of the data set. In this case, the quadratic fit with $n = 2$ seems just right.

Example 8.5 Now we give an example where the approximating class of functions is not polynomial. Consider five data points $(-3,3)$, $(-2,0)$, $(0,4)$, $(1,2)$, $(3,3)$, and let us find a least squares fit by a trigonometric function of the form $y = f(x) = b_0 + b_1 \cos \frac{\pi}{2}x$. The overdetermined system of equations is $f(x_k) = y_k$ for $k = 1,2,3,4,5$.

$$\begin{array}{lllllllll}
f(x_1) & = & b_0 + b_1 \cos \frac{\pi}{2}(-3) & = & b_0 + b_1(0) & = & 3 & = & y_1 \\
f(x_2) & = & b_0 + b_1 \cos \frac{\pi}{2}(-2) & = & b_0 + b_1(-1) & = & 0 & = & y_2 \\
f(x_3) & = & b_0 + b_1 \cos \frac{\pi}{2}(0) & = & b_0 + b_1(1) & = & 4 & = & y_3 \\
f(x_4) & = & b_0 + b_1 \cos \frac{\pi}{2}(1) & = & b_0 + b_1(0) & = & 2 & = & y_4 \\
f(x_5) & = & b_0 + b_1 \cos \frac{\pi}{3}(3) & = & b_0 + b_1(0) & = & 3 & = & y_5,
\end{array}$$

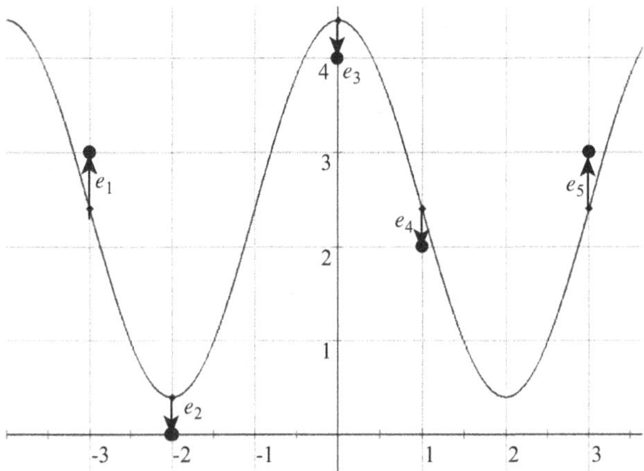

Figure 8.4 Data plot of five points along with the fitted function $\hat{y} = 2.4 + 2\cos\frac{\pi}{2}x$. The five residual vectors are also shown as arrows.

which can be written in matrix form

$$AB = \begin{pmatrix} 1 & 0 \\ 1 & -1 \\ 1 & 1 \\ 1 & 0 \\ 1 & 0 \end{pmatrix} \begin{pmatrix} b_0 \\ b_1 \end{pmatrix} = \begin{pmatrix} 3 \\ 0 \\ 4 \\ 2 \\ 3 \end{pmatrix} = Y.$$

The normal matrix equation is then

$$(A^T A)B = \begin{pmatrix} 5 & 0 \\ 0 & 2 \end{pmatrix} \begin{pmatrix} b_0 \\ b_1 \end{pmatrix} = \begin{pmatrix} 12 \\ 4 \end{pmatrix} = A^T Y,$$

which solves to $b_1 = 2$, $b_0 = 2.4$. The resulting least squares function is thus $\hat{y} = 2.4 + 2\cos\frac{\pi}{2}x$, and is illustrated in Figure 8.4. The computed sum of squares is $SOS = \sum_k e_k^2 = 1.2$.

Exercises for Subsection 8.1.4

8.1.14 In Example 8.3, verify the least squares linear fit $\hat{y} = \frac{1}{35}(53 - 2x)$ for the given four data points. Verify the sum of squares of residuals $SOS = \sum_k e_k^2 = 19.01878$. Why is SOS larger here than for the quadratic fit?

8.1.15 Consider the four data points $(0,4)$, $(1,3)$, $(1,0)$ $(2,1)$ and the family of quadratic polynomials $y = b_0 + b_1x + b_2x^2$.
a. Write the system of equations in matrix form $AB = Y$.
b. Write it in normal matrix form $(A^T A)B = A^T Y$.
c. Row reduce to obtain an augmented matrix in triangular form.
d. Find the least squares solution $\widehat{y} = b_0 + b_1x + b_2x^2$.
e. Compute the sum of squares $SOS = \sum_k (y_k - \widehat{y}_k)^2$ for this solution.

8.1.16 Repeat Exercise 8.1.15 for the family of lines $y = b_0 + b_1x$. Note that the sum of squares is larger here. Why?

8.1.17 For the three data points $(0,2)$, $(2,1)$, $(4,3)$ of Exercise 8.1.7 fit a quadratic $y = b_0 + b_1x + b_2x^2$. Since three points determine a quadratic curve, the least squares solution $\widehat{y} = b_0 + b_1x + b_2x^2$ should fit the three data points exactly. Superimpose the solution on a scatter plot of the data.

8.1.18 Given m distinct points (x_k, y_k) in $\mathbb{R}^2$ for $k = 1, 2, \ldots, m$, the **Lagrange polynomial** is the linear combination given by $L(x) = \sum_{k=1}^{m} y_k l_k(x)$, where for each $k = 1, 2, \ldots, m$, we define

$$\begin{aligned} l_k(x) &= \prod_{j \neq k} \frac{x - x_j}{x_k - x_j} \\ &= \frac{x - x_1}{x_k - x_1} \cdots \frac{x - x_{k-1}}{x_k - x_{k-1}} \cdot \frac{x - x_{k+1}}{x_k - x_{k+1}} \cdots \frac{x - x_m}{x_k - x_m}. \end{aligned}$$

Note that the Lagrange polynomial for distinct m points in $\mathbb{R}^2$ is the $m-1$ degree polynomial that goes through those points. Show that the solution of Exercise 8.1.17 is the same quadratic polynomial as that obtained by Lagrange for the three data points given in the exercise.

8.1.19 For the four data points $(-1,0)$, $(0,1)$, $(1,1)$, $(2,2)$ of Exercise 8.1.9 fit a least squares quadratic solution. Be sure to include the normal matrix equation $(A^T A)B = A^T Y$. Superimpose the solution on a scatter plot of the data.

8.1.20 Given the five data points $(-2,-3)$, $(-1,0)$, $(0,-3)$, $(1,-1)$, $(3,0)$, find the least squares fit using a function of the form $b_0 + b_1 \sin \frac{\pi}{2} x$. Comparing y_k with the fitted $\widehat{y}_k$, for $k = 1, 2, 3, 4, 5$. Would you say that this is a good fit of the data?

8.1.21 Given the four data points $(-2,0)$, $(-1,1)$, $(1,1)$, $(2,1)$, find the least squares fit using a function of the form $b_0 + b_1 \cos \frac{\pi}{2} x$. What is the surprising result? What is the sum of squares of the residuals?

8.1.5 Approximation of Functions

Calculators and computers that have built-in procedures for adding, subtracting, multiplying, and dividing, can easily compute values of rational functions. However, functions such as $\sqrt{x}$, $\log x$, e^x, and $\arctan x$ need to be approximated in order to compute their values. We have already discussed Laguerre, Chebyshev, and Hermite, polynomials (in Sections 7.1–7.2) to find least squares approximations of functions. There are many excellent books on the approximation of functions that explore a variety of methods. Here we consider two examples using the least squares approximation by polynomials.

Example 8.6 Let us find the least squares approximations of the function $\cos t$ on $[0, \pi/2]$ by a linear function of the form

$$p_1(t) = a + bt$$

under the real inner product $\langle f, g\rangle = \int_0^{\pi/2} f(t)g(t)\,dt$. Gram–Schmidt orthogonalization (Theorem 7.17) on the two functions $\{1,\ t\}$ results in the two orthogonal functions

$$w^1(t) = 1$$

$$w^2(t) = t - \frac{\langle t, 1\rangle}{\langle 1, 1\rangle} \cdot 1 = t - \frac{\pi^2/8}{\pi/2} = t - \pi/4.$$

Clearly $\langle w^1, w^2\rangle = 0$ as guaranteed by Gram–Schmidt. By Equation 7.10, the least squares estimate is

$$p_1(t) = \frac{\langle \cos, w^1\rangle}{\langle w^1, w^1\rangle} w^1(t) + \frac{\langle \cos, w^2\rangle}{\langle w^2, w^2\rangle} w^2(t) = 2/\pi + \frac{\pi/4 - 1}{\pi^3/96}(t - \pi/4).$$

So $p_1(t) = a + bt = \frac{4}{\pi}\left(\frac{6}{\pi} - 1\right) + \frac{24}{\pi^2}\left(1 - \frac{4}{\pi}\right)t \approx 1.1585 - 0.6644t$, which has the (least squares estimate) error of

$$\|\cos t - p_1(t)\|_2^2 = \int_0^{\pi/2} \left(\cos t - (1.1585 - 0.6644t)\right)^2 dt = 0.0062.$$

This approximation of $\cos t$ by a linear function $p_1(t) = a + bt$ is mildly tedious when done by hand. It would be more tedious for approximation by a quadratic $p_2(t) = A + Bt + Ct^2$. Yet for a calculator or a computer, such approximations by polynomials of even higher degree are completed in a split second.

Once we have a good approximation of $\cos t$ for the interval $[0, \pi/2]$, we can extend it to all real values of t using identities such as $\cos t = \cos(-t)$, $\cos t = \cos(t - 2\pi)$, and $\cos t = -\cos(\pi - t)$.

Example 8.7 One way to approximate the function $\sqrt{x}$ by polynomials is to first express x in base 16 using the scientific form $x = t \times 16^n$, where n is an integer and $1 \le t < 16$. Then $\sqrt{x} = \sqrt{t} \times 4^n$. The problem thus reduces to an approximation of the function $f(t) = \sqrt{t}$ on the interval $[1, 16)$ by a polynomial $p(t)$, say a linear one $p(t) = a + bt$. Instead of the Gram–Schmidt procedure as used above, this example lends itself to a direct minimization of the least squares error function

$$E(a,b) = \int_1^{16} \left(\sqrt{t} - a - bt\right)^2 dt.$$

Equating the partial derivatives $\frac{\partial}{\partial a}E(a,b)$, $\frac{\partial}{\partial b}E(a,b)$ to zero, results in the equations

$$2\left(\frac{2t^{3/2}}{3} - at - b\frac{t^2}{2}\right)\Bigg|_1^{16} = 0,$$

$$2\left(\frac{2t^{5/2}}{5} - a\frac{t^2}{2} - b\frac{t^3}{3}\right)\Bigg|_1^{16} = 0,$$

or

$$15a + \frac{255}{2}b = 42,$$

$$\frac{255}{2}a + 1365b = \frac{2046}{5}.$$

The solution is $a = \frac{764}{625} = 1.2224$, $b = \frac{116}{625} = 0.1856$ and $E(a,b) = 0.2117$.

Approximation of the same function by a quadratic $p(t) = A + Bt + Ct^2$, using the same method of partial derivatives with respect to A, B, and C, results in the solution $p(t) = 0.8604 + 0.3006t - 0.0068t^2$ with a smaller least squares error

$$\int_1^{16} (\sqrt{t} - A - Bt - Ct^2)^2\, dt = 0.0186.$$

Exercises for Subsection 8.1.5

8.1.22 Repeat Example 8.6 to find a least squares approximation of the function $\cos t$ by a linear function $f(t) = A + Bt$, except do it for the interval $[0, \pi]$. Also find the least squares error.

8.1.23 Follow Example 8.6 to find the least squares approximation of the function $\sin t$ on the interval $[0, \pi]$ by a linear function of the form $f(t) = A + Bt$. Also find the least squares error.

8.1.24 Follow Example 8.7 to find the least squares approximation of the function $\sin t$ on the interval $[0, \pi]$ by a linear function of the form $f(t) = A + Bt$. Also find the least squares error.

8.1.25 Follow Example 8.6 to find the least squares approximation of the function $\sin t$ on the interval $[0, \pi]$ by a quadratic function of the form $f(t) = A + Bt + Ct^2$. Also find the least squares error.

8.1.26 Follow Example 8.7 to find the least squares approximation of the function $\sin t$ on the interval $[0, \pi]$ by a quadratic function of the form $f(t) = A + Bt + Ct^2$. Also find the least squares error.

8.1.27 Follow Example 8.6 to find the least squares approximation of the function $\cos t$ on the interval $[0, \pi]$ by a quadratic function of the form $f(t) = A + Bt + Ct^2$. Also find the least squares error.

8.1.28 Follow Example 8.7 to find the least squares approximation of the function $\cos t$ on the interval $[0, \pi]$ by a quadratic function of the form $f(t) = A + Bt + Ct^2$. Also find the least squares error.

8.1.29 Find the least squares approximation of the function $f(t) = 6+2t+t^2$ on the interval $[0, 1]$ by a linear polynomial $p_1(t) = b_0 + b_1 t$ and by a quadratic polynomial $p_2(t) = A + Bt + Ct^2$. In each case, find the least squares error.

8.2 The Wave Equation

The analysis of the motions of a vibrating string is partly what led to the development of Fourier analysis. We start with simple harmonic motion and then advance to the motion of a vibrating string.

8.2.1 Harmonics

Simple harmonic motion is the idealized motion produced by a mass m sliding horizontally on a frictionless surface under the influence of two identical springs attached to fixed walls as illustrated in Figure 8.5.

By Hook's law, if the mass is displaced from its resting position 0 by a small distance y, the springs induce a force F on the mass that is proportional to the displacement y, say $F = -ky$, where $k > 0$. The negative sign is due to force being toward the origin whenever the point y is to the right of the origin.

The displacement is a function of time $y = y(t)$. By Newton's second law of motion, we have $F = ma$, where $a = y''$ is the acceleration of the mass m. Hence

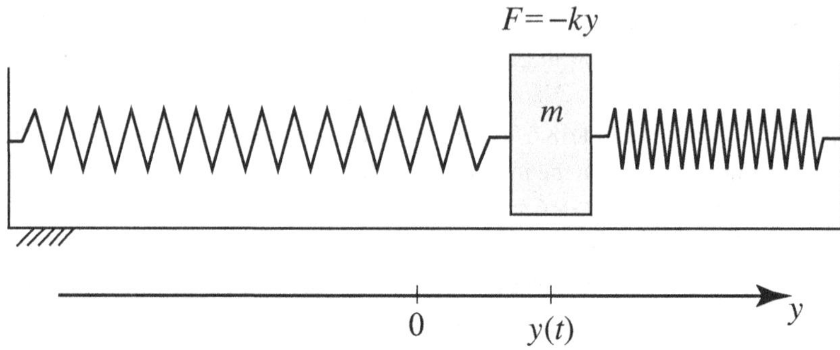

Figure 8.5 A displaced mass under the influence of two springs will produce a simple harmonic motion.

$$F = -ky = my'', \quad \text{or} \quad y'' + c^2 y = 0 \quad \text{with} \quad c = \sqrt{k/m}. \tag{8.4}$$

The solution of this differential equation can be shown (Exercises 8.2.1 and 8.2.2) to be the function of t

$$y = A\sin(ct + \phi), \tag{8.5}$$

where A, c, and ϕ are constants. A function of this form is called a **harmonic** of **amplitude** $|A|$, **angular frequency** c, and **initial phase** ϕ.

A harmonic function is periodic with period $T = 2\pi/c$, as we see here:

$$\begin{aligned} A\sin\left[c(t+T)+\phi\right] &= A\sin\left[c\left(t+\frac{2\pi}{c}\right)+\phi\right] = A\sin\left[(ct+\phi)+2\pi\right] \\ &= A\sin(ct+\phi). \end{aligned}$$

Figure 8.6 gives an example of such a harmonic function. Using the addition formula for the sine function, we have

$$y = A\sin(ct+\phi) = A(\sin\phi\cos ct + \cos\phi\sin ct).$$

Setting the constants $A\sin\phi = a$ and $A\cos\phi = b$, leads to the following form for a harmonic:

$$y = a\cos ct + b\sin ct. \tag{8.6}$$

Conversely, every function of the form (8.6) is a harmonic, as can be derived from the following formulas:

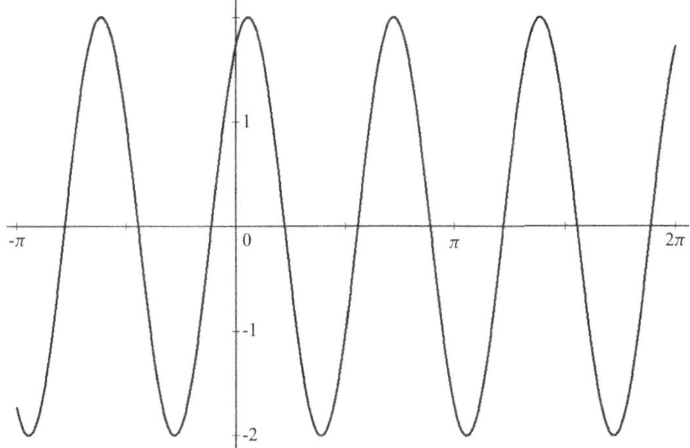

Figure 8.6 A graph of the harmonic $y = 2\sin\left(3t + \frac{\pi}{3}\right) = \sqrt{3}\cos 3t + \sin 3t$ with amplitude 2, angular frequency 3, initial phase $\frac{\pi}{3}$, and period $T = \frac{2}{3}\pi$.

$$A = \sqrt{a^2 + b^2}, \quad \sin\phi = \frac{a}{A} = \frac{a}{\sqrt{a^2+b^2}}, \quad \cos\phi = \frac{b}{A} = \frac{b}{\sqrt{a^2+b^2}}.$$

In terms of the period $T = \frac{2\pi}{c}$, the harmonic function (8.6) may be written as follows:

$$y = a\cos\frac{2\pi t}{T} + b\sin\frac{2\pi t}{T}. \tag{8.7}$$

8.2.2 Sum of Harmonics

Given a fixed T, we can create the following sequence of harmonics

$$a_k\cos\frac{2\pi kt}{T} + b_k\sin\frac{2\pi kt}{T} \quad \text{for} \quad k = 1, 2, 3, \ldots. \tag{8.8}$$

For each k, the harmonic has angular frequency $c_k = 2\pi k/T$ and period $T_k = T/k$. As described in Section 7.5 (Definition 7.12), a finite sum of such harmonics is a **trigonometric polynomial**

$$s_n(t) = \frac{a_0}{2} + \sum_{k=1}^{n}\left(a_k\cos\frac{2\pi kt}{T} + b_k\sin\frac{2\pi kt}{T}\right).$$

Here the constant $a_0/2$ is considered the term for $k = 0$. The entire trigonometric polynomial s_n is of period T (provided that the $k = 1$ term is nonzero).

As described by Formula (7.28) of Definition 7.13, as n tends to ∞, we obtain a **trigonometric Fourier series**

$$\frac{a_0}{2} + \sum_{k=1}^{\infty} \left(a_k \cos \frac{2\pi kt}{T} + b_k \sin \frac{2\pi kt}{T} \right).$$

This will be used below to find the motion of a vibrating string.

8.2.3 Vibrating String

Now let us consider a tightly stretched vibrating string, such as one on a violin or piano. Musicians are aware that a musical instrument will produce, in addition to any fundamental tone, **overtones** or **harmonics**. These are tones which are multiples of the fundamental tone. Overtones are what give the musical instrument its characteristic sound. Even when both a violin and piano are playing the same fundamental tone, you can distinguish between them because of the differences in their overtones.

If the string is displaced from its rest position and then released, it will vibrate. In our analysis, we make some simplifying assumptions.

- The rest position of the string is on the horizontal axis between the points $x = 0$ and $x = L$.
- The string is flexible and has a uniform density.
- Vibrations take place in only one plane, so that each point of the string moves only vertically.
- There is no gravity or air resistance.
- The displacement of any point of the string is small in comparison to the length L.
- At any point, the angle between the string and the horizontal is small.
- The tension forces are the same at all points of the string.

Let the vertical displacement y at the point x and time t be given by the function $y = u(x,t)$. So, at any time t, a point A of the string has coordinates (x, y), where $y = u(x,t)$.

At each point $A = (x, y)$, there are tension forces in equal magnitudes but in opposite directions, as illustrated in Figure 8.7. One force is exerted by the part of the string to the right of A, say with magnitude T. The direction is tangent to the shape of the string. The other tension force $-T$ is exerted by the part of the string to the left of A. Its direction is also tangent to the curve but in the opposite direction.

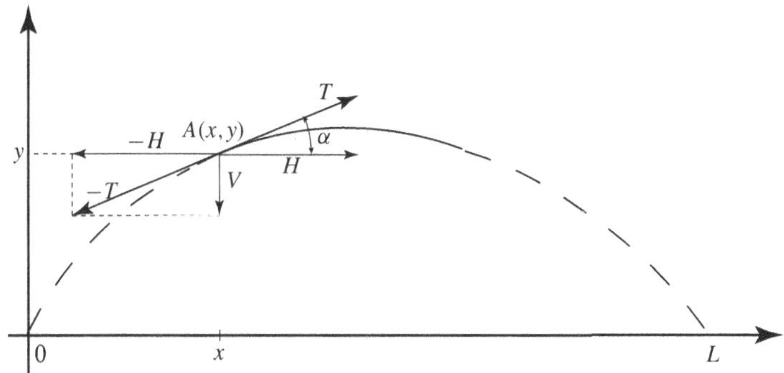

Figure 8.7 An exaggerated view of forces on a string at point $A = (x, y)$.

Because we assume that the displacements at all points of the string are small in comparison to the length L, and the tension forces are the same at all points, we may consider the horizontal component H of the tension force to be constant at all points.

Now consider a part of the string from point $A = (x, y)$ to a point B that is h units along the horizontal axis. Let $V(x,t)$ be the vertical component of the tension force. If α is the angle of inclination of the string at point A at time t, then $\tan\alpha$ is the slope of the line tangent to the curve

$$\frac{-V(x,t)}{H} = \tan\alpha = \frac{\partial}{\partial x}u(x,t).$$

In Figure 8.8, the point A is above the horizontal axis, so $V(x,t) < 0$. If the point A were below the horizontal axis, then we would have $V(x,t) > 0$ and

$$\frac{V(x,t)}{H} = \tan(\pi - \alpha) = -\tan\alpha = -\frac{\partial}{\partial x}u(x,t).$$

In either case, we have

$$V(x,t) = -H\frac{\partial}{\partial x}u(x,t). \tag{8.9}$$

Because of the curvature of the string, the angle of inclination at B is not quite the same as at A. Let β be the angle of inclination of the string at point B. Then a similar analysis gives us

$$\frac{V(x+h,t)}{H} = \tan\beta = \frac{\partial}{\partial x}u(x+h,t)$$

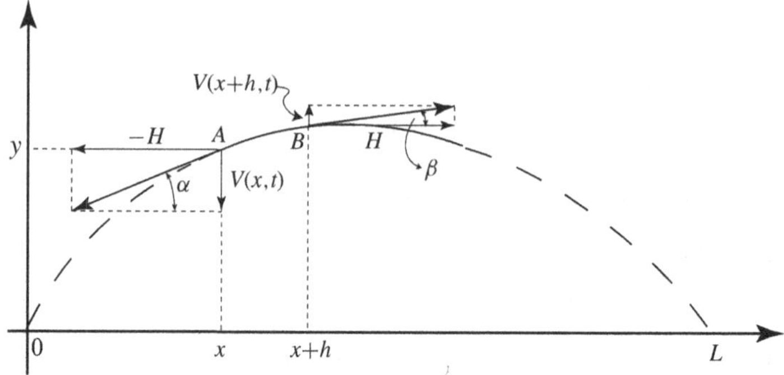

Figure 8.8 Forces on a vibrating string at points A and B.

or

$$V(x+h,t) = H\frac{\partial}{\partial x}u(x+h,t). \tag{8.10}$$

This has the same form as Equation (8.9) except for the reversal of sign. The total force on the segment AB in the vertical directions is thus the sum of (8.9) and (8.10)

$$-H\frac{\partial}{\partial x}u(x,t) + H\frac{\partial}{\partial x}u(x+h,t).$$

Since we assume the string has a constant density, say δ, the mass of the string segment AB is δh. Applying Newton's second law of motion $F = ma$ to the segment, we have

$$F = ma = \delta h\frac{\partial^2}{\partial t^2}u(x,t) = -H\frac{\partial}{\partial x}u(x,t) + H\frac{\partial}{\partial x}u(x+h,t).$$

Dividing by h and letting $h \to 0$, we obtain

$$\frac{\partial^2}{\partial t^2}u(x,t) = \frac{H}{\delta}\lim_{h\to 0}\frac{\frac{\partial}{\partial x}u(x+h,t) - \frac{\partial}{\partial x}u(x,t)}{h} = \frac{H}{\delta}\frac{\partial^2}{\partial x^2}u(x,t).$$

We may write this in the following form:

$$u_{tt}(x,t) = c^2 u_{xx}(x,t) \quad \text{where} \quad c = \sqrt{H/\delta}. \tag{8.11}$$

This is called the **wave equation for a string**. The constant c is called the **velocity of the vibration**.

8.2.4 Solution of the Wave Equation

Given a string with an initial curve $f(x) = u(x,0)$ and initial velocity $v(x) = \frac{\partial}{\partial t}u(x,0)$, the problem is to find the curve $u(x,t)$ at an arbitrary time t. For simplicity of notation, we assume the length L of the string is π; that is, $u(0,t) = u(\pi,t) = 0$. Obviously, there is a trivial solution of *no* motion $u(x,t) \equiv 0$. In the solution below, we assume motion.

Note that the wave equation $u_{tt}(x,t) = c^2 u_{xx}(x,t)$, has differentiation only with respect to t on the left side of the equation and with respect to x on the right. So we look for solutions of separated variables

$$u(x,t) = X(x)T(t). \tag{8.12}$$

Differentiating twice, and using the wave equation, we obtain

$$X(x)T''(t) = c^2 X''(x)T(t)$$

or

$$\frac{1}{c^2}\frac{T''(t)}{T(t)} = \frac{X''(x)}{X(x)}. \tag{8.13}$$

The left-hand side depends only on t and the right-hand side only on x. Hence, both sides are a constant, say $-a$, so that

$$X''(x) = -aX(x) \quad \text{and} \quad T''(t) = -c^2 aT(t). \tag{8.14}$$

Recall that both are forms of the differential equation for a simple harmonic motion observed in Equation (8.4). The constant a must be positive, since a negative value in $T''(t) = -c^2 aT(t)$ will produce no motion. We can thus write $a = \alpha^2$. The solutions of the equations, as given in Equation (8.6), are

$$T(t) = A\cos\alpha ct + B\sin\alpha ct \quad \text{and} \quad X(x) = A'\cos\alpha x + B'\sin\alpha x. \tag{8.15}$$

Since the string is attached at the ends, we have $X(0) = X(\pi) = 0$. Because $X(0) = A' = 0$, the function X reduces to $X(x) = B'\sin\alpha x$. Furthermore, we may assume that $B' \neq 0$, for otherwise the function X would vanish. Additionally, since $X(\pi) = B'\sin\alpha\pi = 0$, dividing by $B' \neq 0$ results in $\sin\alpha\pi = 0$. This can only happen if $\alpha = k$ is an integer. So $X(x) = B'\sin kx$ for some integer k.

We can also assume that $k \neq 0$, for otherwise the function X would vanish. Furthermore, negative values of k will duplicate solutions T and X with other constants since sin is odd and cos is even.

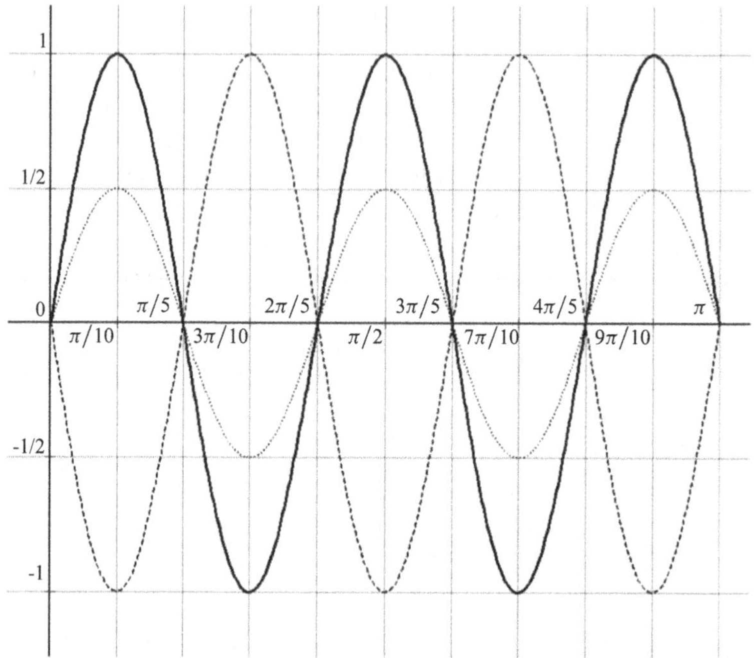

Figure 8.9 The standing wave $u_5(x,t) = X_5(x)T_5(t) = (\sin 5x)T_5(t)$ at amplitude values $T_5(t) = 1, -1$, and $\frac{1}{2}$. The nodes are at $0, \pi/5, 2\pi/2, \ldots, \pi$ and the antinodes are at $\pi/10, 3\pi/10, \ldots, 9\pi/10$.

Incorporating the constant B' into A and B, results in infinitely many solutions of the wave equation of the form $u_k(x,t) = X_k(x)T_k(t)$

$$u_k(x,t) = (\sin kx)(A_k \cos kct + B_k \sin kct) \quad \text{for} \quad k = 1, 2, \ldots. \tag{8.16}$$

We still have to check that the functions $u_k(x,t)$ are solutions of the wave equation (8.11). This is straightforward and is left as Exercise 8.2.4.

Note that each solution (8.16) has the **wave shape** $X_k(x) = \sin kx$ that does not change with time but only oscillates in the amplitude given by $T_k(t) = A_k \cos kct + B_k \sin kct$. Such a solution is called a **standing wave**. Figure 8.9 illustrates an example. The points $x = 0, \frac{\pi}{k}, \frac{2\pi}{k}, \ldots, \pi$ in the interval $[0, \pi]$, where the wave shape has the value $u_k(x,t) = 0$ and stays fixed with t, are called the **nodes**. The points half way between the nodes, where the amplitude oscillates the most, are called the **antinodes**.

Taking $k = 1$ in (8.16) gives us the **fundamental tone** (or **first harmonic**) of the vibrating string

$$u_1(x,t) = (\sin x)(A_1 \cos ct + B_1 \sin ct).$$

Taking $k = 2$ gives us the **first overtone** (or **second harmonic**)

$$u_2(x,t) = (\sin 2x)(A_2 \cos 2ct + B_2 \sin 2ct).$$

And so on, for the other overtones.

The solutions to the wave equation (8.11) are linear. Thus linear combinations of solutions are solutions. If we superimpose solutions, we obtain the general solution

$$u(x,t) = \sum_{k=1}^{\infty} (\sin kx)\big(A_k \cos kct + B_k \sin kct\big). \tag{8.17}$$

Since the initial curve of the string is $f(x) = u(x,0)$ we obtain the trigonometric series

$$f(x) = u(x,0) = \sum_{k=1}^{\infty} A_k \sin kx. \tag{8.18}$$

As was shown in Example 7.13 (Section 7.2), the Fourier coefficients are

$$A_k = \frac{2}{\pi} \int_0^{\pi} f(x) \sin kx \, dx. \tag{8.19}$$

Also, since the initial velocity of the string is $v(x) = \dfrac{\partial}{\partial t} u(x,0)$, we obtain the series

$$v(x) = \sum_{k=1}^{\infty} kcB_k \sin kx, \tag{8.20}$$

which means that

$$B_k = \frac{2}{kc\pi} \int_0^{\pi} v(x) \sin kx \, dx. \tag{8.21}$$

In summary, we have the following theorem.

Theorem 8.2 *The general solution of the wave equation* $u_{tt}(x,t) = c^2 u_{xx}(x,t)$, *on the interval* $[0, \pi]$ *is the sum of harmonics*

$$u(x,t) = \sum_{k=1}^{\infty} \big(\sin kx\big)\big(A_k \cos kct + B_k \sin kct\big), \tag{8.22}$$

where for each k, $X_k(x) = \sin kx$ *is the wave shape of the kth harmonic. The amplitude of the wave shape at time t is* $T_k(t) = A_k \cos kct + B_k \sin kct$*. The coefficients* A_k *and* B_k *are given by equations*

$$A_k = \frac{2}{\pi}\int_0^\pi f(x)\sin kx\,dx \quad \textit{and} \quad B_k = \frac{2}{kc\pi}\int_0^\pi v(x)\sin kx\,dx, \tag{8.23}$$

where $f(x)$ *is the initial curve of the string and* $v(x) = \frac{\partial}{\partial t}u(x,0)$ *is its initial velocity.*

Example 8.8 Plucked string: Let us consider the example where a violin string under tension and of length π is displaced at its center by a distance of h and then let go. The initial velocity is zero, so

$$f(x) = \begin{cases} \frac{2h}{\pi}x & \text{for } 0 \le x \le \frac{\pi}{2} \\ \frac{2h}{\pi}(\pi - x) & \text{for } \frac{\pi}{2} \le x \le \pi \end{cases} \quad \text{and} \quad v(x) \equiv 0.$$

The solution is given by Theorem 8.2. Computing the Fourier coefficients, we obtain

$$A_k = \frac{8h}{\pi^2 k^2}\sin\frac{k\pi}{2} \quad \text{and} \quad B_k = 0.$$

Note that $A_k = 0$ for even k and $A_k = \frac{(-1)^k 8h}{\pi^2 k^2}$ for odd k. The solution is thus

$$u(x,t) = \frac{8h}{\pi^2}\left\{\sin x \cos ct - \frac{1}{9}\sin 3x \cos 3ct + \frac{1}{25}\sin 5x \cos 5ct - \cdots\right\}.$$

Example 8.9 Struck string: In this example, a piano wire under tension and of length π is initially not displaced but has an initial velocity imparted by a hammer of width w striking at the center. Suppose the velocity of the hammer is v_0. So,

$$f(x) \equiv 0 \quad \text{and} \quad v(x) = \begin{cases} v_0 & \text{for } \frac{\pi - w}{2} \le x \le \frac{\pi + w}{2} \\ 0 & \text{elsewhere.} \end{cases}$$

Computing the coefficients, we obtain

$$A_k = 0 \quad \text{and} \quad B_k = \frac{2v_0}{kc\pi}\int_{\frac{\pi-w}{2}}^{\frac{\pi+w}{2}} \sin kx\,dx = \frac{4v_0}{ck^2\pi}\sin\frac{k\pi}{2}\sin kw.$$

The solution for $u(x,t)$ is thus

$$\frac{4v_0}{c\pi}\left\{\sin w \sin x \sin ct - \frac{1}{9}\sin 3w \sin 3x \sin 3ct \right.$$
$$\left. + \frac{1}{25}\sin 5w \sin 5x \sin 5ct - \cdots \right\}.$$

8.2.5 Wave Equation in Higher Dimensions

The wave equation $\frac{\partial^2 u}{\partial t^2} = c^2\frac{\partial^2 u}{\partial x^2}$ can be expressed in higher dimensions. For example in two dimensions, for $u = u(x,y,t)$, it takes the form

$$\frac{\partial^2 u}{\partial t^2} = c^2\left\{\frac{\partial^2 u}{\partial x^2} + \frac{\partial^2 u}{\partial y^2}\right\}, \tag{8.24}$$

which would represent the vibrations of a membrane on a region in the plane. And a wave traveling in 3-space is represented by

$$\frac{\partial^2 u}{\partial t^2} = c^2\left\{\frac{\partial^2 u}{\partial x^2} + \frac{\partial^2 u}{\partial y^2} + \frac{\partial^2 u}{\partial z^2}\right\}. \tag{8.25}$$

In either case, the expression within the brackets on the right is called the **Laplacian** of u. The Laplacian operator of any dimension has the symbol ∇^2, or sometimes simply Δ. In three dimensions the Laplacian operator is

$$\nabla^2 = \frac{\partial^2}{\partial x^2} + \frac{\partial^2}{\partial y^2} + \frac{\partial^2}{\partial z^2}.$$

In any dimension, the wave equation has the form $\dfrac{\partial^2 u}{\partial t^2} = c^2\nabla^2 u$.

We will consider two examples in two dimensions, one with a vibrating membrane stretched over a rectangular region and then another one stretched over a circular region (such as on a drum).

Example 8.10 Consider a membrane $u = u(x,y,t)$ stretched over rectangle $R = [0,a]\times[0,b]$ and satisfying the two-dimensional wave equation (8.24). In this case, let the initial shape and velocity of the membrane be

$$f(x,y) = u(x,y,0) \quad \text{and} \quad v(x,y) = \frac{\partial}{\partial t}u(x,y,0), \quad \text{respectively.}$$

Here the analysis is similar to the one-dimensional case of a vibrating string. The variables separate into three functions

$$u(x,y,t) = X(x)Y(y)T(t).$$

Applying the wave equation as in Equation (8.13) results in

$$\frac{1}{c^2}\frac{T''(t)}{T(t)} = \frac{X''(x)}{X(x)} + \frac{Y''(y)}{Y(y)}. \tag{8.26}$$

This can happen only if the terms are constant, say

$$\frac{X''(x)}{X(x)} = -\alpha^2,\ \frac{Y''(y)}{Y(y)} = -\beta^2,\ \frac{1}{c^2}\frac{T''(t)}{T(t)} = -(\alpha^2+\beta^2).$$

The solution in the one-dimensional case was given by Equation (8.22) in Theorem 8.2. Modifying (8.22) to a string of length a instead of π, it becomes

$$u(x,t) = \sum_{k=1}^{\infty} X_k(x)T_k(t) = \sum_{k=1}^{\infty}\left(\sin\frac{k\pi}{a}x\right)\big(A_k\cos kct + B_k\sin kct\big).$$

Similarly, the general solution on the rectangle $R = [0,a]\times[0,b]$ becomes

$$\begin{aligned} u(x,y,t) &= \sum_{j,k=1}^{\infty} u_{jk}(x,y,t) = \sum_{j,k=1}^{\infty} X_j(x)Y_k(y)T_{jk}(t) \\ &= \sum_{j,k=1}^{\infty}\big(\sin\alpha_j x\big)\big(\sin\beta_k y\big)\big(A_{jk}\cos\gamma_{jk}ct + B_{jk}\sin\gamma_{jk}ct\big), \end{aligned}$$

with constants $\alpha_j = \dfrac{j\pi}{a}$, $\beta_k = \dfrac{k\pi}{b}$, $\gamma_{jk} = \sqrt{\alpha_j^2+\beta_k^2} = \sqrt{\left(\dfrac{j\pi}{a}\right)^2 + \left(\dfrac{k\pi}{b}\right)^2}$.

In the case of a circular membrane, conversion to polar coordinates $x = r\cos\theta,\ y = r\sin\theta$ is convenient.

Theorem 8.3 *For $u\colon \mathbb{R}^2 \longrightarrow \mathbb{R}$, the two-dimensional Laplacian in polar coordinates is*

$$\nabla^2 u = \frac{\partial^2 u}{\partial r^2} + \frac{1}{r}\frac{\partial u}{\partial r} + \frac{1}{r^2}\frac{\partial^2 u}{\partial\theta^2}. \tag{8.27}$$

Proof Using the transformation $x = r\cos\theta$, $y = r\sin\theta$, and the chair rule, we have

$$\frac{\partial u}{\partial r} = \frac{\partial u}{\partial x}\frac{\partial x}{\partial r} + \frac{\partial u}{\partial y}\frac{\partial y}{\partial r} = \frac{\partial u}{\partial x}\cos\theta + \frac{\partial u}{\partial y}\sin\theta.$$

Differentiating again, we obtain

$$\begin{aligned} \frac{\partial^2 u}{\partial r^2} &= \frac{\partial}{\partial r}\left\{\frac{\partial u}{\partial x}\cos\theta + \frac{\partial u}{\partial y}\sin\theta\right\} \\ &= \frac{\partial^2 u}{\partial x^2}\cos^2\theta + \frac{\partial^2 u}{\partial x\partial y}2\sin\theta\cos\theta + \frac{\partial u^2}{\partial y^2}\sin^2\theta. \end{aligned}$$

Similarly, $\dfrac{\partial u}{\partial \theta} = \dfrac{\partial u}{\partial x}\dfrac{\partial x}{\partial \theta} + \dfrac{\partial u}{\partial y}\dfrac{\partial y}{\partial \theta} = -r\dfrac{\partial u}{\partial x}\sin\theta + r\dfrac{\partial u}{\partial y}\cos\theta.$

And differentiating again, we obtain

$$\begin{aligned}\frac{\partial^2 u}{\partial \theta^2} &= r\frac{\partial}{\partial \theta}\left\{-\frac{\partial u}{\partial x}\sin\theta + \frac{\partial u}{\partial y}\cos\theta\right\}\\ &= -r\frac{\partial u}{\partial r} + r^2\frac{\partial^2 u}{\partial x^2}\sin^2\theta - 2r^2\frac{\partial^2 u}{\partial x \partial y}\sin\theta\cos\theta + r^2\frac{\partial u^2}{\partial y^2}\cos^2\theta.\end{aligned}$$

Adding, we obtain $\quad \dfrac{\partial^2 u}{\partial r^2} + \dfrac{1}{r^2}\dfrac{\partial^2 u}{\partial \theta^2} = -\dfrac{1}{r}\dfrac{\partial u}{\partial r} + \dfrac{\partial^2 u}{\partial x^2} + \dfrac{\partial u^2}{\partial y^2}.$

Then transposing $-\dfrac{1}{r}\dfrac{\partial u}{\partial r}$ results in the desired Equation (8.27). ∎

In polar coordinates $u = u(r,\theta,t)$, the wave equation (8.24) becomes

$$\frac{\partial^2 u}{\partial t^2} = c^2\left\{\frac{\partial^2 u}{\partial r^2} + \frac{1}{r}\frac{\partial u}{\partial r} + \frac{1}{r^2}\frac{\partial^2 u}{\partial \theta^2}\right\}. \tag{8.28}$$

The initial shape and velocity of the membrane are then

$$f(r,\theta) = u(r,\theta,0) \quad \text{and} \quad v(r,\theta) = \frac{\partial u}{\partial t}u(r,\theta,0), \quad \text{respectively.}$$

Example 8.11 Now we consider the special case of a circular membrane of radius π where the drum is initially struck and displaced at the center. If the center is at the origin, the initial shape and velocity are independent of the angle θ. Then the wave equation (8.28) reduces to

$$\frac{\partial^2 u}{\partial t^2} = c^2\left\{\frac{\partial^2 u}{\partial r^2} + \frac{1}{r}\frac{\partial u}{\partial r}\right\}$$

with the initial shape and velocity $f(r) = u(r,0)$ and $v(r) = \dfrac{\partial u}{\partial t}u(r,0)$. We separate the variables $u(r,t) = R(r)T(t)$ and differentiate twice to obtain

$$\frac{1}{c^2}\frac{T''(t)}{T(t)} = \left(R''(r) + \frac{1}{r}R'(r)\right)\frac{1}{R(r)}.$$

Following our previous solution of the wave equation, both sides are constant, say $-\alpha^2$. This leads to two equations

$$T''(t) + \alpha^2 c^2 T(t) = 0, \quad \text{and}$$
$$rR''(r) + R'(r) + \alpha^2 r R(r) = 0.$$

As previously shown in Equation (8.15), $T(t)$ is a linear combination of $\cos\alpha ct$ and $\sin\alpha ct$,

$$T(t) = A\cos\alpha ct + B\sin\alpha ct.$$

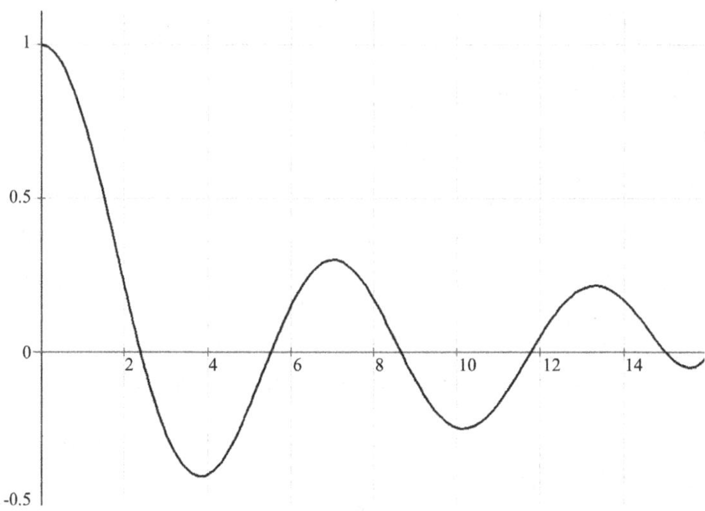

Figure 8.10 A graph of the Bessel function $J_0(x)$. A series expansion is $J_0(x) = \sum_{k=0}^{\infty} \frac{(-1)^k}{(2^k k!)^2} x^{2k} = 1 - \frac{x^2}{2^2} + \frac{x^4}{2^2 4^2} - \frac{x^6}{2^2 4^2 6^2} + \frac{x^8}{2^2 4^2 6^2 8^2} - \cdots$.

The second equation (for $R(r)$) is a differential equation whose solution is called a **Bessel function** of order 0 of the first kind,[2]

$$R(r) = C_1 J_0(\alpha r).$$

Figure 8.10 shows a graph of the Bessel function $J_0(x)$.

The constant C_1, which is nonzero (for otherwise, there would be no motion), can be incorporated into the constants A and B, so that

$$u(r,t) = R(r)T(t) = J_0(\alpha r)\big(A' \cos \alpha ct + B' \sin \alpha ct\big).$$

At the rim of radius π we have $R(\pi) = 0$, so $R(\pi) = J_0(\alpha\pi) = 0$. This means $\alpha\pi$ is one of the infinite roots of J_0, say $\alpha\pi = z_1, z_2, z_3, \ldots$. Then, for $\alpha_k = z_k/\pi$, $k = 1,2,3,\ldots$, the general solution is

$$u(r,t) = \sum_{k=1}^{\infty} J_0(\alpha_k r)\big(A_k \cos \alpha_k ct + B_k \sin \alpha_k ct\big).$$

[2] A differential equation of the form $r^2 R'' + rR' + (\alpha^2 r^2 - m^2)R = 0$ has solutions called Bessel functions of order m, written

$$R(r) = C_1 J_m(\alpha r) + C_2 Y_m(\alpha r),$$

where J_m and Y_m are Bessel functions of order m of the first and second kind, respectively. In our case, $m = 0$. Also it turns out, the Bessel function of the second kind $Y_0(\alpha r)$ is unbounded at $r = 0$, whereas the stretched membrane is not. So we must also have $C_2 = 0$.

Exercises for Section 8.2

8.2.1 Consider the differential equation $y''+c^2y = 0$, as given by Equation (8.4). Show that the function $y = A\sin(ct+\phi)$, (see Equation (8.5)), is a solution.

8.2.2 Show that *every* solution of the differential equation $y''+c^2y=0$ is of the form given by Equation (8.5). Do this by means of the equivalent Equation (8.6), using the functions $f(t) = y(t)\cos ct - c^{-1}y'(t)\sin ct$ and $g(t) = y(t)\sin ct + c^{-1}y'(t)\cos ct$. Differentiate to show that f and g are constants, say $f(t) = a$ and $g(t) = b$.

8.2.3 Consider the simple harmonic motion given by $y = -3\sin(4\pi t + \pi/2)$. Find the amplitude, angular frequency, initial phase, and period of the motion. Find the position at $t = 1/4$. Sketch a graph of the harmonic motion. Find both the maximum speed and the maximum acceleration.

8.2.4 Show that the solution given by Equation (8.16) is actually a solution of the wave equation $\frac{\partial^2 u}{\partial t^2} = c^2\frac{\partial^2 u}{\partial x^2}$.

8.2.5 Verify the coefficients A_k for a plucked string in Example 8.8.

8.2.6 In Example 8.8, explain how the vibrations differ as the tension is increased, for example, when $c = 2$ is increased to $c = 3$.

8.2.7 Repeat Example 8.8 with the string displaced by a distance of h at $x = \frac{\pi}{3}$, instead of $x = \frac{\pi}{2}$.

8.2.8 Calculate the coefficients A_k and B_k in Example 8.9.

8.2.9 Solve the one-dimensional wave equation when

$$c = 1,\ \ f(x) = 2\sin x - \sin 3x,\ \text{ and } v(x) = 0.$$

8.2.10 Solve the one-dimensional wave equation when

$$c = 1,\ \ f(x) = 0,\ \text{ and } v(x) = \sin x + \frac{1}{8}\sin 2x.$$

8.2.11 Solve the one-dimensional wave equation when

$$c = 1,\ \ f(x) = \sin x,\ \text{ and } v(x) = \sin x.$$

8.3 The Heat Equation

This topic starts with the theory of the conduction of heat as described by Jean Fourier in his paper *Théorie Analytique de la Chaleur* in 1822. That same theory also applies to the transfer of energy, motion of fluids, chemical reactions, electricity, economics, Brownian motion, and the modern atomic theory of matter and energy.

8.3.1 Derivation of Heat Equation

Heat is measured in calories. One **calorie** is defined to be the amount of heat required to raise the temperature of one gram of liquid water (equivalently, 1cc of liquid water) by one degree Celsius (or Kelvin).[3]

The **specific heat** (or **heat capacity**) of any material is the amount of calories required to raise the temperature of 1 g of it by one degree C°. The specific heat of water is by definition $\sigma = 1\ (cal/(gm^\circ C))$. But only 0.118 calories are needed to raise one gram of steel by 1°. So the specific heat of steel is $\sigma = 0.118\ (cal/(gm^\circ C))$.

Since anything at 0° Kelvin has no heat energy, the amount of **heat energy** in m grams of a material with specific heat σ and uniform temperature u degrees Kelvin is $m\sigma u$.

Now consider a rod made of a material with uniform density ρ and uniform specific heat σ represented on the x-axis. We will suppose that the temperature u of the rod varies with the position x and time t, say $u = u(x,t)$. The heat energy of an interval (a,b) of the rod at time t is

$$H(t) = \sigma\rho \int_a^b u(x,t)\,dx.$$

Now consider an interval $(a,b) = (x_0, x_0 + h)$, where h is small. Then the heat energy is approximately $H(t) = \sigma\rho u(x_0,t)\cdot h$. The heat flow for the interval is thus approximately

$$\frac{\partial H}{\partial t} = \sigma\rho\frac{\partial u}{\partial t}(x_0,t)\cdot h. \tag{8.29}$$

By the principle of conservation of energy, the heat flow for the interval $(x_0, x_0 + h)$ must be equal to the heat flowing *into* the interval minus the heat flowing *out* of the interval. We assume that the rod is perfectly insulated so that heat can only flow in and out of the interval in the x-direction at the endpoints x_0 and $x_0 + h$.

[3] Normal body temperature is 37 C°. Drinking a half liter (500 g) of cold water at 5 C° will make your body burn $(37-5)\cdot 500 = 16,000$ calories to bring it back to normal temperature. However, food calories are really kilocalories, (1 Cal = 1000 cal), so you lose only 16 Cal. Unfortunately, eating cold ice cream will not make you lose weight. ;-)

Furthermore, Newton's law of cooling states that heat flow is proportional to the difference in temperatures. So

$$\frac{\partial H}{\partial t} = -\kappa \left\{ \frac{\partial u}{\partial x}(x_0,t) - \frac{\partial u}{\partial x}(x_0+h,t) \right\}, \tag{8.30}$$

where κ is called the **thermal conductivity** of the rod.[4] Note that $\kappa > 0$ since heat flows from higher to lower temperatures; that is, if temperature increases with x, then the heat flow is to the left (negative). Combining (8.29) and (8.30) we have

$$\sigma\rho \frac{\partial u}{\partial t}(x_0,t) \cdot h = \kappa \left\{ \frac{\partial u}{\partial x}(x_0+h,t) - \frac{\partial u}{\partial x}(x_0,t) \right\}.$$

Dividing by h, and letting h tend to zero, we obtain the hyphenate **heat equation**

$$\frac{\partial u}{\partial t} = a^2 \frac{\partial^2 u}{\partial x^2} \quad \text{with} \quad a^2 = \frac{\kappa}{\sigma\rho} > 0. \tag{8.31}$$

The constant a^2 is called the **thermal diffusivity coefficient**.

Note the similarity of this heat equation for a rod (8.31), with the wave equation for a string (8.11),

$$\frac{\partial^2 u}{\partial t^2} = c^2 \frac{\partial^2 u}{\partial x^2} \quad \text{for } u = u(x,t).$$

8.3.2 Solutions of the Heat Equation

We can solve the heat equation under various initial conditions with methods similar to those for the wave equation.

As an illustration of the method, we assume dimension 1 and assume the following initial conditions:

1. The rod is of finite length with ends at $x = 0$ and $x = \pi$.
2. At time $t = 0$, the temperature function has the form $u(x,0) = f(x)$.
3. The ends of the rod are held at zero temperature, $u(0,t) = u(\pi,t) = 0$.

As with the wave equation, we separate variables

$$u(x,t) = X(x)T(t), \tag{8.32}$$

and then differentiate the heat equation (8.31) to obtain

$$X(x)T'(t) = a^2 X''(x)T(t), \quad \text{or}$$

[4] The value of κ depends on the composition of the rod. For example, for units in $cal/(sec \cdot cm \cdot C^{\circ})$, $\kappa = 0.92$ for copper, $\kappa = 0.0023$ for glass, and $\kappa = 0.0014$ for water.

$$\frac{1}{a^2}\frac{T'(t)}{T(t)} = \frac{X''(x)}{X(x)}.$$

The left-hand side depends only on t and the right-hand side only on x. Hence, both sides are a constant, say $-\alpha$. Thus

$$\frac{T'(t)}{T(t)} = -a^2\alpha, \text{ and} \tag{8.33}$$

$$X''(x) = -\alpha X(x). \tag{8.34}$$

The differential equation (8.33) has the form $[\ln T(t)]' = -a^2\alpha$ with solution

$$T(t) = T_0 e^{-a^2\alpha t} \quad \text{where} \quad T_0 = T(0).$$

Note that, if $\alpha < 0$, then $T(t)$ increases without bound. Given the above conditions, it is not possible for temperature to increase without bound. We conclude that α is positive, say $\alpha = c^2$. Thus (8.33) has the solution

$$T(t) = T_0 e^{-a^2c^2t}. \tag{8.35}$$

Now to (8.34). Since $\alpha = c^2$, it becomes

$$X''(x) + c^2 X(x) = 0. \tag{8.36}$$

Recall that this is the differential equation for harmonic motion as given by (8.4) in Subsection 8.2.1. The solutions are then shown to be of the form

$$X(x) = A\sin cx + B\cos cx. \tag{8.37}$$

Applying the initial condition $X(0) = 0$, we see that $B = 0$. Applying the initial condition $X(\pi) = 0$, we see that $A\sin c\pi = 0$. If $A = 0$, the entire rod is of zero temperature and nothing changes. Assuming $A \neq 0$, we conclude that $\sin c\pi = 0$, which can happen only when c is an integer, say $c = k$. Equation (8.34) thus has the solutions $X_k(x) = A_k \sin kx$ for $k = 1, 2, \ldots$. As with the wave equation, we obtain the most general solution by adding all possible solutions (the constant T_0 is absorbed by the constant A_k),

$$u(x,t) = \sum_{k=1}^{\infty} e^{-a^2k^2t} A_k \sin kx.$$

Applying the initial condition $u(x,0) = \sum_{k=1}^{\infty} A_k \sin kx = f(x)$, we see that the A_k are Fourier coefficients on $[0,\pi]$,

$$A_k = \frac{2}{\pi}\int_0^{\pi} f(x)\sin kx\,dx.$$

8.3.3 Other Boundary Conditions

The three boundary conditions as given in the Subsection 8.3.2 form just one example of a problem in the theory of the transfer of heat. We could, alternatively, consider the rod to have heat injected at one or both ends or at a point in between, or it could be an infinite rod, or it could be a ring, or the density could be nonuniform.

We will not examine all of these possibilities but we do consider three special cases in the following examples.

Example 8.12 Here, a rod of finite length has the temperature of one end maintained at zero and the other at a constant T. These are the modified conditions:

1. The rod is of finite length with ends at $x = 0$ and $x = \pi$.
2. The initial temperature has the form $u(x,0) = f(x)$.
3. The temperatures at the ends of the rod are constant, with

$$u(0,t) = 0 \quad \text{and} \quad u(\pi,t) = T.$$

There is again a solution in the form of a series. This time it is

$$u(x,t) = \frac{T}{\pi}x + \sum_{k=1}^{\infty} e^{-a^2k^2t} A_k \sin kx,$$

where the A_k are the Fourier coefficients $A_k = \frac{2}{\pi}\int_0^{\pi}\left\{f(x) - \frac{T}{\pi}x\right\}\sin kx\,dx$.

Note that the function $\left\{f(x) - \frac{T}{\pi}x\right\}$ satisfies the conditions of Subsection 8.3.2. This results in the function $\frac{T}{\pi}x$ being added to that solution.

Example 8.13 Next we consider heat flow in a ring of radius 1 and circumference 2π. Following the method of Subsection 8.3.2, we can obtain the solution

$$u(x,t) = \sum_{k=0}^{\infty} e^{-a^2k^2t}\Big\{A_k \sin kx + B_k \cos kx\Big\}.$$

If the initial temperature is given by $f(x) = u(x,0)$, we see that A_k and B_k are the Fourier coefficients

$$A_k = \frac{1}{2\pi}\int_0^{2\pi} f(x)\sin kx\,dx \quad \text{and} \quad B_k = \frac{1}{2\pi}\int_0^{2\pi} f(x)\cos kx\,dx.$$

The complex form of the Fourier series results in the solution

$$u(x,t) = \sum_{k=-\infty}^{\infty} e^{-a^2k^2t}\widehat{f}(k)e^{ikx},$$

which can be written as a convolution $u(x,t) = f * \mathcal{H}_t(x)$, where

$$\mathcal{H}_t(x) = \sum_{k=-\infty}^{\infty} e^{-a^2k^2t} e^{ikx}. \tag{8.38}$$

The family $\{\mathcal{H}_t\}_{(t>0)}$ is called the **heat kernel on the unit circle**. This forms a summability kernel and an approximate identity for 2π-periodic functions f as $t \to 0^+$. This can be confirmed by verifying the conditions **(SK1″)**, **(SK2″)**, and **(SK3″)** as given in Subsection 7.8.3.[5]

Example 8.14 Now we consider heat flow along an infinite rod. Here, the problem is to find a solution $u(x,t)$ of the heat equation (8.31) defined for all $x \in \mathbb{R}$ and $t > 0$ satisfying the initial condition $u(x,0) = f(x)$ for $-\infty < x < \infty$. As shown in Subsection 8.3.2, the solution separates so that

$$u(x,t) = X(x)T(t) \quad \text{with} \quad X''(x) + c^2X(x) = 0 \text{ and } T'(t) + a^2c^2T(t) = 0.$$

These differential equations solve as follows
$X(x) = A\cos cx + B\sin cx$ and $T(t) = t_0e^{-a^2c^2t}$. Thus

$$u(x,t) = X(x)T(t) = \big(A\cos cx + B\sin cx\big)e^{-a^2c^2t}.$$

In the finite rod case, we were able to use the initial conditions $f(0) = f(\pi) = 0$ to make c an integer and thus obtain a series solution. Not so in this case. But we can use the heat kernel on $\mathbb{R}$ as defined in Subsection 7.8.3,

$$H_t(x) = \frac{1}{\sqrt{4\pi a^2 t}} e^{-\frac{x^2}{4a^2t}} \quad \text{to write} \quad f(x) = \lim_{t\to 0^+} H_t(x)$$

and $u(x,t) = f * H_t(x) = \int_{-\infty}^{\infty} f(x-s)H_t(s)\,ds = \int_{-\infty}^{\infty} f(s)H_t(x-s)\,ds$. This solution satisfies the heat equation, as shown in Subsection 7.8.3.

8.3.4 Heat Equation in Higher Dimensions

Recall the definition of the Laplacian operator in two dimensions (Subsection 8.2.5)

$$\nabla^2 = \frac{\partial^2}{\partial x^2} + \frac{\partial^2}{\partial y^2}.$$

The heat equation in two dimensions can thus be written as

$$\frac{\partial u}{\partial t} = a^2\nabla^2 u. \tag{8.39}$$

[5] For a detailed proof, please see Section 1.8.3 of H. Dym & H.P. McKean, *Fourier series and integrals*, Academic Press, New York (1985).

It is the equation for heat flow on the (x, y)-plane. Fourier analytic methods can be applied to two dimensions.

In three dimensions, the Laplacian operator is given by

$$\nabla^2 = \frac{\partial^2}{\partial x^2} + \frac{\partial^2}{\partial y^2} + \frac{\partial^2}{\partial z^2}.$$

Thus for $u(x, y, z, t)$, the heat equation still has the same form (8.39).

Exercises for Section 8.3

8.3.1 Solve the heat equation for a rod with $a = 1$, $u(x,0) = 2\sin x - \sin 3x$, and $u(0,t) = u(\pi,t) = 0$.

8.3.2 Solve the heat equation for a rod with $a = 1$, $u(0,t) = u(\pi,t) = 0$, and

$$u(x,0) = \begin{cases} x & : \text{ if } 0 \leq x \leq \pi/2 \\ \pi - x & : \text{ if } \pi/2 \leq x \leq \pi. \end{cases}$$

8.3.3 Show that the solution of the heat equation for the circle, as given in Example 8.13, actually satisfies the heat equation.

8.4 Harmonic Functions

Definition 8.1 Let G be an open subset of $\mathbb{R}^n$ and let u be a twice continuously differentiable function $u\colon G \longrightarrow \mathbb{R}$. **Laplace's equation** on G is the differential equation

$$\nabla^2 u = 0,$$

where ∇^2 is the Laplacian operator. A solution $u\colon G \longrightarrow \mathbb{R}$ of Laplace's equation is called a **harmonic function** on G.

A solution of a heat equation where $\frac{\partial u}{\partial t} = 0$ is a harmonic function. This would be the case when the heat distribution has reached a steady state and no longer changes with time. In Example 8.12, the linear function $\frac{T}{\pi}x$ is the steady-state solution. In higher dimensions, harmonic functions need not be linear.

The following theorem and its corollary show that a function which is harmonic on a bounded open set G and continuous on its closure, is completely determined by its values on the boundary of G.

Theorem 8.4 *Let G be a bounded open subset of $\mathbb{R}^n$. Suppose that $u\colon G \longrightarrow \mathbb{R}$ is harmonic on G and continuous on the closure of G. If $m \le u(x) \le M$ for all $x = (x_1, x_2, \ldots, x_n)$ on the boundary of G, then $m \le u(x) \le M$ for all $x \in G$.*

Proof We first show that $u(x) \le M$ for all $x \in G$. Since G is a bounded subset of $\mathbb{R}^n$, the closure $\overline{G}$ of G is compact by the Heine–Borel theorem (Theorem 0.23). For notational simplicity consider the case $n = 2$. Let $\epsilon > 0$ be given. Define v to be the function $v = u(x, y) + \epsilon x^2$ on $\mathbb{R}^n = \mathbb{R}^2$. Since v is continuous on the compact set $\overline{G}$, it attains its maximum value at some point $p = (x_0, y_0)$ in $\overline{G}$ (by Theorem 0.27). The point p is either in G or on the boundary of G.

Claim: The point is not in G. Assume that $p \in G$. By elementary calculus, the first order partial derivatives must be zero at p, and the second order partial derivatives cannot be positive; that is, $v_x(p) = v_y(p) = 0$ and $v_{xx}(p) \le 0$, $v_{yy}(p) \le 0$. In particular $v_{xx}(p) + v_{yy}(p) \le 0$. But

$$v_{xx}(p) = u_{xx}(p) + 2\epsilon \quad \text{and} \quad v_{yy}(p) = u_{yy}(p).$$

So $u_{xx}(p) + u_{yy}(p) = v_{xx}(p) - 2\epsilon + v_{yy}(p) \le -2\epsilon < 0$. This violates the condition that u is harmonic on G, which proves the claim.

Thus $p = (x_0, y_0)$ is on the boundary of G. Let $b = \sup\{x \mid (x, y) \in \overline{G}\}$ and let $v = u(x, y) + \epsilon x^2$ as above. We have for any $(x, y) \in G$,

$$v(x, y) \le v(p) = v(x_0, y_0) = u(x_0, y_0) + \epsilon x_0^2 \le M + \epsilon b^2.$$

Hence, for any $(x, y) \in G$, we have

$$u(x, y) = v(x, y) - \epsilon x^2 \le M + \epsilon b^2 - \epsilon x^2 = M + \epsilon(b^2 - x^2).$$

This inequality holds for all $\epsilon > 0$. Taking the infimum we have

$$u(x, y) \le M \quad \text{for all} \quad (x, y) \in G.$$

The same argument can be used to show $m \le u(x, y)$ for all $(x, y) \in G$. ∎

Corollary 8.5 *Let G be a bounded open subset of $\mathbb{R}^n$. Suppose that u and v are harmonic on G and continuous on the closure of G. If $u = v$ on the boundary of G, then $u = v$ on all of G.*

Proof Consider the function $u - v$. It is harmonic on G and continuous on the closure of G and identically zero on the boundary of G. By Theorem 8.4, $u - v$ must be zero on all of G. Thus $u = v$ on G. ∎

Definition 8.2 Consider a bounded open $G \subset \mathbb{R}^n$ and a continuous real-valued function f defined on the boundary of G. The **Dirichlet problem** for G with boundary function f consists of finding a function u that is harmonic on G, continuous on the closure of G, and equal to f on the boundary of G.

Corollary 8.5 shows that any such solution is unique. We give two examples where the solution is found in terms of the Fourier series of f.

Example 8.15 Harmonic functions on a rectangle: Consider the steady-state heat equation $\nabla^2 u = 0$ on a rectangle $R = \{(x,y) \mid 0 \leq x \leq \pi, 0 \leq y \leq 1\}$ where u vanishes on three of the four sides $u(0,y) = u(\pi,y) = u(x,1) = 0$ and is defined by some specified function f on the fourth side $u(x,0) = f(x)$. Using separation of variables, $u(x,y) = g(x)h(y)$, and Laplace's equation, we obtain $g''(x)h(y) + g(x)h''(y) = 0$, which leads to

$$-\frac{g''(x)}{g(x)} = \frac{h''(y)}{h(y)}.$$

One side is a function of x and the other is a function of y. So both must be constant. Following the solutions of the wave equation as given in Equation (8.14), we can show that u has the solution

$$u(x,y) = \sum_{k=1}^{\infty} A_k \left(\frac{\sinh k(1-y)}{\sinh k} \right) \sin kx,$$

where f has the Fourier expansion $f(x) = \sum_{k=1}^{\infty} A_k \sin kx$.

Example 8.16 Harmonic functions on a disk: Consider the steady-state heat equation $\nabla^2 u = 0$ on the unit disk $D = \{(r,\theta) \mid 0 \leq r < 1, 0 \leq \theta < 2\pi\}$ where u is specified by a function $u(1,\theta) = f(\theta)$ on the unit circle $C = \{(r,\theta) \mid r = 1, 0 \leq \theta < 2\pi\}$. Using the polar form of the Laplacian operator (Theorem 8.3), we can separate the variables as

$$r^2 \frac{\partial^2 u}{\partial r^2} + r \frac{\partial u}{\partial r} = -\frac{\partial^2 u}{\partial \theta^2}.$$

Looking for a solution of the form $u(r,\theta) = g(r)h(\theta)$, we have

$$\frac{r^2 X''(r) + g'(r)}{g(r)} = -\frac{T''(\theta)}{h(\theta)}.$$

This must be equal to some constant α. Following the solutions of the wave equation, we can show that α is an integer $\alpha = k$, and h has the solutions

$$h(\theta) = A_k \cos k\theta + B_k \sin k\theta.$$

The only solutions of the differential equation for g are $g(r) = r^k$. The general solution thus comes out to be

$$u(r,\theta) = \sum_{k=1}^{\infty} A_k r^k \sin k\theta + \sum_{k=0}^{\infty} B_k r^k \cos k\theta.$$

For the function f, we have $r = 1$. Hence

$$f(\theta) = u(1,\theta) = \sum_{k=1}^{\infty} A_k \sin k\theta + \sum_{k=0}^{\infty} B_k \cos k\theta,$$

which means that the A_k and B_k are the Fourier coefficients of f.

Exercises for Section 8.4

8.4.1 Show that $u(x, y) = e^x \cos y$ is a harmonic function on $G = \mathbb{R}^2$.

8.4.2 Show that $u(x, y) = x^2 - y^2 + 3x + 4$ is a harmonic function on $G = \mathbb{R}^2$.

8.4.3 Show that $u(x,u) = -\ln(x^2 + y^2)$ is harmonic on $G = \mathbb{R}^2 - \{0\}$.

8.4.4 Find the harmonic function on the rectangle as given in Example 8.15 where f is the constant function $f(x) = 1$.

8.5 Gambler's Ruin and Random Walk

Let us consider a game of tossing a fair coin whereby on heads you win one dollar and on tails you lose a dollar. This is called a fair game because you have an equal chance of winning or losing. It is a common perception that, the more you play a fair game, the more likely you will come out even. In truth (as we will show in Theorem 8.9), the variability of your winnings and losses will *increase* in proportion to the square root of number of games you play. If you start playing with a dollars and your opponent starts with b dollars, then with probability 1, either you or your opponent will run out of money (ruin).

8.5.1 Elements of Discrete Probability

For this game of repeated tosses of a fair coin, let the variable e_k be the outcome of the kth toss. As described above, if the kth toss comes out heads, then you win one dollar ($e_k = 1$), and if tails, you lose one dollar ($e_k = -1$). Because this is a fair game, we have the probabilities

$$Pr(e_k = 1) = Pr(e_k = -1) = 1/2.$$

Your expected winning (or mean value) on the kth play is clearly 0 and can be computed as follows:

$$E(e_k) = (+1)Pr(e_k = 1) + (-1)Pr(e_k = -1) = \frac{1}{2} - \frac{1}{2} = 0. \qquad (8.40)$$

Define your winnings w_n after n plays to be the sum

$$w_n = e_1 + e_2 + \cdots + e_n.$$

Note that the possible values of w_n are $0, \pm 1, \pm 2, \ldots, \pm n$. By linearity, your expected winnings after n plays is zero:

$$E(w_n) = E(e_1 + \cdots + e_n) = E(e_1) + \cdots + E(e_n) = 0 + \cdots + 0 = 0. \qquad (8.41)$$

The variables e_k and w_n are called **discrete random variables** because they are functions that assign real values to outcomes of a discrete random experiment.[6] For a given discrete random variable X, the **expected value** (or **mean value**) $E(X) = \mu_X$ is the weighted average of the possible values of the random variable weighted by the probabilities of those values. This is how we computed $E(e_k) = 0$ in Equation (8.40). For example, if we define a random variable Y to be the square of the outcome e_1, then $Y = e_1^2$ has the value 1, regardless of a win or loss. So the expected value is $E(Y) = E(e_1^2) = (1)Pr(Y = 1) = 1 \cdot 1 = 1$.

Two discrete random variables X and Y are said to be **independent** if

$$Pr(X = x \text{ and } Y = y) = Pr(X = x) \cdot Pr(Y = y).$$

More generally, two random events A and B are **independent** if the probability that **both** A and B happen is the product of the probabilities of each happening. It is clear that tosses of a fair coin are independent:

Theorem 8.6 *Any two tosses of a fair coin are independent.*

The **standard deviation** of a random variable X is defined to be a weighted ℓ_2 norm

$$SD(X) = \sigma_X = \left\{E\Big((X - \mu_X)^2\Big)\right\}^{1/2}.$$

It measures a random variable's variability from its mean μ_X. For example, the random variable $w_1 = e_1$ can attain values of $+1$ and -1. Its variability

[6] An **experiment** is a procedure that results in a measurement or observation. An experiment whose outcome depends upon chance is called a **random experiment**. And if there are only a finite or countably infinite possible outcomes, then its called a **discrete random experiment**.

from its mean 0, is measured by the standard deviation $SD(w_1) = SD(e_1) = \sqrt{E(e_1 - \mu_{e_1})^2} = \sqrt{E(e_1 - 0)^2} = \sqrt{E(e_1^2)} = \sqrt{1} = 1$. Similarly, for any $k = 1, 2, \ldots$, we clearly have $SD(e_k) = 1$.

It is straightforward to show the following formula for the standard deviation of a random variable X,

$$SD(X) = \sqrt{E(X^2) - E^2(X)} = \left\{E(X^2) - \mu_X^2\right\}^{1/2}. \tag{8.42}$$

Theorem 8.7 *If random variables X and Y are independent, then $E(X \cdot Y) = E(X) \cdot E(Y)$.*

Proof Let W be the product random variable $W = X \cdot Y$. The values of W are the products $w = xy$ for all the possible values x of X and y of Y. Given that X and Y are independent, we have

$$Pr(X = x, Y = y) = Pr(X = x) \cdot Pr(Y = y).$$

Thus

$$\begin{aligned}
E(X \cdot Y) = E(W) &= \sum_w w Pr(W = w) \\
&= \sum_{x,y} xy Pr(X = x, Y = y) \\
&= \sum_x \left\{ \sum_y xy Pr(X = x) Pr(Y = y) \right\} \\
&= \sum_x x Pr(X = x) \left\{ \sum_y y Pr(Y = y) \right\} \\
&= \sum_x x Pr(X = x) \{E(Y)\} = E(X) \cdot E(Y).
\end{aligned}$$
■

Theorem 8.8 *If random variables X and Y are independent, then*

$$SD^2(X + Y) = SD^2(X) + SD^2(Y).$$

Proof Let $W = X + Y$. By definition $SD^2(W) = E(W^2) - E^2(W)$. Then

$$\begin{aligned}
E(W^2) = E\big((X + Y)^2\big) &= E(X^2 + 2X \cdot Y + Y^2) \\
&= E(X^2) + 2E(X \cdot Y) + E(Y^2) \\
&= E(X^2) + 2E(X) \cdot E(Y) + E(Y^2).
\end{aligned}$$

The last equality follows from Theorem 8.7. Also

$$\begin{aligned}
E^2(W) = E^2(X + Y) &= \big(E(X) + E(Y)\big)^2 \\
&= E^2(X) + 2E(X) \cdot E(Y) + E^2(Y).
\end{aligned}$$

Subtracting, the middle terms $2E(X) \cdot E(Y)$ cancel, and thus

$$SD^2(X+Y) = E(W^2) - E^2(W) = E(X^2) - E^2(X) + E(Y^2) - E^2(Y)$$
$$= SD^2(X) + SD^2(Y).$$ ∎

Exercises for Subsection 8.5.1

8.5.1 Prove Equation (8.42).

8.5.2 Use the definition of independence to prove Theorem 8.6.

8.5.3 In two tosses of a coin, the possible outcomes are HH, HT, TH, TT. Assuming a fair coin for the two tosses, let Y be the number of heads. Find $E(Y)$ and $SD(Y)$.

8.5.4 Repeat the previous exercise, except for three tosses of a fair coin.

8.5.5 Consider a single roll of fair die. Let Y be the number of dots on the top face. The possible values of Y are $1, 2, 3, 4, 5, 6$. Find $E(Y)$ and $SD(Y)$.

8.5.6 Let X be a discrete random variable and let $Y = aX + b$ for some real numbers a and b. Show that $E(Y) = aE(X) + b$ and $SD(Y) = |a|SD(X)$.

8.5.2 Gambler's Ruin

As promised, we now show that in the fair game of n repeated tosses of a fair coin, although the expected winnings are zero for both players, the variability of the winnings and losses increase with the number of games played; namely, $SD(w_n) = \sqrt{n}$.

Theorem 8.9 *Let w_n be your winnings after n tosses. Then $E(w_n) = 0$ and $SD(w_n) = \sqrt{n}$.*

Proof That $E(w_n) = 0$ was noted before in Equation (8.41). Also recall $SD(e_k) = 1$ for all $k = 1, 2, \ldots$. For a single game, we have $n = 1$ and winnings $w_1 = e_1$. Then $SD(w_1) = SD(e_1) = \sqrt{E(e_1^2) - E^2(e_1)} = \sqrt{E(e_1^2) - 0^2} = \sqrt{1}$, as previously shown. For $n = 2$, we have $w_2 = e_1 + e_2$. Since tosses of a fair coin are independent, we have, by Theorem 8.8,

$$SD(w_2) = \sqrt{SD^2(e_1 + e_2)} = \sqrt{SD^2(e_1) + SD^2(e_2)} = \sqrt{1+1} = \sqrt{2}.$$

Repeating this argument, leads to $SD(w_n) = \sqrt{n}$, for all $n = 1, 2, \ldots$. ∎

In most cases, one proof is enough. However, there is some value in computing $SD(w_n) = \sqrt{n}$ combinatorially by brute force.

Alternative proof: This proof uses Equation (8.42) but not subsequent theorems. For each game, there are 2 possible outcomes of either a win of a dollar or a loss of a dollar. For n games, there are 2^n sequences of wins and losses. These outcomes are all equally likely, so each has a probability of 2^{-n}. In a sequence of n games, if you win a dollar j times (and lose a dollar $n-j$ times), your resulting winnings are $w_n = (j)-(n-j) = 2j-n$ dollars. There are $\binom{n}{j}$ ways this can happen. So the probability of j wins (and $n-j$ losses) in n games is $Pr(w_n = 2j-n) = \binom{n}{j}2^{-n}$. Since $E(w_n) = 0$, by Equation (8.42), we have $SD^2(w_n) = E\left(w_n^2\right)$. Applying the following combinatorial identities,

$$\sum_{j=1}^{n}\binom{n}{j} = 2^n,\ \sum_{j=1}^{n} j\binom{n}{j} = n2^{n-1},\ \text{and}\ \sum_{j=1}^{n} j^2\binom{n}{j} = (n+n^2)2^{n-2}, \tag{8.43}$$

results in

$$\begin{aligned}
SD^2(w_n) = E\left(w_n^2\right) &= \sum_{j=1}^{n}(2j-n)^2 Pr(w_n = 2j-n)\\
&= \sum_{j=1}^{n}(2j-n)^2\binom{n}{j}2^{-n}\\
&= 2^{-n}\left\{n^2\sum_{j=1}^{n}\binom{n}{j} - 4n\sum_{j=1}^{n} j\binom{n}{j} + 4\sum_{j=1}^{n} j^2\binom{n}{j}\right\}\\
&= 2^{-n}\left\{n^2 2^n - 4n^2 2^{n-1} + 4(n+n^2)2^{n-2}\right\} = n.
\end{aligned}$$

■

Exercises for Subsection 8.5.2

8.5.7 Let $n = 4$. List all of the possible values of the random variable w_4. Find the probabilities $Pr(w_4 = k)$ for all $-4 \le k \le 4$. Then use these probabilities to confirm $E(w_4) = 0$ and $SD(w_4) = 2$.

8.5.8 Prove the identities of Equations (8.43).

8.5.9 Player A and player B toss coins, betting one dollar per game, until one of the players is ruined. If each starts with one dollar, clearly after one game, the games stop. Now suppose player A starts with one dollar and player B with two dollars. **(a)** Find the probability that the games stop after one toss. **(b)** After two tosses. **(c)** After three tosses. **(d)** After four tosses.

8.5.10 Repeat Exercise 8.5.9 where each player starts with two dollars.

8.5.3 Random Walk

Now consider an object moving on the x-axis, starting at the origin, so that at times $t = 1, 2, 3, \ldots$ it randomly moves (steps) either to the right (forward) with probability $1/2$ or to the left (backward) one unit with probability $1/2$. Suppose that the outcome at each step is independent of the preceding steps. This is called a **random walk** in one dimension. Let e_k be the movement at the kth step (at time $t = k$); namely, if the object moves forward, then $e_k = 1$, and if the object moves backward, then $e_k = -1$. We have $Pr(e_k = 1) = Pr(e_k = -1) = 1/2$. Since the object starts at the origin, its position at time $t = n$ is

$$w_n = e_1 + e_2 + \cdots + e_n.$$

Obviously, the two examples of coin toss gambling and one-dimensional random walk are abstractly the same. In the gambling example, w_n represents the winnings after n plays; in the random walk example, w_n represents the position of the object after n steps. From Theorem 8.9 we can conclude that, for a random walk, the expected position of the particle is always the starting point, $E(w_n) = 0$, but that the probable position varies with the square root of time, $SD(w_n) = \sqrt{n}$.

In order to study the behavior of w_n, consider the function $f_n \in L_{2\pi}$ defined to have the Fourier coefficients $\widehat{f_n}(k) = Pr(w_n = k)$. That is,

$$f_n(x) = \sum_{k=-\infty}^{\infty} \widehat{f_n}(k)e^{ikx} = \sum_{k=-\infty}^{\infty} Pr(w_n = k)e^{ikx}. \tag{8.44}$$

Theorem 8.10 Pólya[7]**:** $f_n(x) = \cos^n x$.

Proof Function $f_n(x)$ of (8.44) is the expected value of the random variable $e^{iw_n x}$,

$$E(e^{iw_n x}) = \sum_{k=-\infty}^{\infty} Pr(w_n = k)e^{ikx} = f_n(x).$$

[7] George Pólya (1887–1985) was a Hungarian mathematician. He contributed to the theory of combinatorics, number theory, probability, and mathematical physics. A quote: "I thought I am not good enough for physics and I am too good for philosophy. Mathematics is in between." Pólya had a strong interest in pedagogy. He wrote the popular book *How to Solve It*, Princeton University Press (1945), which has been translated into 17 languages.

Since $w_n = e_1+e_2+\cdots+e_n$, we can take all possible values $y_1+y_2+\cdots+y_n$ with $y_1 = \pm 1, \ldots, y_n = \pm 1$. We have

$$\begin{aligned} E(e^{iw_n x}) &= E(e^{i(e_1+\cdots+e_n)x}) \\ &= E(e^{ie_1 x}e^{ie_2 x}\cdots e^{ie_n x}) \\ &= \sum Pr(e_1 = y_1, \ldots, e_n = y_n)e^{iy_1 x}e^{iy_2 x}\cdots e^{iy_n x}, \end{aligned}$$

where the sum is taken of all values of $y_1 = \pm 1, \ldots, y_n = \pm 1$. By independence of the steps (games), we have

$$\begin{aligned} &Pr(e_1 = y_1, \ldots, e_n = y_n) \\ &\quad = Pr(e_1 = y_1)\cdot Pr(e_2 = y_2)\cdots Pr(e_n = y_n) = \frac{1}{2^n}. \end{aligned}$$

Thus $f_n(x) = E(e^{iw_n x}) = \frac{1}{2^n}\sum e^{iy_1 x}e^{iy_2 x}\cdots e^{iy_n x} = \frac{1}{2^n}(e^{ix}+e^{-ix})^n$.
Since $\frac{e^{ix}+e^{-ix}}{2} = \cos x$, we have $f_n(x) = E(e^{iw_n x}) = \cos^n x$. ∎

We can use the formula for Fourier coefficients:

$$Pr(w_n = k) = \widehat{f_n}(k) = \frac{1}{2\pi}\int_{-\pi}^{\pi}(\cos^n x)e^{-ikx}\,dx.$$

In particular, $Pr(w_n = 0) = \frac{1}{2\pi}\int_{-\pi}^{\pi}\cos^n x\,dx$. If $p_n = Pr(w_n = 0)$, then

$$p_n = 0 \text{ for odd } n \text{ and } p_{2n} = \binom{2n}{n}/2^{2n}. \tag{8.45}$$

So $(p_1, p_2, p_3, p_4, \ldots) = \left(0, \frac{2}{4}, 0, \frac{6}{16}, 0, \frac{20}{64}, 0, \frac{70}{256}, 0, \frac{252}{1024}, 0, \ldots\right)$.

If R_n is the event that the object returns to the origin after n steps, $w_n = 0$, then the total number of times that object returns to the origin is $R_1 + R_2 + \cdots$. And the expected number of times that object returns to the origin is thus $p_1 + p_2 + \cdots$. But these probabilities sum to ∞ since, using Abel summation, we have

$$\begin{aligned} \sum_{n=1}^{\infty} p_n &= \lim_{r\to 1^-}\sum_{n=1}^{\infty} r^n p_n \\ &= \lim_{r\to 1^-}\frac{1}{2\pi}\int_{-\pi}^{\pi}\sum_{n=1}^{\infty}(r\cos x)^n\,dx \\ &= \lim_{r\to 1^-}\frac{1}{2\pi}\int_{-\pi}^{\pi}\frac{r\cos x}{1-r\cos x} = \frac{1}{2\pi}\int_{-\pi}^{\pi}\frac{\cos x}{1-\cos x}\,dx = \infty. \end{aligned}$$

We conclude that the object is expected to return to the origin infinitely many times. We have thus proven the following theorem.

Theorem 8.11 *For a random walk in one dimension, starting at the origin, the object returns to the origin infinitely many times.*

Theorem 8.12 *For a random walk in one dimension, starting at the origin, $Pr(w_n = k) = 0$ when $n - k$ is odd. Otherwise, $Pr(w_n = k) = \binom{n}{\frac{n-k}{2}} \Big/ 2^n$.*

The proof is left as Exercise 8.5.14.

In the two-dimensional case, consider a drop of particles placed into a fluid at the origin. As the particles diffuse in the fluid, each particle is bombarded by the movement of the molecules of the fluid and thus experiences a random walk in two dimensions. As the drop diffuses, its radius increases but the pattern remains circular. Although the circle remains centered at the origin, the radius of the circle increases in proportion $\sqrt{t}$. It can be shown that, as in the one-dimensional case, individual particles return to the center infinitely many times.

In three dimensions, the dispersion pattern is spherical of radius proportional to $\sqrt{t}$. However, for three and higher dimension, it can be shown that individual particles no longer return to the center infinitely many times.

Exercises for Subsection 8.5.3

8.5.11 Let $n = 5$. List all of the possible values of the random variable w_5. Find the probabilities $Pr(w_5 = k)$ for all $-5 \leq k \leq 5$. Find $E(w_5)$ and $SD(w_5)$.

8.5.12 Prove Equation (8.45). Hint: In $2n$ steps, there are 2^{2n} different routes which can be taken. Show that $\binom{2n}{n}$ of them come back to $x = 0$.

8.5.13 Use Stirling's formula $n! \approx \sqrt{2\pi n}\left(\frac{n}{e}\right)^n$ to show that $p_{2n} = \frac{\binom{2n}{n}}{2^{2n}} \approx \sqrt{\frac{1}{\pi n}}$. Use this to give an alternative proof that $\sum_{n=1}^{\infty} p_n$ diverges.

8.5.14 Prove Theorem 8.12. Hint: In n steps, there are 2^n different routes that can be taken. Show that $\binom{n}{\frac{n-k}{2}}$ of them end up in position $w_n = k$.

8.5.15 Show that a random walk along the x-axis visits each point $x = k$ infinitely often.

8.6 Atomic Theory of Matter

The theory that matter is formed of atoms can be traced back to the ancient Greek philosophers Democritus and Leucippus. Almost two millennia later, the English chemist John Dalton (1766–1844), noted that elements combine in integral multiples to form compounds. For example, one part oxygen and two parts hydrogen form the compound water. In 1803, in a series of lectures, he declared that this chemical evidence leads to an atomic theory of matter. Yet many still held the contrary view that matter is infinitely divisible.

There was no direct evidence of the existence of atoms until 1905 when Albert Einstein published a paper that explained **Brownian motion**. This motion is named after the Scottish botanist Robert Brown (1773–1858) who, in 1827, observed microscopic particles ejected from pollen grains suspended in water. He noticed that the particles moved in a jittery, irregular, zigzag pattern. The same random motion was observed whenever small particles were suspended in a fluid medium, such as dust and smoke particles in air.

In 1905 Albert Einstein showed that Brownian motion can be explained as a random walk of the suspended particles due to their bombardment by atoms and molecules of the fluid medium.

We limit ourselves to the Brownian motion in a single direction along the x-axis. Let $u(x,t)$ be the density of the particles experiencing Brownian motion at point x and time t. As we saw in Subsection 8.5.3, if we start at position $x = 0$ and time $t = 0$, the suspended particles diffuse in the fluid medium in proportion to $\sqrt{t}$.

Einstein argued that the function u satisfies the differential equation

$$\frac{\partial u}{\partial t} = D\frac{\partial^2 u}{\partial x^2}, \tag{8.46}$$

where D is a constant called the coefficient of diffusion for the fluid. This is called the **diffusion equation**. The proof of (8.46) is not difficult but uses argument of physics outside the scope of this book.

Note that (8.46) is in the form of a heat equation with the thermal diffusivity coefficient a^2 replaced by the coefficient of diffusion D. In Example 8.14, it was shown that $u(x,t)$ is solved by the heat kernel $H_t(x)$ on $\mathbb{R}$ as defined in Equation (7.35). We thus have the following theorem.

Theorem 8.13 *Suppose that Brownian motion on the x-axis starts at position $x = 0$ at time $t = 0$. The diffusion equation* (8.46) *on $\mathbb{R}$, which gives the density $u(x,t)$ of particles at point x and time t, has solution*

$$u(x,t) = \frac{1}{\sqrt{4Dt\pi}} e^{-\frac{x^2}{4Dt}}. \tag{8.47}$$

For each $t > 0$, the function $H_t(x) = u(x,t)$ follows a Gaussian curve (Definition 7.25) with mean $\mu = 0$ and standard deviation $\sigma = \sqrt{2Dt}$. For a given suspension of particles, the standard deviation σ of the dispersion is a measurable quantity. Hence, D is a measurable quantity.

Now we explain how the assumption of the infinite divisibility of matter would lead to no Brownian motion.

A mole of a pure substance is the mass in grams equal to its molecular mass. For example, since the molecular mass of oxygen is 15.9994, a mole of oxygen consists of 15.9994 grams. If a fluid medium is made of molecules, then the number of molecules in a mole is a constant N_A (Avogadro's number). By the use of various gas laws, Einstein showed that the coefficient of diffusion D can be expressed by the equation

$$D = \frac{RT}{6\pi\eta P N_A}, \tag{8.48}$$

where T is the temperature in degrees Kelvin, η is the viscosity, P is the radius of the suspended particles, $R = \frac{\text{(pressure)(volume)}}{T} = 8.1345$ is the universal gas constant, and N_A is Avogadro's number.

The radius of the suspended particles P can be microscopically measured. The temperature T, the viscosity η, and the universal gas constant R are also measurable.

Equation (8.48) shows that N_A is inversely proportional to D. Thus, as the size of the molecules of the fluid medium decrease, N_A increases, and so, D decreases. In the limit, if matter were infinitely divisible, D would be zero and there would be no Brownian motion. The actual existence of Brownian motion shows that matter is discontinuous or "quantized."

Experimental data relating viscosity η and the coefficient of diffusion D could then be used with (8.48) to find Avogadro's number N_A:

$$N_A = \frac{RT}{6\pi\eta PD}. \tag{8.49}$$

At first, experiments seemed to be inconsistent with Einstein's analysis, but the French experimentalist Jean Perrin (1870–1942) finally confirmed Einstein's formulas in 1908. This work of Einstein and Perrin's experimental confirmation became convincing evidence for the existence of atoms and molecules. Perrin received the Nobel Prize for this work in 1926.

8.7 Sampling Theorem

For this section, it is helpful to have read Section 7.11 about the Fourier transform. The human ear cannot hear tones above 20,000 cycles per second.[8] Music signals are thus electronically filtered before being recorded so as to limit frequencies to below 20,000.

Let $f(t)$ represent the amplitude of such a signal at time t. The signal is said to be **band limited** if $\widehat{f}(s) = 0$ for $|s| > N$ for some fixed N, say $N = 20{,}000$. In that case, the sampling theorem below says that we can recover the entire signal by sampling it periodically $2N = 40,000$ times per second.

Theorem 8.14 Sampling theorem: *Suppose that $f(t)$ is continuous and $f \in L^1(\mathbb{R})$. If $\widehat{f}(s) = 0$ for $|s| > N$, then f is completely determined by its values at the points $t_k = \frac{k\pi}{N}$ for $k = 0, \pm 1, \pm 2, \ldots$. In fact,*

$$f(t) = \sum_{k=-\infty}^{\infty} f\left(\frac{k\pi}{N}\right) \frac{\sin(Nt - k\pi)}{Nt - k\pi}. \tag{8.50}$$

Proof Property **(FT3)** (Theorem 7.90) of the Fourier transform says for $f \in L^1(\mathbb{R})$, we have $\widehat{f} \in L^2(\mathbb{R})$. Since $\widehat{f}$ has support on the interval $[-N, N]$, we actually have $\widehat{f} \in L^2[-N, N]$. Then $\widehat{f}$ can be extended to a $2N$-periodic function and has a convergent Fourier series expansion (see Section 7.7)

$$\widehat{f}(s) = \sum_{k=-\infty}^{\infty} c_k e^{i(\frac{\pi}{N})ks}, \tag{8.51}$$

with coefficients c_k given by

$$c_k = \frac{1}{2N}\int_{N}^{N} \widehat{f}(s)e^{-i\frac{\pi}{N}ks}\, ds = \frac{1}{2N}\int_{-\infty}^{\infty} \widehat{f}(s)e^{-i\frac{\pi}{N}ks}\, ds. \tag{8.52}$$

Since $\widehat{f}(s) = 0$ for $|s| > N$, Fourier inversion (Theorem 7.92) results in

$$f(t) = \frac{1}{2\pi}\int_{-\infty}^{\infty} \widehat{f}(s)e^{ist}\, ds = \frac{1}{2\pi}\int_{-N}^{N} \widehat{f}(s)e^{ist}\, ds. \tag{8.53}$$

In particular, $\frac{\pi}{N} f\left(\frac{-k\pi}{N}\right) = \frac{\pi}{N}\frac{1}{2\pi}\int_{-N}^{N} \widehat{f}(s)e^{i\left(\frac{-k\pi}{N}\right)s} ds$. This is equal to c_k by (8.52). Thus (8.51) becomes

$$\widehat{f}(s) = \sum_{k=-\infty}^{\infty} \frac{\pi}{N} f\left(\frac{-k\pi}{N}\right) e^{i(\frac{\pi}{N})ks}.$$

[8] Nor can tones be heard below 20 cycles per second. However, a loud tone below 20 cycles can sometimes be felt as a vibration.

Inserting this into (8.53), we obtain

$$f(t) = \frac{1}{2\pi} \int_{-N}^{N} \left[\sum_{k=-\infty}^{\infty} \frac{\pi}{N} f\left(\frac{-k\pi}{N}\right) e^{i(\frac{\pi}{N})ks} \right] e^{ist}\, ds.$$

Here we are taking the inner product in $L^2[-N, N]$ of the Fourier series of $\widehat{f}$ with the function e^{ist}. We can integrate termwise,

$$\begin{aligned} f(t) &= \frac{1}{2N} \sum_{k=-\infty}^{\infty} f\left(\frac{-k\pi}{N}\right) \int_{-N}^{N} e^{i(\frac{\pi}{N})ks+ist}\, ds \\ &= \sum_{k=-\infty}^{\infty} f\left(\frac{-k\pi}{N}\right) \frac{\sin(Nt + k\pi)}{Nt + k\pi}. \end{aligned}$$

Finally, replacing k by $-k$ in the series yields the desired result. ∎

Because the range of human hearing does not extend beyond 20,000 cycles per second, music for CD recordings are typically sampled a little bit past twice that, to 44,100 times per second.

The expansion (8.50) unfortunately does not converges rapidly. For that reason, sampling is often done at a higher frequency. This is called **oversampling**. For example, professional video recording equipment samples audio at a rate of 48,000 times per second; and DVDs and Blu-ray audio double this rate further to 96,000 times per second.

Telephone signals, on the other hand, have a frequency band of 300–3400 cycles per second (or a bandwidth of 3,100 cycles per second). They are typically sampled 8,000 times per second. In the silent gaps between sampled points, more voice signals can be inserted. Actually, up to 8,000 phone conversations have been transmitted on a single fiber optic cable.

Similarly, in video transmission, if the signal is band limited, only a small portion of an image needs to be transmitted in order to reconstruct an entire high definition image. The sampling theorem is one of the reasons space probes are able to transmit breathtaking pictures from deep space.

Theorem 8.15 *Let C_N be the space of $L^1(\mathbb{R})$ functions f whose Fourier transforms $\widehat{f}$ vanish outside of $[-N, +N]$. Define the sequence of functions $w_k(t) = \frac{\sin(Nt-k\pi)}{Nt-k\pi}$ for $k = 0, \pm 1, \pm 2, \ldots$. The sequence w_k is an orthonormal basis of C_N and Equation (8.50) is the Fourier series of a function $f \in C_N$.*

Exercise for Section 8.7

8.7.1 Prove Theorem 8.15.

8.8 Uncertainty Principle

The duality between a function f and its Fourier transform $\widehat{f}$ gives rise to a theorem dual to the sampling theorem. Recall that a function $f(t)$ is **band limited** if there exists $N > 0$ for which $\widehat{f}(s)$ vanishes outside of $[-N, N]$. The dual concept would be case where there exists $L > 0$ such that $f(t)$ vanishes outside of $[-L, L]$. Such a function is said to be **time limited. Wavelets**, which we will consider in Section 8.9, are examples of time limited signals.

Because of the Fourier inversion theorem (Theorem 7.92), there is a dual formulation of the sampling theorem for time limited functions.

Theorem 8.16 *Suppose that a continuous function $f \in L^2(\mathbb{R})$ is time limited; that is, $f(t) = 0$ whenever $|t| > L$, for some $L > 0$. Then $\widehat{f}$ is completely determined by its values at the points $s_k = \frac{k\pi}{N}$ for $k = 0, \pm 1, \pm 2, \ldots$. In fact,*

$$\widehat{f}(s) = \sum_{k=-\infty}^{\infty} \widehat{f}\left(\frac{k\pi}{N}\right) \frac{\sin(Ns - k\pi)}{Ns - k\pi}. \tag{8.54}$$

The proof is left as Exercise 8.8.1.

Theorem 8.17 *A signal cannot be both band limited and time limited.*

We will not prove this theorem but instead we prove a general uncertainly principle, which says that it is not possible for both f and $\widehat{f}$ to be essentially localized. That is, if f is mostly confined to a small interval, then $\widehat{f}$ will be widely dispersed, and vice versa. To make this principle precise, we formally define the dispersion of a function.

Definition 8.3 Suppose f is a function in both $L^1(\mathbb{R})$ and $L^2(\mathbb{R})$. Define the **dispersion** of f about a point a to be

$$\Delta_a f = \frac{\displaystyle\int_{-\infty}^{\infty} (t-a)^2 |f(t)|^2 \, dt}{\displaystyle\int_{-\infty}^{\infty} |f(t)|^2 \, dt}. \tag{8.55}$$

If the dispersion $\Delta_a f$ is small, it means that f is concentrated near the point a. The uncertainty principle below says it is not possible for the dispersion of a function f and the dispersion of its Fourier transpose $\widehat{f}$ to both be very small.

Theorem 8.18 Uncertainty principle: *Suppose the function f is in both $L^1(\mathbb{R})$ and $L^2(\mathbb{R})$ and has a continuous derivative. For any $a, b \in \mathbb{R}$, we have*

$$(\Delta_a f)(\Delta_b \widehat{f}\,) \geq \frac{1}{4}. \tag{8.56}$$

Proof We begin with the case $a = b = 0$. If $tf(t)$ is not in $L^2(\mathbb{R})$, then by definition, $\Delta_0 f = \infty$, which makes (8.56) true. So we may assume $tf(t)$ in $L^2(\mathbb{R})$. Then if $f'(t)$ is not in $L^2(\mathbb{R})$, it turns out that $\Delta_0 \widehat{f} = \infty$ (see Exercise 8.8.3), which again makes (8.56) true. We can thus assume $tf(t)$ and $f'(t)$ are both in $L^2(\mathbb{R})$. Integrating by parts, we obtain

$$\int_{-\infty}^{\infty} |f(t)|^2\,dt = t|f(t)|^2\Big|_{-\infty}^{\infty} - \int_{-\infty}^{\infty} t\frac{d}{dt}|f(t)|^2\,dt.$$

Since $tf(t) \in L^2(\mathbb{R})$, we clearly have $\lim_{t\to\pm\infty} t|f(t)|^2 = 0$. Then

$$\begin{aligned}\int_{-\infty}^{\infty} |f(t)|^2\,dt &= 0 - \int_{-\infty}^{\infty} t\frac{d}{dt}|f(t)|^2\,dt = -\int_{-\infty}^{\infty} t\frac{d}{dt} f(t)\overline{f(t)}\,dt \\ &= -\int_{-\infty}^{\infty}\left(tf'(t)\overline{f(t)} + t\overline{f'(t)}f(t)\right)dt,\end{aligned}$$

which leads to $\int_{-\infty}^{\infty} |f(t)|^2\,dt \leq 2\int_{-\infty}^{\infty} |t|\,|f(t)|\,|f'(t)|\,dt$.

Using the Cauchy–Schwarz inequality (Theorem 7.5), we obtain the inequality

$$\left(\int_{-\infty}^{\infty} |f(t)|^2\,dt\right)^2 \leq 4\left(\int_{-\infty}^{\infty} t^2|f(t)|^2\,dt\right)\left(\int_{-\infty}^{\infty} |f'(t)|^2\,dt\right).$$

By the Plancherel identity (Theorem 7.94), we have $\int_{-\infty}^{\infty} |f'(t)|^2\,dt = \frac{1}{2\pi}\int_{-\infty}^{\infty} \left|\widehat{f'}(s)\right|^2 ds$. But $\widehat{f'}(s) = is\widehat{f}(s)$. Thus

$$\int_{-\infty}^{\infty} |f'(t)|^2\,dt = \frac{1}{2\pi}\int_{-\infty}^{\infty} \left|is\widehat{f}(s)\right|^2 ds = \frac{1}{2\pi}\int_{-\infty}^{\infty} s^2\left|\widehat{f}(s)\right|^2 ds. \tag{8.57}$$

Inserting Equation (8.57) into the above inequality, we get

$$\left(\int_{-\infty}^{\infty} |f(t)|^2\,dt\right)^2 \leq 4\left(\int_{-\infty}^{\infty} t^2|f(t)|^2\,dt\right)\left(\frac{1}{2\pi}\int_{-\infty}^{\infty} s^2\left|\widehat{f}(s)\right|^2 ds\right). \tag{8.58}$$

By Plancherel again, the left-hand side of (8.58) becomes

$$\left(\int_{-\infty}^{\infty} |f(t)|^2\,dt\right)^2 = \left(\int_{-\infty}^{\infty} |f(t)|^2\,dt\right)\left(\frac{1}{2\pi}\int_{-\infty}^{\infty} \left|\widehat{f}(t)\right|^2 dt\right).$$

Dividing Equation (8.58) by these two factors and by 4 results in $\frac{1}{4} \leq (\Delta_0 f)(\Delta_0 \widehat{f})$. This proves the case $a = b = 0$.

The case for general $a, b \in \mathbb{R}$ follows by replacing the function $f(t)$ with $F(t) = e^{-itb} f(t+a)$. Then $\Delta_a f = \Delta_0 F$ and $\Delta_b \widehat{f} = \Delta_0 \widehat{F}$. ∎

This theorem is what we will use to prove the famous Heisenberg inequality (Theorem 10.15). This inequality describes the position-momentum uncertainty of quantum theory, covered later in Section 10.4.

Exercises for Section 8.8

8.8.1 Prove Theorem 8.16. It is essentially the same as the proof of the sampling theorem (Theorem 8.14).

8.8.2 Suppose the function f is in both $L^1(\mathbb{R})$ and $L^2(\mathbb{R})$ and has a continuous derivative. Use integration by parts to prove $\widehat{f'}(s) = is\widehat{f}(s)$.

8.8.3 Suppose the function f is in both $L^1(\mathbb{R})$ and $L^2(\mathbb{R})$ and has a continuous derivative. Assume $tf(t)$ is in $L^2(\mathbb{R})$ but $f'(t)$ is not. Follow the argument of Theorem 8.18 and use Equation (8.57) to prove that $\Delta_0 \widehat{f} = \infty$.

8.8.4 Complete the proof of Theorem 8.18 by showing that $\Delta_a f = \Delta_0 F$ and $\Delta_b \widehat{f} = \Delta_0 \widehat{F}$, when $F(t) = e^{-itb} f(t+a)$.

8.8.5 Show that the uncertainty inequality (8.56) is an equality when $f(t) = e^{-At^2}$ with $A > 0$.

8.9 Wavelets

Unlike the oscillations of the wave functions sine and cosine which are periodic forever, a **wavelet** is a time limited (Section 8.8) burst of oscillations. It still has the wave-like characteristic in that it oscillates between positive and negative values, but its amplitude begins at zero, alternately increases and decreases, and then fades back to zero in a short period of time. An example of a wavelet is a musical tone of short duration, say a middle C on a piano that dampens to zero. Another example is a transient heart irregularity recorded by an electrocardiogram. Yet another is a seismic rumbling (tremor) of the earth. The application of wavelets to engineering originated with seismic signal processing in the oil exploration industry.

As we have seen, the **orthonormal trigonometric basis** $\frac{1}{2\pi} e^{ikt}$ $(k = 0, \pm 1, \pm 2, \ldots)$ is an excellent tool for the analysis of periodic functions. On the other hand, the defining characteristics of orthonormal bases used for the analysis of wavelets are as follows.

Definition 8.4 A **orthonormal wavelet basis** for $L^2(\mathbb{R})$ is a family of functions

$$\psi_{jk}(t) = 2^{\frac{j}{2}}\psi(2^j t - k) \quad \text{for} \quad j,k = 0, \pm 1, \pm 2, \ldots, \tag{8.59}$$

generated by a single fixed **mother wavelet** $\psi \in L^2(\mathbb{R})$ such that:

(1) The mother wavelet oscillates equally between positive and negative values in the sense that $\int_{-\infty}^{\infty} \psi(t)\,dt = 0$.
(2) The family ψ_{jk} forms an orthonormal basis for $f \in L^2(\mathbb{R})$, with series

$$f(t) = \sum_{j,k} \langle f, \psi_{jk} \rangle\, \psi_{jk}(t), \quad \text{where} \quad \langle f, g \rangle = \int_{-\infty}^{\infty} f(t)\overline{g(t)}\,dt. \tag{8.60}$$

One advantage of a wavelet basis, as compared to a trigonometric basis, is that each term $\langle f, \psi_{jk} \rangle\, \psi_{jk}$ in the expansion (8.60) has an effect only on the short-time support of the term ψ_{jk}. Furthermore, that support decreases with j because of the dilation factor 2^j.

An arbitrary function $\psi \in L^2(\mathbb{R})$ does *not* in itself guarantee that the family $\{\psi_{jk}\}_{k,\,j\in\mathbb{Z}}$ forms an orthonormal wavelet basis. There are, however, many examples of orthonormal wavelet bases. Figure 8.11 shows four examples of commonly used mother wavelets that do lead to orthonormal wavelet bases. We begin with an extension of the square wave Haar functions as discussed in the Chapter 7.

8.9.1 Haar Wavelets

The Haar wavelet system starts with the Haar orthonormal system $\{h_{jk}\}$ as discussed in Subsection 7.10.3 and then extends the functions from the Hilbert space $L^2[0,1]$ to the Hilbert space $L^2(\mathbb{R})$.

Let $H: \mathbb{R} \longrightarrow \mathbb{C}$ be the indicator (characteristic) function of the interval $(0,1]$. It is often called the **father wavelet** of the Haar system. Define the **mother wavelet** h on $\mathbb{R}$ to be

$$h(t) = \begin{cases} 1 & : \text{ if } 0 < t \le 1/2 \\ -1 & : \text{ if } 1/2 < t \le 1 \\ 0 & : \text{ otherwise.} \end{cases}$$

Then define the other **Haar wavelets** $h_{jk}: \mathbb{R} \longrightarrow \mathbb{C}$ as in (8.59) so that:

$$h_{jk}(t) = 2^{\frac{j}{2}} h(2^j t - k) \quad \text{for} \quad j,k \in \mathbb{Z}.$$

Recall that the Haar *functions* were confined to the unit interval $(0,1]$. For Haar *wavelets*, the dilation factors 2^j are extended to negative values of j, and the translation terms $-k$ extends them to cover all integral values. Note that the

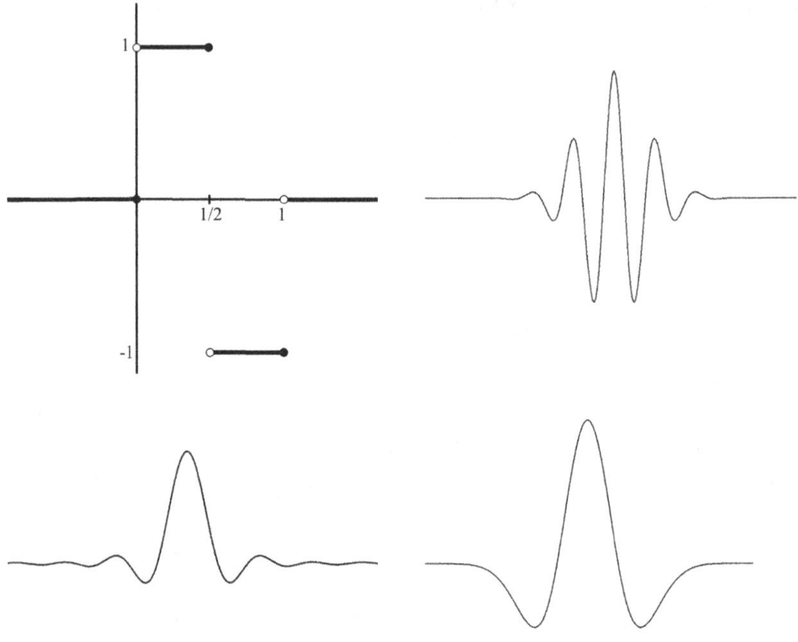

Figure 8.11 (Top) The Haar and Morlet wavelets. (Bottom) The Meyer and Mexican hat wavelets.

mother wavelet is also $h = h_{00}$. Figure 8.12 shows graphs of Haar wavelets h_{jk} for various values of j and k.

The mother Haar wavelet $h(t)$ has support $[0, 1]$ of width 1.

The dilations	$\ldots, h(2^{-2}t), h(2^{-1}t), h(2^{0}t), h(2^{1}t), h(2^{2}t), \ldots$	
have supports	$\ldots, [0, 2^2],\ [0, 2^1],\ [0, 1],\ [0, 2^{-1}],\ [0, 2^{-2}], \ldots,$	
of widths	$\ldots,\ 2^2,\ 2^1,\ 1,\ 2^{-1},\ 2^{-2}, \ldots,$	respectively.

Then the terms $-k$ translate the dilated wavelets to cover the entire real axis.

Theorem 8.19 *The Haar wavelets $h_{jk}(t) = 2^{j/2}h(2^j t - k)$ for $j, k \in \mathbb{Z}$ form a complete orthonormal set in the Hilbert space $L^2(\mathbb{R})$ under the inner product*

$$\langle f, g \rangle = \int_{-\infty}^{\infty} f(t)\overline{g(t)}\, dt.$$

Proof That the family $\{h_{jk}\}$ (for $j, k \in \mathbb{Z} = \{0, \pm 1, \pm 2, \ldots\}$) is orthonormal is easy to show (see proof of Theorem 7.86).

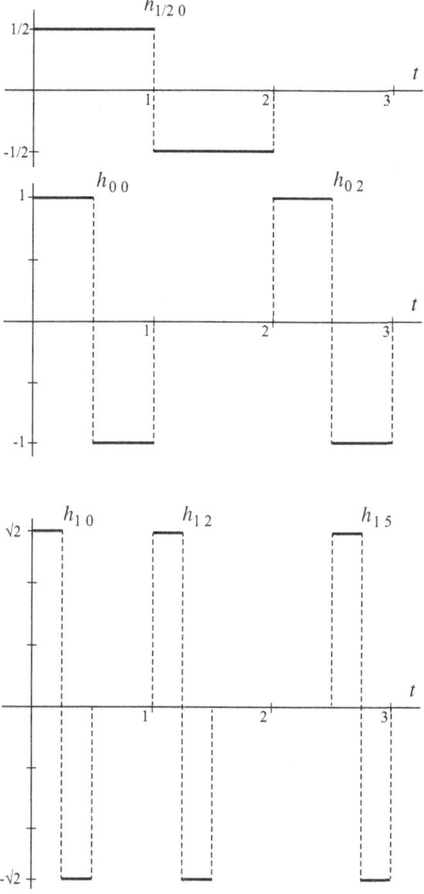

Figure 8.12 Haar wavelets h_{jk} for various j and k.

Let V_0 consist of the step functions[9] that are constant on the intervals $(k, k+1]$ for $k \in \mathbb{Z}$. Clearly, V_0 is a subspace of $L^2(\mathbb{R})$. It is also clear that the functions $H_{0k}(t) = H(t-k)$ for $k \in \mathbb{Z}$, which are the indicator functions of the intervals $(k, k+1]$, form an orthonormal basis on V_0.

[9] By definition (Definition 2.5), a step function has only a finite number of distinct values which occur on intervals.

For each fixed $j \in \mathbb{Z}$, define V_j, the **dyadic step functions of scale j,** to be those step functions that are constant on the intervals $(a_{jk}, b_{jk}]$, where $a_{jk} = 2^j k$ and $b_{kj} = 2^j(k+1)$ for $k \in \mathbb{Z}$. In each V_j, define the functions $H_{jk}(t) = 2^{\frac{j}{2}} H(2^j t - k)$ for $k = 0, \pm 1, \pm 2, \ldots$. The following six statements are not difficult to prove:

(1) For each $j \in \mathbb{Z}$, V_j is a linear subspace of $L^2(\mathbb{R})$.
(2) In each V_j, the sequence $\{H_{jk}(t)\}_{k \in \mathbb{Z}}$ forms an orthonormal basis.
(3) $\cdots \subset V_{-2} \subset V_{-1} \subset V_0 \subset V_1 \subset V_2 \subset \cdots$.
(4) For each $j \in \mathbb{Z}$, $f(t) \in V_j$ if and only if $f(2^{-j}t) \in V_0$.
(5) $\bigcap_j V_j = \{0\}$.
(6) $\bigcup_j V_j$ is a dense subspace of $L^2(\mathbb{R})$.

Consider the inclusion $V_0 \subset V_1$. As noted, the sequence $\{H_{0k}(t)\}_{k \in \mathbb{Z}}$ forms a orthonormal basis for V_0. Direct calculations show that the Haar wavelets $\{h_{0,k}(t)\}_{k \in \mathbb{Z}}$ are orthonormal in V_1. It is left as Exercise 8.9.2 to show that the sequence $\{H_{0k}\}_{k \in \mathbb{Z}}$ along with the sequence $\{h_{0k}\}_{k \in \mathbb{Z}}$ form an orthonormal basis in V_1. If we define W_0 to be the span of the Haar wavelets $\{h_{0k}\}_{k \in \mathbb{Z}}$, then we have $V_0 \oplus W_0 = V_1$.

Similarly, for any fixed $j \in \mathbb{Z}$, consider $V_j \subset V_{j+1}$. We can show that the set of functions $\{H_{jk}\}_{k \in \mathbb{Z}}$ forms a total orthonormal sequence in V_j and that the set $\{H_{jk}, h_{jk}\}_{k \in \mathbb{Z}}$ is a total orthonormal sequence in V_{j+1}. Defining W_j to be the span of $\{h_{jk}\}_{k \in \mathbb{Z}}$, we have $V_j \oplus W_j = V_{j+1}$. The details are left as Exercise 8.9.3. Combining the case $j = 0$ and $j = 1$, we obtain $V_0 \oplus W_0 \oplus W_1 = V_1 \oplus W_1 = V_2$ and the set $\{H_{0k}, h_{0k}, h_{1k}\}_{k \in \mathbb{Z}}$ forms an orthonormal basis for V_2.

Continuing this way, we can show that for any $-\infty < j' < j < \infty$, the set $\{H_{j'k}, h_{j'k}, \ldots, h_{jk}\}_{k \in \mathbb{Z}}$ is an orthonormal basis of V_{j+1} and

$$V_{j'} \oplus W_{j'} \oplus W_{j'+1} \oplus \cdots \oplus W_{j-1} \oplus W_j = V_{j+1}. \tag{8.61}$$

Letting $j' \to -\infty$, and using statement (5), for each fixed j, the set $\{h_{j'k}\}_{k \in \mathbb{Z}, -\infty < j' \leq j}$ forms an orthonormal basis in V_{j+1}. Letting $j \to \infty$, and using statement (6), completes the proof. ∎

Corollary 8.20 *Every $f \in L^2(\mathbb{R})$ can be written as a series expansion of Haar wavelets*

$$f(t) = \sum_{j=-\infty}^{\infty} \sum_{k=-\infty}^{\infty} c_{jk} h_{jk}(t) \quad \text{with} \quad c_{jk} = \langle f, h_{jk} \rangle. \tag{8.62}$$

Exercises for Section 8.9

8.9.1 Show that V_j is a linear space for each fixed $j \in \mathbb{Z}$.

8.9.2 Prove the claim made in the proof of Theorem 8.19 that the sequence $\{H_{0,k}\}_{k\in\mathbb{Z}}$ along with the sequence $\{h_{0,k}\}_{k\in\mathbb{Z}}$ form a total orthonormal sequence in V_1.

8.9.3 Prove the claim made in the proof of Theorem 8.19 that for any fixed $j \in \mathbb{Z}$, the set of functions $\{H_{jk}\}_{k\in\mathbb{Z}}$ forms a total orthonormal sequence in V_j. Then show that the set $\{H_{jk}, h_{jk}\}_{k\in\mathbb{Z}}$ is a total orthonormal sequence in V_{j+1}, and that $V_j \oplus W_j = V_{j+1}$.

8.9.4 Consider Equation (8.61). Suppose $-\infty < j' < j < \infty$ and $f \in V_{j+1}$. Show that if $f \perp H_{j'k}$ (for all $k \in \mathbb{Z}$) and $f \perp h_{lk}$ (for all $k \in \mathbb{Z}$ and all $j' \le l \le j$), then $f = 0$.

8.9.5 Suppose $f \in L^2(\mathbb{R})$ and $f \perp V_j$ and $f \perp W_{j'}$ for all $j' \ge j$. Show that $f = 0$. This shows that the set $\{H_{jk}, h_{jk}, h_{j+1k}, h_{j+2k} \cdots\}$ for all $k \in \mathbb{Z}$ is an orthonormal basis for $L^2(\mathbb{R})$.

8.10 A Continuous but Nowhere Differentiable Function

In the early nineteenth century it was commonly thought that a continuous function is differentiable everywhere except in a set of isolated points. It thus shocked the mathematics community when Karl Weierstrass presented to the Royal Academy of Sciences in Berlin in 1872 an example of a function that is continuous but differentiable *nowhere*. The function was

$$f(x) = \sum_{k=0}^{\infty} a^k \cos b^k x, \tag{8.63}$$

with $0 < a < 1$, $b > 1$ an odd integer, and $ab > 1 + \frac{3}{2}\pi$. Since the series is uniformly convergent on $\mathbb{R}$, it is continuous on $\mathbb{R}$. The graph is self-similar (fractal) in the sense that magnifications of the graph at any point will produce shapes similar to the original.

The first publication of the result with proof that $f'(x)$ does not exist was by Paul du Bois-Reymond.[10] That proof finds, for each $x \in \mathbb{R}$, sequences $y_k < x < z_m$ that converge to x but

[10] Paul du Bois-Reymond (1831–1889) was a German mathematician who worked on the theory of functions and the field of mathematical physics. His interests included Fourier series, integral equations, Sturm–Liouville theory, and variational calculus.

$$\limsup_{k\to\infty}\frac{f(y_k)-f(x)}{y_k-x}<\liminf_{k\to\infty}\frac{f(z_k)-f(x)}{z_k-x}.$$

Since then, there have been many other examples discovered of continuous nowhere differentiable functions. For example, Weierstrass's function has been extended to the case where $0 < a < 1$, $b > 1$, and $ab \geq 1$. These examples are not really special. Indeed, in Example 6.15 we showed that the set of functions in $C[0,1]$ differentiable at some point $x \in [0,1]$ is of the first category, whereas $C[0,1]$ is of second category. This clearly shows that "most" continuous functions are nowhere differentiable.

In this section we prove that a specific Weierstrass function (with $a = 1/2$, $b = 2$) is continuous but nowhere differentiable.

Theorem 8.21 *The function* $f(x) = \sum_{k=0}^{\infty} \frac{1}{2^k}\cos 2^k x$ *is continuous but not differentiable anywhere.*

Proof By the Weierstrass M-test, we have uniform convergence of the series. Hence, f is continuous. In complex form of a Fourier series, the function is $f(x) = \sum_{k=-\infty}^{\infty} \frac{1}{2^{k+1}} e^{i2^k x}$. So $\widehat{f}(\pm 2^n) = \frac{1}{2^{n+1}}$ and $\widehat{f}(k) = 0$ otherwise.[11]

We first consider differentiability at the point $x_0 = 0$. **Assume that $f'(0)$ exists.** By changing the function to $f(x) - f(0)\cos x - f'(0)\sin x$, we may assume $f(0) = f'(0) = 0$. This changes only the value of $\widehat{f}(2^0)$ and $\widehat{f}(-2^0)$.

Since $\widehat{f}(k) = 0$ for $2^{n-1} < k < 2^{n+1}$ and $k \neq 2^n$, we have

$$\frac{1}{2^{n+1}} = \widehat{f}(2^n) = \frac{1}{2\pi}\int_{-\pi}^{\pi} f(x)e^{-i2^n x}\,dx = \frac{1}{2\pi}\int_{-\pi}^{\pi} f(x)P_n(x)e^{-i2^n x}\,dx$$

for any trigonometric polynomial $P_n = \sum_{k=-N}^{N} \alpha_k e^{ikx}$ with $\alpha_0 = P_n(0) = 1$ and $N < 2^{n-1}$. Recall the Fejér kernel (Definition 7.20),

$$F_n(x) = \frac{D_0(x) + D_1(x) + \cdots + D_n(x)}{n+1} = \sum_{k=-n}^{n}\left(1 - \frac{|k|}{n+1}\right)e^{ikx}.$$

Note that $\|F_n\|_{L^2}^2 = \sum_{k=-n}^{n}\left(1 - \frac{|k|}{n+1}\right)^2 > \frac{n}{2}$. Also Inequality (7.30) states $0 \leq F_n(x) \leq \frac{1}{(n+1)\sin^2(\delta/2)} \leq \frac{\pi^2}{(n+1)\delta^2}$ for $0 < \delta \leq |x| \leq \pi$.

This shows $0 \leq \left(\frac{F_n(x)}{\|F_n\|_{L^2}}\right)^2 \leq \frac{2\pi^4}{n^3\delta^4}$ for $0 < \delta \leq |x| \leq \pi$.

Now let $P_n = \left(F_{2^{n-3}}/\|F_{2^{n-3}}\|_{L^2}\right)^2$. This is a trigonometric polynomial with $\alpha_0 = P_n(0) = 1$ and of degree $< 2^{n-1}$. Also $\|P_n\|_{L^1} = 1$.

[11] Such a sequence $\widehat{f}$ where there are increasingly larger gaps between nonzero entries $\widehat{f}(\alpha_n)$ with $\frac{\alpha_{n+1}}{\alpha_n} > c > 1$, is called a **lacunary** sequence. Here $\alpha_n = 2^n$ with $\frac{\alpha_{n+1}}{\alpha_n} = 2$.

Let $\epsilon > 0$ be given. If $f(0) = f'(0) = 0$, we may choose $\delta > 0$ so that $\left|\frac{f(x)-f(y)}{x-y}\right| < \epsilon$ for $|x|, |y| < \delta$. Choose n sufficiently large so that $\pi/2^{n/2} < \delta/2$ and $\|f - T_{\pi/2^n} f\|_1 < \epsilon$. By Equation (7.29), we have

$$\begin{aligned}
\frac{1}{2^{n+1}} = |\widehat{f}(2^n)| &= \left|\frac{1}{2}\left\{\widehat{f - T_{\pi/2^n} f}\right\}(2^n)\right| \\
&= \left|\frac{1}{2\pi}\int_{-\pi}^{\pi}\Big(f(x) - f(x - \pi/2^n)\Big)P_n(x)\,dx\right| \\
&\leq \frac{1}{2^{n+1}}\int_{|x|<\delta}\left|\frac{f(x) - f(x-\pi/2^n)}{\pi/2^n}\right|P_n(x)\,dx \\
&\quad + \frac{1}{2\pi}\int_{\delta\leq|x|\leq\pi}\Big|f(x) - f(x-\pi/2^n)\Big|P_n(x)\,dx \\
&\leq \frac{\epsilon}{2^{n+1}}\int_{|x|<\delta}P_n(x)\,dx \\
&\quad + \frac{2\pi^4}{(2^{n-3})^3\delta^4}\|f - T_{\pi/2^n} f\|_{L^1} \\
&\leq \frac{\epsilon}{2^{n+1}}\left(1 + 2^7\right).
\end{aligned}$$

That is, $\frac{1}{2^{n+1}} \leq \frac{\epsilon}{2^{n+1}}\left(1+2^7\right)$. But this is not true for $\epsilon < 1/2^8$.

So the assumption that $f'(0)$ exists is false.

For differentiability at a general point x_0 we make a transformation $T_{x_0} f(x) = f(x - x_0)$. In this case, $\widehat{T_{x_0} f}(n) = e^{-nx_0}\widehat{f}(n)$. Differentiability of f at x_0 is equivalent to differentiability of $T_{x_0} f$ at 0. The same argument works for the transformed function since the absolute values of the Fourier coefficients are the same. ∎

Exercise for Section 8.10

8.10.1 Show that any series (8.63) defining the Weierstrass functions is uniformly convergent in $\mathbb{R}$.

8.11 Group Structure and Fourier Series

It is not by accident that Fourier series are used to represent functions. Nor is the use of the orthonormal sequence $\left(e^{ikx}\right)_{k\in\mathbb{Z}}$. There is a deep relationship between Fourier series and Lebesgue integration which we explore in this section. We show how the topological group structure of $\mathbb{R}$ leads to the study of 2π-periodic functions and Fourier series and the Lebesgue integral. These connections presented here are deep and beautiful. It is helpful to have some familiarity of elementary group theory.

A function f defined on an interval $[a, b]$ can be extended to a periodic function of period $c = b - a$ defined on all of $\mathbb{R}$, by way of

$$f(x) = f(x + kc) \quad \text{for all} \quad k = 0, \pm 1, \pm 2, \ldots .$$

It is convenient to use the period $c = 2\pi$ because of the natural periodic mapping of $\mathbb{R}$ onto the unit circle $\{z \in \mathbb{C} \mid |z| = 1\}$ given by $p_0(x) = e^{ix}$. Clearly, a function f defined on $\mathbb{R}$ is 2π-periodic if and only if there exists a function f_0 on the unit circle such that

$$f(x) = f_0(e^{ix}) = f_0\big(p_0(x)\big) \quad \text{for all} \quad x \in \mathbb{R}.$$

8.11.1 Elementary Group Theory

Definition 8.5 An **abelian group** is a set G with a binary operation, which we denote by addition $+$, satisfying the following properties for all $x, y, z \in G$

(AG1) $x + y = y + x$.
(AG2) $x + (y + z) = (x + y) + z$.
(AG3) There exists an identity $0 \in G$ such that $x + 0 = x$.
(AG4) There exists $-x$, such that $x + (-x) = 0$.

Definition 8.6 An abelian group G is a **topological group** if it has a topology (which could be defined by a norm, a metric, etc.) that satisfies the the following additional properties:

(TG1) For all $a \in G$, the function $x \longrightarrow a + x$ is continuous.
(TG2) The function $x \longrightarrow -x$ is continuous.
(TG3) The function $(x, y) \longrightarrow x + y$ is continuous.
We will consider only abelian groups that are topological groups.

A nonempty subset $F \subset G$ is said to be a **subgroup** of G if F is also an abelian group under the same operator $+$ of G (and the topology induced by G). The critical properties for a subgroup F are **algebraic closure** (if $x, y \in F$, then $x + y \in F$) and **inverse** (if $x \in F$, then $-x \in F$).

Example 8.17 • The set $\mathbb{R}$ is a topological group under the usual operation of addition $+$ and the usual norm given by the absolute value $|\cdot|$.
• The sets $\mathbb{Z}$ and $\mathbb{Q}$ are two examples of subgroups of $\mathbb{R}$.

• The function spaces $C[0,1]$, L, $L^2_{2\pi}$ are all topological groups under the usual function addition $f + g$ and topologies given by their usual norms.
• The group $C_{2\pi}$ is a subgroup of $L^2_{2\pi}$, which in turn is a subgroup of $L_{2\pi}$.

Theorem 8.22 *Other than the trivial* $\{0\}$ *and* $\mathbb{R}$, *the only (topologically) closed subgroups of* $\mathbb{R}$ *are of the form*

$$c\mathbb{Z} = \{0, \pm c, \pm 2c, \pm 3c, \ldots\} \quad \textit{for some} \quad c \in \mathbb{R}.$$

Proof Showing that $\{0\}$, $\mathbb{R}$, for $c \in \mathbb{R}$ are actually subgroups is routine. Conversely, suppose F is a closed subgroup of $\mathbb{R}$. Define $F^+ = \{x \in F \mid x > 0\}$.

Case $F^+ = \emptyset$**:** Since $-x \in F$ whenever $x \in F$, it is clear that $F = \{0\}$. If F^+ is not empty, let $c = \inf\{x \mid x \in F^+\}$.

Case $c = 0$**:** Let $\epsilon > 0$ be given. Since 0 is the infimum of F^+, there exists $d \in F^+$ such that $0 < d < \epsilon$. All multiples of d are in F. For each $x \in \mathbb{R}$, there exists a multiple nd such that $nd \leq x \leq (n+1)d$; that is, there exists $y \in F$ such that $|x - y| < \epsilon$. This means that F is dense in $\mathbb{R}$. Since F is closed in $\mathbb{R}$, we have $F = \overline{F} = \mathbb{R}$.

Case $c > 0$**:** In this case, it is clear that $F^+ = \{c, 2c, 3c, \ldots\}$ and that $F = c\mathbb{Z} = \{0, \pm c, \pm 2c, \pm 3c, \ldots\}$. ∎

For convenience, we choose $c = 2\pi$ and denote such a closed subgroup by $2\pi\mathbb{Z}$.

Definition 8.7 If F is a subgroup of G, then a **coset of** F **with respect to** G is a subset of the form $x + F$, for some $x \in G$.

The following three theorems are elementary results in group theory. Proofs are omitted.

Theorem 8.23 *Suppose* G *is an abelian group with a subgroup* F. *Two cosets* $x + F$ *and* $y + F$ $(x, y \in G)$ *are either identical or disjoint. That is, the cosets of* F *form a partition of the group* G.

Theorem 8.24 *We can define addition of cosets by the rule* $(x+F)+(y+F) = (z + F)$ *for* $z = x + y \in G$. *Under this definition, the cosets form a group called the quotient group of* G *with respect to* F. *The quotient group is denoted by* G/F. *The identity element of* G/F *is* $0 + F$.

Theorem 8.25 *Let* F *be a subgroup of the topological group* G. *The map* $p_0 \colon G \longrightarrow G/F$ *defined by* $p_0(x) = x + F$ *is a* ***homomorphism****; that is,*

$$p_0(x + y) = p_0(x) + p_0(y) \quad \textit{for all} \ x, y \in G.$$

Furthermore, p_0 is continuous if and only if F is a closed subgroup of G. The inverse image of the identity element of G/F is F.

Definition 8.8 The quotient group $\mathbb{R}/2\pi\mathbb{Z}$ is called the **circle group**. The elements of $\mathbb{R}/2\pi\mathbb{Z}$ can be identified with points on the unit circle. The map $p_0(x) = e^{ix}$ is continuous since $2\pi\mathbb{Z}$ is closed in $\mathbb{Z}$. The circle group is compact.

8.11.2 Characters

Definition 8.9 A **character** of a topological group G is a complex valued function $\chi : G \longrightarrow \mathbb{C}$ satisfying the following three conditions:

(Cha1) $\chi \neq 0$.

(Cha2) $\chi : G \longrightarrow \mathbb{C}$ is bounded and continuous.

(Cha3) $\chi(a+b) = \chi(a) \cdot \chi(b)$ for all $a, b \in G$.

Boundedness of the function χ means that $|\chi(na)| = |\chi^n(a)| = |\chi(a)|^n \leq M$ for some $M > 0$, which can only happen if $|\chi(a)| \leq 1$ for all $a \in G$. Also $|\chi(-a)| = \frac{1}{|\chi(a)|} \leq 1$ leads to $|\chi(a)| \geq 1$. Thus $|\chi(a)| = 1$ for all a. Note that $\chi(0) = \chi(0+0) = \chi(0)\chi(0)$, so it follows that $\chi(0) = 1$. We can write all this as a fourth property of characters:

(Cha4) $|\chi(a)| = 1$ for all $a \in G$ and $\chi(0) = 1$.

If G is compact, as in the case $G = \mathbb{R}/2\pi\mathbb{Z}$, then continuity implies boundedness, so then the boundedness condition in **(Cha2)** may be omitted.

Example 8.18 The **principal character** χ_0 of a topological group is defined by $\chi_0(a) = 1$ for all $a \in G$. The characters of G form an abelian group under the multiplication operation $\cdot$ defined by $(\chi_1 \cdot \chi_2)(a) = \chi_1(a)\chi_2(a)$. The principal character is the identity element of the group of characters. For each χ, the inverse character is defined by $\chi^{-1}(a) = \chi(-a)$. So $(\chi^{-1} \cdot \chi)(a) = \chi_0(a) = 1$.

Theorem 8.26 *The characters of the abelian group $\mathbb{R}$ are precisely the functions*

$$\chi_r(x) - e^{irx} \quad \textit{for} \quad r \in \mathbb{R}.$$

The function for $r = 0$ is the principal character, $\chi_0(x) = 1$.

Proof To show that all of the functions e^{irx} are actually characters on $\mathbb{R}$ is routine. Now suppose χ is a character of $\mathbb{R}$. Since χ is continuous, it is integrable. Let $f(x) = \int_0^x \chi(t)\,dt$. Then

$$
\begin{aligned}
f(x+b) - f(x) &= \int_0^{x+b} \chi(t)\,dt - \int_0^x \chi(t)\,dt \\
&= \int_x^{x+b} \chi(t)\,dt = \int_0^b \chi(x+t)\,dt \\
&= \chi(x) \int_0^b \chi(t)\,dt = \chi(x) \cdot f(b).
\end{aligned}
$$

Since $\chi(0) = 1$ and χ is continuous, there exists b for which $f(b) \neq 0$. Thus $\chi(x) = \big(f(x+b) - f(x)\big)/f(b)$. By the fundamental theorem of calculus (Section 3.3), the function f is differentiable. Hence, χ is differentiable and so

$$
\begin{aligned}
\chi'(x) &= \lim_{h \to 0} \frac{\chi(x+h) - \chi(x)}{h} \\
&= \lim_{h \to 0} \frac{\chi(h) - \chi(0)}{h} \cdot \chi(x) = \chi'(0) \cdot \chi(x).
\end{aligned}
$$

The character χ thus satisfies $\chi' = c\chi$ with $c = \chi'(0)$. The solution of this differential equation is $\chi(x) = Ae^{cx}$. By **(Cha4)** we have $\chi(0) = 1$, so $A = 1$, and $|\chi(x)| = 1$, so $\chi(x) = e^{cx}$ is on the unit circle of $\mathbb{C}$. Thus c must be pure imaginary; that is, $c = ir$ with $\chi(x) = e^{irx}$ for some $r \in \mathbb{R}$. ∎

Corollary 8.27 *The characters of the circle group $\mathbb{R}/2\pi\mathbb{Z}$ are precisely the functions*

$$\chi_k(x) = e^{ikx} \quad \textit{for} \quad k \in \mathbb{Z}.$$

Proof To show that the functions χ_k are characters on $\mathbb{R}/2\pi\mathbb{Z}$ is routine. Now suppose χ is a character of $\mathbb{R}/2\pi\mathbb{Z}$. If $p_0(x) = e^{ix}$, then $\chi \circ p_0$ is clearly a character on $\mathbb{R}$. By the Theorem 8.26, $\chi \circ p_0(x) = e^{irx}$ for some $r \in \mathbb{R}$. But $\chi \circ p_0$ is 2π-periodic. From $e^{irx} = e^{ir(x+2\pi)}$, we conclude $e^{2\pi ir} = (-1)^{2r} = 1$. This can only happen when r is an integer $r = k \in \mathbb{Z}$. ∎

8.11.3 Translates Operators

Here are basic properties of translation operators $T_a \colon C(\mathbb{R}) \longrightarrow C(\mathbb{R})$ defined by $T_a(f)(x) = f(x-a)$ for all $a, x \in \mathbb{R}$.

(1) $T_a(f+g) = T_a(f) + T_a(g)$, $\forall a \in \mathbb{R}$.

(2) $T_0 = T_{2\pi} = I$.

(3) $T_{a+b} = T_a \circ T_b$, $\forall a, b \in \mathbb{R}$.

(4) $T_a^{-1} = T_{-a}$. $\forall a \in \mathbb{R}$.

Instead of translations on the space $C(\mathbb{R})$, we could just as well consider translations on other function spaces such as $L^2_{2\pi}$, $L^2(\mathbb{R})$, $C_{2\pi}$, or $L_{2\pi}$.

Definition 8.10 An **invariant subspace** V of $C(\mathbb{R})$ under translations is one for which $T_a V \subset V$ for all $a \in \mathbb{R}$. Two trivial examples are $V = \{0\}$ and $V = C(\mathbb{R})$. We want to find nontrivial **minimal invariant subspaces.** One-dimensional invariant subspaces are certainly minimal.

For each $f \in C(\mathbb{R})$, let V_f be the smallest invariant subspace containing f,

$$V_f = \text{span}\{T_a f \mid a \in \mathbb{R}\}.$$

Theorem 8.28 *The one-dimensional minimal invariant subspaces of* $C(\mathbb{R})$ *are the spaces*

$$V_r = \text{span}\{e^{irx}\}, \textit{ for } r \in \mathbb{R}.$$

Proof If $f = 0$, then we obtain the trivial $V_f = \{0\}$ of dimension 0. So let $f \neq 0$. We want to find functions f such that dimension of V_f is 1. Then for each a, the translate $T_a(f)$ is a multiple of f, say the multiple c_a.

$$T_a(f)(x) = f(x-a) = c_a \cdot f(x). \tag{8.64}$$

The assigned multiple c_a of f is a character of $\mathbb{R}$, say $c_a = \chi(a)$; proof of that is left as Exercise 8.11.10. By Theorem 8.26, $c_a = e^{ira}$ for some $r \in \mathbb{R}$. Letting $x = 0$, we have

$$T_a(f)(0) = f(-a) = c_a \cdot f(0) = f(0)e^{ira} \text{ for some } r \in \mathbb{R} \text{ and all } a \in \mathbb{R}.$$

Changing a to the variable $-x$, we have $f(x) = f(0)e^{-irx}$, for some $-r \in \mathbb{R}$, and all x. This proves $V_f = \text{span}\{e^{irx}\}$ for some $r \in \mathbb{R}$. ∎

Corollary 8.29 *The one-dimensional minimal invariant subspaces of* $C_{2\pi}$ *are the spaces*

$$V_k = \text{span}\{e^{ikx}\}, \textit{ for } k = 0, \pm 1, \pm 2, \ldots.$$

Details of the proof are left as Exercise 8.11.12.

8.11.4 Spectral Analysis and Synthesis

For a given $f \in C_{2\pi}$ (or $L^2_{2\pi}$), the problem of finding the corresponding components of f lying in the invariant subspaces V_k is called **harmonic analysis** or **spectral analysis**. We have seen in the theory of Fourier series (Chapter 7) that the sequence $(e^k)_{k\in\mathbb{Z}}$, where $e^k(x) = e^{ikx}$ for $k \in \mathbb{Z}$, forms a complete orthonormal sequence in $L^2_{2\pi}$ with $\widehat{f}(k) = \langle f, e^k \rangle$.

This is summarized in the following theorem.

Theorem 8.30 Harmonic (or spectral) analysis: *Given $f \in C(\mathbb{R})$, the component of f lying in the subspace V_k, for $k \in \mathbb{Z}$, is $\widehat{f}(k)e^k$ where $e^k(x) = e^{ikx}$, and*

$$\widehat{f}(k) = \langle f, e_k \rangle = \frac{1}{2\pi} \int_{-\pi}^{\pi} f(x) e^{-ikx}\, dx.$$

Given the Fourier series $\sum_{k=-\infty}^{\infty} \widehat{f}(k)e^{ikx}$, the problem of reconstituting $f(x)$ from its harmonic components is called **harmonic synthesis** or **spectral synthesis**. This is *harder* than spectral analysis. As we have seen in Chapter 7, the series may not converge pointwise to the function f, or it may converge only under special conditions on f. And, as discussed in Section 7.8, the problem can involve approximate identities and summability.

8.11.5 Haar Integral

Definition 8.11 A topological group G is said to be **locally compact** if every $x \in G$ has a compact neighborhood. That is, for every $x \in G$ there exists a compact set K and open set U, such that $x \in U \subset K \subset G$.

Example 8.19 The topological group $\mathbb{R}$ is locally compact. More generally, $\mathbb{R}^n$ is locally compact for all $n = 1, 2, \ldots$. Every compact group G is locally compact. Therefore, $\mathbb{R}/2\pi\mathbb{Z}$ is locally compact.

Definition 8.12 Let G be a locally compact group and let $C_c(G)$ the space of continuous complex valued functions on G with compact support. A **Haar**[12] **integral** on G is a linear functional $I : C_c(G) \longrightarrow \mathbb{C}$ satisfying the following:

(h1): $I(f) > 0$ whenever f is real valued and $f > 0$.

(h2): $I(T_a F) = I(f)$ for all $f \in C_c(G)$ and $a \in G$.

Remark Here is why functions are restricted to those with compact support. Except in the case where G is itself compact, *no* linear functional $I(f)$ can be finite valued for *all* continuous $f \in C(G)$. Recall that the Lebesgue integral was extended to permit infinite valued for some measurable nonnegative real valued functions. But restricted to continuous functions of compact support $C_c(\mathbb{R}^n)$, the Lebesgue integral $\int f\, dm$ is a linear functional with finite complex values.

Theorem 8.31 *The Lebesgue integral $I(f) = \int f\, dm$ is a Haar integral on $G = \mathbb{R}^n$.*

[12] Alfréd Haar (1885–1933), Hungary.

Proof Every continuous function f of compact support on $\mathbb{R}^n$ is measurable (Corollary 2.3) and bounded, and thus dominated by a simple function with the same support. This ensured that f is integrable. That the Lebesgue integral is a linear functional is Theorem 2.35. Property **(h1)** is a standard property for continuous functions (see Exercise 5.1.14). Property **(h2)** says that translating a function does not change the value of the Lebesgue integral over $\mathbb{R}^n$. ∎

Corollary 8.32 *The Lebesgue integral $I(f) = \int f\, dm$ is a Haar integral on $\mathbb{R}/2\pi\mathbb{Z}$.*

The proof is left as Exercise 8.11.14.

Theorem 8.33 *The Haar integral I on $\mathbb{R}/2\pi\mathbb{Z}$ is unique up to multiplication by a positive constant. If we normalize I so that $I(\mathbf{1}) = 1$ for the identity function $\mathbf{1}$ on $\mathbb{R}/2\pi\mathbb{Z}$, then the Haar integral I can only be the Lebesgue integral*

$$I(f) = \frac{1}{2\pi}\int_{-\pi}^{\pi} f\, dm \quad \text{for all continuous } f \text{ on } \mathbb{R}/2\pi\mathbb{Z}.$$

Proof Suppose I is a Haar integral on $\mathbb{R}/2\pi\mathbb{Z}$. By Property **(h1)** and linearity, if $f \le g$, then $I(f) \le I(g)$. Since $-(\sup f)\mathbf{1} \le f \le (\sup f)\mathbf{1}$, this leads to

$$|I(f)| \le I(\mathbf{1}) \cdot \sup f. \tag{8.65}$$

Claim: If g is continuous and $\frac{1}{2\pi}\int_{-\pi}^{\pi} g(x)\, dx = 0$, then $I(g) = 0$.

Let $f(x) = \int_0^x g(t)\, dt$. By the fundamental theorem of calculus f is continuously differentiable with $f' = g$. By compactness of $\mathbb{R}/2\pi\mathbb{Z}$, the limit

$$\lim_{h\to 0} \frac{T_h f - f}{h} = -f'$$

holds uniformly. By Property **(h2)** and (8.65), this implies $I(f') = 0$. Thus $I(g) = I(f') = 0$. This completes the proof of the claim.

Now choose a nonnegative function h_0 such that $\frac{1}{2\pi}\int_{-\pi}^{\pi} h_0(x)\, dx = 1$. For any continuous function h, define g by

$$g(x) = h(x) - h_0(x)\frac{1}{2\pi}\int_{-\pi}^{\pi} h(t)\, dt. \tag{8.66}$$

Then g is continuous and $\frac{1}{2\pi}\int_{-\pi}^{\pi} g(x)\, dx = 0$. We conclude that $I(g) = 0$. Applying this to (8.66) results in

$$I(h) = I(h_0) \cdot \frac{1}{2\pi}\int_{-\pi}^{\pi} h(x)\, dx.$$

This established uniqueness up to multiplication by the positive $I(h_0)$. ∎

Corollary 8.34 *The Haar integral I on $\mathbb{R}^n$ is unique up to multiplication by a positive constant.*

The proof is left as Exercise 8.11.15.

Remarkably, Haar extended this result in 1933 to *every* locally compact group, as stated in Theorem 8.35. We omit the proof.

Theorem 8.35 Haar: *For every locally compact group G, a Haar integral exists and it is unique up to multiplication by a positive constant.*

Definition 8.13 Given the Haar integral I for a locally compact G, we can define the **Haar measure** of sets. For example, for each compact K, we can define $m_H(K) = \inf\{I(f) \mid f \in C_c(G) \text{ such that } f(x) = 1 \; \forall x \in K\}$. Then, just as we extended the Lebesgue measure from the Euclidean measure of rectangles in $\mathbb{R}^n$ to the most general class of Lebesgue measurable set $\mathcal{L}(\mathbb{R}^n)$ (in Chapter 1), we can do the same for the Haar measure. That is, we can extend the Haar measure of compact subsets of G, as defined above, to the most general Haar measurable sets $\mathcal{H}(\mathbb{R}^n)$ of G.

Theorem 8.36 *The Haar measure on $\mathbb{R}^n$, normalized such that the measure of the unit cube is 1, is the Lebesgue measure on $\mathbb{R}^n$. The corresponding Haar integral is then the Lebesgue integral,*

$$I(f) = \int f \, dm.$$

Exercises for Section 8.11

8.11.1 Prove Theorem 8.23.

8.11.2 Prove Theorem 8.24. Be sure to show that coset addition is well defined; that is, if $x + F = x' + F$ and $y + F = y' + F$, then $(x + y) + F = (x' + y') + F$.

8.11.3 Show that the circle group is a compact topological group.

8.11.4 Prove that the principal character of a group G is actually a character.

8.11.5 Prove that the characters of an abelian group form a group under the multiplication operation $\cdot$ as described in Example 8.18.

8.11.6 Consider $G = \mathbb{Z}$. Show that it is an abelian group under $+$. If χ is a character of $\mathbb{Z}$, show that it is of the form $\chi(n) = (c)^n$ for $c = \chi(1) \in \mathbb{C}$.

8.11.7 Show that a function $\chi(x) = e^{irx}$ for $r \in \mathbb{R}$ is actually a character on $\mathbb{R}$.

8.11.8 Show that $\chi(x) = e^{ikx}$ $(k = 0, \pm 1, \pm 2 \ldots)$ are all characters on $\mathbb{R}/2\pi\mathbb{Z}$.

8.11.9 Prove the four basic properties of translation operators $T_a(f)(x) = f(x - a)$ as given at the beginning of Subsection 8.11.3.

8.11.10 Prove that function $\chi : a \longrightarrow c_a$ in Equation (8.64) is a character. Show that the conditions **(Cha1)–(Cha3)** are satisfied.

8.11.11 Show that $\widehat{f}(k) \neq 0$ if and only if $V_k \cap V_f \neq \{0\}$.

8.11.12 Prove Corollary 8.29.

8.11.13 Show that the subgroup $\mathbb{Z}$ of $\mathbb{R}$ is locally compact but that the subgroup $\mathbb{Q}$ of $\mathbb{R}$ is not locally compact.

8.11.14 Prove Corollary 8.32.

8.11.15 Prove Corollary 8.34.

9
Sequence Spaces

The infinite-dimensional spaces easiest to understand are the sequence spaces, such as $\omega, c, c_0, \ell_p, cs, bs, bv_0$, and bv. Although sequence spaces are of interest in themselves, they are also the gateway to more general infinite-dimensional spaces, such as function spaces. For example, we have seen in previous chapters that the function space $L_{2\pi}$ can be identified with the sequence spaces $\widehat{L_{2\pi}}$.

This chapter is devoted to properties that are special to sequence spaces. There are many unsolved problems in this part of functional analysis. Here are three examples.

- The sequences of Fourier coefficients $\widehat{f}$ of $f \in L^2_{2\pi}$ are characterized by the property $\widehat{f} \in \ell_2$. What about characterizing Fourier coefficients $\widehat{f}$ for $f \in L_{2\pi}$ instead of $f \in L^2_{2\pi}$? It is known that every convex null sequence belongs to $\widehat{L_{2\pi}}$. Such a containment statement given in terms of properties of sequences is called an **integrability condition**. Is there an integrability condition that characterizes the entire sequence space $\widehat{L_{2\pi}}$?
- Summability fields consist of sequences that are summed by some summability method (a precise definition is given in Section 10.2). What characteristic distinguishes these sequence spaces from other sequence spaces?
- Which sequence spaces have approximate identities?

Definition 9.1 Recall that for each k the sequence $e^k = (0,\ldots,0,1,0,\ldots)$ has 1 in the kth position and 0 elsewhere. Also, $\Phi = \text{span}\{e_1, e_2, \ldots\}$ is the sequence space with finitely many nonzero entries. All sequence spaces E considered here will be assumed to contain Φ; that is, we freely assume all e^k belong to E.

9.1 K-spaces

Definition 9.2 A normed sequence space E is called a **K-space** if for all $k = 1, 2, \ldots$, the coordinate functionals $f_k \colon x \longrightarrow x_k$ are continuous (or equivalently, the seminorms $p_n(x) = |x_k|$ are continuous). That is, for each $k = 1, 2, \ldots$, there exists M_k such that $|x_k| \leq M_k \|x\|$ for all $x \in E$.

As a consequence, a normed sequence space E is a K-space if and only if, whenever a sequence (x^n) converges to x, then the sequence x_k^n converges to x_k as $n \to \infty$ for each coordinate k; that is, $\lim_{n\to\infty} x_k^n = x_k$ for $k = 1, 2, \ldots$.

Sequence spaces of interest are usually K-spaces. There is something unnatural about spaces that are not K-spaces, as the following example shows.

Example 9.1 Here we construct a sequence space that is *not* a K-space. Let E be the subspace of ℓ consisting of all sequences $x = (x_1, x_2, x_3, \ldots)$ for which $x_1 = \sum_{k=2}^{\infty} x_k$. Clearly E is a linear space and it is clear that $\|x\| = \sup_{k\geq 2} |x_k|$ defines a norm on E. However, E is not a K-space since there is no bound M_1 with

$$|x_1| \leq M_1 \|x\| \; \forall x \in E,$$

as required by the definition. This can be seen by considering the sequence $x^n = (n, 1, 1, \ldots, 1, 0, \ldots) \in E$ with values 1 in positions 2 through $n + 1$. Then $n = |x_1^n| \leq M_1 \|x^n\| = M_1$ fails for all bounds M_1.

Example 9.2 The space $\widehat{L_{2\pi}}$ is a K-space because $|\widehat{f}(k)| \leq \|f\|_1$ (Theorem 7.46).

9.2 BK-spaces

Definition 9.3 A **BK-space** is a K-space which is a Banach space.

Theorem 9.1 *Suppose that T is a linear operator* (*Definition* 6.1) *from a Banach space E into a BK-space F. Then T is continuous if and only if for all k, the linear functionals $f_k(Tx) = (Tx)_k$ are continuous.*

Proof The proof of ($\Rightarrow$) is clear. For a proof of ($\Leftarrow$), we use the closed graph theorem (Theorem 6.36). Suppose that (x^n, y^n) is a sequence in the graph of T (that is, $Tx^n = y^n$, for all n) that converges to (x, y) in $E \times F$. Because of the definition of the norm on $E \times F$, $\|(x, y)\| = \|x\| + \|y\|$, clearly $Tx^n \to y$. Since F is a K-space, we have coordinatewise convergence, $(Tx)_k = y_k$ for all k, which means that $Tx = y$. Thus (x, y) is on the graph of T, which makes the graph closed. This means that T is continuous. ∎

You are certainly familiar with $m \times n$ matrices which map $\mathbb{R}^n$ into $\mathbb{R}^m$. Here we extend that notion to infinite matrices.

Definition 9.4 Let $A = (a_{nk})$, with $n, k = 1, 2, \ldots$, be an infinite matrix and let E, F be K-spaces. Suppose that, for each $x \in E$, the sums

$$y_n = (Ax)_n = \sum_{k=1}^{\infty} a_{nk} x_k \text{ exist for all } n = 1, 2, \ldots. \tag{9.1}$$

We write $T_A x = y = (y_1, y_2, \ldots)$ for the sequence given by (9.1). If $T_A x \in F$ for all $x \in E$, then T_A is a linear operator from E to F and is called a **matrix map**. We usually write Ax instead of $T_A x$.

Theorem 9.2 *If T_A is a matrix map between BK-spaces E and F, then it is continuous.*

Proof Showing the linearity of T_A is routine. By Theorem 9.1, it is sufficient to show that $f_n \colon Ax \longrightarrow (Ax)_n$ is continuous for all n. We have

$$f_n(Ax) = y_n = \sum_{k=1}^{\infty} a_{nk} x_k, \tag{9.2}$$

which is assumed to converge by definition of a matrix map. Define $T_m(x) = \sum_{k=1}^{m} a_{nk} x_k$ for each m. Since each T_m is a finite linear combination of coordinates, each T_m is continuous. Furthermore, $\lim_{m \to \infty} T_m x = f_n(Ax)$. By the Banach–Steinhaus theorem (Theorem 6.33), the functionals defined by (9.2) are continuous for all n. ∎

The fact that all matrix maps between BK-spaces are continuous makes it important to characterize matrix maps between sequence spaces. For some cases of BK-spaces E and F, all continuous linear operators between them are matrix maps. For example, we will show in Example 11.5 (Section 11.4) that T is a continuous linear operator from c_0 to c_0 if and only if it is a matrix map for a matrix $A = (a_{nk})$ with columns that converge to zero and $\sup_n \sum_{k=1}^{\infty} a_{nk} = M < \infty$.

Definition 9.5 The **product** of sequences $x = (x_k)$ and $y = (y_k)$ is defined to be the sequence $z = (z_k)$ for which $z_k = x_k y_k$ for all k.

Definition 9.6 Suppose that E, F are BK-spaces and, for some sequence $x \in \omega$, we have

$$xy = (x_k y_k) \in F \quad \text{for all} \quad y \in E.$$

The map $T_x \colon y \longrightarrow xy$ is then called a **multiplier map** from E to F.

Theorem 9.3 *All multiplier maps between BK-spaces are continuous.*

Proof The multiplier map $T_x : y \longrightarrow xy$ is a matrix map defined by the matrix

$$A = \begin{pmatrix} x_1 & 0 & 0 & 0 & \cdots \\ 0 & x_2 & 0 & 0 & \cdots \\ 0 & 0 & x_3 & 0 & \cdots \\ 0 & 0 & 0 & x_4 & \cdots \\ \vdots & \vdots & \vdots & \vdots & \ddots \end{pmatrix}.$$

■

Corollary 9.4 *If E, F are BK-spaces with $E \subset F$, then the identity map $I : E \longrightarrow F$ is continuous.*

Proof The map I is a matrix map with respect to the identity matrix. ■

Corollary 9.5 *Every BK-space E has a unique topology. That is, if a sequence space E has two norms, each of which make it a BK-space, then the two norms are equivalent.*

Proof Let E_1 and E_2 be the BK-spaces E under the two norms. The identity map $I : E_1 \longrightarrow E_2$ is continuous and the same goes for the other direction. The proof is then a consequence of the definition of equivalent norms. ■

Exercises for Section 9.2

9.2.1 Show that in a K-space E, for each k, there exists M_k such that

$$|x_k| \leq M_k ||x||, \forall\, x \in E.$$

9.2.2 A sequence space E is said to be **symmetric** if for all $x \in E$, x_π is in E for all permutations π of the coordinates. Give an example of a symmetric BK-space. Give an example of a BK-space which is not symmetric.

9.2.3 Find a sequence x which is in ℓ_∞ but not in $\bigcup_{p=1}^{\infty} \ell_p$.

9.2.4 Find a sequence x which is in $\bigcap_{p>1} \ell_p$ but not in ℓ_1.

9.2.5 Find a sequence x which is in $\bigcap_{p>q} \ell_p$ but not in ℓ_q.

9.2.6 Find a sequence x which is in ℓ_q but not in $\bigcup_{p<q} \ell_p$.

9.3 Examples of BK-spaces

Example 9.3 The space c_0 is a BK-space. That it is a Banach space under the supremum norm

$$\|x\|_\infty = \sup_k |x_k|,$$

was shown as Example 5.38 (Section 5.8). Since we have $|x_k| \leq \|x\|_\infty$, for all $k = 1, 2, \ldots$ and all $x \in c_0$, the space is a K-space.

Example 9.4 The space ℓ_∞ is a BK-space under the supremum norm. One can follow the pattern of the proof that c_0 is a BK-space. This is left as Exercise 9.3.1.

Example 9.5 The space c is a BK-space. Having the above result that ℓ_∞ is a BK-space, we need only show that c is a closed subspace. This was observed in Exercise 5.8.8 in Section 5.8.

Example 9.6 The spaces bs and cs are BK-spaces under the norm $\|x\|_{bs} = \sup_n |\sum_{k=1}^n x_k|$. Consider the matrix

$$\Sigma = \begin{pmatrix} 1 & 0 & 0 & 0 & 0 & \cdots \\ 1 & 1 & 0 & 0 & 0 & \cdots \\ 1 & 1 & 1 & 0 & 0 & \cdots \\ 1 & 1 & 1 & 1 & 0 & \cdots \\ 1 & 1 & 1 & 1 & 1 & \cdots \\ \vdots & \vdots & \vdots & \vdots & \vdots & \ddots \end{pmatrix}$$

and its inverse

$$\Delta = \begin{pmatrix} 1 & 0 & 0 & 0 & 0 & \cdots \\ -1 & 1 & 0 & 0 & 0 & \cdots \\ 0 & -1 & 1 & 0 & 0 & \cdots \\ 0 & 0 & -1 & 1 & 0 & \cdots \\ 0 & 0 & 0 & -1 & 1 & \cdots \\ \vdots & \vdots & \vdots & \vdots & \vdots & \ddots \end{pmatrix}.$$

The matrix maps $\Sigma\colon bs \longrightarrow \ell_\infty$ and $\Delta\colon \ell_\infty \longrightarrow bs$ are clearly isometries between ℓ_∞ and bs. For $x \in bs$, $y \in \ell_\infty$, we have $\|x\|_{bs} = \|\Sigma x\|_\infty$ and $\|y\|_\infty = \|\Delta y\|_{bs}$. These matrix maps are also isometries between cs and c.

Example 9.7 For all $1 \leq p < \infty$, the sequence spaces ℓ_p are BK-spaces

The case $p = \infty$: this was done in Example 9.4.

The case $1 \leq p < \infty$: For all k, it is clear that $|x_k| \leq M_k \|x\|_p$, with $M_k = 1$. Hence, ℓ_p is a K-space. To show that ℓ_p is complete, let x^n be a Cauchy sequence. Since ℓ_p is a K-space, for each k, x_k^n is Cauchy in $\mathbb{R}$ and hence converges, say to a_k. Let $a = (a_1, a_2, \ldots)$.

Claim 1: $\lim_n x^n = a$. Let $\epsilon > 0$ be given. There exists $N > 0$ such that for each integer $M > 0$,

$$\left(\sum_{k=1}^{M} |x_k^n - x_k^m|^p\right)^{1/p} \le \|x^n - x^m\| < \epsilon \text{ whenever } n, m > N.$$

Letting $m \to \infty$, and then afterwards $M \to \infty$, we have,

$$\left(\sum_{k=1}^{\infty} |x_k^n - a_k|^p\right)^{1/p} \le \epsilon \text{ whenever } n > N. \tag{9.3}$$

This proves Claim 1 (note that, in the limit, $< \epsilon$ is changed to $\le \epsilon$).

Claim 2: $a \in \ell_p$. Let $\epsilon = 1$. Then for a sufficiently large n, and each positive integer M, we have,

$$\left(\sum_{k=1}^{M} |a_k|^p\right)^{1/p} \le \left(\sum_{k=1}^{\infty} |x_k^n - a_k|^p\right)^{1/p} + \|x^n\| \le 1 + \|x^n\| < \infty. \tag{9.4}$$

Letting $M \to \infty$, we see that $a \in \ell_p$. This proves Claim 2.

Finally, (9.3) shows that $\|x^n - a\|_p \to 0$, so the ℓ_p spaces are BK-spaces.

Example 9.8 The spaces $\widehat{L_{2\pi}^p}$ for all $1 \le p \le \infty$ with the norms $\|\widehat{f}\| = \|f\|_p$ are all BK-spaces. The isomorphism between $\widehat{L_{2\pi}^p}$ and $L_{2\pi}^p$, and the fact that the $L_{2\pi}^p$ spaces are Banach spaces (Theorems 6.43 and 6.44) shows that the $\widehat{L_{2\pi}^p}$ spaces are Banach spaces. That $\widehat{L_{2\pi}}$ is a K-space was noted in Example 9.2. That the other $\widehat{L_{2\pi}^p}$ spaces are K-spaces follow from the inequality $\|\cdot\|_1 \le \|\cdot\|_p$ of Corollary 6.38 (Section 6.10).

Example 9.9 The nth Cesàro mean of a sequence x with partial sums $s_k = x_1 + \cdots x_k$ is defined to be

$$\sigma_n = \frac{s_1 + \cdots + s_k + \cdots + s_n}{n}. \tag{9.5}$$

The space

$$\sigma s = \{x \in \omega \mid \lim_{n\to\infty} \sigma_n \text{ exists }\} \tag{9.6}$$

is a BK space under the norm $\|x\|_{\sigma s} = \sup_n |\sigma_n|$.

Theorem 9.6 *For σs as defined in* (9.6), *we have $cs \subset \sigma s$. Furthermore, if $\lim_{n\to\infty} s_n = L$, then $\lim_{n\to\infty} \sigma_n = L$.*

Proof Let $\epsilon > 0$ be given. Suppose $|s_n - L| < \epsilon/2$ for $n > N$, and let $n > N$. Then

$$\begin{aligned}|\sigma_n - L| &= \left|\frac{(s_1 - L) + \cdots + (s_n - L)}{n}\right| \\ &\leq \frac{|s_1 - L| + \cdots + |s_N - L|}{n} + \frac{|s_{N+1} - L| + \cdots + |s_n - L|}{n} \\ &\leq \frac{|s_1 - L| + \cdots + |s_N - L|}{n} + \frac{\epsilon}{2}.\end{aligned}$$

The last quantity can be made $< \epsilon$ for sufficiently large n. ∎

Definition 9.7 The kth forward difference of a sequence x is $\Delta x_k = x_k - x_{k+1}$. The space bv of sequences of bounded variation is defined by

$$bv = \left\{ x \in \omega \mid \sum_{k=1}^{\infty} |\Delta x_k| < \infty \right\}.$$

Theorem 9.7 *We have the following properties of the sequence space bv:*

(a) *The space bv is a subset of the space of convergent sequences c.*
(b) *The space bv contains all real convergent monotonic sequences.*
(c) *The span of the set of convergent monotonic sequences is bv.*

Proof **(a):** Let $x \in bv$. For $m > n$, $\sum_{k=n}^{m-1} \Delta x_k$ telescopes to $x_n - x_m$. Thus

$$|x_n - x_m| = \left| \sum_{k=n}^{m-1} \Delta x_k \right| \leq \sum_{k=n}^{m-1} |\Delta x_k| \to 0 \text{ as } n, m \to \infty.$$

This shows that x_n is a Cauchy sequence in $\mathbb{R}$. By the completeness property of $\mathbb{R}$, x_n is a convergent sequence.

(b): If x is monotonic, then Δx_k is always of the same sign. Thus the partial sums telescope:

$$\sum_{k=1}^{n} |\Delta x_k| = \left| \sum_{k=1}^{n} \Delta x_k \right| = |x_1 - x_{n+1}|.$$

(c): In the real case, first we will write each $x \in bv$ as a difference of two convergent monotonic sequences (a_n) and (b_n). For each n, obtain a_n by adding up the nonnegative Δx_k terms for $k \geq n$. Then

$$a_n = \frac{1}{2} \sum_{k=n}^{\infty} \big(|\Delta x_k| + \Delta x_k\big).$$

Clearly $0 \leq a_n$ is monotonically decreasing and bounded, therefore convergent. To show that $b_n = a_n - x_n$ is also convergent and monotonically increasing is routine and left as Exercise 9.3.3. In the complex case, the real and imaginary parts of $x \in bv$ are each a difference of convergent monotonic sequences. ∎

Theorem 9.8 *The sequence space bv is a BK-space under the norm*

$$\|x\|_{bv} = \sum_{k=1}^{\infty} |\Delta x_k| + \sup_k |x_k|.$$

Proof The proof follows the pattern of Theorem 9.7. In place of Inequality (9.3), we use

$$\sum_{k=1}^{\infty} |\Delta x_k^n - \Delta a_k| + \sup_k |x_k - a_k| \leq \epsilon.$$

And in place of Inequality (9.4), we use

$$\sum_{k=1}^{M} |\Delta a_k| + \sup_{1 \leq k \leq M} |a_k| \leq \sum_{k=1}^{\infty} |\Delta x_k^n - \Delta a_k| + \sup_k |x_k - a_k| + \|x^n\|.$$ ∎

Corollary 9.9 *Every $x \in bv$ is a sum of an element $y \in bv_0$ and a multiple of $e = (1, 1, 1, \ldots)$. Actually we have the direct sum $bv = bv_0 \oplus \operatorname{span}\{e\}$.*

Corollary 9.10 *The closure of the space of finite sequences Φ in bv is bv_0.*

Corollary 9.11 *The space bv_0 is a BK-space with the same norm as bv.*

The result is clear if we observe that bv_0 is a closed subspace of bv.

Remark An equivalent norm of bv is $\|x\|'_{bv} = |x_1| + \sum_{k=1}^{\infty} |\Delta x_k|$.

Definition 9.8 The second (forward) difference is defined for all k by

$$\Delta^2 x_k = \Delta x_k - \Delta x_{k+1} = x_k - 2x_{k+1} + x_{k+2}.$$

A real sequence is called **convex** if $\Delta^2 x_k \geq 0$ for $k = 1, 2, \ldots$.
The sequence space q of **quasiconvex sequences** is defined by

$$q = \left\{ x \in \omega \;\middle|\; \|x\|_q = \sum_{k=1}^{\infty} k |\Delta^2 x_k| + \sup_k |x_k| < \infty \right\}.$$

We also define the space $q_0 = q \cap c_0$.

Theorem 9.12 *For every $x \in q$, we have the following.*

(a) $\Delta x_n = \sum_{k=n}^{\infty} \Delta^2 x_k$.

(b) $n\Delta x_n \to 0$ *as* $n \to \infty$.

(c) $\sum_{k=1}^{\infty} |\Delta x_k| \leq \sum_{k=1}^{\infty} k|\Delta^2 x_k|$.

(d) $x \in bv$.

Proof **(a):** Since $\sum_{k=1}^{\infty} k|\Delta^2 x_k| < \infty$, it is true that $\sum_{k=1}^{\infty} |\Delta^2 x_k| < \infty$, and thus $\sum_{k=1}^{\infty} \Delta^2 x_k$ converges. But the sum telescopes:

$$\sum_{k=n}^{m-1} \Delta^2 x_k = (\Delta x_n - \Delta x_{n+1}) + \cdots + (\Delta x_{m-1} - \Delta x_m) = \Delta x_n - \Delta x_m. \tag{9.7}$$

This shows that $\lim_{m\to\infty} \Delta x_m$ exists. If it were nonzero, then x_m would be unbounded. Thus $\lim_{m\to\infty} \Delta x_m = 0$. The result then follows from (9.7).

(b): From **(a)** we obtain

$$|\Delta x_n| = \left|\sum_{k=n}^{\infty} \Delta^2 x_k\right| \leq \sum_{k=n}^{\infty} \frac{1}{k} k|\Delta^2 x_k| \leq \frac{1}{n}\sum_{k=n}^{\infty} k|\Delta^2 x_k|.$$

Hence, $n|\Delta x_n| \leq \sum_{k=n}^{\infty} k|\Delta^2 x_k| \to 0$ as $n \to \infty$.

(c): From **(a)** we obtain

$$\sum_{k=1}^{\infty} |\Delta x_k| \leq \sum_{k=1}^{\infty}\sum_{k=n}^{\infty} |\Delta^2 x_k| = |\Delta^2 x_1| + 2|\Delta^2 x_2| + 3|\Delta^2 x_3| + 4|\Delta^2 x_4| + \cdots .$$

(d): From **(c)** we obtain $\|\cdot\|_{bv} \leq \|\cdot\|_q$. Thus $x \in q$ implies $x \in bv$. ∎

Lemma 9.13 Abel's transformation: *Consider two sequences (a_n) and (b_n) with $A_n = \sum_{k=1}^{n} a_k$. Then $\sum_{k=1}^{n} a_k b_k = \sum_{k=1}^{n-1} A_k(b_k - b_{k+1}) + A_n b_n$.*

The proof is left as Exercise 9.3.5.

Corollary 9.14 *Consider a sequence (a_n) and a nonnegative nonincreasing sequence (b_n). Then*

$$\left|\sum_{k=1}^{n} a_k b_k\right| \leq b_1 \cdot \sup_{1\leq k\leq n} |A_k|.$$

Theorem 9.15 *The following hold for quasi-convex sequences:*

(a) *Every bounded convex sequence is quasi-convex.*
(b) *Every real $x \in q$ is the difference of two real bounded convex sequences.*
(c) *The linear space q is the span of the set of convex sequences.*

Proof **(a):** Suppose x is a bounded sequence with $\Delta^2 = \Delta x_k - \Delta x_{k+1} \geq 0$ for all k. This means that $\Delta x_k \downarrow$.

Claim 1: $\Delta x_k \geq 0$ for all k. Assume $\Delta x_n < 0$, for some n. Then, for all $k \geq n$, we have $\Delta x_k < 0$ and $|\Delta x_k| \geq |\Delta x_n|$. And for any $m > n$, we have

$$x_m - x_n = (x_m - x_{m-1}) + \cdots + (x_{n+1} - x_n) = -\sum_{k=n}^{m-1} \Delta x_k \geq (m-n)|\Delta x_n|. \tag{9.8}$$

But this implies that $x_m \to \infty$, which contradicts the hypothesis.

Claim 2: $\lim_{k\to\infty} x_k$ exists. By the previous claim, x is decreasing. Since x is also bounded, it must be convergent.

Claim 3: $n\Delta x_n \to 0$ as $n \to \infty$. This follows from (9.8) above since we now know that x is a convergent sequence.

Claim 4: $\sum_{k=1}^{n} k|\Delta^2 x_k| = \sum_{k=1}^{n} k\Delta^2 x_k < \infty$. If x is bounded and convex then using Abel's transformation (Lemma 9.13) we have,

$$x_1 - x_{n+1} = \sum_{k=1}^{n} 1 \cdot \Delta x_k = \sum_{k=1}^{n} k\Delta^2 x_k + n\Delta x_n. \tag{9.9}$$

Since x_{n+1} converges and $n\Delta x_n \to 0$, the claim follows. This proves **(a)**.

(b): To show that every real $x \in q$ is the difference of two real bounded convex sequences, we follow the proof of Theorem 9.7(**c**). Let a_n be the sum of the nonnegative terms $(k-n+1)\Delta^2 x_k$ for $k \geq n$:

$$a_n = \frac{1}{2}\sum_{k=n}^{\infty}(k-n+1)\big(|\Delta^2 x_k| + \Delta^2 x_k\big).$$

Clearly a_n is nonnegative and decreasing, hence bounded. By an easy calculation we can show that $\Delta a_k = a_n - a_{n+1}$ is the sum of the nonnegative terms $\Delta^2 x_k$ for $k \geq n$. Thus Δa_k is positive and decreases to zero, which means that a_k is convex. Similarly we can show that

$$b_n = \frac{1}{2}\sum_{k=n}^{\infty}(k-n+1)\big(|\Delta^2 x_k| - \Delta^2 x_k\big)$$

has the same properties. Routine calculations will show that $x_n = a_n - b_n$.

In the complex case, following the proof of part **(b)**, we see that both the real and imaginary parts of $x \in q$ can be written as the difference of two bounded convex sequences.

(c): This part clearly follows from **(b)**. ∎

Here are two lemmas that will be useful later in proving Theorem 9.30.

Lemma 9.16 *Let $0 < m_1 < m_2 \cdots$ be an increasing sequence of real numbers, and let $1 = N_1 < N_2 < \cdots$ be an increasing sequence of integers such that for $k = 2, 3, \ldots$ we have*

$$N_{k+1}(m_k - m_{k-1}) \geq N_k(m_{k+1} - m_{k-1}) - N_{k-1}(m_{k+1} - m_k).$$

Let $y = (y_n)$ be given such that for each k we have $y_{N_k} = m_k$ and y is linear between N_k and N_{k+1}. Then the sequence y is concave downward; that is, $\Delta^2 y \leq 0$ for each n.

Proof Since y_n is linear between N_k and N_{k+1}, we have $\Delta^2 y_n = 0$ unless $n = N_k - 1$. But for $n = N_k - 1$, the values of the N_k are chosen so as to make the slopes of the line segments $M_k = \frac{m_k - m_{k-1}}{N_k - N_{k-1}}$ decreasing. We have

$$\begin{aligned}
\Delta^2 y_n &= -(M_k - M_{k+1}) \\
&= -\frac{N_{k+1}(m_k - m_{k-1}) - N_k(m_{k+1} - m_{k-1}) + N_{k-1}(m_{k+1} - m_k)}{(N_{k+1} - N_k)(N_k - N_{k-1})} \\
&\leq 0.
\end{aligned}$$

∎

Lemma 9.17 *Let $0 < m_1 < m_2 \cdots$ and $1 = N_1 < N_2 < \cdots$ be given as in Lemma 9.16. Then the sequence $\lambda = (\lambda_n)$ given by $\lambda_n = 1/y_n$ is a convex null sequence.*

Proof Let M_k be defined as in the proof of Lemma 9.16. For $n = N_k - 1$, we have

$$\begin{aligned}
\Delta^2 \lambda_n &= (m_k - M_k)^{-1} - 2(m_k)^{-1} + (m_k + M_{k+1})^{-1} \\
&= \frac{m_k(M_k - M_{k+1}) + 2M_k M_{k+1}}{m_k(m_k - M_k)(m_k + M_{k+1})} > 0
\end{aligned}$$

since $M_k - M_{k+1} \geq 0$ by Lemma 9.16 and $M_k M_{k+1} > 0$.

For $n \neq N_k - 1$, we have

$$\begin{aligned}
\Delta^2 \lambda_n &= (y_{n+1} - M_k)^{-1} - 2(y_{n+1})^{-1} + (y_{n+1} + M_{k+1})^{-1} \\
&= \frac{2M_k^2}{y_{n+1}(y_{n+1} - M_k)(y_{n+1} + M_{k+1}))} > 0.
\end{aligned}$$

∎

Exercises for Section 9.3

9.3.1 Complete Example 9.4 by showing that ℓ_∞ is a BK-space.

9.3.2 Prove that the space σs, as defined in Example 9.9, is a BK-space.

9.3.3 Complete the proof of part (**c**) of Theorem 9.7 by showing that the constructed sequence (b_n) is convergent and monotonically decreasing.

9.3.4 Prove the claim that an equivalent norm of bv is given by

$$\|x\|'_{bv} = |x_1| + \sum_{k=1}^{\infty} |\Delta x_k|.$$

9.3.5 Prove Lemma 9.13.

9.3.6 Prove its Corollary 9.14.

9.4 AD-spaces

Definition 9.9 A normed sequence space E is said to be an **AD-space** (or to have the **property AD**) if the space of finitely nonzero sequences Φ is dense in E.

Example 9.10 Every AD-space E is separable since the subset of Φ, consisting of the finitely nonzero sequences with rational entries, is dense in Φ and hence also dense in E. However, not every separable K-space is an AD-space. For example c is *not* an AD-space since the span$\{\Phi\} = c_0 \not\subset c$. However, c is separable since, by adjoining $e = (1, \ldots, 1, \ldots)$, we have $c = \text{span}\{e, e^1, e^2, e^3, \}$.

Example 9.11 The spaces c_0, cs, and ℓ_p for $1 \le p < \infty$ are AD-spaces. Also the spaces $\widehat{L^p_{2\pi}}$ for $1 \le p < \infty$ are AD-spaces. However, ℓ_∞ and c are not.

Definition 9.10 For each continuous linear functional f on a sequence space E, we define the sequence $y^f = (f(e^1), f(e^2), f(e^3), \ldots)$. Then define the sequence space

$$E^f = \{y^f \mid f \in E'\}. \tag{9.10}$$

If $f \longrightarrow y^f$ is a one-to-one correspondence between E' and E^f, we write

$$E' \simeq E^f.$$

Theorem 9.18 *A BK-space E is an AD-space if and only if $E' \simeq E^f$.*

Proof Clearly $E' \simeq E^f$ if and only if every $f \in E'$ is determined by its values on $\{e^1, e^2, \ldots\}$, and hence on $\Phi = \text{span}\{e^1, e^2, \ldots\}$. This is property AD. ∎

9.5 Solid Spaces

Definition 9.11 A sequence space E is said to be **solid** if for every $x \in E$, we have also $y = (y_1, y_2, \dots) \in E$ whenever $|y_k| \le |x_k|$ for all k.

Example 9.12 The spaces c_0 and ℓ_p for $1 \le p \le \infty$ are solid.
But the spaces c, cs, bv, q, and $\widehat{L_{2\pi}}$ are not.

Definition 9.12 The **solid hull** of a sequence space E is

$$\ell_\infty \cdot E = \{xy \in \omega \mid x \in \ell_\infty,\ y \in E\}.$$

A sequence space E is ℓ_∞**-invariant** if $E = \ell_\infty \cdot E$.

Theorem 9.19 *The solid hull of E is a solid sequence space, and it is the smallest solid sequence space containing E.*

The proof is left as Exercise 9.5.1.

Corollary 9.20 *A sequence space E is solid if and only if it is ℓ_∞-invariant.*

Exercises for Section 9.5

9.5.1 Prove Theorem 9.19.

9.5.2 Show that the solid hull of bv is ℓ_∞.

9.5.3 Show that the solid hull of bv_0 is c_0.

9.6 AK-spaces

Definition 9.13 A BK-space E is called an **AK-space** (or E has the **property AK**) if

$$P^n x = \sum_{k=1}^{n} x_k e^k \to x \quad (\text{as} \quad n \to \infty) \quad \text{for all} \quad x \in E;$$

that is, $\sum_{k=1}^{\infty} x_k e^k = x$ for all $x \in E$. The sequence $P^n x$ is called the n**th section** of x. It is important to note that, for sequences defined on the integers $\mathbb{Z}$, we sum from $-n$ to n; in this case, $P^n x = \sum_{k=-n}^{n} x_k e^k$.

Clearly an AK-space must have the property AD. Thus the spaces c and ℓ_∞ do not have the property AK.

Example 9.13 The spaces c_0, cs, and ℓ_p for $1 \le p < \infty$ have the property AK. The proof for the space cs is left as Exercise 9.6.1. That the others have the property AK, is clear from their norms.

Example 9.14 The space bv_0 has the property AK but the space q_0 does not. The proofs are left as Exercises 9.6.2 and 9.6.6.

Definition 9.14 Consider a normed space V. A **Schauder basis** is a sequence $g^1, g^2, \ldots$ of elements of V such that each $x \in V$ can be uniquely expressed as a series

$$x = \sum_{k=1}^{\infty} x_k g^n, \text{ where } x_1, x_2, \ldots \text{ are scalars.}$$

Convergence of the series is with respect to the norm of V, that is,

$$\lim_{n\to\infty} \left\| x - \sum_{k=1}^{n} x_k g^k \right\| = 0.$$

If V has a Schauder basis, then V can be represented as a sequence space with the property AK under the association $x \longrightarrow \widehat{x} = (x_1, x_2, \ldots)$. Here $g^1 \longrightarrow e^1 = (1, 0, \ldots)$, $g^2 \longrightarrow e^2 = (0, 1, 0, \ldots)$, ...

Theorem 9.21 **(a)** *The sequence space $\widehat{L^2_{2\pi}}$ is an AK-space.*

(b) *The sequence space $\widehat{L_{2\pi}}$ is* not *an AK-space.*

Proof Recall that sequences of Fourier coefficients range over all of $\mathbb{Z}$.

(a) By Corollary 7.34 (Section 7.3), we have $\widehat{L^2_{2\pi}} = \ell_2$, which is an AK-space.

(b) Theorem 7.64 (Subsection 7.8.1) shows that $\widehat{L_{2\pi}}$ is not an AK-space. ∎

Theorem 9.22 shows that every continuous linear operator from an AK-space E to a BK-space F is a matrix map.

Theorem 9.22 *Suppose that E is a BK-space with the property AK, and that F is a BK-space. If $T : E \longrightarrow F$ is a continuous linear operator, then there exists an infinite matrix $A = (a_{nk})$ such that, for all $x \in E$,*

$$\sum_{k=1}^{\infty} a_{nk} x_k \quad \textit{converges and} \quad Tx = Ax.$$

Proof Let $T : E \longrightarrow F$ be linear and continuous, and let $x \in E$. Since $x = \sum_{k=1}^{\infty} x_k e^k$, we have $Tx = \sum_{k=1}^{\infty} x_k T e^k$. For each k, let $Te^k = (a_{1k}, a_{2k}, \ldots)$. Since F has continuous coordinate functionals, we have for each $n = 1, 2, \ldots,$

$$(Tx)_n = \sum_{k=1}^{\infty} x_k (Te^k)_n = \sum_{k=1}^{\infty} a_{nk} x_k.$$

If A is the matrix with entries (a_{nk}), then $Tx = Ax$ for all $x \in E$. ∎

Definition 9.15 Let E be a K-space. The ***β* dual** of E is defined to be the space

$$E^\beta = \{x \in \omega \mid xy \in cs \ \ \forall\, y \in E\}.$$

Recall the definition of E^f given above by (9.10).

Theorem 9.23 *Consider a BK-space E with the property AD. Then E has the property AK if and only if for every $f \in E'$, we have $y^f \in E^\beta$ and*

$$f(x) = \sum_{k=1}^{\infty} x_k y_k^f \quad \textit{for all} \quad x \in E.$$

In this case, we have $E^f = E^\beta$.

Proof ($\Rightarrow$): Suppose E has the property AK. Then for all $x \in E$, $x = \sum_{k=1}^{\infty} x_k e^k$. Thus for $f \in E'$, we have

$$f(x) = \sum_{k=1}^{\infty} x_k f(e^k) = \sum_{k=1}^{\infty} x_k y_k^f \quad \text{for all} \quad x \in E.$$

This means that $y^f \in E^\beta$, which shows that $E^f \subset E^\beta$. The other inclusion follows from Theorem 9.3 which states that all multiplier maps are continuous and the fact that the sum is continuous on cs.

($\Leftarrow$): Consider the continuous linear operators $T_n : E \longrightarrow E$ defined by $T_n x = \sum_{k=1}^{n} x_k e^k$. By hypothesis, for each $f \in E'$, $y^f \in E^\beta$, and

$$|f(T_n x)| = \left| \sum_{k=1}^{n} x_k y_k^f \right| \le \sup_n \left| \sum_{k=1}^{n} x_k y_k^f \right| = \|xy^f\|_{cs}.$$

By the Banach–Mackey theorem (Corollary 6.31), the set $B = \{T_n x \mid n = 1, 2, 3, \dots\}$ is bounded. That is, there exists $M > 0$ such that $\|T_n x\| \le M\|x\|$ for all $x \in V$. Furthermore, for each $y \in \Phi$, we have $T_n y \to y$ as $n \to \infty$. Since Φ is dense in E, by the corollary of the Banach–Steinhaus theorem (Corollary 6.34), we have that $T_n x = P^n x$ converges. Since E is a K-space, it must converge to x. ∎

Example 9.15 The topological dual of c_0 is ℓ_1.

The topological dual of ℓ_p is ℓ_q when $1 \le p < \infty$ and $1/p + 1/q = 1$.

The topological dual of cs is bv.

The topological dual of bv_0 is bs.

Definition 9.16 Let E and F be sequence spaces. We say that E is F**-invariant** if $E = F \cdot E$.

Theorem 9.24 Garling: *A BK-space E is an AK-space if and only if it is bv_0-invariant.*

Proof ($\Leftarrow$): Suppose $E = bv_0 \cdot E$ and let $x = \lambda y \in E$ for some $\lambda \in bv_0$ and $y \in E$. We want to show that

$$P^n x = \sum_{k=1}^{n} x_k e^k \to x \quad \text{as} \quad n \to \infty.$$

Consider the multiplier map $T_y : bv_0 \longrightarrow E$ defined by $T_y \lambda = \lambda y$ for $\lambda \in bv_0$ and $y \in E$. Since bv_0 has the property AK, and multiplier maps are continuous, we have

$$P^n x = P^n \lambda y = T_y(P^n \lambda) \to T_y(\lambda) = \lambda y = x \quad \text{as} \quad n \to \infty.$$

($\Rightarrow$): Suppose E has the property AK. Let $x \in E$ and $\lambda \in bv_0$. Using Abel's transformation (Lemma 9.13), we have

$$\begin{aligned} P^m \lambda x - P^n \lambda x &= \sum_{k=n}^{m} \lambda_k x_k e^k = \sum_{k=n}^{m} \lambda_k (P^k x - P^{k-1} x) \\ &= \sum_{k=n}^{m} (\lambda_k - \lambda_{k+1}) P^k x + \lambda_{m+1} P^m x - \lambda_n P^{n-1} x. \end{aligned}$$

Thus

$$\|P^m \lambda x - P^n \lambda x\| \le \sum_{k=n}^{m} |\lambda_k - \lambda_{k+1}| \|P^k x\| + |\lambda_{m+1}| \|P^m x\| + |\lambda_n| \|P^{n-1} x\|.$$

Since $\sum_{k=1}^{\infty} |\lambda_k - \lambda_{k+1}| < \infty$, and $\lambda \in c_0$, and $\sup_k \|P^k x\| < \infty$, we have

$$\|P^m \lambda x - P^n \lambda x\| \to 0 \quad \text{as} \quad n, k \to \infty.$$

Thus $P^k \lambda x$ converges. Since E is a K-space, it can only converge to its coordinatewise limit λx. This shows that $\lambda x \in E$ and hence $bv_0 \cdot E \subset E$.

To show the opposite inclusion, let $x \in E$. We can find $0 < N_1 < N_2 < \cdots$ such that $\|x - P^n x\| < 1/4^k$ for $n > N_k$. Define the sequence λ by

$$\lambda_n = 1 \quad \text{for} \quad 1 \le n \le N_1 \quad \text{and} \quad \lambda_n = 1/2^k \quad \text{for} \quad N_k < n \le N_{k+1}.$$

Since λ decreases monotonically to zero, by Theorem 9.7 we have $\lambda \in bv_0$. We now show that the sequence $z = (z_k) = (x_k/\lambda_k)$ is in E. Since $x = \lambda z$, this implies that $E \subset bv_0 \cdot E$.

$$\begin{aligned}\|P^{N_j}z - P^{N_k}z\| &= \left\|\sum_{i=k}^{j-1}(P^{N_{i+1}}z - P^{N_i}z)\right\| \\ &\le \sum_{i=k}^{j-1} 2^i \, \|(P^{N_{i+1}}x - P^{N_i}x)\| \\ &< 2\sum_{i=k}^{j} 1/2^i < 1/2^{k-2}.\end{aligned}$$

Thus $P^{N_k}z$ is Cauchy. It converges coordinatewise to z. Thus $z \in E$. ∎

Corollary 9.25 *A solid BK-space has the property AK if and only if it is c_0-invariant.*

Proof The result follows easily from the above theorem if we recognize that c_0 is the solid hull of bv_0. See Exercise 9.6.3. ∎

Corollary 9.26 *Every solid BK-space E with the property AD also has the property AK.*

Proof Since E is solid, $E = \ell_\infty \cdot E$. For each $y \in E$ consider the multiplier map $T_y \colon \ell_\infty \longrightarrow E$ defined by $T_y\lambda = \lambda y$. For each $\lambda \in \ell_\infty$, the sections $P^n\lambda$ are bounded in ℓ_∞. Because multiplier maps are continuous, the sections $T_y(P^n\lambda) = P^n(\lambda y)$ are bounded in E. Since every $x \in E$ is of the form $x = \lambda y$, the sections $P^n x$ are bounded for every $x \in E$.

By the uniform boundedness principle (Theorem 6.30), the maps $T_n \colon E \longrightarrow E$ defined by $T_n x = P^n x$ are uniformly bounded. Since $\lim_{n\to\infty} T_n x = x$ for all $x \in \Phi$ and since Φ is dense in E, the Banach–Steinhaus theorem (Theorem 6.32) shows that $\lim_{n\to\infty} P^n x = \lim_{n\to\infty} T_n x = x$ for all $x \in E$. ∎

Example 9.16 $\boldsymbol{cs' \simeq cs^f = bv}$**:** The proof that cs has the property AK is left as Exercise 9.6.1. Garling's theorem shows that $bv_0 \subset cs^\beta = cs^f$. Since $e \in cs^\beta$, we have also $bv \subset cs^\beta = cs^f$. By Theorem 9.18, we have $E' \simeq E^f$. It remains to be shown that $cs^f \subset bv$.

Consider $f \in cs'$ and $b = y^f$. Fix $n = 1, 2, \ldots$. We construct $a = (a_k) \in cs$ as follows. Let $a_1 = \operatorname{sgn}(\Delta b_1)$ and let a_k by obtained by making $A_k = \sum_{j=1}^{k} a_j = \operatorname{sgn}(\Delta b_k)$ for $k = 1, \ldots, n$ and $A_k = 0$ for $k > n$. Then $\|a\|_{cs} = \sup_k |A_k| = 1$. Using Abel's transformation (Lemma 9.13), we have

$$f(a) = \sum_{k=1}^{n} a_k b_k = \sum_{k=1}^{n} A_k(b_k - b_{k+1}) + A_{n+1} b_{n+1} = \sum_{k=1}^{n} |\Delta b_k|.$$

Thus for all n, $\|f\| = \sup_{\|x\|_{cs} \leq 1} |f(x)| \geq \sum_{k=1}^{n} |\Delta b_k|$. This shows that $y^f = b \in bv$.

Exercises for Section 9.6

9.6.1 Prove that the space cs with the usual norm has the property AK as claimed in Example 9.13.

9.6.2 Show that the space bv_0 has the property AK but that bv does not.

9.6.3 Show that bv_0 is bv-invariant but that bv is not bv_0-invariant.

9.6.4 Show that an AD space has the property AK if and only if it is bv-invariant.

9.6.5 Show that the space bs does not have the property AK.

9.6.6 Show, as claimed in Example 9.14, that q_0 does not have the property AK.

9.6.7 Show that the β dual of a K-space is actually a linear space.

9.7 σK-spaces

Definition 9.17 The Cesàro sections of a sequence $x \in \omega$ are defined to be

$$\sigma^n x = \frac{P^1 x + P^2 x + \cdots + P^n x}{n} \quad \text{for} \quad n = 1, 2, \ldots.$$

Definition 9.18 We say that a BK-space E has the **property σK** (or E is a **σK-space**) if

$$\sigma^n x \to x \quad (\text{as} \quad n \to \infty) \quad \text{for all} \quad x \in E.$$

Theorem 9.27 *If a BK-space E has the property AK then it has the property σK. The converse is not true in general.*

Proof The proof is similar to that of $cs \subset \sigma s$ (Theorem 9.6). Let $\epsilon > 0$ be given. Suppose $\lim\limits_{n \to \infty} \|P^k x - x\| = 0$. If $\|P^k x - x\| < \epsilon/2$ for $k > N_1$, choose $N > N_1$ so that

$$\frac{N_1}{N}\sup_k \|P^k x - x\| < \epsilon/2.$$

Then for $n > N$, we have

$$\begin{aligned}
\|\sigma^n x - x\| &= \left\| \frac{P^1 x + P^2 x + \cdots + P^n x}{n} - x \right\| \\
&\le \frac{\|P^1 x - x\| + \cdots + \|P^n x - x\|}{n} \\
&\le \frac{N_1}{n} \sup_{k \le N_1} \|P^k x - x\| + \frac{\|P^{N_1+1} x - x\| + \cdots + \|P^n x - x\|}{n} \\
&\le \frac{N_1}{N} \sup_k \|P^k x - x\| + \frac{\epsilon}{2} < \frac{\epsilon}{2} + \frac{\epsilon}{2} = \epsilon.
\end{aligned}$$

That the converse is not true in general can be seen from the example of q_0. That q_0 is not an AK-space was observed in Example 9.14. That this space has the property σK is shown below. ∎

Example 9.17 q_0 is a σK-space: For each sequence $x \in q_0$, direct calculations show that

$$x - \sigma^n x = \left(0, \frac{1}{n}x_2, \frac{2}{n}x_3, \ldots, \frac{n-2}{n}x_{n-1}, \frac{n-1}{n}x_n, x_{n+1}, x_{n+2} \ldots\right). \quad (9.11)$$

Using brute force on this sequence, we obtain

$$\|x - \sigma^n x\|_q = S_1 + S_2 + S_3 + S_4,$$

where $S_1 = \frac{1}{n}\sum_{k=1}^{n-1} k\left|\Delta^2(k-1)x_k\right|$, $S_2 = n\left|\frac{n-1}{n}x_n - 2x_{n+1} + x_{n+2}\right|$, $S_3 = \sum_{k=n+1}^{\infty} k\left|\Delta^2 x_k\right|$, and S_4 is the supremum norm of the sequence (9.11). Clearly $S_4 \to 0$ as $n \to \infty$ since $q_0 \subset c_0$. Also, $S_3 \to 0$ by definition of the norm of q.

Next, $S_2 = n\left|\Delta^2 x_n - \frac{1}{n}x_n\right| \le n|\Delta^2 x_n| + |x_n| \to 0$ since $x \in q$ and $x \in c_0$.

Finally, $S_1 = \frac{1}{n}\sum_{k=1}^{n-1} k\left|(k-1)\Delta^2 x_k - 2\Delta x_{k+1}\right| \le \frac{1}{n}\sum_{k=1}^{n-1} k\big(k|\Delta^2 x_k| + 2|\Delta x_{k+1}|\big)$.

The sum $S_1 \to 0$ because of both $q \subset bv$ and Theorem 9.28, whose proof is Exercise 9.7.1.

Theorem 9.28 *If* $\sum_{k=1}^{\infty} |y_k| < \infty$, *then* $\lim_{n\to\infty} \frac{1}{n}\sum_{k=1}^{n} k\left|y_k\right| = 0$.

Example 9.18 $\widehat{L_{2\pi}}$ is a σK-space. This result follows from Fejér's theorem (Theorem 7.60).

Theorem 9.29 *Suppose E is a BK-space with the property σK under the norm $\|\cdot\|$. Then an equivalent norm for E is given by*

$$\|x\|' = \sup_n \|\sigma^n x\| \quad \textit{for all} \quad x \in E.$$

Proof It is routine to verify that $\|\cdot\|'$ is a norm on E. Since $\lim\limits_{n\to\infty} \|\sigma^n x\| = \|x\|$ for all $x \in E$, we have the inequality

$$\|x\| \le \sup_n \|\sigma^n x\| = \|x\|' \quad \text{for all} \quad x \in E.$$

Consider the linear operators $T^n x = \sigma^n x$ on E. By the uniform bounded principle (Theorem 6.30), $\sup_n \|T_n\| = M < \infty$. Thus, for each $x \in E$, we have

$$\|\sigma^n x\| = \|T^n x\| \le \|T^n\| \cdot \|x\| \le M\|x\|.$$

So $\|x\| \le \|x\|' \le M\|x\|$ for all $x \in E$. ∎

Theorem 9.30 *A BK-space E is an σK-space if and only if it is q_0-invariant.*

Proof ($\Leftarrow$): Suppose $E = q_0 \cdot E$. Consider the multiplier map $T_y\colon q_0 \longrightarrow E$ defined by $T_y\lambda = \lambda y$ for $\lambda \in q_0$ and $y \in E$. Since q_0 has the property σK by Example 9.17, and multiplier maps are continuous, we have

$$\sigma^n \lambda y = T_y(\sigma^n \lambda) \to T_y(\lambda) = \lambda y \quad \text{as} \quad n \to \infty.$$

Since every sequence $x \in E$ is of the form $x = \lambda y$, the result follows.

($\Rightarrow$): Suppose E has the property σK and let $x \in E$. From Theorem 9.29 we have $\|x\| = \lim_n \|\sigma^n x\| \le \sup_n \|\sigma^n x\| \le M\|x\|$. Let $\lambda \in q_0$. Using Abel's transformation (Lemma 9.13) twice we can show that for any sequence y we have

$$P^m \lambda y = \sum_{k=1}^{m-1} k\Delta^2\lambda_k \sigma^k y + (m-1)\Delta\lambda_m \sigma^{m-1} y + \lambda_m P^m y.$$

Let $\epsilon > 0$ be given. Choose $M > 0$ such that $\sum_{k=M}^{\infty} k|\Delta^2\lambda_k| < \epsilon$. Let $y = \sigma^r x - \sigma^s x$ with $r > s$ sufficiently large that $\|\sigma_k y\| < \epsilon$ for each $k \le M$. By properties of q_0, as shown in Theorem 9.12, we have

$$(m-1)|\Delta\lambda_m| \sup_m \|\sigma^{m-1} y\| + |\lambda_m| \sup_{m\le r} \|P^m y\| \to 0 \quad \text{as} \quad m \to \infty.$$

Now let $m \to \infty$. Then

$$\begin{aligned}\|\sigma^r \lambda x - \sigma^s \lambda x\| &\leq \sum_{k=1}^{M-1} k|\Delta^2 \lambda_k| \cdot \|\sigma^k y\| + \sum_{k=M}^{\infty} k|\Delta^2 \lambda_k| \cdot \|\sigma^k y\| \\ &\leq \epsilon \|\lambda\|_q + \epsilon \sup_k \|\sigma^k y\| \\ &\leq \epsilon\big(\|\lambda\|_q + M\|\sigma^r x - \sigma^s x\|\big) \leq \epsilon\big(\|\lambda\|_q + 2M \sup_n \|\sigma^n x\|\big).\end{aligned}$$

Thus $\sigma^n \lambda x$ is a Cauchy sequence in E, which must converge to $\lambda x \in E$. This shows that $q_0 \cdot E \subset E$.

It remains to be shown that $E \subset q_0 \cdot E$. Let $x \in E$. Construct $\lambda \in q_0$ and $y \in E$ following Lemma 9.16 and Lemma 9.17. Using $m_k = 2^k$ in Lemma 9.16, choose $1 = N_1 < N_2 < \cdots$ (for $k = 2, 3, 4, \ldots$) such that

$$N_{k+1} \geq 3N_k - 2N_{k-1} \quad \text{and} \quad \|\sigma^r x - \sigma^s x\| < 1/4^k \quad \text{for} \quad r > s > N_k.$$

The sequence λ constructed with Lemma 9.17 is a convex null sequence. Hence, it is in q_0. We are done if can we show that $yx = (x_k/\lambda_k) \in E$.

Let $s = N_{i+1}$ and $r = N_i$ and $z = \sigma^s x - \sigma^r x$. Then

$$P^s yx = \sum_{k=1}^{s-1} k\Delta^2 y_k \sigma^k x + (s-1)\Delta y_s \sigma^{s-1} x + y_s P^s x.$$

$$\begin{aligned}\|P^s yz\| &= \|\sigma^s yx - \sigma^r yx\| \\ &\leq \sum_{k=1}^{s-1} k|\Delta^2 y_k| \cdot \|\sigma^k z\| + (s-1)|\Delta y_s| \cdot \|\sigma^{s-1} z\| + |y_s| \cdot \|P^s z\| \\ &\leq \left(\sum_{k=1}^{s-1} k|\Delta^2 y_k| + (s-1)|\Delta y_s| + |y_s|\right) M\|z\| \\ &\leq (-y_1 + 2y_s)M\|z\| = (-2 + 2 \cdot 2^{k+1})(M/4^k) < M2^{-k+2}.\end{aligned}$$

For general $s = N_m > r = N_i$, we have

$$\|\sigma^s yx - \sigma^r yx\| \leq \sum_{k=i}^{\infty} M2^{-k+2} = M2^{-i+3}.$$

This shows that the sequence $\sigma^{N_1} yx, \sigma^{N_2} yx, \sigma^{N_3} yx, \ldots$ is Cauchy in E, which must converge to $yx \in E$. ∎

Corollary 9.31 *Suppose E is a BK-space with the property AK. Then E is q_0-invariant.*

Example 9.19 $\widehat{L_{2\pi}}$ **is** q_0 **invariant:** Since $\widehat{L_{2\pi}}$ is a sequence space on the integers $\mathbb{Z}$, we need q_0 to consist of even sequences $\lambda_k = \lambda_{-k}$ for which $(\lambda_k)_{k=0}^{\infty}$ is quasiconvex. Actually, using the Abel transformation twice, as in Theorem 9.30, it can be shown that $q_0 \subset \widehat{L_{2\pi}}$. From Property **(C4)** of Theorem 7.42, it follows that

$$\widehat{L_{2\pi}} = q_0 \cdot \widehat{L_{2\pi}} \subset \widehat{L_{2\pi}} \cdot \widehat{L_{2\pi}} \subset \widehat{L_{2\pi}}.$$

Thus $\quad L_{2\pi} = L_{2\pi} * L_{2\pi}.$

Many results for AK-spaces also hold for the more general σK-spaces. We state the following two without proof because they are similar to those for the property AK.

Definition 9.19 Let E be a K-space. The $\boldsymbol{\sigma}$ **dual** of E is defined to be

$$E^{\sigma} = \{x \in \omega \mid xy \in \sigma s \ \ \forall\, y \in E\}.$$

Since $cs \subset \sigma s$ (Theorem 9.6), we have $E^{\beta} \subset E^{\sigma}$.

Theorem 9.32 *Consider a BK-space E with the property AD. Then E has the property σK if and only if for every $f \in E'$, we have $y^f \in E^{\sigma}$ and*

$$f(x) = \lim_{n\to\infty} \sigma^n x y^f \quad \text{for all} \quad x \in E.$$

In this case, we have $E^f = E^{\sigma}$.

Corollary 9.33 *Suppose E is a BK-space with the property AK. Then*

$$E^f = E^{\beta} = E^{\sigma}.$$

Theorem 9.34 *Suppose that E is a BK-space with the property σK and let F be a BK-space. If $T : E \longrightarrow F$ is continuous and linear, then there exists an infinite matrix $A = (a_{nk})$ such that, for all $x \in E$,*

$$\sum_{k=1}^{\infty} a_{nk} x_k \quad \text{converges and} \quad Tx = Ax.$$

Exercises for Section 9.7

9.7.1 Prove Theorem 9.28.

9.7.2 Show that the space q_0 has the property σK.

9.7.3 Show that the space q does not have the property σK.

9.7.4 Show that the space q_0 is q-invariant but that q is not q_0-invariant.

9.7.5 Show that the space bv is q-invariant.

9.7.6 Show that the space q is not bv-invariant.

9.7.7 Show that the space bv_0 is q_0-invariant.

9.7.8 Show that the space cs has the property σK.

9.7.9 Show that the space bs does not have the property σK.

9.7.10 Show that an AD space has the property σK if and only if it is q-invariant.

10

Matrix Maps, Multipliers, and Duality

10.1 FK-spaces

We are by now familiar with normed vector spaces. In this section we define more general spaces that, instead of a single norm, have a sequence of seminorms $p_1, p_2, p_3, \ldots$ satisfying the following property:

$$\text{for all } x \in V,\ x = 0 \iff p_n(x) = 0\ \forall\, n = 1, 2, \ldots. \tag{10.1}$$

Theorem 10.1 *If V is a vector space with seminorms $p_1, p_2, p_3, \ldots,$ satisfying* (10.1)*, then V is a metric space under the metric*

$$d(x, y) = \sum_{n=1}^{\infty} \frac{p_n(x-y)}{2^n\big(1 + p_n(x-y)\big)}. \tag{10.2}$$

Proof We can use the inequality of Lemma 5.1 (Section 5.1) to obtain the triangle inequality. The other requirements for a metric are easily verified. ∎

Definition 10.1 A **seminorm space** is a metric linear space defined by a sequence of seminorms $p_1, p_2, p_3, \ldots$ satisfying condition (10.1) with a metric given by Equation (10.2) above.

We may specify the defining seminorms to be increasing by using seminorms $q_1 = p_1, q_2 = p_1 + p_2, q_3 = p_1 + p_2 + p_3, \ldots, q_n = p_1 + \cdots + p_n, \ldots.$ Although the q_k result in a different metric (10.2), both sets of seminorms result in the same open sets.

Theorem 10.2 *In a seminormed space V, a sequence (x^k) converges to $x \in V$ if and only if*

$$p_n(x^k - x) \to 0 \quad as \quad k \to \infty \quad for\ each \quad n = 1, 2, \ldots. \tag{10.3}$$

Proof The proof of ($\Rightarrow$) is easy. Conversely, suppose condition (10.3) is satisfied. Let $\epsilon > 0$ be given. First, choose $M > 0$ such that $\frac{1}{2^M} = \sum_{n=M+1}^{\infty} \frac{1}{2^n} < \frac{\epsilon}{2}$. Then choose $N > 0$ such that $p_n(x^k - x) < \frac{\epsilon}{2}$ for $n = 1, 2, \ldots, M$ whenever $k > N$. Then

$$d(x_n, x) = \sum_{n=1}^{M} \frac{p_n(x_n - x)}{2^n(1 + p_n(x_n - x))} + \sum_{n=M+1}^{\infty} \frac{p_n(x_n - x)}{2^n(1 + p_n(x_n - x))}$$
$$\le \sum_{n=1}^{M} \frac{\epsilon/2}{2^n} + \sum_{n=M+1}^{\infty} \frac{1}{2^n} < \frac{\epsilon}{2} + \frac{\epsilon}{2} = \epsilon.$$

∎

Similarly, we can show that a sequence (x^k) is Cauchy in a seminorm space V if and only if for each $n = 1, 2, \ldots$, we have

$$p_n(x^k - x^j) \to 0 \quad \text{as} \quad k, j \to \infty.$$

There is no notion of open spheres $S(z, r)$ in a seminorm space. However, every open neighborhood of z contains a finite intersection of sets

$$U_n(z, r) = \{x \in V \mid p_n(x - z) < r\} \quad \text{for} \quad r > 0 \quad \text{and} \quad n = 1, 2, 3, \ldots. \tag{10.4}$$

Definition 10.2 A **Fréchet space** is a complete vector space whose topology is generated by a countable number of seminorms as given by the metric (10.2).

Definition 10.3 Recall that a **BK-space** is a K-space that is also a Banach space. An **FK-space** is a K-space that is also a Fréchet space.

Example 10.1 The space ω with coordinate seminorms $P_k(x) = |x_k|$ is an FK-space. The proof is left as Exercise 10.1.5.

FK-spaces were introduced by Karl Zeller[1] in 1951. They have many properties in common with BK-spaces. We will show that matrix domains are not generally BK-spaces but are always FK-spaces.

All of the results mentioned for BK-spaces in Section 9.2 hold for FK-spaces as well. The proofs are essentially the same, except that, instead of a norm, they involve a sequence of increasing seminorms. We list these results below but we do not repeat the proofs.

[1] Karl Zeller (1924–2006) was a German mathematician. He introduced functional analytic methods to summability theory.

Theorem 10.3 *Suppose that T is a linear transformation from FK-space E into FK-space F. Then T is continuous if and only if for all k, the linear functionals $f_k(Tx) = (Tx)_k$ are continuous.*

Theorem 10.4 *Let A be an infinite matrix. If A defines a matrix map between FK-spaces E and F, then the map is a continuous linear operator.*

Theorem 10.5 *All multiplier maps between FK-spaces are continuous.*

Corollary 10.6 *If E and F are FK-spaces with $E \subset F$, then the identity map $I\colon E \longrightarrow F$ is continuous.*

Corollary 10.7 *If a sequence space E is an FK-space, then its topology is unique.*

Example 10.2 The space ω with defining coordinate seminorms $P_k(x) = |x_k|$ is an FK-space but is not a BK-space. This can be shown as follows. If the FK-topology of ω were defined by a norm, then that norm would have to be continuous by Corollary 10.6. The continuous seminorms on ω are the linear combinations of the defining seminorms $P_1, P_2, P_3, \ldots$. But every such linear combination $p(x) = \sum_{n=1}^{N} a_n|x_n|$ will have zero value $p(x) = 0$ for some nonzero sequence x; for example, $x = e^{N+1} = (0, \ldots, 0, 1, 0, \ldots)$. This is not possible for a norm, so a norm cannot be continuous on ω.

Exercises for Section 10.1

10.1.1 Let V be a seminorm space defined by seminorms $p_1, p_2, \ldots$. Show that a sequence (x^k) is Cauchy in V if and only if for each $n = 1, 2, \ldots$, we have

$$p_n(x^k - x^j) \to 0 \quad \text{as} \quad k, j \to \infty.$$

10.1.2 Suppose V is a seminorm space defined by seminorms $p_1, p_2, \ldots$. For each n, define $r_n(x) = \max\{p_1(x), p_2(x), \ldots, p_n(x)\}$. Show that $r_1, r_2, \ldots$ are increasing seminorms on V defining a space with the same topology as that defined by $\{p_1, p_2, \ldots\}$.

10.1.3 Suppose V is a seminorm space defined by seminorms $p_1, p_2, \ldots$. Show that if $x \neq y$ in V, then there exists n such that $p_n(x) \neq p_n(y)$.

10.1.4 Suppose V is a seminorm space defined by seminorms $p_1, p_2, \ldots$. Show that $\bigcap_k \{x \in V \mid p_k(x) = 0\} = \{0\}$.

10.1.5 Prove that ω is an FK-space as described in Example 10.1

10.1.6 Suppose E is an FK-space defined by seminorms $p_1, p_2, \ldots$. Show that the defining seminorms are continuous.

10.1.7 Let V and W be FK-spaces with a continuous linear operator $T: V \longrightarrow W$. Prove the open mapping theorem for T. See Theorem 6.35 in Section 6.8.

10.1.8 Let V and W be FK-spaces with a linear operator $T: V \longrightarrow W$. Prove the closed graph theorem for T. See Theorem 6.36 in Section 6.9.

10.1.9 Prove the Banach–Steinhaus theorem (Theorem 6.32 in Section 6.7) for FK-spaces. Let V be an FK-space, W a normed space, and $T_1, T_2, \ldots,$ a sequence of continuous linear operators from V into W. If

$$\lim_{k\to\infty} T_k(x) = T(x)$$

for each $x \in V$, then T is a continuous linear operator from V to W.

10.2 Matrix Maps Between FK-spaces

Definition 10.4 Let A be an infinite matrix and let E be a K-space. The **matrix domain of A relative to the space E**, or simply the **E-domain of the matrix A**, is

$$E_A = \{x \in \omega \mid Ax \in E\}.$$

If $E = \omega$, then ω_A is called the **domain** of matrix A. If $E = c$, then c_A is called the **summability field** of A.

Theorem 10.8 *Let A be an infinite matrix and let E be an FK-space defined by the seminorms $p_1, p_2, \ldots$. The E-domain of A is an FK-space with defining seminorms*

$$P_n(x) = |x_n|, \quad q_n(x) = p_n(Ax), \quad \text{and } r_n(x) = \sup_m \left|\sum_{k=1}^{m} a_{nk}x_k\right| \quad \text{for each } n.$$

Proof Clearly E_A is a linear space. Because the seminorms $P_1, P_2, P_3, \ldots$ are included, it is also a K-space. The space E_A has a countable number of seminorms, so it remains to be proven that E_A is complete.

Let $(x^m) = (x^1, x^2, \ldots)$ be a Cauchy sequence in E_A. Since E_A is a K-space, there exists a coordinatewise limit $x \in \omega$. For the nth row of the matrix, the sequence

$$y^m = \left(a_{n1}x_1^m, a_{n2}x_2^m, a_{n3}x_3^m, \ldots\right)$$

forms a Cauchy sequence in cs. Since cs is a BK-space, this sequence converges to an element of cs, which must be

$$y_n = (a_{n1}x_1, a_{n2}x_2, a_{n3}x_3, \ldots) = (Ax)_n.$$

It is sufficient to show that $y = Ax \in E$. Since $q_n(Ax^m) = p_n(Ax^m)$ is Cauchy, we have that (Ax^m) is a Cauchy sequence in E, which converges to some element of E. If we can show that this element is y, we are done. But it is only necessary to show that (Ax^m) converges coordinatewise to y. The nth coordinate of (Ax^m) is $\sum_{k=1}^{\infty} a_{nk}x_k^m$, which is the sum of the convergent sequence y^m. Since the sum is a continuous linear functional on cs, and y^m converges to y in cs, the result follows. ∎

Corollary 10.9 *Let A be an infinite matrix. The summability field c_A is an FK-space with defining seminorms*

$$P_n(x) = |x_n|, \ q(x) = \sup_n \left| \sum_{k=1}^{\infty} a_{nk}x_k \right|, \ r_n(x) = \sup_m \left| \sum_{k=1}^{m} a_{nk}x_k \right| \ \textit{for each } n.$$

If A is row-finite (that is, the rows belong to Φ), then the seminorms r_n are superfluous, since $r_n \leq q$ for all n.

Example 10.3 The space ω_A: The domain of a matrix A is an FK-space with seminorms

$$P_n(x) = |x_n| \text{ and } r_n(x) = \sup_m \left| \sum_{k=1}^{m} a_{nk}x_k \right| \quad \text{for each } n.$$

The seminorms $q_n(x) = p_n(Ax)$ may be left out since $q_n \leq r_n$ for all n.

Theorem 10.10 *Suppose A is a matrix that defines a one-to-one map. Then c_A is a BK-space with norm*

$$\|x\| = \sup_n |(Ax)_n|.$$

Proof By the open mapping theorem (Theorem 6.35), the inverse of the map $A : x \longrightarrow Ax$ is continuous. Thus the map is an isometry between c_A and c. So

$$\|x\|_{c_A} = \|Ax\|_{\infty}.$$

This means that the other seminorms of Corollary 10.9 are superfluous. ∎

Caution A matrix A may have an inverse without being one-to-one. For example,

$$A = \begin{pmatrix} 2 & 1 & 0 & 0 & 0 & \cdots \\ 0 & 2 & 1 & 0 & 0 & \cdots \\ 0 & 0 & 2 & 1 & 0 & \cdots \\ 0 & 0 & 0 & 2 & 1 & \cdots \\ 0 & 0 & 0 & 0 & 2 & \cdots \\ \vdots & \vdots & \vdots & \vdots & \vdots & \ddots \end{pmatrix}$$

has a two-sided inverse:

$$A^{-1} = \begin{pmatrix} \frac{1}{2} & -\frac{1}{4} & \frac{1}{8} & -\frac{1}{16} & \frac{1}{32} & \cdots \\ 0 & \frac{1}{2} & -\frac{1}{4} & \frac{1}{8} & -\frac{1}{16} & \cdots \\ 0 & 0 & \frac{1}{2} & -\frac{1}{4} & \frac{1}{8} & \cdots \\ 0 & 0 & 0 & \frac{1}{2} & -\frac{1}{4} & \cdots \\ 0 & 0 & 0 & 0 & \frac{1}{2} & \cdots \\ \vdots & \vdots & \vdots & \vdots & \vdots & \ddots \end{pmatrix}.$$

Yet if $x = (-\frac{1}{2}, \frac{1}{4}, -\frac{1}{8}, \frac{1}{16}, -\frac{1}{16}, \ldots)$, then $Ax = 0$. Thus A is not one-to-one.

10.3 Duality and Multipliers

Many properties of linear spaces depend on properties of their topological dual spaces. It is thus important to be able to find dual spaces of FK-spaces. In this connection, α and β duality were introduced in 1934 by Gottfried Köthe[2] and Otto Toeplitz.[3]

Definition 10.5 Let E and F be K-spaces. The **multiplier space** from E to F is defined as follows

$$(E \longrightarrow F) = \{x \in \omega \mid xy \in F \;\; \forall\, y \in E\}.$$

Sometimes it is convenient to use the notation $E^F = (E \longrightarrow F)$.

[2] The German mathematician Gottfried Köthe (1905–1989) wrote "That I became a mathematician, is almost by chance. At school I had two interests which I pursued rather intensively. One was chemistry, the other philosophy. At the university I began with the study of chemistry. A meeting with the Innsbruck philosopher Kastil brought philosophy back into the foreground. Since I was fascinated by epistemology and logic, in particular the paradoxes of set theory, it seemed best to give up chemistry and to study mathematics and philosophy instead. It turned out that mathematics attracted me more strongly than philosophy; in mathematics I found the precision and certainty which I had sought in philosophy, but failed to find there. However, I have always retained an interest in the questions which lie at the boundaries between mathematics and philosophy."

[3] Otto Toeplitz (1881–1940) was a German mathematician who worked in the theory of infinite-dimensional spaces. He criticized Banach's work as being too abstract.

If $F = \ell_1$, the multiplier space is called the **α dual** space of E:

$$E^\alpha = (E \longrightarrow \ell_1).$$

If $F = cs$, the multiplier space is called the **β dual** space of E:

$$E^\beta = (E \longrightarrow cs).$$

If $F = \sigma s$, the multiplier space is called the **σ dual** space of E:

$$E^\sigma = (E \longrightarrow \sigma s).$$

Theorem 10.11 *If $X = \alpha, \beta, \sigma$ or F, define $E^{XX} = (E^X)^X$ and $E^{XXX} = (E^{XX})^X$. Then*

(a) *if $E_1 \subset E_2$, then $E_2^X \subset E_1^X$ and $E_1^{XX} \subset E_2^{XX}$,*

(b) *$E \subset E^{XX}$ for any sequence space E, and*

(c) *$E^X = E^{XXX}$ for any sequence space E.*

The proofs are left as Exercises 10.3.2, 10.3.3, and 10.3.4.

Theorem 10.12 *Let E be a sequence space.*

(a) *The α dual of E is solid.*

(b) *The β dual of E is bv-invariant.*

(c) *The σ dual of E is q-invariant.*

These results follow from the fact that **(a)** ℓ_1 is solid, **(b)** cs is bv invariant, and **(c)** σs is q invariant.

Theorem 10.13 *If E is solid, then $E^\alpha = E^\beta = E^\sigma$.*

Proof If E is solid, then so is E^β (this is Exercise 10.3.1). Then for any $y \in E^\beta$, the sequence xy is absolutely summable for all $x \in E$. Thus every $y \in E^\beta$ is in E^α. The proof that $E^\beta = E^\sigma$ follows from Corollary 9.33 (Section 9.7). ∎

Theorem 10.14 **(a)** *For any FK-space E, we have $\Phi \subset E^\alpha \subset, E^\beta \subset E^\sigma \subset E^f$.*

In the other direction:

(b) *If E is a solid AD-space, then $E^\alpha = E^\beta = E^\sigma = E^f$.*

(c) *If E has the property AK, then $E^\beta = E^\sigma = E^f$.*

(d) *If E has the property σK, then $E^\sigma = E^f$.*

Proof **(a):** The first three inclusions are obvious. By Theorem 10.4 (p. 428), every $y \in E^\sigma$ defines a continuous linear map from E to σs. The Cesàro sum is a continuous linear functional on σs. The composite f_y is a linear functional on E with the property $f_y(e_k) = y_k$ for all k.

(c): This follows from Theorem 9.23 (Section 9.6).

(d): This follows from Theorem 9.32 (Section 9.7).

(b): A solid AD-space is an AK-space by Corollary 9.26 (Section 9.6). By Theorem 10.13 above and part **(c)**, we have $E^\alpha = E^\beta$. ∎

Example 10.4 Here we list some α, β, and σ duals.

$$
\begin{array}{rclrclrcl}
\omega^\alpha & = & \Phi & \omega^\beta & = & \Phi & \omega^\sigma & = & \Phi \\
\Phi^\alpha & = & \omega & \Phi^\beta & = & \omega & \Phi^\sigma & = & \omega \\
\ell_p^\alpha & = & \ell_q & \ell_p^\beta & = & \ell_q & \ell_p^\sigma & = & \ell_q \\
cs^\alpha & = & \ell_1 & cs^\beta & = & bv & cs^\sigma & = & bv \\
c_0^\alpha & = & \ell_1 & c_0^\beta & = & \ell_1 & c_0^\sigma & = & \ell_1 \\
bs^\alpha & = & \ell_1 & bs^\beta & = & bv_0 & bs^\sigma & = & bv_0 \\
bv^\alpha & = & \ell_1 & bv^\beta & = & cs & bv^\sigma & = & cs \\
bv_0^\alpha & = & \ell_1 & bv_0^\beta & = & bs & bv_0^\sigma & = & bs \\
\sigma s^\alpha & = & \ell_1 & \sigma s^\beta & = & bv_0 & \sigma s^\sigma & = & q_0 \\
q^\alpha & = & \ell_1 & q^\beta & = & cs & q^\sigma & = & \sigma s \\
q_0^\alpha & = & \ell_1 & q_0^\beta & = & bs & q_0^\sigma & = & \sigma b
\end{array}
$$

Exercises for Section 10.3

10.3.1 Show that the β dual of a solid sequence space is solid.
10.3.2 Prove part **(a)** of Theorem 10.12.
10.3.3 Prove part **(b)** of Theorem 10.12.
10.3.4 Prove part **(c)** of Theorem 10.12.
10.3.5 Verify the α duals given in Example 10.4.
10.3.6 Verify the β duals given in Example 10.4.
10.3.7 Verify the σ duals given in Example 10.4.

10.4 Application: Matrix Mechanics

The atomic model of Niels Bohr[4] says that an atom consists of negatively charged electrons spinning around a small nucleus. The nucleus contains protons, which are positively charged, and neutrons, which carry no electrical

[4] Niels Bohr (1885–1962) was a Danish physicist. He won the Nobel Prize in 1922 for his work on the structure of atoms.

charge. The number of protons determines the element of the atom. It is called the atomic number of the element. The protons and neutrons give the atom most of its mass. For example, the element carbon has atomic number 6, which means that each nucleus has 6 protons. And most atoms of carbon also contain 6 neutrons but a few (isotopes) contain either 7 or 8 neutrons. The average atomic mass of carbon comes out to be about 12.0107. Similarly, the atomic number of oxygen is 8 and it has average atomic mass of 15.9994. The analysis of properties and motions of atoms and subatomic particles is called **quantum mechanics**.

The electrons have discrete orbits with certain energy levels. An electron in orbit level 1 has the lowest energy. For each value, $n = 1, 2, \ldots$, there corresponds an orbit. If an atom is bombarded with light energy, the energy is absorbed by the transitioning of the electrons to higher orbits. When this energy is later released as light, the electrons transition back to lower orbits. Regardless of the energy and frequency of the light absorbed by the bombardment, the released light comes in certain energies and frequencies, depending on which orbits change. These energies and frequencies follow the Planck–Einstein equation

$$\frac{E}{\nu} = nh, \tag{10.5}$$

where E is the energy, ν is the frequency of the light, n in an integer, and h is **Planck's constant**.[5] That is, $\frac{E}{\nu}$ is quantized in units of Planck's constant. The smallest possible amount of absorbed or emitted light is a photon, corresponding to $n = 1$. A single photon has energy and frequency given by

$$E = \nu h. \tag{10.6}$$

The possible frequencies and intensities, which depend on the orbits of the electrons, characterize the element, like a fingerprint. This distribution of frequencies along with their intensities is called the **spectrum** of the element. Observing the spectrum is a way that the makeup of stars can be determined. For example, energized iron will glow in a certain distribution of frequencies and intensities. If this spectrum is observed in starlight, along with the spectra of other elements, the star is then known to contain iron.

[5] This relationship was discovered in 1900 by Max Planck (1858–1947), a German physicist. He calculated h to be $6.55 \times 10^{-34} m^2 kg/s$. Most recent measurements give $h = 6.62606957 \times 10^{-34} m^2 kg/s$. Planck received the 1918 Nobel prize in physics for his work on quantum theory.

Let E_n be the energy of an electron in orbit n. From Equation (10.5) we see that, if an electron emits a photon when it transitions from orbit n to k, the energy of the photon is $E_n - E_k$ and its frequency is

$$\omega_{nk} = (E_n - E_k)/h.$$

Suppose that the position of an electron in orbit n about a nucleus is given by $f_n(t)$. Because the motion is periodic, say of period T_n, we may expand it as a Fourier series

$$f_n(t) = \sum_{k=-\infty}^{\infty} x_{nk} e^{2\pi ikt/T_n}. \tag{10.7}$$

Since f_n has only real values, the coefficient must satisfy

$$x_{nk} = \overline{x_{nj}} \quad \text{for} \quad j = -k.$$

Bohr suggested that the radiation emitted in a transition from level $n = 1, 2, \ldots$ to level $n - k$ is represented by the kth harmonic $x_{nk}e^{2\pi ikt/T_n}$ of this Fourier series, with the amplitude given by the coefficient x_{nk} and frequency $\omega_{nk} = 2\pi k/T_n$. So we may write

$$f_n(t) = \sum_k x_{nk} e^{i\omega_{nk}t}. \tag{10.8}$$

Similarly, the functions $v_n(t)$ giving the velocity of a particle in orbit n has a similar representation $v_n(t) = \sum_k v_{nk} e^{i\omega_{nk}t}$. Differentiating (10.8), we arrive at $v_{nk} = ix_{nk}\omega_{nk}$.

Yet, the theory of the Bohr atom posed many mysteries. And a lot of things did not make sense. A breakthrough in this analysis came to Werner Heisenberg,[6] when he realized that the focus should be on what can be observed. The position $f_n(t)$ and velocity $v_n(t)$ of an electron in orbit about a nucleus *cannot* in principle be observed. Such concepts should not be used in computations of energies. However, the emission amplitudes x_{nk} at the frequencies ω_{nk} are **observable magnitudes**. So instead of the function $f_n(t)$ giving the actual position, one would represent the position by the observables x_{nk} only. And, instead of the velocity function $v_n(t)$, one represents the concept of velocity by the observables

$$v_{nk} = ix_{nk}\omega_{nk}. \tag{10.9}$$

[6] Werner Paul Heisenberg (1901–1976) was a German theoretical physicist. He said "We cannot observe electron orbits inside the atom.... Now, since a good theory must be based on directly observable magnitudes, I thought it more fitting to restrict myself to these, treating them, as it were, as representatives of the electron orbits." In 1932, Heisenberg won the Nobel Prize in physics for the creation of quantum mechanics.

Now in a transition from orbit n to orbit k, the electron could first transition to some orbit j before jumping from j to k. In so doing, the principle of conservation of energy has to hold. That is, the energies of emissions for the jump n to j and then j to k should add up to that from n to k. His calculations showed that, when two observables a_{nk} and b_{nk} are multiplied, the product is an observable c_{nk} with

$$c_{nk} = \sum_{j=-\infty}^{\infty} a_{nj} b_{jk}.$$

From the point of view of Fourier series, this makes sense since Theorem 7.50 (Section 7.5) shows that the product of Fourier series

$$\sum_{k=-\infty}^{\infty} a_{nk} e^{2\pi ikt/T_n} \quad \text{and} \quad \sum_{k=-\infty}^{\infty} b_{nk} e^{2\pi ikt/T_n}$$

is the Fourier series $\sum_{k=-\infty}^{\infty} c_{nk} e^{2\pi ikt/T_n}$ with coefficients c_{nk} give by the convolution $c_{nk} = \sum_{j=-\infty}^{\infty} a_{nj} b_{jm}$.

When Heisenberg showed this work to Max Born,[7] Born recognized that this "quantum multiplication" of Heisenberg was matrix multiplication between infinite matrices $A = (a_{nk})$ and $B = (b_{km})$.

This theory became known as **matrix mechanics**. In 1925 Werner Heisenberg, along with Max Born and Pascual Jordan,[8] set forth in a paper the matrix formulation of quantum mechanics.

A striking characteristic of these operations was that products of observables are not commutative because matrix multiplication is not commutative. So AB is not the same as BA. Actually, Heisenberg obtained the following relationship for Planck's constant h:

$$h = 4\pi m \sum_j \left(x_{nj} x_{jn} \omega_{jn} - x_{nj} x_{jn} \omega_{nj} \right), \tag{10.10}$$

where m is the mass of an electron and $X = (x_{nk})$ is the position observable. Using Equation (10.9), $V = (v_{nk}) = (i x_{nk} \omega_{nk})$, we obtain

$$\sum_j \left(x_{nj} v_{jn} - v_{nj} x_{jn} \right) m = \frac{ih}{2\pi}. \tag{10.11}$$

[7] Max Born (1882–1970), a German-British physicist, who had studied mathematics at the University of Göttingen. He was awarded the 1954 Nobel Prize in physics (shared with Walther Bothe) for the statistical interpretation of quantum mechanics. His gravestone has the inscription $pq - qp = \frac{h}{2\pi} i$, which is based on Equation (10.13). One of his grandchildren is the popular singer Olivia Newton-John.

[8] Ernst Pascual Jordan (1902–1980) was a student of Max Born and worked with Heisenberg and Born to create matrix mechanics. The nonassociative Jordan algebras are named after him.

Since momentum is mass times velocity, we can write this in terms of the momentum observable $P = mV = (p_{nk})$ as

$$(XP - PX)_{nn} = \frac{ih}{2\pi}. \tag{10.12}$$

This gives the *diagonal* of the matrix $XP - PX$. Born showed that the other entries of the matrix are zero. So we have

$$XP - PX = \frac{ih}{2\pi} I, \tag{10.13}$$

where I is the identity matrix.

Considering the one-dimensional case of a particle, say an electron, moving along the horizontal axis. A **state function** or **wave function** $f(x,t) \in L^2(\mathbb{R}^2)$ is a function where $|f(x,t)|^2$ is the probability density of its position. We fix t, so we will write $f(x)$ instead of $f(x,t)$. Thus the probability that the particle is in the interval $[a,b]$ at a fixed time t is given by

$$Pr(a < x < b) = \int_a^b |f(x)|^2\, dx.$$

We, of course, must have probability 1 for the entire real axis $(-\infty, \infty)$. Thus

$$\int_{-\infty}^{\infty} |f(x)|^2\, dx = \|f\|_2^2 = 1.$$

In classical mechanics, the momentum of a particle with mass m and velocity v is $p = mv = mx'$. In quantum mechanics, the state function $f_p(x)$ for the momentum p is simply the Fourier transform of the state function for position $\widehat{f}(x)$, provided we use units of measurement that give us $\frac{h}{2\pi} = 1$.[9] That is, the probability that the momentum p is in the interval $[c,d]$ at a fixed time t is given by

$$Pr(c < p < d) = \frac{1}{2\pi} \int_c^d |\widehat{f}(x)|^2\, dx.$$

To show that the integral over the entire real axis (∞, ∞) is 1, we use the Plancherel theorem (Theorem 7.94 in Section 7.11) to obtain $\frac{1}{2\pi}\int_{-\infty}^{\infty} |\widehat{f}(x)|^2\, dx = \|f\|_2^2 = 1$.

[9] To obtain the state function in metric units, we can perform a change of variable to give us the state function $f_p(u) = \frac{1}{\sqrt{h}} \widehat{f}\left(\frac{2\pi}{h} u\right)$ in original units. For more detail about momentum as a Fourier transform, please see Chapter 2 of *Fourier series and integrals* by H. Dym and H.P. McKean. Academic Press (1985).

We can use this to derive the **Heisenberg inequality**, which states that a definite position and momentum of a particle cannot be simultaneously measured. Heisenberg used matrix mechanics, such as equations (10.10) through (10.13), and quantum notions of the position and mass of an electron to derive his inequality. Parts of his derivations have been called magic because formulas appeared seemingly with the wave of his hand.[10] Instead, we rely on the uncertainty principle (Theorem 8.18 in Section 8.8).

Recall from Section 8.8, the dispersion of a function f about a point a is given by $\Delta_a f$ (see Equation (8.55)).

Theorem 10.15 Heisenberg's inequality: *Let f be the state function of a particle moving along the x-axis. Suppose $|f|^2$ is in both $L^1(\mathbb{R})$ and $L^2(\mathbb{R})$ and has a continuous derivative. Let $\overline{x} = \int_{-\infty}^{\infty} x|f(x)|^2\,dx$ and $\overline{p} = \int_{-\infty}^{\infty} x|f_p(x)|^2\,dx$ be the expected (average) position and momentum of the particle, respectively. Then*

$$(\Delta_{\overline{x}} f)(\Delta_{\overline{p}} f_p) \geq \frac{h^2}{16\pi^2}.$$

Proof The uncertainty principle (Theorem 8.18) says that $(\Delta_a f)(\Delta_b \widehat{f}\,) \geq \frac{1}{4}$ for any $a, b \in \mathbb{R}$. This proves the theorem for $a = \overline{x}$ and $b = \overline{p}$, provided we use units for which $\frac{h}{2\pi} = 1$. To obtain it in metric units, we make the substitution $u = \frac{h}{2\pi}x$, as follows:

$$\Delta_b f_p = \int_{-\infty}^{\infty} (u-b)^2 \left|f_p(u)\right|^2 du = \frac{h^2}{4\pi^2}\int_{-\infty}^{\infty} \left(x - \frac{2\pi}{h}b\right)^2 \left|\widehat{f}(x)\right|^2 dx,$$

which is equal to $\frac{h^2}{4\pi^2}\Delta_{\frac{2\pi}{h}b}\widehat{f}$. Thus, with $a = \overline{x}$ and $b = \overline{p}$, we obtain Heisenberg's inequality in original units,

$$\left(\Delta_{\overline{x}} f\right)\left(\Delta_{\overline{p}} f_p\right) = \frac{h^2}{4\pi^2}\left(\Delta_a f\right)\left(\Delta_{\frac{2\pi}{h}b}\widehat{f}\,\right) \geq \frac{h^2}{4\pi^2}\cdot\frac{1}{4}. \qquad \blacksquare$$

Just as the observables position and momentum have a Born matrix relationship,

$$XP - PX = \frac{ih}{2\pi}I,$$

10 How did he do it? Details of calculations that Heisenberg may have used can be found in the paper *Understanding Heisenberg's 'magical' paper of July 1925: a new look at the calculational details,* by Ian J. R. Aitchison, David A. MacManus, and Thomas M. Snyder. American Journal of Physics. Vol. **72** (2004), pp. 1370–1380.

other observables such as velocity, energy, and time, have state functions and may be represented by Born-type matrix relationships. If we have the matrix relationship

$$AB - BA = icI$$

with state functions f_a, f_b, respectively, and $c > 0$, we can derive a Heisenberg-type inequality

$$(\Delta_{\overline{a}} f_a)(\Delta_{\overline{b}} f_b) \geq c^2 \cdot \frac{1}{4}.$$

If, on the other hand, the matrices commute, $AB - BA = 0$, then $c = 0$, and we have no uncertainty inequality. Then the two observables can be measured simultaneously with arbitrary precision. Observables in our macroscopic world of measurements lead to matrices that commute.

Louis de Broglie[11] suggested that an electron moving with the momentum p is associated with a wave of length $\lambda = \frac{h}{p}$. This was supported experimentally by observations of diffraction patterns formed by electrons of momentum p. This view of an electron as a wave, inspired Erwin Schrödinger[12] in 1925 to found a **wave mechanics** explanation of quantum theory. In 1926, Shrödinger showed that matrix mechanics and wave mechanics are equivalent. Schrödinger's approach has now become the more standard way of explaining quantum theory.

[11] Louis de Broglie (1892–1987) was a French mathematical physicist who studied the particle-wave duality of the electron. He was awarded the Nobel prize in 1929.

[12] Erwin Schrödinger (1887–1961) was an Austrian theoretical physicist. He was awarded the Nobel prize in 1933.

11
Summability

11.1 Introduction

The main task of **summability** is to assign a sum to a divergent series or a limit to a divergent sequence. We saw with Fourier series the importance of Cesàro summability. There are situations that arise with differential equations where one deals with methods of summing divergent series.

Suppose that we have such a summability method; call it W. The set of all sequences that have limits by the method W is called the summability field of W and denoted c_W. We have already studied summability fields of matrix summability methods in Section 10.2.

The scalar L assigned to a series $x_1 + x_2 + \cdots$ by a method W is denoted $\sum x_k\ (W) = L$. If we consider the sequence of partial sums $(s_1, s_2, s_3, \ldots)$ then we write $\lim_W s_n = L$.

If W is a method of summing sequences, there are certain assumptions.

(S1) The summability method W is a linear functional; that is,
if $\sum x_k\ (W) = A$ and $\sum y_k\ (W) = B$, then

$$\sum (\alpha x_k + \beta y_k)\ (W) = \alpha A + \beta B \text{ for all scalars } \alpha,\ \beta.$$

(S2) For every finite sequence $x = \sum_{k=1}^{n} x_k e^k$, we have $\sum x_k\ (W) = \sum_{k=1}^{n} x_k$.

The summability method W is said to be **regular** if we also have the following condition **(S3)**.

(S3) If $x \in cs$, then it will be summed by the method W and

$$\sum x_k\ (W) = \sum_{k=1}^{\infty} x_k.$$

Or **(S3)** if $\lim_{n\to\infty} s_n$ exists, then so does $\lim_W s_n$ and $\lim_W s_n = \lim_{n\to\infty} s_n$.

Definition 11.1 A summability method W_1 is said to be **stronger** than a method W_2 if the summability field of W_2 is a subset of that of W_1. A regular method W is thus stronger than the ordinary method of summability.

11.2 Cesàro Method

The Cesàro method (of order 1) of summability $(C, 1)$ sums sequences as follows. For a sequence x, take the partial sums

$$s_1 = x_1, \quad s_2 = x_1 + x_2, \quad s_3 = x_1 + x_2 + x_3, \quad s_4 = x_1 + x_2 + x_3 + x_4, \ldots.$$

Then, for each n, take the average of the first n partial sums

$$\sigma_n = \frac{s_1 + s_2 + \cdots + s_n}{n}.$$

If $\lim_{n\to\infty} \sigma_n = L$, then we say that L is the Cesàro sum of the sequence

$$\sum x_k \; (C, 1) = \lim_{n\to\infty} \sigma_n = L.$$

Example 11.1 Consider the series $1 - 1 + 1 - 1 + 1 - \cdots$. The partial sums of $x = (1, -1, 1, -1, \ldots)$ are

$$\begin{aligned}
s_1 &= 1 \\
s_2 &= 1 - 1 = 0 \\
s_3 &= 1 - 1 + 1 = 1 \\
s_4 &= 1 - 1 + 1 - 1 = 0 \\
&\vdots
\end{aligned}$$

The partial sums fail to converge. Yet the Cesàro method takes the average of these partial sums.

$$\begin{aligned}
\sigma_1 &= s_1 = 1 \\
\sigma_2 &= \frac{s_1 + s_2}{2} = \frac{1 + 0}{2} = \frac{1}{2} \\
\sigma_3 &= \frac{s_1 + s_2 + s_3}{3} = \frac{1 + 0 + 1}{3} = \frac{2}{3} \\
\sigma_4 &= \frac{s_1 + s_2 + s_3 + s_4}{4} = \frac{1 + 0 + 1 + 0}{4} = \frac{2}{4} = \frac{1}{2} \\
\sigma_5 &= \frac{s_1 + s_2 + s_3 + s_4 + s_5}{5} = \frac{1 + 0 + 1 + 0 + 1}{5} = \frac{3}{5}
\end{aligned}$$

and so on; $s_{2n} = \dfrac{n}{2n} = \dfrac{1}{2}$ and $s_{2n+1} = \dfrac{n+1}{2n+1}$. Thus

$$\lim_{n\to\infty} s_n = \frac{1}{2}.$$

We see that the Cesàro sum of the sequence x is $\sum x_k\ (C,1) = \frac{1}{2}$.

11.3 Matrix Methods

Definition 11.2 Let $A = (a_{nk})$ be an infinite matrix, and let $x = (x_1, x_2, \ldots)$ be a sequence with partial sums $s = (s_1, s_2, x_3, \ldots)$. If

(a) $\sum_{k=1}^{\infty} a_{nk} s_k = (As)_n$ exists for each $n = 1, 2, \ldots$, and

(b) $\lim_{n\to\infty} (As)_n = L$ exists,

then we say that the ***A*-limit** of s is L (or the ***A*-sum** of x is L) and write

$$\lim_A s = L \qquad \text{or} \qquad \textstyle\sum_A x = L.$$

Example 11.2 The Cesàro method is a matrix method with matrix

$$C_1 = \begin{pmatrix} 1 & 0 & 0 & 0 & 0 & \cdots \\ \frac{1}{2} & \frac{1}{2} & 0 & 0 & 0 & \cdots \\ \frac{1}{3} & \frac{1}{3} & \frac{1}{3} & 0 & 0 & \cdots \\ \frac{1}{4} & \frac{1}{4} & \frac{1}{4} & \frac{1}{4} & 0 & \cdots \\ \vdots & \vdots & \vdots & \vdots & \vdots & \ddots \end{pmatrix}.$$

11.4 Silverman–Toeplitz Theorem

The Silverman–Toeplitz theorem was first proved by Otto Toeplitz in 1911 and Louis Silverman[1] in 1913. The original proof was very difficult, using classical analysis (also called "hard" analysis). Functional analytic methods were first applied to the theory of matrix maps by Stefan Banach and his student Stanislaw Mazur. Banach gave a short proof of the Silverman–Toeplitz

[1] Louis Lazarus Silverman (1884–1967) worked in the area of divergent series. He was born in what is now Lithuania, and immigrated with his family to the United States when he was eight years old.

theorem in 1932 using the Banach–Steinhaus theorem. Matrices that satisfy the conditions of the theorem are called **Toeplitz matrices**.

Definition 11.3 A matrix method is said to be **regular** if

(a) $c \subset c_A$ (that is, every convergent sequence s has an A-limit), and

(b) for all $s \in c$, we have $\lim_A s = \lim\limits_{n\to\infty} s_n$.

Theorem 11.1 Silverman–Toeplitz: *A matrix method given by the matrix $A = (a_{nk})$ is regular if and only if the following three conditions are satisfied:*

(1) $\sup_n \sum |a_{nk}| < \infty$.

(2) $a_{nk} \to 0$ *as* $n \to \infty$, *for* $k = 0, 1, 2, \ldots$ *(the columns converge to 0).*

(3) $\sum_k a_{nk} \to 1$ *as* $n \to \infty$ *(the row sums tend to 1).*

Proof ($\Rightarrow$): If A is regular, the sequence e^n with $\lim_a e^n = 0$ shows that the columns of A must converge to 0. The sequence $e = (1, 1, 1, \ldots)$ with $\lim_A e = 1$ shows that condition **(3)** must hold. To show that condition **(1)** is necessary, consider the matrix map $A : c \longrightarrow c$ defined by $As = \big((As)_1, (As)_2, (As)_3 \ldots\big)$. Since matrix maps are continuous, we have

$$\|As\|_c = \sup_n \left|\sum_k a_{nk}s_k\right| \le \|A\| \cdot \|s\|_c.$$

Choose any n and let $s_k = \operatorname{sgn} a_{nk}$. Then for any $N > 0$, the Nth section $P^N s = (s_1, s_2, \ldots, s_N, 0, 0, \ldots)$ of s is in c and $\|P^N s\|_c = 1$. Thus

$$\sum_{k=1}^{N} |a_{nk}| \le \|A\| \cdot \|P^N s\|_c = \|A\|.$$

Letting N tend to ∞, we have condition **(1)**.

($\Leftarrow$): Suppose matrix A satisfies the three conditions of the theorem. By condition **(3)** we have for each $s \in c$,

$$|(As)_n| = \left|\sum_k a_{nk}s_k\right| \le \sup_n \sum |a_{nk}| \cdot \|s\|_c.$$

This shows that the linear functionals on c defined by $f_n(x) = (Ax)_n$ are uniformly bounded.

Notice that $\lim_A e = 1$ by condition **(3)**, and $\lim_A e^k = 0$ exists for each k by condition **(2)**. Thus $\lim_A x = x$ on all of the sequences e^k and e. Since the span of $\{e, e^1, e^2, e^3, \ldots\}$ is dense in c we have by Corollary 6.34 to the Banach–Steinhaus theorem (Theorem 6.33) that $\lim_A s = \lim s$ on all of c. ∎

Clearly the Cesàro matrix of Example 11.2 is a regular matrix. Thus we have the following result.

Corollary 11.2 *If a sequence x has a convergent series with sum $\sum_k x_k = L$, then the Cesàro sum of x exists and $\sum_k x_k \ (C,1) = L$.*

Corollary 11.3 *The sequence space c is a subset of the summability field of the Cesàro matrix.*

Theorem 11.4 *If matrices A and B are regular, then so is the product AB.*

The proof is left as Exercise 11.4.2.

Example 11.3 Cesàro and Hölder methods: By Theorem 11.4, powers of the Cesàro matrix C_1 are defined and regular. Whereas, the Cesàro method defined by the matrix $H_1 = C_1$ takes averages of partial sums, the square $H_2 = C_1^2$ is a matrix method that takes averages of the averages of partial sums. And so on, for the higher powers. The Hölder method of order j is defined by the matrix $H_j = C_1^j$, for $j = 1, 2, \ldots$. As the orders increase, the methods get stronger.

The Cesàro method of order j is equivalent to the Hölder methods of order j, in the sense that they both sum the same sequences, and to the same sums. However, the Cesáro method of order j is defined in a more sophisticated way by the lower-triangular matrix $C_j = (c_{nk})$ with entries

$$c_{nk} = \frac{\binom{n-k+j-1}{n-k}}{\binom{n+j-1}{n-1}} \quad \text{if} \quad k \le n, \quad \text{and} \quad c_{nk} = 0 \quad \text{otherwise.}$$

For example the matrix C_2 is

$$C_2 = \begin{pmatrix} 1 & 0 & 0 & 0 & 0 & \cdots \\ \frac{2}{3} & \frac{1}{3} & 0 & 0 & 0 & \cdots \\ \frac{3}{6} & \frac{2}{6} & \frac{1}{6} & 0 & 0 & \cdots \\ \frac{4}{10} & \frac{3}{10} & \frac{2}{10} & \frac{1}{10} & 0 & \cdots \\ \frac{5}{15} & \frac{4}{15} & \frac{3}{15} & \frac{2}{15} & \frac{1}{15} & \cdots \\ \vdots & \vdots & \vdots & \vdots & \vdots & \ddots \end{pmatrix}.$$

Example 11.4 Using methods similar to the Silverman–Toeplitz theorem one can show that a matrix $A = (a_{nk})$ maps c into c if and only if

(1) $\sup_n \sum_{k=1}^{\infty} |a_{nk}| < \infty$,
(2) all of the columns of A converge, and
(3) $\lim_{n\to\infty} \sum_{k=1}^{\infty} a_{nk}$ exists.

Example 11.5 Using methods similar to the Silverman–Toeplitz theorem one can show that a matrix $A = (a_{nk})$ maps c_0 into c_0 if and only if

(1) $\sup_n \sum_{k=1}^{\infty} a_{nk} < \infty$ and
(2) the columns of A converge to zero.

Exercises for Section 11.4

11.4.1 Define the Borel matrix $B = (b_{nk})$ by $b_{nk} = \dfrac{n^{k-1}}{e^n(k-1)!}$.
Show that B satisfies the Silverman–Toeplitz conditions.

11.4.2 Prove Theorem 11.4. Be sure to show that the product AB is defined when both matrices satisfy the Silverman–Toeplitz conditions.

11.4.3 Prove the statement of Example 11.4.

11.4.4 Prove the statement of Example 11.5.

11.4.5 Show that a matrix A maps cs into c if and only if
(1) $\sup_n \sum_{k=1}^{n} |a_{n,k+1} - a_{nk}| < \infty$, and
(2) all the columns of A converge.

11.4.6 Continuing with Exercise 11.4.5, show that the matrix map from cs to c assigns to each convergent series its usual sum if, in addition, **(3)** the column limits are all equal to 1.

11.5 Abel Summability

Definition 11.4 Here we consider sequences $x = (x_0, x_1, x_2, \ldots)$ whose index starts with $k = 0$. A sequence $x \in \omega$ is said to be **Abel summable** if the function $f(r) = \sum_{k=0}^{\infty} x_k r^k$ is defined for all $0 < r < 1$, and

$$\lim_{r\to 1^-} f(r) = \lim_{r\to 1^-} \sum_{k=0}^{\infty} x_k r^k \quad \text{exists.}$$

In this case, we write $\sum x_k \; (A) = \lim_{r\to 1^-} \sum_{k=0}^{\infty} x_k r^k$.

Theorem 11.5 Abel: *If* $f(r) = \sum_{k=0}^{\infty} x_k r^k$ *is defined for some* $r > 0$, *then* $f(t) = \sum_{k=0}^{\infty} x_k t^k$ *exists for all* $0 \le t < r$ *and*

$$f(r) = \lim_{t \to r^-} f(t) = \lim_{t \to r^-} \sum_{k=0}^{\infty} x_k r^k. \tag{11.1}$$

Proof The case $r = 1$: In this case, $f(1) = \sum_{k=0}^{\infty} x_k$ exists. Let $\epsilon > 0$ be given and let $0 \le t \le 1$. Then there exists $N > 0$ such that

$$|s_m - s_n| = |x_{n+1} + \cdots + x_m| < \epsilon \quad \text{whenever} \quad N < n < m.$$

Let $s_n(t) = \sum_{k=0}^{n} x_k t^k$ for some $0 \le t \le 1$. Then the sequence (t^n) is nonincreasing. By an Abel transformation (Corollary 9.14, Section 9.3), we have

$$|s_m(t) - s_n(t)| = |x_{n+1} t^{n+1} + \cdots + x_m t^m| \le t^{n+1} \epsilon \le \epsilon.$$

This shows that $f(t)$ exists for all $0 \le t \le 1$. Since N does not depend on t, the convergence is uniform. This means that $f(t)$ is continuous on $[0, 1]$. Hence, $\lim_{t \to 1^-} f(t) = f(1)$.

In the general case, use the case $r = 1$ on the sequence $y_k = x_k r^k$. ∎

Corollary 11.6 *The Abel method is regular.*

Lemma 11.7 *If* x *is Abel summable, then for every* $0 < r < 1$, *we have for partial sums* s^n,

$$\lim_{n \to \infty} r^n s_n = 0.$$

Proof Let $0 < r < t < 1$. Then

$$|r^n s_n| \le \left(\frac{r}{t}\right)^n \sup_k |x_k t^k| \sum_{k=0}^{n} t^{n-k} = \left(\frac{r}{t}\right)^n \sup_k |x_k t^k| \left(\frac{1 - t^{n+1}}{1 - t}\right).$$

Since x is Abel summable, the terms $x_k t^k$ of $\sum_{k=0}^{\infty} x_k t^k$ tend to zero. Hence $\sup_k |x_k t^k| < M < \infty$, for some bound M. Then we have

$$|r^n s_n| \le M \left(\frac{r}{t}\right)^n \left(\frac{1 - t^{n+1}}{1 - t}\right) \quad \text{which tends to zero as } n \to \infty.$$ ∎

Lemma 11.8 *A sequence* x *is Abel summable to* L *if and only if we have*

$$\lim_{r \to 1^-} (1 - r) \sum_{k=0}^{\infty} s_k r^k = L \text{ for partial sums } s^n.$$

Proof Let $0 < r < 1$. Then for any n, we have

$$\begin{aligned}\sum_{k=0}^{n} x_k r^k &= s_0 r^0 + (s_1 - s_0)r^1 + \cdots + (s_n - s_{n-1})r^n \\ &= s_0(r^0 - r^1) + s_1(r^1 - r^2) + \cdots + s_{n-1}(r^{n-1} - r^n) + s_n r^n \\ &= (1-r)\sum_{k=0}^{n-1} s_k r^k + s_n r^n.\end{aligned}$$

Clearly, if $s_n r^n$ tends to zero, then for all $0 < r < 1$.

$$\sum_{k=0}^{\infty} x_k r^k = (1-r)\sum_{k=0}^{\infty} s_k r^k. \tag{11.2}$$

If x is Abel summable, then by Lemma 11.7, $s_n r^n$ tends to zero. Conversely, if the right side of (11.2) exists, $s_n r^n$ tends to zero as well. ∎

Theorem 11.9 *The Abel method is more powerful than any of the Cesàro methods.*

Proof We prove this for the Cesàro method of order 1. The higher orders we leave as Exercise 11.5.1. Suppose that x is Cesàro summable. We may change the value of x_1 to make the Cesàro sum 0. Let $\epsilon > 0$ be given and suppose that $|\sigma^n| < \epsilon/2$ for $n > N$.
Noting that $\sum_{k=0}^{\infty} kr^k = \frac{r}{(1-r)^2}$, we have

$$\begin{aligned}(1-r)^2\sum_{k=0}^{\infty} k\sigma_k r^k &\le (1-r)^2\sum_{k=0}^{n} k\sigma_k r^k + (1-r)^2\sum_{k=n+1}^{\infty} k\sigma_k r^k \\ &\le (1-r)^2\sum_{k=0}^{n} k\sigma_k r^k + \frac{\epsilon}{2}(1-r)^2\frac{r}{(1-r)^2} \\ &\le (1-r)^2\sum_{k=0}^{n} kr^k \sup_k|\sigma_k| + \frac{\epsilon}{2}.\end{aligned}$$

This expression can be made to be less than ϵ by making r sufficiently close to 1. We conclude that

$$\lim_{r\to 1^-}(1-r)^2\sum_{k=0}^{\infty} k\sigma_k r^k = 0. \tag{11.3}$$

Noting $s_k = k\sigma_k - (k-1)\sigma_{k-1}$, we can repeat the argument of Lemma 11.8 to obtain

$$\sum_{k=0}^{\infty} x_k r^k = (1-r)\sum_{k=0}^{\infty} s_k r^k = (1-r)^2 \sum_{k=0}^{\infty} k\sigma_k r^k \tag{11.4}$$

provided that $k\sigma_k \to 0$ as $k \to \infty$. But (11.3) implies $k\sigma_k \to 0$. So we conclude that $\lim_{r\to 1^-} \sum_{k=0}^{\infty} x_k r^k = 0$. ∎

We state the following without proof. For a proof, please see the articles *Sur les méthodes continues de limitation* (I) & (II), by L. Włodarski in Studia Mathematica. Vol. **14** (1955), pp. 161–187, 188–199.

Theorem 11.10 *The summability field of the Abel method is an FK-space with defining seminorms*

$$p_0(x) = \sup_{0<r<1} \left| \sum_{k=1}^{\infty} x_k r^k \right|$$

$$p_n(x) = \sup_k |x_k| \left(\frac{n}{n+1}\right)^k \quad \textit{for} \quad n = 1, 2, \ldots .$$

Exercises for Section 11.5

11.5.1 Prove Theorem 11.9 for the Cesàro method of orders $j > 1$.

11.5.2 Show that $1-2+4-8\cdots$ is not summable by either the (C,1) method or the Abel method.

11.5.3 Show that $1-2+3-4\cdots$ sums to $1/4$ by the Abel method.

Index

SYMBOLS

A

O

P

Q

R

S

Printed in the United States
by Baker & Taylor Publisher Services